PHYSICS OF THE SUN
Volume II: The Solar Atmosphere

GEOPHYSICS AND ASTROPHYSICS MONOGRAPHYS

PHYSICS OF THE SUN

Volume II: The Solar Atmosphere

Edited by

PETER A. STURROCK

Center for Space Science and Astrophysics,
Stanford University, Stanford, California, U.S.A.

Associate Editors:

THOMAS E. HOLZER

High Altitude Observatory,
National Center for Atmospheric Research, Boulder, Colorado, U.S.A.

DIMITRI M. MIHALAS

High Altitude Observatory,
National Center for Atmospheric Research, Boulder, Colorado, U.S.A.

ROGER K. ULRICH

Astronomy Department, University of California,
Los Angeles, California, U.S.A.

D. REIDEL PUBLISHING COMPANY

A MEMBER OF THE KLUWER ACADEMIC PUBLISHERS GROUP

DORDRECHT / BOSTON / LANCASTER / TOKYO

Library of Congress Cataloging-in-Publication Data

Main entry under title:

Physics of the sun.

    (Geophysics and astrophysics monographs)
    Includes bibliographies and index.
    Contents: v. 1. The solar interior − v. 2. The solar atmosphere.
    1. Sun. I. Sturrock, Peter A. (Peter Andrew). II. Series.
GB521.P48 1985   559.9′4   85-20996
ISBN 90-277-1823-7 (set)
ISBN 90-277-1860-1 (v. 1)
ISBN 90-277-1861-X (v. 2)

Published by D. Reidel Publishing Company,
P.O. Box 17, 3300 AA Dordrecht, Holland.

Sold and distributed in the U.S.A. and Canada
by Kluwer Academic Publishers
190 Old Derby Street, Hingham, MA 02043, U.S.A.

In all other countries, sold and distributed
by Kluwer Academic Publishers Group.
P.O. Box 322, 3300 AH Dordrecht, Holland.

Printed in the Netherlands

TABLE OF CONTENTS

PREFACE

This volume, together with its two companion volumes, originated in a study commissioned by the United States National Academy of Sciences on behalf of the National Aeronautics and Space Administration. A committee composed of Tom Holzer, Dimitri Mihalas, Roger Ulrich and myself was asked to prepare a comprehensive review of current knowledge concerning the physics of the sun. We were fortunate in being able to persuade many distinguished scientists to gather their forces for the preparation of 21 separate chapters covering not only solar physics but also relevant areas of astrophysics and solar-terrestrial relations.

In proved necessary to divide the chapters into three separate volumes that cover three different aspects of solar physics. Volumes I and III are concerned with "The Solar Interior" and with "Astrophysics and Solar-Terrestrial Relations." This volume, devoted to "The Solar Atmosphere," covers not only the chromosphere and corona but also the principal phenomena usually referred to as "solar activity." The emphasis is on identifying and analyzing the relevant physical processes, but each chapter also contains a great deal of descriptive material.

In preparing our material, the authors and editors benefited greatly from the efforts of a number of scientists who generously agreed to review individual chapters. I wish therefore to take this opportunity to thank the the following individuals for this valuable contribution to our work: S. K. Antiochos, E. H. Avrett, J. N. Bahcall, C. A. Barnes, G. Bicknell, D. Black, M. L. Blake, P. Bodenheimer, F. H. Busse, R. C. Canfield, T. R. Carson, J. I. Castro, J. Christensen-Dalsgaard, E. C. Chupp, A. N. Cox, L. E. Cram, P. R. Demarque, L. Fisk, W. A. Fowler, D. O. Gough, L. W. Hartmann, J. W. Harvey, R. F. Howard, P. Hoyng, H. S. Hudson, G. J. Hurford, C. F. Kennel, R. A. Kopp, A. Krueger, R. M. Kulsrud, R. B. Larson, H. Leinbach, R. E. Lingenfelter, J. L. Linsky, D. B. Melrose, M. J. Mitchell, A. G. Newkirk, F. W. Perkins, R. Roble, R. T. Rood, R. Rosner, B. F. Rozsynai, S. Schneider, E. C. Shoub, B. Sonnerup, H. Spruit, R. F. Stein, M. Stix, J. Tassoul, G. Van Hoven, G. S. Vaiana, A. H. Vaughan, S. P. Worden, R. A. Wolf, and J. B. Zirker.

On behalf of the editors of this monograph, I wish to thank Dr. Richard C. Hart of the National Academy of Sciences, Dr. David Larner of Reidel Publishing Company, and Mrs. Louise Meyers of Stanford University, for the efficient and good-natured support that we received from them at various stages of the preparation of this volume.

Stanford University, June, 1985   P. A. STURROCK

CHAPTER 8

RADIATION OUTPUT

R. G. ATHAY

1. Introduction

The following discussion of the Sun's radiation output is more in the nature of an essay than a review. Among the reasons for adopting such an approach are: (1) a balanced review would require more discussion than is appropriate for this volume; (2) the desired goal of the chapter is to focus attention on the nature and importance of the problems rather than describe the attempts to solve particular problems; and (3) current and rather extensive reviews already exist in such volumes as *The Solar Output and its Variations* (ed. O. R. White, 1977). *The Sun as a Star* (ed. S. Jordan, 1981), *Solar Flares* (ed. P. A. Sturrock, 1980), and *Solar Active Regions* (ed. F. Q. Orrall, 1981).

The topic of radiation output has many aspects ranging from the formalism of radiative transfer and diagnostic methods to the structure and energy balance of the solar atmosphere itself. The radiation not only carries in its spectral signature the information by which we diagnose the physical properties of the atmosphere but also a major role in determining certain of those physical properties. This chapter selects from the broad set of problems a subset that is representative of the problems currently undergoing active development. Because the selected topics are undergoing development, there are differences of opinion as to the methods and directions of development and as to the level of progress achieved.

The essay that follows is biased towards the author's point of view. Such bias is reflected in both the choice of topics and in the methods chosen to illustrate the nature of the problems.

Much work of current interest and many important historical developments are omitted from the following discussion. This is a regrettable but unavoidable restraint imposed by the conditions mentioned in the first paragraph.

Viewed from the perspective of the past 30 years, solar physics appears to be evolving from an essentially 'discovery mode' to what is better described as an 'understanding mode'. The discovery mode was characterized by exploratory observations in new regions of the spectrum and with new instrumental techniques for observing motions and magnetic fields. Virtually each new observing technique and each new spectral region revealed new phenomena and new aspects of previously known phenomena.

Most of the accessible spectrum has now been explored, and the rate of discovery of new phenomena appears to be slowing down. The rapid discovery of new phenomena over the past 30 years has impacted solar physics in many ways. Two of the more important impacts are represented on the one hand by the inescapable conclusion that the Sun is more complex than we had previously allowed ourselves to believe and, on the other hand, and perhaps more importantly, the growing conviction that we now have the potential to understand the physical processes involved in solar phenomena. These seemingly contradictory conclusions result from a newly recognized degree of order and correlation among solar phenomena that both restricts the range of plausible models and, in some cases, even suggests the nature of the model itself. It is the quest for under-standing the basic physical processes and the attempts to build physical models that shape and motivate most of today's efforts in solar physics.

The effort to understand the physical processes involved in the diverse phenomena of the solar atmosphere requires that we determine the cospatial thermodynamic, velocity, and magnetic field structures together with their temporal evolution. In addition, we must be able to identify and quantify the major energy and momentum processes. Experience indicates that such a goal cannot be accomplished without a combination of precise observations coupled with careful diagnostics based on realistic physical models. For the most part, solar physics has outgrown its heyday when crude observations and zeroth-order interpretation could produce worthwhile new results.

One of the major current limitations in the study of solar phenomena is the lack of adequate spatial resolution. Abundant evidence indicates that many solar phenomena have spatial scales considerably smaller than the best resolution currently attainable. In XUV regions of the spectrum we are further limited by insufficient spectral and temporal resolution. These are engineering and resource problems and can be improved upon with new ground and space facilities such as the Solar Optical Telescope (SOT).

Increased precision in spectroscopic data must be accompanied, of course, by refined diagnostic techniques. Such refinement often requires more realistic physical models, which, in turn, are mostly likely to come from better and more complete observations. Thus, the prospect for improved diagnostics is closely coupled to the prospect for im-proved observations.

The diagnostic methods discussed in this chapter are based on plane-parallel models. However, considerable work has been done in multidimensional radiative transfer (cf. review by Jones and Skumanich, 1980; Rybicki, 1967; Avery and House, 1968; Jones and Skumanich, 1973; Cannon, 1973; Stenhölm, 1977), but more remains to be done, particularly in the case of diagnostics for application to data with high spatial resolution. Many of the important features of the solar atmosphere, such as magnetic flux tubes, granules, intergranular lanes, spicules and fibrils, have characteristic dimensions of an arcsec (725 km), or less. In all such features, the multidimensional structure plays an important role in determining their interaction with and their contributions to the emergent radiation field.

Parametric studies of multidimensional geometries (cf. review by Jones and Skumanich, 1980) demonstrate that photon transfer is governed primarily by the nearest 'border' (defined by optical distance). In a plane-parallel atmosphere, the nearest border is always in the vertical direction. However, in multidimensional media a discrete structure may have 'borders' in the horizontal directions that are much nearer than the borders in the vertical direction. In addition, a discrete structure may be irradiated by other nearby structures of quite different radiation properties. Thus the 'incident' radiation becomes an important factor in the problem.

Spectral diagnostics is a rather subtle art even for plane-parallel geometries. Additional and important complexities arise in multidimensional cases, and reliable diagnostics become even more subtle. The changes that occur, however, are more quantitative than qualitative. The following discussion of diagnostics for plane-parallel geometries should be regarded, therefore, as illustrative of the qualitative effects. Accurate, quantitative diagnostics of high-resolution data will very probably require the development of suitable diagnostics for multidimensional media.

One of the primary strengths of solar physics is the ability to resolve many of the structural features of the solar atmosphere. From our present perspective, it is clear that many important physical phenomena occur on scales that are a tiny fraction of the solar radius. In fact, the solar chromosphere and corona, as well as important phenomena of the photosphere, exhibit many aspects of highly localized gas dynamic and MHD phenomena, which, in turn, appear to play a fundamental, and perhaps controlling, role in determining even the large-scale physical properties of the atmosphere. This leaves solar physicists with no alternative but to attempt to achieve spatial resolution commensurate with the scale of the hydrodynamic and MHD phenomena.

The need in solar physics to pursue detailed, precise observations and to seek refined diagnostic methods beyond those normally used in astrophysics may appear to those uninitiated in solar physics as perhaps a loss of perspective or as a hindrance to rapid progress. However, without such details much of solar physics would merely stagnate on a shaky framework of unanswered questions, unsolved problems, and uncertain conclusions. It was the pursuit of unexplained details in line profiles by a number of workers following the pioneering work of Jeffries and Thomas (1958) that led to a workable formulation of non-LTE methods in radiative transfer. Subsequent refinements and additions to non-LTE diagnostics and their applications to solar and stellar problems have profoundly altered the conclusions derived from analysis of line profiles.

Similarly, it was the pursuit of details in solar magnetograph data by several workers that led finally to the conclusion that solar magnetic fields are mostly confined to small flux tubes with diameters of a fraction of an arcsecond, which is much below the resolution limit of the magnetographs. This surprising conclusion has totally reshaped the interpretation of solar magnetic field phenomena.

These are but two of many examples in solar physics where careful observations and careful attention to theoretical detail have overturned previous concepts and replaced them with radically new ones. The process is by no means complete. We are still forced to rely heavily on primitive concepts that are known to be inadequate or incorrect, but which represent the best we can do at the time. Diagnostic methods for velocity and magnetic fields are particular cases in point.

In the following discussion, it is necessary to omit much of the detail referred to above. The reader should bear in mind, however, that it is just these details in both observation and theory that provide the basis for progress.

Section 2 deals with some concepts and definitions in radiative transfer that are useful for subsequent discussion. Those already familiar with radiative transfer theory may wish to skip this rather elementary discussion. Section 3 discusses diagnostic methods and associated problems. Section 4 considers the role of radiation in determining the thermodynamic properties of the atmosphere. Sections 5 and 6 deal briefly with certain aspects of the nonradial structure of the atmosphere and the temporal fluctuations of the radiation output. These topics are discussed in broader context in other chapters of this volume. Finally, Section 7 contains some comments on future expectations and needs in solar physics.

2. Basic Concepts and Definitions

2.1. FORMATION OF SPECTRA

With but few exceptions, quantitative studies of the Sun are dependent upon the interpretation of spectroscopic data. Whether such data are collected at radio, visual, or X-ray wavelengths, their interpretation rests upon our ability, first, to identify the process by which the radiation is produced, second, to interpret the radiation in terms of local plasma conditions at the place of origin, and, third, to organize the different plasma regimes observed at different wavelengths in proper relationship to each other.

To illustrate the problems involved in each of these steps, we consider a spectrum giving specific intensity I_λ as a function of wavelength. We represent the functional dependence of the spectrum in the symbolic form:

$$I_\lambda(r_i, S_i, \varphi_{\lambda i}, \kappa_{0i}, \Delta z_i, \Delta\lambda, \Delta\theta, \Delta t),$$

where r_i denotes the different radiation processes (usually two or more, such as line and continuum), S_i denotes the source function for photons of species i, $\varphi_{\lambda i}$ is the absorption profile, κ_{0i} is the absorption coefficient, Δz_i denotes the spread in the line-of-sight distance within which the observed radiation originates and $\Delta\lambda$, $\Delta\theta$, and Δt, respectively, are spectral, spatial, and temporal resolution. The latter three quantities are instrumental parameters and, within limitations, are set by the observer. On the other hand, the five quantities r_i, S_i, κ_{0i}, $\varphi_{\lambda i}$, and Δz_i are determined by the solar atmosphere, and these are the five quantities that convey most of the information about the atmosphere.

Even under optimum conditions for $\Delta\theta$ and $\Delta\lambda$, the radiation intensity I_λ at a given value of λ arises from a spread in distance that is often larger than the characteristic lengths for changes in thermodynamic, fluid dynamic, and magnetic field variables. This has the consequence that a given I_λ cannot be unambiguously associated with specific values of such variables as temperature, fluid velocity, and magnetic fidle strength. Instead, one must interpret I_λ in terms of a model atmosphere in which both the local plasma parameters and their gradients are given. To accomplish this, it is necessary to consider values of I_λ over a number of different wavelengths with different representative values of Δz.

A further consequence of the inability to assign a unique value of z to I_λ is that the average value, $\langle z \rangle$, may depend upon the particular parameter extracted from I_λ. For example, there is no *a priori* reason for the value of $\langle z \rangle$ associated with a temperature inferred from the source function to correspond to that value of $\langle z \rangle$ associated with a fluid velocity inferred from φ_λ. Also, the value of $\langle z \rangle$ associated with a fluid velocity may be quite different from the $\langle z \rangle$ associated with a magnetic field intensity, even though both are inferred from φ_λ.

The preceding comments illustrate the inherent difficulty of properly organizing the plasma variables in terms of relative values of z. This, however, is only part of the problem. An equally, and perhaps more, difficult problem is in the complicated transport of photons through the atmosphere and the consequent complexity of the information content in I_λ. Although all of the photons in I_λ can be represented as being emitted within the range Δz, it does not follow that the photon energy in I_λ is extracted from the interval energy within Δz. In fact, much, and perhaps most, of the photon energy within Δz may simply be a result of photons diffusing through the atmosphere, i.e. scattering. Scattering photons often react passively with the local plasma with no essential communication between the photon and the internal energy of the plasma. An electron raised to a higher level through absorption of a photon merely returns to its original level and re-emits the photon.

In addition to photons diffusing in depth, they diffuse in wavelength. Thus, photon energy at wavelength λ contributing to I_λ may have originated at a depth far removed from Δz and at a wavelength quite different from λ. What this means in a practical sense is that of the total photons in I_λ only a fraction, say, ρI_λ are produced within the depth internal Δz and it is only this fraction that is strongly coupled to the thermodynamic state of the plasma within Δz. In many cases of interest in solar physics, ρ is very small and must be determined carefully. This requires accurate solutions of the radiative transfer equations.

As difficult as it is to diagnose properly the solar spectrum, such diagnoses provide the primary means of probing the physical state of the atmosphere. It is essential, therefore, that we understand the complexities of the problem in order that we treat it with proper care.

Studies of the Sun, of course, often utilize the resolved structure on the solar disk and at the limb to aid in determining the relative locations of different solar features and different plasma regimes. Features that are high in the atmosphere, for example, are usually more readily visible at the limb whereas features that are low in the atmosphere tend to be obscured at the limb. Similarly, limb brightening or darkening gives a relatively unambiguous indication of the sign of the intensity gradient. Even such rudimentary knowledge provides a strong and useful supplement to the spectroscopic data.

2.2. ENERGY BALANCE

Radiation from the Sun not only provides the primary means for diagnosing the solar atmosphere; it plays a major role in determining the physical state of the atmosphere. In the photosphere, radiation provides both the main source of energy and the main loss of energy. Higher in the atmosphere, in the chromosphere and corona, the plasma is heated to high temperature by nonradiative sources and, in turn, eventually loses almost all of this energy as radiation. The fraction lost in the solar wind and in energetic particles may dominate in the outer corona and in such places as coronal holes, but, as a global average, the radiation losses dominate.

The problem of ascertaining the magnitude of the radiation terms in the local energy balance is no different, in essence, from that of performing the diagnostics for the thermodynamic variables. In both cases one needs to determine from the total photon ensemble the fraction of the energy that is produced locally. However, whereas the diagnostics might be performed with only a small subset of the spectrum, the energy balance requires an examination of all relevant parts of the spectrum. In some cases, it is not clear *a priori* just which spectral features are most relevant. For this reason, relatively less progress has been made with the energy balance problem. In the photosphere and low chromosphere, we still rely on approximate treatments of questionable validity even though the thermodynamic structure in these layers is known better than in the higher layers. The techniques for solving these problems with higher accuracy are known, but the sheer magnitude of the problem is intimidating.

2.3. NOTATION

Extended discussions of spectral diagnostics are given in such texts as Thomas (1965), Jeffries (1968), Athay (1972), and Mihalas (1978). Here we review only certain aspects of the problems that are particularly germane to the present discussion. For the most part, the emphasis will be on concepts rather than the formal mathematics of the problem, which are beyond the intended scope of this chapter. This may tend to convey a false illusion of simplicity since we will avoid the harsh reality of dealing with quantitative solutions.

We assume that the reader is familiar with the definitions of the customary parameters of radiative transfer, including the following:

I_ν = specific intensity
J_ν = mean intensity over angle
H_ν = net outward flux
μ = $\cos \theta$, θ measured from the normal
τ_ν = optical depth
Φ_ν = absorption profile normalized so that $\int_{-\infty}^{\infty} \Phi_\nu \, d\nu = 1$
φ_ν = $M_0 \Phi_\nu$ = absorption profile normalized to unity at $\nu = \nu_0$
Ψ_ν = emission profile normalized so that $\int_{-\infty}^{\infty} \Psi_\nu \, d\nu = 1$
S_ν = source function
B_ν = Planck function.

In addition, we wish to make use of a number of concepts that are best discussed in terms of other quantities, which we now define.

2.4. ESCAPE PROBABILITY AND SCATTERING DEPTH

We define the single-step escape probability, P_e, for a photon ensemble at depth τ_0 ($\nu = \nu_0$) and between frequencies ν_1 and ν_2 as

$$P_e = \int_{\nu_1}^{\nu_2} \int_0^1 \Psi_\nu \exp(-\tau_\nu/\mu) \, d\mu \, d\nu, \qquad (2.1)$$

i.e. escape means the arrival of a photon at $\tau = 0$. Since Ψ_ν is normalized such that $\int_{-\infty}^{\infty} \Psi_\nu \, d\nu = 1$, P_e is the average probability that a photon is emitted in the interval ν_1 to ν_2 and reaches the surface $\tau_0 = 0$ without interacting with the gas by scattering or absorption.

The context in which we shall use P_e is illustrated through the integral form of the transfer equation. The specific intensity at the surface $\tau_0 = 0$ due to radiation between $\tau_0 = 0$ and a finite value τ_0 is given by

$$I_\nu(0) = \int_0^{\tau_0} S_\nu \exp(-\tau_\nu/\mu) \, \frac{\mathrm{d}\tau_\nu}{\mu} \, . \tag{2.2}$$

After multiplying by μ and averaging over μ, we obtain

$$H_\nu(0) = \int_0^{\tau_0} S_\nu M_0 \int_0^1 \Phi_\nu \exp(-\tau_\nu/\mu) \, \mathrm{d}\mu \, \mathrm{d}\tau. \tag{2.3}$$

We next integrate over frequency, assuming $S_\nu \Phi_\nu = S\Psi_\nu$, where S is independent of frequency, and obtain

$$H(0) = \int_0^{\tau_0} S P_e M_0 \, \mathrm{d}\tau_0. \tag{2.4}$$

Thus, $P_e M_0 S$ gives the contribution to $H(0)$ of photons emitted in the optical depth interval between τ_0 and $\tau_0 + \mathrm{d}\tau_0$.

Note that P_e is a function only of τ_0, Ψ_ν and φ_ν and can be evaluated as a function of τ_0 without solving the transfer equation. Near the surface in an optically thick atmosphere P_e approaches $1/2$.

Since P_e is defined as the average probability of reaching $\tau_0 = 0$ in a single step, we define a scattering depth, N, by setting

$$NP_e = 1/2. \tag{2.5}$$

Note that N approaches unity at the surface $\tau_0 = 0$. The dependence of N on τ_0 varies with the form of Φ_ν and Ψ_ν. For $\Psi_\nu = \Phi_\nu = \text{const}$, $N \propto E_2(\tau_0)$, where E_2 is the second exponential integral, and for a Gaussian Φ_ν, $N \propto \tau_0(\ln\tau_0)^{1/2}$. As defined, N is the average number of photon emissions required in order for one photon at τ_0 to reach the surface $\tau_0 = 0$ by direct flight. In the following, we will sometimes use N rather than τ_0 as a depth variable as a matter of convenience.

2.5. ESCAPE COEFFICIENT

We now define an escape coefficient (sometimes called the net radiative bracket, Thomas (1961)) that is different from the escape probability and therefore useful in a different context. The equation of radiative transfer

$$\mu \, \frac{\mathrm{d}I_\nu}{\mathrm{d}\tau_\nu} = I_\nu - S_\nu \tag{2.6}$$

can be averaged over angle to give

$$\frac{\mathrm{d}H_\nu}{\mathrm{d}\tau_\nu} = J_\nu - S_\nu \tag{2.7}$$

and integrated over frequency to give

$$dH = \left(\int J_\nu \Phi_\nu \, d\nu - S \right) M_0 \, d\tau_0. \tag{2.8}$$

Equation (2.8) may be rewritten as

$$dH = -\rho S M_0 \, d\tau_0. \tag{2.9}$$

where ρ is the escape coefficient defined by

$$\rho = 1 - \frac{\int J_\nu \Phi_\nu \, d\nu}{S}. \tag{2.10}$$

We note from Equation (2.9) that the total flux at $\tau_0 = 0$ is given by

$$H(0) - H(\tau_0) = \int_0^{\tau_0} \rho S M_0 \, d\tau_0, \tag{2.11}$$

This is parallel to Equation (2.4) with the exception that P_e is replaced by ρ and there is an additional term, $H(\tau_0)$. However, the interpretation of the two equations is quite different. In Equation (2.9) $\rho S M_0$ is the increment in H at τ_0, whereas in Equation (2.4) $P_e S M_0$ represents an increment in $H(0)$ at $\tau_0 = 0$. Thus, we interpret ρ as the probability that a photon emitted at τ_0 adds to the net flux at τ_0. In other words, it is the probability that a photon at τ_0 is not a result of diffusion and, therefore, must be a photon created locally.

Note that even at $\tau_0 = 0$, ρ and P_e are different. As noted earlier, P_e depends only upon τ_0 and the form of φ_ν and Ψ_ν and not upon the local value of J_ν. On the contrary, ρ depends on $\int J_\nu \Phi_\nu \, d\nu/S$ and can be evaluated only through an explicit and careful solution of the transfer equation. In many cases of interest, ρ is of the order of 10^{-2} or smaller, near $\tau_0 = 0$. The only cases where ρ approaches unity at the surface $\tau_0 = 0$ are those for which the total optical thickness of the radiating layer is much less than unity and in which $H(\tau_0)$ is small. Whenever the total optical thickness is large enough that $H(0)$ becomes of the order of S near $\tau_0 = 1$, ρ is likely to be small.

2.6. CREATION AND DESTRUCTION PROBABILITIES

The terms photon creation and destruction are used in reference to energy exchange between the photon field and the plasma, between photons in different spectral regions and between different radiation processes such as line and continuum. A line photon absorbed in a continuum transition, for example, is destroyed from the line. Similarly, a photon emitted at the wavelength of the line by a continuum transition represents a creation of a line photon. Electrons excited from the lower level to the upper level of a line transition by collisional interactions followed by spontaneous decay represent creations of line photons from the thermal energy of the plasma. Also, cascading electrons from recombinations to excited levels or from excitation of highly excited states create new photons in the spectral lines involved in the cascade. Each of these processes has an inverse through which line photons can be destroyed.

For purposes of this discussion, we need not be concerned with specific details of all the various photon processes. Rather we shall simply express the total probabilities for photon creation, P_c, and destruction, P_d, as

$$P_c = P_c^{\text{th}} + P_c^{\text{nth}} \qquad (2.12)$$

and

$$P_d = P_d^{\text{th}} + P_d^{\text{nth}}, \qquad (2.13)$$

where the superscripts 'th' and 'nth' represent, respectively, the processes, such as collisions and continuum absorptions, that couple photon energy directly to the thermal energy of the plasma and the basically nonthermal processes in which photon energy at wavelength λ is merely redistributed to photons at other wavelengths.

We will further make use of the notation

$$P_c^{\text{th}} = \epsilon B/S, \qquad (2.14)$$

$$P_c^{\text{nth}} = \delta/S, \qquad (2.15)$$

$$P_d^{\text{th}} = \epsilon, \qquad (2.16)$$

and

$$P_d^{\text{nth}} = \epsilon^*, \qquad (2.17)$$

but we will not give explicit expressions for ϵ, ϵ^*, and δ. For such expressions the reader is referred to more extended discussions of the transfer problem, such as referenced earlier or as given by Athay (1981a). In the notation adopted, ϵ is at most a slow function of temperature, and ϵ^* and δ may or may not be functions of temperature. S is the frequency independent form of the line source function.

2.7. THERMALIZATION AND DEGRADATION LENGTHS

Two length scales of interest in the transport of photons through an atmosphere are the typical length a photon migrates before being converted back to thermal energy and the typical length a photon migrates before being lost from a spectral line by any process including thermalization. We define these length scales in terms of the scattering distance N.

The thermalization distance, N_{th}, is defined as

$$N_{\text{th}} = \frac{1}{P_d^{\text{th}}} = \frac{1}{\epsilon}, \qquad (2.18)$$

and the degradation distance, N_{deg}, is defined as

$$N_{\text{deg}} = \frac{1}{P_d} = \frac{1}{\epsilon + \epsilon^*}. \qquad (2.19)$$

For many lines in the solar spectrum, particularly resonance lines, $N_{th} \approx N_{deg}$. On the other hand, for most subordinate lines $N_{th} \gg N_{deg}$.

2.8. INTRA-ATMOSPHERE EXCHANGE PROBABILITY

Photon exchange between different depths in an atmosphere are important in the net photon balance. We define the probability of a photon at $\tau_0 = \tau_1$ traveling to depth $\tau_0 = \tau_2 > \tau_1$ in a single step as

$$P_{\tau_1}^{\tau_2} = \int_0^\infty \int_{-1}^0 \Psi_{\nu_1} \exp\left[(\tau_{\nu_2} - \tau_{\nu_1})/\mu\right] d\mu \, d\nu, \qquad \mu < 0, \qquad (2.20)$$

and the reverse probability as

$$P_{\tau_2}^{\tau_1} = \int_0^\infty \int_0^1 \Psi_{\nu_2} \exp\left[-(\tau_{\nu_2} - \tau_{\nu_1})/\mu\right] d\mu \, d\nu, \qquad \mu > 0, \qquad (2.21)$$

where Ψ_{ν_1} and Ψ_{ν_2} are the emission profiles at τ_1 and τ_2 and τ_{ν_1} and τ_{ν_2} are the frequency dependent optical depths at τ_1 and τ_2.

Note that $P_{\tau_1}^{\tau_2} \neq P_{\tau_2}^{\tau_1}$ unless $\Psi_{\nu_1} = \Psi_{\nu_2}$. Gradients in Ψ_ν can be produced by gradients in temperature, velocity, or magnetic field strength. When such gradients are present, as they always are in the solar atmosphere, photon migration between depths τ_1 and τ_2 is asymmetric with respect to the plus and minus directions in μ. Thus, photons may be preferentially trapped in some regions and preferentially drained from other regions.

2.9. THE SOURCE FUNCTION

The line source function, neglecting stimulated emissions, can be written in the rather general form (cf. Jeffries, 1968; Athay, 1972; Mihalas, 1978)

$$S_\nu = \left(\frac{\int\!\int J_\nu \Phi_\nu \, d\nu + \epsilon B + \delta}{1 + \epsilon + \epsilon^*}\right) \frac{\Psi_\nu}{\Phi_\nu}. \qquad (2.22)$$

In the following, we assume $\Psi_\nu \equiv \Phi_\nu$, which renders S_ν independent of frequency. The frequency independent form, S, can be rewritten by means of Equation (2.10) as

$$S = \frac{\epsilon B + \delta}{\rho + \epsilon + \epsilon^*}. \qquad (2.23)$$

and this, in turn, can be solved for ρ to obtain

$$\rho = \epsilon \frac{B}{S} + \frac{\delta}{S} - \epsilon - \epsilon^*. \qquad (2.24)$$

From Equations (2.12)–(2.17), we find

$$\rho = P_c - P_d. \qquad (2.25)$$

Solutions to the coupled radiative transfer and statistical equilibrium equations for S normally begin by specifying B and ϵ and at least some of the elements in δ and ϵ^*. Sometimes such specifications can be given as functions of τ_0, but more generally they are given as functions of geometrical depth. However, in cases where δ and ϵ^* are important, these terms frequently involve values of ρ for transitions other than the one in question. Thus, the solutions to the coupled set of transfer equations must specify a self-consistent set of ρ's for each of the relevant transitions.

2.10. THE TRANSFER EQUATION

The normal equation of radiative transfer, expressed as Equation (2.6), treats the net gain and loss of photons for a specific frequency and direction cosine. For purposes of computing S/B, this represents far more information than is needed. Also, it is difficult to relate the normal transfer equation to some of the useful concepts discussed in the preceding paragraphs.

It is more practical for computation of S/B to use the transfer equation in a form that deals with just those photons that directly influence S/B. Equations (2.8) and (2.9) are such equations. By combining Equations (2.9) and (2.25), we obtain

$$\frac{1}{M_0} \frac{dH}{d\tau_0} = (P_d - P_c)S. \tag{2.26}$$

Also, we note that Equation (2.8) may be rewritten as

$$\frac{1}{M_0} \frac{dH}{d\tau_0} = \int \Phi_\nu \Lambda_\nu(S_\nu)\, d\nu - S, \tag{2.27}$$

where we use $\Lambda_\nu(S_\nu)$ to denote the standard lambda transform of S_ν. A primary characteristic of the lambda transform of a function $f(T)$ is that it returns $f(T)$ as a leading term. Thus, we may replace the full lambda transform of S_ν by

$$\Lambda_\nu(S_\nu) = S_\nu + \lambda_\nu(S_\nu), \tag{2.28}$$

where $\lambda_\nu(S_\nu)$ is the difference between J_ν and S_ν. Equation (2.27) then becomes

$$\frac{1}{M_0} \frac{dH}{d\tau_0} = \int \Phi_\nu \lambda_\nu(S_\nu)\, d\nu \tag{2.29}$$

and this combines with Equation (2.26) to give

$$\int \Phi_\nu \lambda_\nu(S_\nu)\, d\nu = (P_d - P_c)S. \tag{2.30}$$

We now have an equation for S that contains only the photon interactions that have a direct influence on S.

It is still convenient to convert Equation (2.30) to a differential form. To illustrate this for the case of a spectral line formed in regions of low continuum opacity, we assume S_ν

to be independent of frequency. We further assume that $S(\tau)$ can be represented by a series of steps in each of which S is constant. Thus, at depth τ_i we take S to be constant at a value S_i between the limits $t_i \leq \tau_i \leq t_{i+1}$, where t_i is given by

$$t_i = (\tau_i \tau_{i-1})^{1/2}. \tag{2.31}$$

This results in the following expansion at depth τ_i for the integral in Equation (2.30) (the reader should consult a standard text in radiative transfer):

$$\int \Phi_\nu \lambda_\nu(S) \, d\nu = - \sum_{j=1}^{i-1} S_j(P_j^i - P_{j+1}^i) +$$

$$+ \sum_{i+1}^{m} S_j(P_j^i - P_{j+1}^i) - S_i(P_i^i + P_{i+1}^i), \tag{2.32}$$

where P_j^i is the intra-atmosphere exchange probability from depth t_j to depth τ_i as defined by Equations (2.20) and (2.21).

We now rewrite Equation (2.32) in the form

$$\int \Phi_\nu \lambda_\nu(S) \, d\nu = - \sum_{j=1}^{i-2} S_j(P_j^i - P_{j+1}^i) - S_{i-1}(P_{i-1}^i - P_i^i) -$$

$$- S_i(P_i^i + P_{i+1}^i) + S_{i+1}(P_{i+1}^i - P_{i+2}^i) +$$

$$+ \sum_{j=i+2}^{m} S_j(P_j^i - P_{j+1}^i). \tag{2.33}$$

The terms S_{i-1} and S_{i+1} can be replaced by

$$S_{i-1} = S_i - \frac{\mathrm{d}S_i}{\mathrm{d}\tau_i} (\tau_i - \tau_{i-1}) \tag{2.34}$$

and

$$S_{i+1} = S_i + \frac{\mathrm{d}S_i}{\mathrm{d}\tau_i} (\tau_{i+1} - \tau_i). \tag{2.35}$$

After collecting terms and substituting into Equation (2.30), we obtain

$$[(P_{i+1}^i - P_{i+2}^i)(\tau_{i+1} - \tau_i) - (P_i^i - P_{i-1}^i)(\tau_i - \tau_{i-1})] \frac{\mathrm{d}S_i}{\mathrm{d}\tau_i}$$

$$= [P_d - P_c + P_{i-1}^i + P_{i+2}^i] S_i + \sum_{j=1}^{i-2} S_j [P_j^i - P_{j+1}^i] -$$

$$- \sum_{i+2}^{m} S_j [P_j^i - P_{j+1}^i]. \tag{2.36}$$

Aside from the minor approximation of replacing $S(T)$ by a series of step functions, which can be made arbitrarily accurate, Equation (2.36) is exact. The quantities P_c and P_d are given by Equations (2.14)–(2.17) and the exchange probabilities P_j^i are readily computed from Equations (2.20) and (2.21).

The advantage of working with a transfer equation that includes only the photon events that directly influence S is the ease with which the equation can be solved. Equation (2.36), for example, is readily solved by a simple iterative technique beginning from an initial solution for $S(T)$ that may be far from the final solution.

Equations even simpler than Equation (2.36) have been developed by several authors. These simpler equations, however, are restricted to cases in which Φ_ν does not vary with depth and, even then, are only approximate. The two most notable examples are given by Frisch and Frisch (1975) as

$$2P \frac{dS}{dP} = -S + P_d \frac{dB}{dP} \tag{2.37}$$

and by Scharmer (1981) as

$$(P_d + 2P_e) \frac{dS}{dP_e} = -\left(1 + \frac{dP_d}{dP_e}\right)S + \frac{d(P_d B)}{dP_e} . \tag{2.38}$$

The quantity P in Equation (2.37) is defined by

$$P = P_d + (1 - P_d)P_e. \tag{2.39}$$

The most notable difference between Equation (2.36) and the two approximate equations, (2.37) and (2.38), is the use of the escape probability, P_e, to replace the exchange probabilities. All of the exchange effects are approximated by photons escaping from lower depths.

Equation (2.36) is readily converted to a differential equation in P_e or N rather than τ using Equations (2.1) and (2.4). As far as solving the equation is concerned, these substitutions make no substantial difference. However, it is convenient to discuss the solutions in terms of N.

Solutions to the transfer equation for constant ϵ, ϵ^* and δ and for B either constant or increasing monotonically with N give $S \approx B$ for $N \gg N_{\text{th}}$ ($P_e \ll P_d^{\text{th}}$) and $S \propto (\epsilon N)^{1/2}$ for $N \ll N_{\text{th}}$ ($P_e \gg P_d^{\text{th}}$). The depth at which S saturates to B is near $N = N_{\text{th}}$. These properties are readily demonstrated from Equations (2.37) and (2.38).

In spectral diagnostics, one sometimes uses inversions of Equation (2.2) to determine $S_\nu(\tau_0)$ from observed values of $I_\nu(0)$. If this is done successfully, the empirical values of S_ν can provide strong clues to the nature of the remaining parameters in Equation (2.36), such as B. Such approaches have been used very successfully in analyses of continuum data. On the other hand, such analyses are extremely suspect in the case of spectral lines. The function Φ_ν varies rapidly with distance from line center and, as a result, τ_ν is very sensitive to both the characteristic width of ϕ_ν and to local Doppler shifts of ϕ_ν. This has the consequence, in some cases, of making $S_\nu(\tau_0)$ an extremely rapid function of τ_0, and of grossly distorting the mapping of S_ν into $I_\nu(0)$ to such a degree that empirical inversions of Equation (2.2) can be virtually meaningless.

3. Spectral Diagnostics

3.1. TEMPERATURE DIAGNOSTICS

In optically thick cases, temperature information resides essentially in the Planck function, B, in Equation (2.36). It is of interest, therefore, to investigate the response of S to chromospheric temperature increases of the form (Jeffries and Thomas, 1958)

$$B_c = B_0 \left(1 + A \exp(-N/N_c)\right). \tag{3.1}$$

The ratio of the solution using B_c to the solution using B_0 is given by (Athay, 1976)

$$\frac{S_c}{S_0} = 1 + \frac{(\epsilon N_c)^{1/2}}{(1 + \epsilon N_c)^{1/2}} A, \tag{3.2}$$

which has the form

$$\frac{S_c}{S_0} = 1 + A, \qquad \epsilon N_c \gg 1, \tag{3.3}$$

and

$$\frac{S_c}{S_0} = 1 + (\epsilon N_c)^{1/2} A, \qquad \epsilon N_c \ll 1. \tag{3.4}$$

It follows that a single spectral line provides a good temperature diagnostic only if the quantity $(\epsilon N_c)^{1/2} A$ is not small compared to unity. Plots of S_c/B_0 for a range of values of A and ϵN_c are shown in Figure 1.

One can maximize N_c by selecting very strong lines, such as Ca II H and K, Mg II, k and h or Lyman-α. For these lines, however, ϵ tends to be very small. On the other hand, one can maximize ϵ by selecting forbidden lines, or, in some cases, subordinate lines, but these selections tend to minimize N_c. The quantity A, which measures the response of the Planck function to an increase of temperature, of course, increases at shorter wavelengths. Thus, lines formed in the extreme ultraviolet are better for temperature diagnostics than lines formed in the visual and infrared regions. Table I contains approximate solar values of A, N_c, and ϵ for a few lines of interest.

TABLE I

Approximate value of A, N_c, and ϵ for representative solar lines in the low chromosphere

Line	A	N_c	ϵ
Hα	4	10^2	5×10^{-4}
Ca II K	10	3×10^4	2×10^{-4}
Si II (1816.93)	10^2	10^3	3×10^{-3}

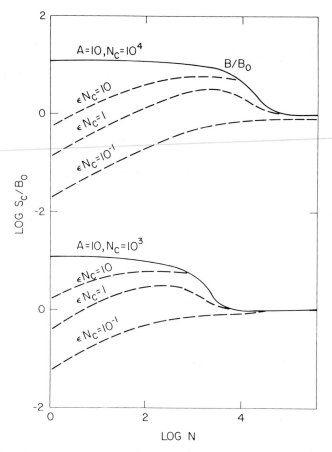

Fig. 1. An illustration of the ratio of the chromospheric source function to the Planck function without the chromospheric component (Equation (3.7)) for $A = 10$ and a range of values of ϵ and N_c.

corona, and transition region, one can be reasonably certain that the total thickness of the radiating layer is less than N_{th}. This leads to a useful approximation in which P_d can be ignored and ρ is consequently given by P_c (Equation (2.25)). If it is also known that $P_c^{th} \gg P_c^{nth}$, which is often true, then Equation (2.14) gives

$$\rho S = \epsilon B \tag{3.5}$$

and Equation (2.11) gives

$$H(0) - H(\tau_0) = \int_0^{\tau_0} M_0 \, \epsilon B \, d\tau_0. \tag{3.6}$$

In the case of collisional excitations and no incident flux $(H(\tau_0) = 0)$, Equation (3.6) reduces to the well-known relation for emission lines.

$$H(0) = \frac{h\nu}{4\pi} \int_0^{z_0} N_L \, C_{L\nu} \, dz, \tag{3.7}$$

where N_L is the population density of the lower level of the transition, C_{LU} is the collisional excitation rate between the lower and upper levels, and z is the geometrical height in the atmosphere.

Equation (3.7) is simply the statement that each collisional excitation generates a photon, and the photon escapes the atmosphere before being lost from the line. This latter condition is referred to as 'effectively thin'. The only requirement is that the photon eventually escapes, which requires only that the total scattering depth $N \ll N_{th}$. For many lines in the solar spectrum, $N_{th} \gtrsim 10^4$, and Equation (3.7) can still be valid even for strong lines.

The collision term in Equation (3.7) can be written $C_{LU} = QN_e f(T)$, where Q is a collision strength, N_e is electron density, and $f(T)$ is an exponential function of temperature of the form $\exp(-h\nu_0/kT)$. Similarly, N_L can be written $N_L = A_{el}N_H g(T)$, where A_{el} is the abundance of the element relative to hydrogen, N_H is the hydrogen density, and $g(T)$ is a mildly peaked function of temperature. Substitution of these expressions in Equation (3.7) given

$$H(0) = \frac{h\nu}{4\pi} QA_{el} \int_0^{z_0} N_e N_H F(T) \, dz, \qquad (3.8)$$

where $F(T) = f(T)g(T)$. The function $F(T)$ is sometimes sharply peaked in temperature and sometimes only mildly peaked. More typically, $F(T)$ remains near its maximum for $\Delta T/T \approx 1$.

Equation (3.8) is most useful as a temperature diagnostic when used in combination with several spectral lines. One must remember, however, that in making use of lines for which $N \ll N_{th}$ all information on the depth of line formation has been lost. Thus, the only direct information one obtains from Equation (3.8) is the integral of $N_e N_H$ over the depths where $F(T)$ is close to its maximum value. When several lines are used one gets only a distribution function for the emission measure as a function of temperature. In order to organize this distribution function into a depth distribution, other information, such as center-limb variations, must be used.

Diagnostics of free–bound and free–free continua are essentially similar to those for spectral lines. The continua have the advantage, however, in that φ_ν is less affected by local atmospheric parameters and is, therefore, better known. Also, in the case of the H^- continuum the effective ϵ is large relative to strongly permitted lines. Thus, the thermalization depth is smaller, and S is more closely coupled to B than in the case of the lines.

Free–bound continua that are optically thin ($N = 1$) have the interesting property that d ln I_ν/dλ is a function of temperature only. Thus, if one can isolate the continuum emission the slope gives a unique measure of the local temperature. Unfortunately, the free–bound continua are not usually intense enough to give an accurate value for d ln I/dλ unless they have appreciable optical thickness. When this occurs the slope d ln I_ν/dλ depends upon the temperature gradient as well as the temperature.

3.2. DENSITY DIAGNOSTICS

Density information in Equation (2.36) is contained mainly in ϵ and N emerging in the effectively thin limit with the combined dependence as shown in Equation (3.8). This latter form is useful, however, only when N_e and N_H can be related to T (e.g. through hydrostatic equilibrium) or when the temperature dependence is slight as is the case near the heads of free–bound continua. The high Balmer lines and continuum, for example, have been much used in this context.

A more popular density diagnostic results from the use of certain pairs of lines whose flux ratios are dependent on density (Gabriel and Jordan, 1969; Doschek and Feldman, 1978; Dere et al., 1979; Dere and Mason, 1981). To illustrate this case, we write the flux in the line in the form

$$H(0) = \frac{h\nu}{4\pi} \int_0^{z_0} A_{UL} N_U \, \mathrm{d}z. \tag{3.9}$$

where A_{UL} is the spontaneous transition probability and N_U is the population of the upper level. Equation (3.9) is a special case of Equation (3.8) for which $C_{LU} N_L = A_{UL} N_U$, and is generally valid only for optically thin conditions, $N = 1$. Both equations are restricted to values of z_0 such that $N \ll N_{th}$.

Consider two spectral lines with upper levels 2 and 3 and with a common lower level 1. One of the lines (2–1) is supposed to be strongly permitted and the other (3–1) weakly permitted or forbidden so that the transition probabilities A_{21} and A_{31} have a large ratio. Also, transitions from 3 to 2 are forbidden so that the collision rate between 3 and 2 exceeds the radiative rate. Such line pairs are representative of the permitted resonance line of He-like ions paired with either the intersystem or forbidden line. In these circumstances, level 3 will build up a large population relative to level 2 and will be depopulated mainly by collisional transitions to levels 2 and 1 and spontaneous transitions to level 1. The population ratio N_3/N_2 will then be density dependent, and it follows from Equation (3.9) that the ratio of line fluxes will be density dependent.

The equilibrium ratio for levels N_3/N_2 can be expressed in the form

$$\frac{N_3}{N_2} = \frac{R_{23} R_{11} + R_{21} R_{13}}{R_{32} R_{11} + R_{31} R_{12}}, \tag{3.10}$$

where the R_{ij}'s represent total transition rates from level i to level j and R_{11} represents the total rate out of level 1. We ignore all photo-excitations and consider $A_{32} = 0$ and $A_{21} \gg C_{21}$. Equation (3.10) can then be written

$$\frac{N_3}{N_2} = \frac{C_{23}(C_{12} + C_{13}) + A_{21} C_{13}}{C_{32}(C_{12} + C_{13}) + (A_{31} + C_{31}) C_{12}}. \tag{3.11}$$

We now suppose that levels 2 and 3 have approximately the same energy so that C_{12} and C_{13} have the same temperature dependence and C_{23} and C_{32} are essentially independent of temperature. The collision rates C_{ij} can then be replaced by $Q_{ij} N_e$ where Q_{ij} is a collision strength. Equation (3.11) then becomes

$$\frac{N_3}{N_2} = \frac{Q_{23} N_e + (1 - \sigma) A_{21}}{(Q_{32} + \sigma Q_{31}) N_e + \sigma A_{31}}. \tag{3.12}$$

where

$$\sigma = \frac{Q_{12}}{Q_{12} + Q_{13}}. \tag{3.13}$$

Although we have represented C_{23} and C_{32} as functions of electron density, in practice heavy particle collisions may be equally important when the energy difference between levels 2 and 3 is small.

It is evident from Equation (3.12) that the ratio N_3/N_2 is density dependent in the range of densities such that $C_{23} \ll (1 - \sigma)A_{21}$ and $C_{32} + \sigma C_{31} \gg \sigma A_{31}$. Thus, there is both an upper and lower limit on densities at which the ratio is density dependent. For the solar case, only the low-density limit is of interest.

The line ratio method for determining densities is not restricted to lines close together in wavelength nor to lines of the same ion. All that is required is for the population ratio to be expressable in terms of known quantities and for the two lines to be formed in the same temperature regime. If the lines are well separated in wavelength or come from different ions, the line ratio will be temperature dependent as well as density dependent and, thus, the temperature must be known. Also, if two different atomic species are represented their relative abundances must be known. If the lines are formed in even mildly different temperature regimes, the integral in Equation (3.9) will have a somewhat different centroid for the two lines, and this introduces an added element of uncertainty in the density diagnostic. In practice, it is usually necessary to use lines for which the ratio is temperature dependent, and some authors have used line pairs from different atomic species. Such line pairs are useful, but the results can sometimes be misleading.

A final density diagnostic useful in the corona and in isolated solar features such as prominences and spicules in Thompson scattering of photospheric radiation by free electrons. In this case the scattering intensity varies linearly with electron density and is independent of temperature. The method has been widely and successfully used in the corona. In prominences and spicules, care must be taken to exclude line emission, which is not easily accomplished.

3.3. VELOCITY AND MAGNETIC FIELD DIAGNOSTICS

Velocity and magnetic field information reside almost entirely in φ_ν with only secondary effects in S_ν. We note from Equation (2.2) that

$$\frac{dI_\nu(0)}{dS_\nu} = \int_0^{\tau_0} \varphi_\nu \exp(-\tau_\nu/\mu) \frac{d\tau_0}{\mu}. \tag{3.14}$$

where

$$\frac{dI_\nu(0)}{d\varphi_\nu} = \int_0^{\tau_0} \left(S_\nu + \varphi_\nu \frac{dS_\nu}{d\varphi_\nu} - \frac{\varphi_\nu \tau_0}{\mu} S_\nu\right) \exp(-\tau_\nu/\mu) \frac{d\tau_0}{\mu}. \tag{3.15}$$

It is immediately evident, therefore, that the response of $I_\nu(0)$ to local changes in φ_ν induced by local changes in velocity or magnetic field strength can be quite different from the response to local changes in S_ν. Such effects have been demonstrated repeatedly by numerical modeling of line profiles in the presence of velocity gradients (Athay, 1970a, b; Gouttebroze, 1977; Gouttebroze and Leibacher, 1980). The most commonly used velocity diagnostics are wavelength shifts in line centers and in line widths. Strictly speaking, such diagnostics are valid only in the condition of zero velocity gradient. Whenever there is a velocity gradient on a scale that is less than the depth range over which the line is formed, the line will broaden as well as Doppler shift (Kulander, 1968; Athay, 1970b; Sedlmayer, 1976). Since even in the optically thin case the thickness of the layer over which the line is formed can easily exceed the density scale height, it is probable that the effects of velocity gradients are present even in optically thin lines. For optically thick lines, the thickness of the line-forming layer may cover many density scale heights, and it is virtually certain that velocity gradients are influencing the line shape.

Classical velocity diagnostics of single line profiles in terms of microturbulence, macroturbulence, and constant streaming velocity are still useful, provided it is recognized that the relative errors $\delta v/v$ may be substantial. A proper velocity diagnostic should incorporate observations in several lines whose overall range of formation is large compared to the range for any one line (Tripp et al., 1978; Vernazza et al., 1981). The profiles should then by synthesized with a model atmosphere that includes the velocity gradients. Within the photosphere one can select lines for which S is closely equal to B, which avoids the necessity of computing S independently. In the chromosphere and higher layers, however, one should compute S for the model atmosphere with the velocity gradients included.

It appears that every identifiable structural feature of the solar atmosphere from subarcsec to the order to 100 arcsec in size is either moving itself or is associated with fluid motion. Also, the global P-mode waves (Chapter 7) have appreciable velocity amplitude, and it seems likely that other types of waves are also present. There is no question, therefore, that fluid motions are universally present on a hierarchy of length scales, lifetimes, and velocity amplitudes. The effect of these motions appears as line broadening, line asymmetry and Doppler shift.

At the present time there are no fully adequate diagnostics for extracting the velocity distribution functions from observational data. The most recent methods of analysis, such as those discussed by Gail et al. (1974), Frisch (1975), Durrant (1979), Stahlberg (1979), and Froeschle and Frisch (1980) introduce a two-parameter distribution characterized by the r.m.s. velocity and a correlation length. This approach is clearly an improvement over the older method of micro- and macroturbulence, and, in fact, includes micro- and macroturbulence as special cases. Although the new methods hold much promise, they have not yet been widely applied to spectroscopic data.

From what we already know of solar velocity fields, and as is evident from the highest resolution spectroscopic data, the velocity distribution is very probably more complex than can be described by a two-parameter function. Thus, the problem of velocity diagnostics will require continued improvement well beyond its present level.

All of the preceding remarks concerning velocity diagnostics probably apply equally to magneitc field diagnostics. In the case of the magnetic fields, however, there has been little or no numerical modeling of the effects of gradients in magnetic field strength. Thus, we can only infer from the numerical modeling of the effects of velocity gradients that the results will be similar.

3.4. ABUNDANCE DIAGNOSTICS

Equations (2.36) and (3.6) depend upon the elemental abundances through $d\tau_0$, and in Equation (3.3) the abundance, A_{el}, appears explicitly. Equation (2.36) applies to both absorption and emission lines, and it or an equivalent form should be used for all strong lines. Equations (3.6) and (3.3) are appropriate for emission lines. Abundances are best determined using weak or moderately strong lines

In the case of weak Fraunhofer lines, one can make use of the Eddington–Barbier relation

$$I_\nu(0) = S_\nu(\tau_\nu = 1) \tag{3.16}$$

to obtain

$$I_c(0) - I_\nu(0) = B(\tau_c = 1) - S_\nu(\tau_c + \varphi_\nu \tau_0 = 1), \tag{3.17}$$

where B and τ_c are the continuum source function and optical depth. For simplicity we set $S_\nu = B$ and $d\tau_c/d\tau_0 = \tau_0 = \text{const}$ and assume that B is of the form

$$B = B_0(1 + \beta\tau_c). \tag{3.18}$$

We then obtain

$$\frac{I_c(0) - I_\nu(0)}{I_c(0)} = \frac{\beta\varphi_\nu\tau_0}{1 + \beta} \tag{3.19}$$

and, after integrating over frequency, we obtain

$$W(\nu) = \frac{\beta}{1 + \beta} M_0 \tau_0, \tag{3.20}$$

where $W(\nu)$ is the equivalent width of the line in frequency units. In wavelength units and for a Gaussian absorption profile, Equation (3.20) reduces to

$$\frac{W(\lambda)}{\lambda} = \frac{\beta}{1 + \beta} \frac{\pi e^2}{mc^2} \lambda f N_a, \tag{3.21}$$

where f is the oscillator strength and N_a is the column density of absorbing atoms.

In order to obtain an abundance from Equation (3.21), N_a must be replaced by

$$N_a = A_i \int_0^{z_0} G(T)N_H \, dz, \tag{3.22}$$

where $G(T)$ gives the ratio of N_a to the total concentration of the atomic species. Note that with this substitution for N_a, Equation (3.21) is very similar to Equation (3.8). The extra factor of N_e in Equation (3.8) comes because we have used $N_L C_{LU} = N_U A_{UL}$, which gives $N_U/N_L \propto N_e$. In deriving Equation (3.21), however, we assumed $S = B$ for the line source function, which results in N_U/N_L being independent of N_e. This latter assumption is appropriate for many weak photospheric lines formed near $\tau_c = 1$. On the other hand, it is inappropriate for a weak helium line, such as $\lambda 10830$ or D_3, formed in the chromosphere where $\tau_c < 10^{-5}$. Such lines require solution of the transfer equation to evaluate S relative to B and to evaluate τ_0.

Unlike Equation (3.8), which deals with emission lines and is valid as long as $N \ll N_{th}$, Equation (3.20) is valid only on the linear portion of the absorption curve of growth, which requires $\beta_1 \tau_0/(1 + \beta) \ll 1$. This difference arises because W is defined relative to a fixed continuum level and saturation effects set in as soon as the residual intensity approaches zero. From Equation (3.19), we note that this happens as $\beta \tau_0/(1 + \beta)$ approaches unity. In the photosphere, β ranges from unity near $\lambda 7000$ to 4.5 near $\lambda 4000$. Thus, saturation sets in for τ_0 between 1 and 2. There is no corresponding saturation effect in Equation (3.8) because the emission flux is not limited to a fixed level. Thus, the linear portion of the emission curve-of-growth continues until photon destruction competes with photon escape.

Regardless of whether one deals with absorption or emission lines, one obtains a reasonably accurate measure of the product fN_a for each transition. The problem of deriving an abundance is then threefold: (1) obtaining correct values for f; (2) converting N_a to N_{el} (N_{el} = concentration of the element): and (3) relating N_{el} to N_H. Major difficulties exist in all three areas. For many atoms and ions, f-values are only crudely known. In many cases, the excitation and ionization equilibrium is difficult to determine accurately, which means that N_{el} is not readily derivable from N_a. Also, we can properly relate N_{el} to N_H only if we understand the actual distribution of temperature and matter. In other words, we must know the centroids of the integrals giving N_a relative to the total mass and temperature distributions.

In the deeper layers of the photosphere the temperature and mass structure are regular enough and the excitation and ionization processes are close enough to local thermodynamic equilibrium for many ions that problems (2) and (3) are reasonably straightforward. In fact, few major inconsistencies in abundances are found other than those that can be attributed to errors in oscillator strengths. There are small but notable differences in relative abundances found in the photosphere in going from disk center to the limb and in using different ionization stages. These discrepancies could possibly reflect uncertainties in the photospheric model, particularly those associated with the granular structure. In the chromosphere and corona, however, the temperature and mass distribution is highly structured and, in addition, the correlation between temperature and mass density is often not known. Thus, the problems of relating N_a to N_{el} and N_{el} to N_H are often severe and the uncertainties can be large. The more severe of the two problems is relating N_{el} to N_H because of the difficulty of measuring N_H locally in the corona. For several heavy elements such as iron, magnesium, and silicon the coronal spectrum is moderately rich with several ion stages represented. In such cases, the temperature ranges represented by the spectra of one atomic species, such as iron, sufficiently overlaps the temperature ranges represented by the spectra of other atomic species that relative values of N_{el} can be derived with an accuracy comparable to those in the photosphere (Pottasch, 1984; Dupree and Goldberg, 1967). Thus even though the abundances relative to hydrogen are not well determined, abundances of several elements are known relative to each other with reasonable accuracy.

Available evidence suggests that the solar atmosphere is probably well mixed. Within the inherent limitations on accuracy in abundance determinations, there appears to be no major remaining conflicts between coronal and photospheric abundances (Dupree and Goldberg, 1967). The one major discrepancy for the iron abundance discovered by Pottasch (1964) was resolved after redetermination of the f-values used for Fe I lines in the photosphere revealed major errors. There are still discrepancies between coronal and photospheric abundances for individual elements of the order of a factor of 2, as reported by some authors. However, different authors using different data do not find the same discrepancies, and it appears likely that all the discrepancies can be attributed to the inaccuracies inherent in the analysis and in the spectroscopic data.

4. The Role of Radiation in Determining Atmospheric Properties

4.1. PHOTOSPHERE

Most of the energy carried by convection in the convection zone is converted to radiation beneath the visible layers of the photosphere. Of the portion that remains, some is used to heat the chromosphere and corona but the majority is probably dissipated in the photosphere itself. The small remnant that heats the chromosphere and corona produces a dramatic increase in temperature and leads as well to a variety of MHD phenomena. The larger portion dissipated in the photosphere produces little apparent effect, however.

To understand why this is so, we consider the increase in net flux at $\tau_0 = 0$ produced by an increment ΔB in the continuum source function between optical depths τ_1 and $\tau_2 > \tau_1$. The result is well known in radiative transfer theory (cf. Kourganoff, 1963) and is given by

$$\Delta H(0) = \frac{\Delta B}{2} \left[E_3(\tau_1) - E_3(\tau_2) \right], \tag{4.1}$$

where $E_3(\tau)$ is the exponential integral of order 3 and for small τ is given by $1/2 - \tau$. Thus, at small τ Equation (4.1) can be written

$$\Delta H(0) = \frac{\Delta B}{2} (\tau_2 - \tau_1). \tag{4.2}$$

Since any mechanical energy added to the atmosphere must be radiated away, $\Delta H(0)$ is proportional to the added mechanical energy. Thus for a given amount of mechanical energy added ΔB is proportional to $(\tau_2 - \tau_1)^{-1}$. It follows that a given increase in B requires 1000 times more mechanical energy input at $\tau_c = 10^{-1}$ than at $\tau_c = 10^{-4}$. Conversely, a given mechanical energy input produces a 1000-fold larger increase in ΔB at $\tau_c = 10^{-4}$ than at $\tau_c = 10^{-1}$. It is not difficult to understand, therefore, that mechanical energy dissipation in the deeper photosphere produces only minor changes in B from the radiative equilibrium values, and that to a good approximation the photosphere may be treated as being in radiative equilibrium.

To further illustrate the effects of adding nonradiative energy to the atmosphere, consider a case in which a constant amount of nonradiative energy, δ, is added per unit optical depth in an atmosphere with a grey absorption coefficient. Equation (2.7) then gives

$$J - B = -\delta, \tag{4.3}$$

where J and B are the frequency-integrated mean intensity and source function. We replace J and B with

$$J = J_0 + J' \tag{4.4}$$

and

$$B = B_0 + B', \tag{4.5}$$

where J_0 and B_0 are the radiative equilibrium values and J' and B' are the incremental values produced by the nonradiative energy input. It follows that

$$J' - B' = -\delta \tag{4.6}$$

since $J_0 = B_0$. We next differentiate Equation (4.6) to obtain

$$\frac{dJ'}{d\tau} - \frac{dB'}{d\tau} = 0. \tag{4.7}$$

Since energy is being added to the atmosphere, $dJ'/d\tau$ must be negative, which requires that $dB'/d\tau$ be negative. As long as the gradients $dJ_0/d\tau$ and $dB_0/d\tau$ are moderately large, the small added terms $dJ'/d\tau$ and $dB'/d\tau$ need not change the sign of $dJ/d\tau$ and $dB/d\tau$. However, at depths where $dJ_0/d\tau$ and $dB_0/d\tau$ approach zero, the small negative terms $dJ'/d\tau$ and $dB'/d\tau$ will cause both $dJ/d\tau$ and $dB/d\tau$ to be negative. In other words, at sufficiently small optical depth, B (and therefore T) will increase with height. This requires that B (also T) has a minimum. Thus, the existence of a temperature minimum is a consequence of decreasing τ and in no way proves, by itself, that the mechanical energy input begins at the location of minimum temperature. On the other hand, the region below the temperature minimum is clearly not dominated by nonradiative energy input. Otherwise the terms $dJ'/d\tau$ and $dB'/d\tau$ would be large enough to reverse the gradients.

The conclusion that the photospheric structure is governed mainly by the flow of radiation does not warrant the further conclusion that the structure is easily determined. In order for a spectral line to form in radiative equilibrium photon energy must be exchanged between the line and the continuum. An absorption line in the local mean intensity, for example, raises the continuum level, and, at the same time makes $dH_c/d\tau_c > 0$. The effect is a local cooling relative to the same situation without the absorption line. Conversely, a local emission line produces heating. The degree of heating and cooling depends upon the strength of the line and upon the coupling between the radiation and the thermal environment. In the upper photosphere and beyond, this coupling becomes rather weak and the problem must be treated with great care.

4.2. LINE BLANKETING AND COOLING

To illustrate the nature of the complex line-continuum interplay in radiative equilibrium computations, we consider a single spectral line mixed with a grey continuum. The results are readily generalized to the case of many spectral lines by summing the line effects individually.

For the case of one line plus continuum we may write

$$S_\nu = \frac{\varphi_\nu}{\varphi_\nu + r_0} S + \frac{r_0}{\varphi_\nu + r_0} B, \tag{4.8}$$

$$d\tau_\nu = d\tau_c + \varphi_\nu \, d\tau_0 = d\tau_c \left(1 + \frac{\varphi_\nu}{r_0}\right), \tag{4.9}$$

and

$$J_\nu = J_\nu^0 - \eta_\nu J_0^c, \tag{4.10}$$

where J_0^c is the extrapolated value of J_ν^c at line center and η_ν is the depression in the line measured positively for an absorption line in J_ν. Substituting Equations (4.8), (4.9), and (4.10) into Equation (2.7), integrating over frequency and collecting like terms, we obtain (Athay, 1981a)

$$\frac{1}{J_0^c} \frac{dH_c}{d\tau_c} = \left(wr_0 + w^* + \frac{S}{J_0^c} - 1\right)\frac{M_0}{r_0}. \tag{4.11}$$

The quantities w and w^* are defined by

$$w = \frac{1}{M_0} \int \eta_\nu \, d\nu, \tag{4.12}$$

and

$$w^* = \frac{1}{M_0} \int \varphi_\nu \eta_\nu \, d\nu. \tag{4.13}$$

Thus, w is the reduced equivalent width in frequency units and w^* is a weighted equivalent width of order w for weak lines and much less than w for saturated lines.

All of the quantities r_0, w, w^*, and S are interdependent, and, in the general case of non-LTE, they must be computed by solving the coupled transfer and statistical equilibrium equations. This is a rather formidable task in view of the many thousands of solar lines involved. Fortunately, throughout most of the photosphere the non-LTE effects do not produce substantial changes in $dH_c/d\tau_c$ from the LTE case. The LTE model computed by Kurucz (1974), for example, does not differ markedly from the non-LTE model computed by Athay (1970c) for $\tau_c \gtrsim 10^{-4}$ (Figure 2). We can with some degree of confidence, therefore, consider Equation (4.11) for the LTE case.

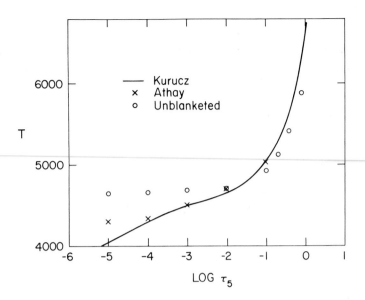

Fig. 2. A comparison of three radiative equilibrium models of the solar photosphere with and without line-blanketing effects.

In LTE, $S = B_0$ and $S/J_0^c - 1 = B/J_0^c - 1$. For B of the form given by Equation (3.18), $B > J_0^c$ for $\beta < 1.5$ and $B < J_0^c$ for $\beta > 1.5$. In the solar photosphere $\beta < 1.5$ for $\lambda > 6000$ Å and $\beta > 1.5$ for $\lambda < 6000$ Å. It follows that $B/J_0^c - 1$ tends to be positive for $\lambda > 6000$ Å and negative for $\lambda < 6000$ Å.

The parameters w and w^* in Equation (4.11) are positive for $J_\nu < J_0^c$, i.e. for absorption lines. Many lines that are in absorption ($\eta_\nu > 0$) near the surface of the Sun are in emission ($\eta_\nu < 0$) deeper inside the atmosphere, and the net effect of the line may be to heat the atmosphere rather than to cool it.

Within the line, Equation (3.18) can be written (for $B_0 = 1$)

$$B = 1 + \beta^*(\tau_c + \tau_\nu),\tag{4.14}$$

where τ_ν is the added optical depth due to the line and where

$$\beta^* = \beta \, \frac{\tau_c}{\tau_c + \tau_\nu} = \beta \, \frac{r_0}{r_0 + \varphi_\nu}.\tag{4.15}$$

Two opposing effects are influencing J_ν within the line: (a) the increased optical depth tends to increase J_ν and (b) the decrease of $dB/d\tau$ from β to β^* tends to decrease J_ν. The combined effects tend to produce emission lines inside the atmosphere when β is large but to produce absorption lines when β is small. Thus, in the violet end of the spectrum the lines reverse to emission lines inside the atmosphere, and w, w^*, and $B/J_0^c - 1$ are each negative. The resultant negative values of $dH_c/d\tau_c$ is the equivalent of heating, and the effect is referred to as backwarming. The same lines becomes absorption lines near the surface and $dH_c/d\tau_c$ becomes positive with resultant surface cooling.

It is possible for the effective value of β to increase within the line rather than decrease. This happens when r_0 increases with increasing τ_c, which is the tendency for neutral metal lines because of the decreased concentration of neutral relative to singly ionized metals in the deeper layers. As a result, the tendency for absorption lines to reverse to emission lines inside the atmosphere is stronger for lines of neutral metals than for ions and the neutral metals are therefore more effective in producing backwarming.

Most of the backwarming inside the atmosphere is produced by the multitude of weak lines in the solar spectrum. Surface cooling, however, is controlled mainly by the lines that are near optical depth unity in the surface layers, i.e. the lines that have total optical thickness at $\tau_c = 1$ of the order of $10^3 - 10^4$ at line center. LTE is a poor approximation for lines formed in the upper photosphere and may also be a poor approximation for the continuum. In non-LTE, S, r_0, w, and w^* are each different from their LTE value, and the problem of $dH_c/d\tau_c$ from Equation (4.11), or ite equivalent, is vastly more complicated.

Figure 2 illustrates radiative equilibrium models computed without lines, with lines formed in LTE, and with lines formed in non-LTE.

At $\tau_c = 10^{-4}$, radiative equilibrium models for both non-LTE (Athay, 1970c) and LTE (Kurucz, 1974) give a temperature close to 4300 K, which is some 400 K cooler than radiative equilibrium models without lines. The agreement between the line models, however, is misleading because of uncertain non-LTE effects, and, in fact, the temperature is not known to an accuracy better than about ±150 K.

The total reduced equivalent width of all Fraunhofer lines is about 0.12. Thus, the line-continuum interchange deals with total changes in $\Delta H_c/H$ of the order of 10^{-1}, which far exceeds the expected energy input from nonradiative sources. Most of the line-continuum interchange, of course, occurs in the layers near $\tau_c = 1$ and only a small fraction of the total occurs near $\tau_c = 10^{-4}$. On the other hand, the quantity of interest in connection with local changes in $\Delta H_c/H$ is the fraction of the total equivalent width formed locally. In the Sun, there are about 50 lines whose centers are formed near $\tau_c = 10^{-4}$ or smaller. To estimate W' we set $W' = 50\pi^{1/2} \Delta\lambda_D$, where $\Delta\lambda_D$ is an average Doppler width. Thus, for $\Delta\lambda_D = 20$ mÅ, we obtain $W' \approx 1.7$ Å. By comparison the product of $\tau_c W_c$ for the continuum at $\tau_c = 10^{-4}$ is only about 0.4 Å, where W_c is the equivalent width of the total continuum flux in units of the maximum flux near 4500 Å. Thus, τW is larger for the lines than for the continuum. At such small values of τ_c, even a small input to $\Delta H_c/H$ from the lines can be important in the model.

The uncertainty in ΔT of ±150 K near $\tau_c = 10^{-4}$ results in an uncertainty in $\Delta H_c/H$ given by (Osterbrock, 1961)

$$\frac{\Delta H_c}{H} = \frac{16T^3}{T_{\text{eff}}^4} \Delta T \tau_c \approx 10^{-3} \Delta T \tau_c \approx \pm 1.5 \times 10^{-5}. \qquad (4.16)$$

where T_{eff} is the solar effective temperature. By comparison the estimated mechanical energy flux needed to heat the chromosphere and corona is $3-5 \times 10^6$ erg cm^{-2} s^{-1} (Vernazza et al., 1981; Athay, 1981a), which is equivalent to $\Delta H_c/H = 5-8 \times 10^{-5}$. Since the uncertainty given by Equation (4.16) is an appreciable fraction of the equivalent energy needed for the chromosphere and corona, the current radiative equilibrium models are not accurate enough to assess the relative importance of mechanical heating in the temperature minimum region.

In many discussions of mechanical heating in the solar atmosphere it is assumed that the energy is deposited mainly in the regions above the temperature minimum. There is neither theoretical nor empirical evidence to compel such an assumption. It seems more likely, in fact, that the location of the temperature minimum is determined, in part, by the decreasing opacity rather than solely by the sudden onset of mechanical heating. Before we can properly describe the mechanical heating, therefore, it will be necessary that we improve considerably both the empirical and the theoretical radiative equilibrium models in the upper photosphere and temperature minimum region.

4.3. THE CHROMOSPHERE

The solar chromosphere has been defined in a variety of ways by different authors. For purposes of this discussion, it is the region below $T = 2.5 \times 10^4$ K in which $dT/d\tau$ is negative, i.e. in which T increases outward. The most recent model for the mean chromosphere is shown in Figure 3. In the definition of the chromosphere used here,

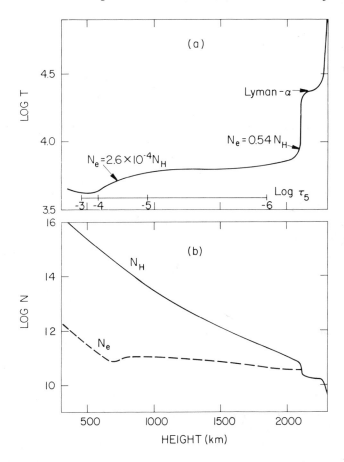

Fig. 3. Plots of temperature and density versus height in a recent chromospheric model (Vernazza *et al.*, 1981). The significance of the points indicated by arrows in the upper panel is discussed in the text.

the base of the chromosphere is at the temperature minimum. Nearly all current models agree in placing the temperature minimum near $\tau_c = 10^{-4}$. The visual limb occurs near $\tau_c = 3 \times 10^{-3}$, which places the limb well below the base of the chromosphere. At total eclipse, the spectrum of the region between the limb and the base of the chromosphere exhibits numerous emission lines from a variety of low-excitation sources including molecules, rare earth elements, and neutral metals, and a continuum mainly from the negative hydrogen ion. Higher excitation features such as Balmer lines and continuum, helium and Ca II lines come mainly from the chromosphere.

Two alternative definitions that lead to approximately the same location for the lower boundary of the chromosphere are: (1) the region in which the dominant energy input changes from radiation to mechanical energy dissipation, and (2) the region in which the dominant radiation output shifts from the continuum to the lines. The first of these is synonymous with the definition of $dT/d\tau < 0$; the second may reflect simply a fortuitous agreement.

An important additional characteristic of the chromosphere is that it is effectively thin ($N < N_{th}$, $\tau < \tau_{th}$) in nearly all the spectral lines. The only major exceptions are the hydrogen Lyman lines. Thus, virtually all of the photon energy generated in the chromosphere by collisional excitations escapes without being rethermalized. In addition to the radiation flux generated in the chromosphere, there is scattering of photospheric radiation. Only a small portion of the total emission from the chromosphere, in fact, represents an energy drain from those layers. This makes it difficult to quantify the radiation loss from the chromosphere empirically.

In the case of emission lines, such as occur in the EUV spectral regions, the problem is less ambiguous. Such lines cannot form by scattering of photospheric radiation and the energy flux in the lines can be safely identified with energy loss from the chromosphere. The same is true of the emission components in the Ca II and Mg II lines. On the other hand, for the latter lines it does not follow that the observed flux in the emission cores represents all the chromospheric energy loss in these lines. Much of the energy generated by collisions in these transitions diffuses into the line wing before escaping the chromosphere. Thus, the true chromospheric radiation loss includes some fraction of the radiation in the wings as well as the emission cores. Similarly, there are many lines that contain appreciable chromospheric radiation but do not show emission cores. These are lines for which the quantity $(\epsilon N_c)^{1/2} A$ in Equation (3.2) is less than unity. Notable examples of this type are the hydrogen Balmer lines, the Ca II infrared triplet, Na D lines, Mg b lines and strong lines of Fe I and Fe II. In fact, all lines on the wing portion of the curve-of-growth very likely have a chromospheric contribution. In addition, high excitation lines such as the Balmer series of hydrogen and lines of He I and He II have strong chromospheric contributions. Figure 4 indicates the depths where some of the prominent spectral features and formed.

Some authors have speculated that the initial temperature rise in the low chromosphere is due to the Cayrel mechanism (Cayrel, 1963; Praderie and Thomas, 1972), which is a non-LTE effect in the continuum. In order for the Cayrel mechanism to be effective, the continuum source function must become dependent on density as well as temperature. Also, this must occur deep enough in the atmosphere that the following two conditions are satisfied: (i) the atmosphere must still be in radiative equilibrium, and (ii) the main contribution to the total net flux, ΔH, from a temperature perturbation ΔT must be in

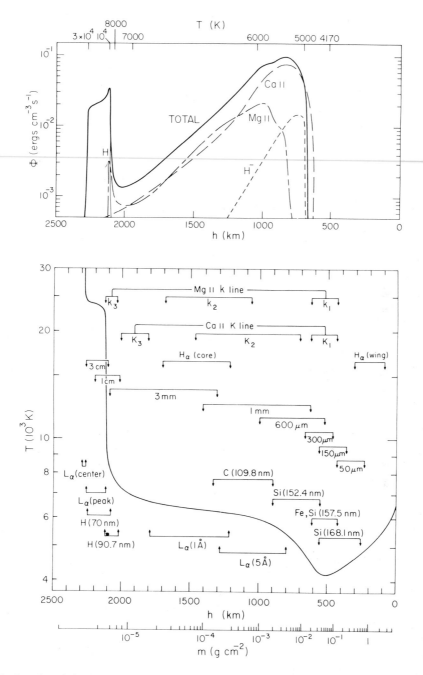

Fig. 4. A replot of the temperature curve in Figure 3 showing the regions of formation of some prominent spectral features, and a plot of the rates of radiation loss from H⁻, Ca II, Mg II, and hydrogen (from Avrett, 1981).

the continuum rather than the lines. Under these conditions, the continuum source function at small optical depths must remain essentially constant. However, since the source function is found to decrease with decreasing density (cf. Gebbie and Thomas, 1970) and since density is decreasing outwards there must be an offsetting rise in temperature.

As noted in the discussion by Gebbie and Thomas (1970), the Cayrel mechanism is self-limiting; both the degree of departure from LTE in the source function and the subsequent rise in temperature are limited by the radiation field. Since the atmosphere is assumed to have no energy source other than the radiative flux, the temperature rise is limited to $T = T_c$, where T_c is the color temperature of the radiation. This limitation results simply from the fact that the free electrons receive all their energy from the photons that photoionize the H⁻ ion and is analogous to a similar effect in circumstellar nebulae.

The relevance of the Cayrel mechanism to the sun depends upon whether the two conditions (i) and (ii) are both satisfied. Current evidence, in fact, suggests that neither is satisfied. If, for example, the continuum source function were controlling the temperature near the temperature minimum, the minimum temperature should be near the classical limit for the Sun of 4700 K. Instead, the minimum temperature is near 4300 K in theoretical line-blanketed models and even lower in current empirical models (Vernazza et al., 1981). The drop in temperature below 4700 K clearly is due to cooling by the spectral lines. In the theoretical models, therefore, the cooling due to lines at the temperature minimum is already important, and the balance shifts rapidly in favor of lines at smaller continuum optical depths. This clearly suggests that the Cayrel mechanism is not the main cause of the temperature rise in the low chromosphere.

The structure and energy balance of the chromosphere provide a continuing challenge to solar physicists. Although much progress has been made in constructing one-dimensional models representative of the average Sun, some features of the model are still unclear, particularly with respect to energy balance.

Since energy loss from the chromosphere is dominated by radiation in spectral lines and since the energy loss in the lines is a complicated function of temperature, there are certain distinctive features of the chromospheric temperature structure associated with changes in radiation loss. It is of interest, therefore, to examine the nature of the chromospheric radiation loss in relation to the temperature structure.

4.4. CHARACTERISTICS OF CHROMOSPHERE RADIATION LOSS

Radiation loss in spectral lines, as expressed by Equation (3.8), is the result of electron–atom and electron–ion collisions. Thus, the loss rate is critically dependent upon both the material density and electron density as well as the temperature. Current empirical models (Vernazza et al., 1981) indicate that the energy loss rate decreases at approximately the same rate as the material density. This requires that the product $N_e F(T)$ in Equation (3.8) remain relatively constant with height.

At the temperature minimum, electrons come mainly from metal ionization, with hydrogen ionization being of little consequence. However, in order for $N_e F(\tau)$ to be constant with height the temperature must first increase rapidly to the point where the electrons come mainly from hydrogen. The vast reservoir of electrons bound to hydrogen nuclei will then become the governing factor in the temperature rise.

To illustrate the effect of hydrogen ionization on the radiation loss, we rewrite Equation (3.8) in the form

$$4\pi \ dH/dz = QA_{el}N_e N_H f(T)g(T), \tag{4.17}$$

where

$$f(T) = \exp(-X_u/kT) \tag{4.18}$$

is the temperature dependence of the collisional excitations to a level at energy X_u and

$$g(T) = \exp(\pm A/k)T \tag{4.19}$$

is the temperature dependence of the ratio N_L/N_H. This latter expression is valid over restricted ranges of T with an appropriate choice for the parameter A for a particular atom or ion. The $\pm$ sign is chosen according to whether N_L/N_H is increasing or decreasing with temperature.

The product $f(T)g(T)$ is the same for all lines originating on a common upper level or on upper levels of close to the same excitation energy. In the case of Ca II, for example, the infrared triplet and the H and K doublet all have essentially common values for $f(T)g(T)$.

The electron density, N_e, in Equation (4.17) can be expressed as a function of N_H and T through an appropriate ionization equation. This, then, permits the derivation of the temperature gradient from Equation (4.17) as a function of the gradients of n_H and the radiative loss rate. The results give (Athay, 1981c)

$$\frac{d \ln T}{dz} = \frac{kT}{X_u \pm A} \left(\frac{d \ln L}{dz} - 2 \frac{d \ln N_H}{dz} \right), \ N_p \ll N_e \tag{4.20}$$

and

$$\frac{d \ln T}{dz} = \frac{kT}{X_2 + (1 + \alpha)(X_u \pm A)} \left[(1 + \alpha) \frac{d \ln L}{dz} - \right.$$

$$\left. - (2 + \alpha) \frac{d \ln N_H}{dz} - \frac{d \ln (1 - \gamma)}{dz} \right], N_p \approx N_e \tag{4.21}$$

L is the loss rate per cm^{-3}, dH/dz, X_2 is the excitation energy of the second principal quantum level in hydrogen, γ is the fractional ionization of hydrogen defined as

$$\gamma = \frac{N_p}{N_p + N_H^0} , \tag{4.22}$$

where N_H^0 is the density of neutral hydrogen, and α is parameter varying from unity when the Lyman continuum optical depth τ_{LC}, is large to zero when τ_{LC} approaches unity (cf. Thomas and Athay, 1961).

We supposed that $(d \ln L)/dz$ is fixed by the energy input and that both $(d \ln L)/dz$ and $(d \ln N_H)/dz$ are slowly varying in regions of low temperature gradient. The derivative $(d \ln(1 - \gamma))/dz$ is neglibible when hydrogen is mainly neutral ($\gamma \ll 1$) but may become large when hydrogen ionization (γ) approaches unity. Equations (4.20) and (4.21) then suggest three distinct regimes for $(d \ln T)/dz$: one for $N_p \ll N_e$, a second for $N_p \approx N_e$ but $\gamma \ll 1$, and a third for $\gamma \approx 1$. The first of these corresponds to the case where N_e comes mainly from metal ionization, the second to the case where N_e comes mainly from hydrogen and, consequently, is a strong function of temperature, and the third where hydrogen is sufficiently ionized that N_e is no longer a strong function of temperature.

The preceding discussion in this section serves, in a qualitative sense, to identify three possible changes in temperature gradient arising purely from the radiative properties of astrophysical plasmas and essentially unrelated to the details of the energy input. The three changes occur at the depths where $d\tau$ switches from mainly continuum processes to line processes, where electrons begin to come mainly from hydrogen, $N_e \approx N_p$, and where hydrogen ionization approaches completeness. As we shall indicate, these conditions are pertinent, in the order given, to the temperature minimum, a slightly higher region near the base of the first chromospheric temperature plateau, and the top of the first plateau, as illustrated in Figure 3.

In astrophysical plasmas and for stars of near solar type, the main radiation loss in the low chromosphere comes in lines for which $X_u \pm A \ll X_2$. Thus, the smallest temperature gradient comes in the second case where $\gamma \ll 1$ but $N_e \approx N_p$. The largest temperature gradient is expected when γ approaches unity forcing $(d \ln(1 - \gamma))/dz$ to become large in absolute value.

A further characteristic of radiation loss in stars of solar type can produce a marked change in temperature gradient. It arises because the condition $\gamma \approx 1$ sets in before the Lyman-α line of hydrogen becomes effectively thin, and thus before Lyman-α photons escape readily. At a given pressure, the potential for radiation loss in Lyman-α (Figure 5) is larger than for any group of lines from any other ion, because of the low relative abundances of other elements. Thus, at the level where γ approaches unity the strongest potential radiation loss is still waiting to be tapped. Strong losses in Lyman-α will begin as soon as τ_0 in Lyman-α drops below τ_{th}.

In the regions where Lyman-α photons escape readily $g(T)$ in Equation (4.17) is proportional to $\exp(X_2/kT)$ ($A = X_2$) and $f(T)$ is proportional to $\exp(-X_2/kT)$ ($X_u = X_2$). Also, $N_e \approx N_H$ and, at constant pressure ($N_e T = P_0$), Equation (4.17) is of the form

$$dH = \frac{h\nu}{4\pi} Q P_0^2 T^{-2} \, dz \tag{4.23}$$

which gives

$$\frac{d \ln T}{dz} = -\frac{1}{2} \frac{d \ln L}{dz}. \tag{4.24}$$

This implies a second region of relatively low-temperature gradient identified with $\tau_0 < \tau_{th}$ in Lyman-α.

We now seek to identify these expected changes in temperature gradient more explicitly

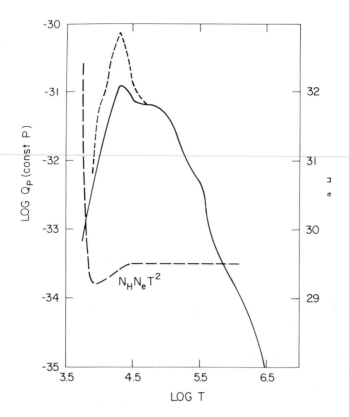

Fig. 5. The radiation loss coefficient plotted versus temperature for constant gas and electron pressures ($Q_p N_H N_e T^2$ gives the loss rate in erg cm^{-3} s^{-1}). A plot of $N_H N_e T^2$ in the solar atmosphere is included. For log $T > 4.6$, the emission coefficient is taken from Summers and McWhirter (1979). The dashed curve is a continuation of their results for log $T < 4.6$. The solid curve for log $T < 4.6$ is a plot of Equations (4.31) and (4.32).

with the empirical temperature structure and to examine the implications for the over-lying corona.

4.5. TEMPERATURE MINIMUM

In the Sun, N_e comes almost exclusively from metals at the temperature minimum and the region of temperature rise just above the temperature minimum corresponds to the first (d ln T)/dz regime identified by Equation (4.21). For stars of earlier spectral type, minimum temperature will be higher than in the Sun and the condition $N_e \approx N_p$ may already be satisfied at the temperature minimum. This would result in a smaller value for (d ln T)/dz corresponding to Equation (4.24) with $\gamma \ll 1$. Stars of spectral type later than the Sun, however, should have a lower minimum temperature and should require an even larger increase in temperature above the minimum value in order to achieve the condition $N_e \approx N_p$. Thus, the region of initial steep temperature rise should be more pronounced in such stars.

4.6. The First Temperature Plateau

The existence of even a modest temperature gradient in the regime where hydrogen is weakly ionized will cause both N_p and N_e to increase in the direction of increasing temperature. Eclipse observations in the Balmer continuum (cf. Athay, 1976) where the intensity of radiation is proportional to $N_e N_p$ show $N_e N_p$ increasing with height above the limb to a maximum value near 400–500 km (750–850 km above $\tau_c = 1$). The empirical model in Figure 3 shows a similar effect with maximum N_e and N_p near 800 km above $\tau_c = 1$. The level where $N_e = 2.6 \times 10^{-4} N_H$ marked in the figure is slightly lower. It is clear from the figure that the knee in the temperature curve separating the initial temperature rise from the first temperature plateau and marking the second substantial change in the temperature gradient is near the level where $N_e \approx N_p$, as expected.

We have also indicated, in Figure 3 the level at which $N_e = 0.54 N_H$ ($\gamma \approx 1/2$). The agreement between the location of $\gamma = 1/2$ and the top of the temperature plateau, marking the third sharp break in the temperature gradient, is striking.

It appears, therefore, that the first temperature plateau in the solar chromosphere conforms closely to the predicted constraints imposed by hydrogen ionization. The plateau is a result of the large reservoir of available electrons bound to hydrogen nuclei and ends when the reservoir begins to be depleted. Since these constraints are relatively independent of any particular property of the Sun, the same constraints should apply to stars of nearby spectral type.

It is now appropriate to ask what would happen to the first temperature plateau if the energy input to this region of the chromosphere were changed. Using a compilation of results from detailed computations by several authors (Shine et al., 1975; Milkey and Mihalas, 1974) for the H and K doublet and infrared triplet of Ca II and the Mg II resonance doublet, we find a combined radiation loss given by (Athay, 1981a)

$$4\pi \frac{dH}{dz} = hcN_e N_H \, [8.4 \times 10^{-3} \exp(-3.66 \times 10^4/T) +$$

$$+ 1.7 \times 10^{-7} \exp(-5.16 \times 10^4/T)]. \tag{4.25}$$

with the exponential terms in the order quoted for the two ions. The bracketed term in Equation (4.25) increases by approximately a factor of 5 from 6000 to 8000 K. More importantly, N_e increases relative to N_H by approximately 10^3, and the combined increase in dH/dz for a fixed N_H approaches four orders of magnitude. Thus, even large changes in the heat input to the low chromosphere can be accommodated by modest changes in temperature. The temperature is limited only by the condition $\gamma \ll 1$, and regardless of the temperature within the plateau, the plateau will end when γ approaches unity.

In the solar chromosphere $4\pi \, dH/dz$ is of the order 10^{-1} erg cm^{-3} s^{-1} near the base of the first plateau and of order 10^{-3} erg cm^{-3} s^{-1} near the top of the plateau (cf. Vernazza et al., 1981). The hydrogen density decreases by a similar factor, which means that the energy input is very nearly proportional to the total density. The total loss from the plateau is of the order of 2×10^6 erg cm^{-2} s^{-1}. Most of the loss is in the Ca II lines in the lower chromosphere. Near the top of the plateau Mg II losses compete with the Ca II loss.

4.7. THE SECOND TEMPERATURE PLATEAU

The second temperature plateau forms within the region where Lyman-α is effectively thin, which requires only that $\tau_0 \ll \tau_{th}$. We may write τ_0 as $\tau_0 \propto N_H^0 h_0$, where h_0 is the scale height for neutral hydrogen. The neutral hydrogen density, N_H^0, is a strong function of T for $T \lesssim 5 \times 10^4$ K. It follows that both h_0 and N_H^0 can vary by large factors depending upon T and dT/dz. In particular, we note that the constraint $\tau_0 \ll \tau_{th}$ does not necessarily constrain Δz, the geometrical thickness of the plateau.

As noted by Equation (4.23), the radiative loss rate from the second plateau is proportional to T^{-2}. The total loss, H, integrated across the plateau is proportional to $T^{-2} \Delta z$ when the width of the plateau, Δz, is less than $((d \ln N_e N_H)/dz)^{-1} = h_\rho/2$. When Δz exceeds $h_\rho/2$ the total loss is proportional to $T^{-2} h_\rho/2$. However, since $h_\rho \propto T$ this latter result reduces to $H \propto T^{-1}$.

We have the rather curious condition, therefore, that dH/dz can be varied only over a limited range. On the other hand, H can be varied over a considerable range provided $\Delta z < h_\rho/2$. This suggests that the second plateau is of a much different nature than the first plateau in which both H and dH/dz can be varied by large factors without regard for the thickness of the plateau.

The limited ability of dH/dz to adjust to changes in energy input in the second plateau makes it unlikely that this plateau is in static energy equilibrium. However, the large potential value for H and the ability of H to take on a wide range of values insures that a plateau will tend to form.

The rate of energy loss from the Lyman-α plateau can be written in the form (Athay, 1981a; Athay *et al.*, 1975)

$$4\pi \frac{dH}{dz} = hcN_e N_H (1 - \gamma)[1.9 \times 10^{-3} \exp(-1.18 \times 10^5/T)], \qquad (4.26)$$

which is convenient for determining dH/dz from the empirical model given N_e, N_H, γ, and T even though the product of $(1 - \gamma)$ and the exponential temperature term is approximately constant. For the Sun, $4\pi \, dH/dz \approx 2 \times 10^{-2}$ and $\Delta z \approx 150$ km. The total loss is $H \approx 3 \times 10^5$ erg cm^{-2} s^{-1}. However, we note that the value of $h_\rho/2$ at the mean temperature of the plateau is about 600 km, which means that the Sun is capable of radiating much more energy flux in Lyman-α than is observed, on the average.

If the energy supply for the Lyman-α flux were coming directly from the external mechanical energy flux, it is difficult to see why the second plateau is not thicker since the overlying corona attests to the fact that the energy supply is not exhausted in the upperchromosphere. However, this problem disappears if it is assumed that the energy is first dissipated in the corona then carried by thermal conduction and/or enthalpy flow into the second chromospheric temperature plateau where it is radiated as Lyman-α flux. This places a fixed limit on the Lyman-α flux that is set by conditions in the corona rather than in the temperature plateau itself. Evidence that this is the case in the Sun is provided by the estimated thermal conduction flux at the base of the corona of about 6×10^5 erg cm^{-2} s^{-1} (Gabriel, 1976; Athay, 1981b). Only about 3×10^5 erg cm^{-2} s^{-1} can be accounted for by radiation from the transition region, and an energy balance is achieved only if the Lyman-α flux is included as part of the energy balancing the total thermal conduction from the corona.

Still further evidence is provided by comparing the relative values of dH/dz between the two temperature plateaus. As already noted in the preceding, $4\pi\ dH/dz \approx 10^{-3}$ erg cm^{-3} s^{-1} near the top of the first plateau and jumps to 2×10^{-2} erg cm^{-3} s^{-1} in the second plateau. This sudden jump by a factor of 20 in dH/dz between the two plateaus again clearly suggests a discontinuity in the source of energy.

The simplest explanation of the distribution of radiation losses in the chromosphere is to assume that the sites of primary energy input to the solar atmosphere are concentrated in the low chromosphere and in the corona with the corona, in turn, supplying the energy for the transition region and upper chromosphere. In this picture, the temperature structure of the second temperature plateau is set by the corona and is not directly related to the temperature of the first plateau. Also, it seems likely that the plateau concept may be misleading in the Lyman-α region and, instead, is indicative only of the amount of material near the mean temperature of the regions emitting Lyman-α. Observations of Lyman-α structure at high spatial resolution, as shown in Figure 6 below (Bonnet et al., 1980), are not suggestive of a horizontally stratified model.

4.8. THE TRANSITION REGION

Energy balance in the transition region is perhaps more complicated than anywhere else in the solar atmosphere. The initial suggestion of a simple stratified transition region driven by thermal conduction in balance with radiation losses soon gave way to the notion that the conduction is channeled along magnetic field lines concentrated in the chromospheric network and fanning out to fill the overlying corona (Feldman et al., 1979; Kopp and Kuperus, 1968). This picture, in turn, has been modified to recognize that a major portion of the input energy flux may be carried as enthalpy in the return flow of spicule material setting back into the network whence it originated (Pneuman and Kopp, 1978).

The enthalpy flux is given by

$$F_{en} = 5kTF_p, \tag{4.27}$$

where F_p is the average flux of hydrogen atoms estimated at 1.2×10^{15} cm^{-2} s^{-1} (cf. Athay, 1981b). Downflow in the network has been observed up to temperatures of the order of 2×10^5 K (Doschek et al., 1976) and could extend to higher temperatures. At $T = 2 \times 10^5$ K, $F_{en} = 1.6 \times 10^5$ erg cm^{-2} s^{-1}, which is already a significant factor in the energy balance. If the spicule material is heated to as high as 5×10^5 K, the maximum enthalpy flux increases to 4×10^5 erg cm^{-2} s^{-1}, which is then a major energy term exceeding the Lyman-α radiation loss.

The downward enthalpy flux, of course, must be initially supplied as heat input to the spicule material (Athay, 1981b). In addition the spicule matter acquires potential energy as it is lifted, and this potential energy possibly provides a substantial secondary source of heat energy (Athay and Holzer, 1982). It could be, for example, that the downfalling spicule material converts potential energy into heat, which then becomes the source of the enthalpy downflow. A second plausible possibility is direct heating from the same mechanism that produces the spicule motion. In any case, the enthalpy flux is an important component of the energy balance in the transition region and one should place little confidence in models that ignore it.

Radiation loss from the transition region itself (excluding Lyman-α) appears, in the case of the sun, to be considerably smaller than the conduction flux into the top of the transition region. In addition, the downward enthalpy flux and the heat flux generated through release of potential energy in the downflowing matter (Athay and Holzer, 1982) may each exceed the radiation flux from the transition region. When one allows for such possibilities, there is no longer a unique relationship between the conduction flux and the radiation loss from the transition region *per se* (cf. Athay, 1981b).

On the other hand, none of the energy processes identified with rising and falling spicule material remove energy from the atmosphere as a whole. Energy supplied to the transition region from the corona or that is supplied as heat or potential energy in the elevated spicule material must eventually be radiated away or carried away in the solar wind. For the average Sun, the latter process is of little consequence, and radiation provides the dominant energy release. Thus, the total radiation loss from the corona, transition region, and upper chromosphere provides a good measure of the total energy input to the corona.

4.9. THE CORONA

Energy balance in the million degree solar corona involves a chain of events, some parts of which are not well understood. Mechanical energy dissipated as heat in the corona must ultimately be carried into outer space as radiation and the solar wind (including outward thermal conduction and waves). However, only a small fraction of the total energy loss from the corona is in the solar wind and in the radiation flux from the million degree temperature regime. Instead, the bulk of the energy loss from the corona appears to be by thermal conduction, and possibly enthalpy flow, into the cooler temperature regime of the transition region and upper chromosphere. This inward transport of heat energy is then disposed of by radiation from the lower temperature regime, thus completing the cycle.

This chain of events in which heat energy is deposited in the corona and carried by conduction and advection into lower temperature regimes where it is then radiated into outer space is deduced entirely from the corona and transition region radiation. The process of arriving at this conclusion involves the use of Equation (3.8) together with observed radiation fluxes in coronal and transition region spectral lines and theoretical computations of the temperature ranges, ΔT, in which the different lines are formed.

Equation (3.8) can be rewritten as

$$H(0) = \frac{h\nu}{4\pi} \, QA_{el}\kappa \int_{T_1}^{T_2} \frac{P_0^2 \, T^{1/2}}{F_c} \, F(T) \, \mathrm{d}T, \qquad (4.28)$$

where the conduction flux, F_c, is given by

$$F_c = \kappa T^{5/2} \, \frac{\mathrm{d}T}{\mathrm{d}z} \, , \qquad (4.29)$$

$P_0 = N_e T$ and $N_e = N_H$. T_1 and T_2 are the limits within which a given spectral line receives most of its contribution. It is then assumed that P_0 and F_c are constant between

T_1 and T_2, which permits the completion of the integral in Equation (4.28). One then applies Equation (4.28) to several spectral lines and obtains F_c as a function of T (cf. Elwert and Raju, 1972; Dupree and Goldberg, 1967; Athay, 1976, Chap. VII).

This process of arriving at an estimate of the conduction flux depends on a knowledge of Q, A_{el}, P_0, and $F(T)$, and, in addition, on the assumption that the radiating plasma is more or less stratified in such a way that the inferred value of dT/dz can be interpreted as a temperature gradient. Strictly speaking, the inferred value of dT/dz is a measure only of the volume of space between the temperature limits T_1 and T_2, and in a highly inhomogeneous atmosphere the ratio $\Delta z/T_2 - T_1$ may not be representative of the mean temperature gradient. In addition, Q, A_{el}, P_0, and $F(T)$ are each subject to uncertainties of the order of 50% or somewhat more.

These uncertainties lead us to have little confidence in any one determination of F_c. On the other hand, F_c has now been determined for many spectral lines and from several independent sets of data with relatively consistent results. Near $T = 10^6$ K, the inferred values of F_c are consistently near 6×10^5 erg cm^{-2} s^{-1} (cf. Elwert and Raju, 1972; Gabriel, 1976; Athay, 1976). Such a value is entirely consistent with the observed radiation loss from the corona, transition region, and Lyman-α region of the chromosphere. Thus, the proposed energy balance chain from the corona to the transition region and upper chromosphere is consistent in broad outline with the observational data.

On the other hand, the proposed scenario for the energy balance has major problems. No one has yet been able to construct a physical model based on this scenario that even approximately explains the largest bulk of the radiation flux it is supposed to account for. The basic problem is that thermal conduction and enthalpy flow are extremely efficient at temperatures near 10^6 K but very inefficient at temperatures near 10^5 K and lower, whereas the bulk of the radiation is at temperatures below 10^5 K (Gabriel, 1976; Athay, 1981b). In order to produce the observed radiation flux at these lower temperatures it is necessary to have values of Δz that effectively reduce the conduction flux to zero. Also, if the energy flux is carried as enthalpy, the required flow velocities at these lower temperatures are unacceptably large. Thus, the picture is seriously incomplete. Either the geometry is so complex (Feldman et al., 1979) that we are improperly interpreting $\Delta z/\Delta T$ or the major energy transport processes are not yet identified.

Efforts to balance thermal conduction and enthalpy flow from the corona against radiation losses encounter additional objections based on plausibility arguments. In the case of minimum flux models for stellar coronae (Hearn, 1975), it is assumed that radiation is balanced only by thermal conduction and that increased energy input to coronae results in both increased thermal conduction flux and increased P_0. Part of the increase in radiation flux required to match the increased energy input is provided by the increased P_0 and the remainder is provided by extending the height of maximum temperature in the corona, thereby increasing the radiating volume. The latter requirement is objectionable on the grounds that it imposes requirements on the distribution of mechanical energy input that may be unrealistic (Manageney and Souffrin, 1979; Endler et al., 1979), and the former is objectionable in that the gas pressure, P_0, at least in the solar case, is fixed by conditions at the top of the first temperature plateau in the chromosphere and appears to be unresponsive to coronal heating (Athay, 1981c).

The objection to energy balance invoking enthalpy flow is that the downflow can only be the result of a corresponding upflow, and, therefore, the upflow rather than coronal heating controls the rate of enthalpy downflow. A possible way around this difficulty is to make the upflow itself the source of coronal heating; as recently suggested by Athay and Holzer (1982). An earlier suggestion by Kuperus and Athay (1967) that spicule upflow is produced by the downward conduction flux merely adds another link to the energy chain without completing the energy cycle.

In summary, there seems little reason to doubt that most of the coronal energy initially leaves the corona as thermal conduction and enthalpy flow, which then ultimately appears as radiation from the lower transition region and upper chromosphere. However, the process by which the initial downflow of energy is carried into the proper temperature regimes is not at all understood. Perhaps, as Jordan (1980) has suggested, the energy transport is associated with the large amplitude, small-scale motions identified with nonthermal broadening of spectral lines in these temperature regimes.

5. Nonradial Structure

5.1. FLUID MOTIONS AND MAGNETIC FIELDS

Nonradial structure in the Sun is associated with fluid motions and magnetic fields and their mutual interactions. The observed structures are often classed in a variety of ways according to size, lifetime, fluid velocity, or correlations with other features. However, our concern in this chapter is limited to those structural features known to be essential to a proper interpretation of the radiation output. In particular, we are interested only in the rather pronounced structures that are so prevalent as to invalidate the assumption of spherical symmetry. This viewpoint overlooks many important hydrodynamical and magnetohydrodynamical aspects of the structures, and, in addition, overlooks entire classes of phenomena such as the small-scale structure and short-lived events.

In the photosphere, the structure is dominated by convective motions with typical scales of 1 arcsec (granulation) and 40 arcsec (supergranule cells). The granule structure is readily recognized in the intensity distribution in the visual continuum whereas the supergranules are recognized mainly by their velocity pattern of outflow from the cell center to the border. The magnetic flux, apart from that in sunspot groups, is concentrated at the borders of the supergranules apparently as a result of the converging flow from adjacent supergranules.

Beginning in the upper photosphere, and extending through the chromosphere and transition region, the borders of the supergranules, defined by the concentration of vertical magnetic field, are seen in certain spectral lines as a bright pattern of network forming a mosaic over the solar disk. As the gas pressure decreases with height, the network expands to cover a larger fraction of the total area. As it expands, it first becomes increasingly distinct, reaching maximum contrast in the lower transition region, then it fades and becomes indistinguishable in the corona.

The network itself contains much fine structure, including spicules. Also, the cell interiors contain a variety of fine structures. At some level of analysis, most of the fine structure will become important to the interpretation. For the present purposes, however, we shall restrict attention to the general network structure and its associated verical features consisting mainly of spicules and legs of magnetic loops.

The structure of the corona is largely governed by magnetic fields associated with active regions and larger quiet Sun regions of a dominant magnetic polarity. On a finer scale, the corona appears to consist largely of magnetic loops and rays selectively heated and selectively filled with matter.

The approximation of spherical symmetry becomes progressively worse from the photosphere to the corona. Unfortunately, our knowledge of the true structure and its role in the total energy output also becomes progressively poorer.

5.2. PHOTOSPHERE AND TEMPERATURE MINIMUM REGION

At the present level of sophistication in photospheric models, spherical symmetry is still a useful approximation. The effects of energy carried by convection in the low photosphere and of uncertain opacity in the ultraviolet, including line opacity, are still areas of major concern to be resolved before much progress can be made with multicomponent models.

The eventual need for multicomponent models has been recognized for many years and is receiving renewed emphasis from recent observations of the high photosphere and temperature minimum regions. Figure 6 is a photograph of the temperature minimum

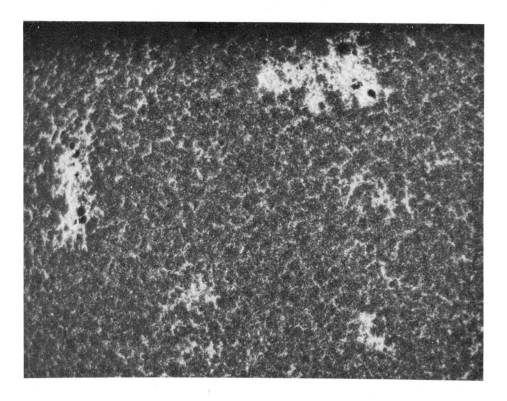

Fig. 6. Photograph of a portion of the solar disk in a 100 Å band centered at 1600 Å (courtesy Bonnet *et al.*, 1982). Sunspots and plage areas are readily recognized as are the bright network veins. Note the numerous bright dots within the supergranule cells.

region made in a 100 Å band centered at 1600 Å (Bonnet *et al.*, 1982). The entire disk is seen to be covered by small bright points enclosed in a well-defined network pattern. Sunspot regions are surrounded by bright plages similar to those seen in the chromosphere. Although photometric details are not available at the time of writing, it is clearly evident that this region of the atmosphere is highly structured and that spherical symmetry probably is a poor approximation.

In Section 4.2, the estimated uncertainty in temperature, ΔT, at the temperature minimum is given as ±150 K. This represents a factor of ±2 in the continuum radiation intensity at λ1600 if the continuum were formed in LTE. However, the continuum near λ1600 is known to depart substantially from LTE (Cuny, 1971), which increases the values of ΔT required to produce a given change in brightness.

Spectroheliogram observations in spectral lines showing inhomogeneous structure in the high photosphere and temperature minimum regions similar to that shown in Figure 6 have been available for many years. However, observations in the continuum, such as those in Figure 6, are relatively new in solar physics and relatively little effort has been made to investigate their influence on the model atmosphere. In particular, it is not known to what extent the energy balance is changed.

The association of enhanced emission in the network with magnetic fields raises many questions. Does the increased brightness at higher photospheric and chromospheric levels results from additional energy input or from changes in opacity? What are the characteristic lifetimes of the different features? If there is additional energy input, does it come directly from the magnetic energy or does the presence of the magnetic field merely enhance other forms of energy input, such as dissipation of sound waves? Such questions probably cannot be answered until the spectral properties and dynamical characteristics of the bright regions are much better known than at present.

5.3. CHROMOSPHERE AND TRANSITION REGION

Structural properties of the chromosphere have been much discussed (cf. Bray and Loughhead, 1974; Athay, 1976) and will not be discussed in detail in this chapter. However, certain aspects of chromospheric and transition region structure have a strong influence on the radiative output from these layers.

In the lowest layers of the chromosphere there is enough similarity between models based on disk and limb data that one need not reject completely the use of a mean model. However, beginning in the upper chromosphere and continuing into the corona, the concept of a mean model is, in fact, of questionable value. There is now ample evidence that temperature and geometrical height are not necessarily related in any simple way. In some general sense, temperature increases outwards to the temperature maximum. On the other hand, cool magnetic loops apparently rise above much hotter material. Similarly, cool jets of spicule matter penetrate into the million degree corona.

Figures 7 and 8 are high resolution Lyman-α filtergrams of the Sun obtained by Bonnet *et al.* (1980). The structure on the disk reveals a prominent network pattern, which on close inspection appears to consist of a multitude of fine filaments of more or less horizontal orientation. At the limb, the structure appears to be dominated by spicules, which are predominantly vertical. This is strongly reminiscent of the fibril and spicule structure observed in Balmer-α. The dominantly horizontal fibrils appear to be

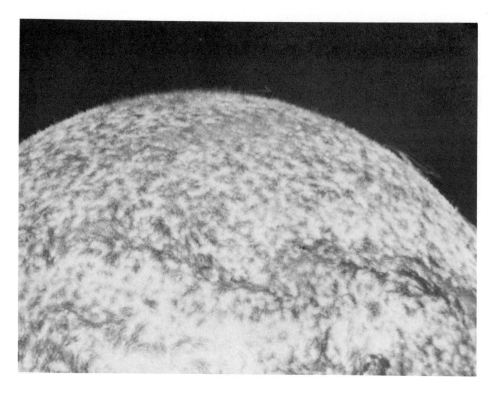

Fig. 7. A disk photograph in the Lyman-α line (courtesy Bonnet *et al.*, 1980). Note the bright diffuse
network and the widespread evidence of loop and filamentary structure.

low-lying magnetic loops (or at least flux tubes with a strong horizontal component),
whereas the spicules apparently rise in either the legs of large loops or in open field flux
tubes.

   A tantalizing study by Schmabl and Orrall (1979) reveals that high-temperature
transition region lines observed on the disk of the Sun show strong evidence of absorption
by the Lyman continuum of hydrogen. This curious result requires that a canopy of
neutral hydrogen overlies much of the transition region, which itself is at temperatures
much too high to permit enough neutral hydrogen to produce observable absorption
effects in the Lyman continuum. Evidently, the fibril structure, which is ubiquitous in
both Lyman-α and Blamer-α, and therefore relatively cool ($T \lesssim 20\,000$ K). Rises above
transition region material at a few hundred thousand degrees.

   Similar effects are observed at the limb (Withbroe, 1970). and were originally attributed to absorption by spicules. It is clear from the results of Schmahl and Orall,
however, that the absorption is much too widespread to be due to spicules alone.

   It is widely recognized that the magnetic fields in the network must undergo rapid
expansion between the photosphere where the gas pressure is of the order of 10^5 dyne
cm^{-2} and the transition region where the pressure is of the order of 10^{-1} dyne cm^{-2}
(cf. Kopp and Kuperus, 1986; Gabriel, 1976). The fibril structure could be associated,
at least in part, with the outer segments of the expanding field, where the field lines are

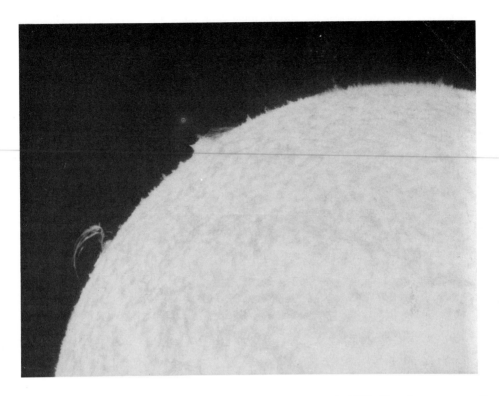

Fig. 8. A limb photograph in the Lyman-α line (courtesy Bonnet *et al.*, 1980). Note the appearance of vertical spicule structure at the limb together with larger prominences.

near horizontal. This provides an alternative to the loop picture, which requires nearby regions of opposite magnetic polarity.

Whatever the interpretation of the fibril structure, it is clear from high-resolution observations at the limb that much of the radiation produced at temperatures characteristic of the upper chromosphere arises at elevations much above the mean height of the chromosphere. Part of this effect is due to spicules, but a more important part may be due to the fibrils.

Within the chromosphere proper, network structure is pronounced and accounts for a large component of the radiation flux. The fraction of the surface area covered by network increases to about 45% in the upper chromosphere and transition region and accounts for about 60–65% of the observed emission in the upper chromosphere and for about 60% in the lower transition region (Reeves, 1976). The chromospheric network model of Vernazza *et al.* (1981) is a few hundred degrees hotter and at slightly higher pressure than the average Sun model.

5.4. CORONA

Prior to the era of observations of the solar corona from space, knowledge of the corona was limited mainly to limb observations and low-resolution observations at radio wavelengths. With such restrictive observational data, there was a widespread belief that the corona could be adequately described in terms of a more-or-less prevalent, spherically symmetric, hydrostatic amblent with embedded condensations associated with active regions and prominences. However, photographs of the corona viewed against the solar disk obtained with imaging X-ray telescopes quickly dispelled the notion that any substantial portion of the corona could be described as spherically symmetric. Instead, the observed structure consists of rather sharply defined large-scale features, typified by coronal holes (Bohlin, 1977; Kreiger, 1977) and active regions (Webb, 1981), a sprinkling of small 'bright points' (Golub *et al.*, 1974) scattered somewhat randomly over the disk, and extended diffuse regions of intermediate brightness surrounding the active regions. The active regions and surrounding diffuse regions appear to consist primarily of magnetic loops in organized but rather complex patterns (Webb, 1981).

Coronal loops are a heterogeneous collection representing a wide range in brightness in a particular wavelength band. Although most of the coronal loops are at temperatures near 2×10^6 K (Webb, 1981), loop structure at coronal heights is frequently observed in lines formed at transition region temperatures of the order of 10^5 K. There is little hard evidence that either temperature or gas pressure varies appreciably within a given loop (Webb, 1981). Thus, the temperature differences occur mainly between different sets of loops rather than within single loops. It is not uncommon, however, that the 'cool' and 'hot' loops exist in close proximity to each other.

Taken as a global average, radiation loss from the corona is small compared to the apparent inward conduction loss. However, because the radiation loss is concentrated in features making up but a small fraction of the coronal volume, radiation loss probably plays a major role in the energy balance in all of the brighter loops.

Current efforts to model the solar corona concentrate almost exclusively on models of individual loops, of active regions, or of coronal holes. Thus, the concept of an 'average' coronal model has almost ceased to exist.

6. Temporal Fluctuations

In its integrated radiation flux, H, the Sun is remarkably steady with no regularly occurring fluctuation, as large as 10^{-3}. In certain specific spectral bands, however, the radiation flux varies substantially. The most evident fluctuations are those associated with solar activity. These fluctuations give rise to short-term sporadic changes, to modulation by solar rotation, and to a roughly sinusoidal rise and fall with the solar activity cycle. The largest fluctuations are in the extreme ends of the spectrum and the smallest are in the visual continuum.

During the period from sunspot maximum in 1969–70 to the following solar minimum in 1975–76, the X-ray flux in the 0–8 Å band decreased by a factor of 500 (Kreplin *et al.*, 1977). Near solar maximum the modulation amplitude of the flux in this spectral band due to solar rotation is about an order of magnitude and individual flare events temporarily increase the flux by factors in excess of 10^2 (Kreplin *et al.*, 1977). The solar Lyman-α flux at 1216 Å decreases by somewhat less than a factor of 2 from sunspot maximum to minimum, it is modulated by no more than 30% with solar rotation, and increases by only about 10–20% with major flares (Vidal-Madjar, 1977). At still longer wavelengths, the flux in a 1 Å band including the K_3, K_2 and portions of K_1 components of the Ca II K line increased by about 18% from solar minimum in 1975–76 to solar maximum in 1974–80 (White and Livingston, 1981). During this same time perdod, the equivalent with of a C I line at 5380 Å decreased by 2.3% and the equivalent width of a Si I line at 10 827 Å remained constant to within 0.2% (Livingston et al., 1981). The Ca II variations correlate well with plage area and appear to be due entirely to the localized emission in these regions.

Studies of fluctuations in integrated solar flux from ground-based observations have been hampered by inadequate observational accuracy due to terrestrial atmospheric effects. These problems can be and are being overcome by space observations, but the data base is not yet large enough for accurate studies of long-term trends.

Observations during the first 153 days of operation of an active cavity irradiance monitor on the SMM satellite launched in February 1980 (Willson et al., 1981) show typical fluctuations over intervals of several days of ±0.05%. Two large dips lasting 7–10 days with amplitudes of 0.1–0.15% were observed in association with central meridian passage of large, recently developed sunspot groups and in agreement with a predicted curve based upon observed sunspot area. There is little apparent reason to doubt the validity of these important new results.

In an earlier study of solar constant data from the Smithsonian Astrophysical Observatory obtained between 1923 and 1952, Foukal and Vernazza (1979) found evidence of recurrent peaks in the autocorrelation index due to solar rotation with an amplitude of the order of 0.07%. Cross-correlation studies indicated that the solar flux increased in correlation with bright photospheric faculae and decreased in correlation with projected sunspot area. These results appear to be consistent with the more precise but limited results of Willson et al. (1981).

The demonstration that sunspots reduce the total irradiance and that faculae increase the total irradiance suggests that the radiation balance adjusts to surface features of this type on the time-scale of at least several days. As noted by Foukal and Vernazza (1979), this provides information on the depths to which these features modify the energy transport, which could be quantified through careful observations of the time delays.

7. Challenges for the Future

7.1. RADIATION DIAGNOSTICS

Much remains to be done before we can reliably extract from spectroscopic data all of the information content. Ideally, we would like to be able to describe both the vector velocity field and the vector magnetic field at known values of temperature and gas pressure. All we can do at present, however, is obtain estimates of two or three parameters of the velocity fields, a coarse estimate of the line-of-sight component of the magnetic field, and only a crude estimate of the transverse component, none of which can be closely related to temperature or pressure. These limitations were not very troubling as long as we were concerned (and still are to some extent) with only the general structural properties of the solar atmosphere. However, as soon as one begins to ask questions about the basic physical processes that determine the observed properties of the atmosphere –

such as magnetic field topology or local energy balance — the limitations described are entirely unacceptable.

Although considerable progress has been made with new velocity diagnostics, our current knowledge of line broadening is still based on the catch-all concepts of micro- and macroturbulence, whose velocity amplitudes typically exceed the streaming velocity and whose meaning is exceedingly obscure. We measure magnetic field strengths using very limited observational data interpreted according to assumptions about spectral line formation and line widths that are known to be inadequate and that fail often rather badly when applied to more complete data on the full Stokes profiles of the lines (Heasley *et al.*, 1981). The errors that result from such limited diagnostics are unknown but, by analogy with velocity field diagnostics where it has been demonstrated repeatedly that similarly limited diagnostics can give large errors, we must assume that the errors in magnetic field strength and configuration may also be large.

Better magnetic field diagnostics are possible utilizing known techniques in radiative transfer and known, but subtle, properties of atomic structure applied to more realistic models of line formation. The work has not yet been done mainly because it requires a wide range of expertise in what are normally rather diverse topics and because it requires a major commitment of time and computer resources. The atomic physics requires a careful treatment of level-crossing effects in strong fields, and the treatment of the transport of polarized radiation may require three-dimensional geometry. Such effects can be handled, but they considerably enlarge the scope of the problem.

Better diagnostics for velocity fields are possible, also. In this case, however, we are still dealing with conceptual problems. What is the nature of the velocity fields represented by microturbulence? Is it unresolved wave motion (Deubner, 1976), velocity gradients in laminar flow, true aerodynamic turbulence, or a combination of these effects? Most likely it is the latter. What seems to be needed is systematic analysis of each of the known or suspected categories of motion with a careful treatment of their influences on line profiles. Once each of the known effects, such as waves and convective motions, has been evaluated the residual line widths may reveal the necessary clues for advancing the next step.

As noted in Section 1, radiation diagnostics must eventually account for the multi-dimensional nature of the phenomena being studied. There is still much improvement to be made in the plane-parallel case, however.

7.2. INFLUENCE OF RADIATION ON ATMOSPHERIC PROPERTIES

Radiation either dominates or plays a vital role in determining the structure in all layers of the solar atmosphere. Throughout much of the atmosphere, however, the precise role of the radiation is still somewhat obscure. In the extreme upper photosphere and temperature minimum, for example, it is believed that radiative equilibrium is still a valid approximation, but we have no acceptable radiative equilibrium model that agrees with the low temperatures found in empirical models. Nor do we have a satisfactory theoretical model for sunspots.

At temperatures between about 10^4 and 3×10^5 K in the upper chromosphere and transition region, we have no clear understanding of the relationship between the emergent radiation and the atmospheric structure. Current models span the range from assuming that virtually all the emission is in horizontally stratified layers to those claiming that virtually all the emission is in isolated vertical structures, such as spicules. Similarly, the corona outside coronal holes is now envisioned as primarily a collection of magnetic loops acting as more or less independent entities and taking on a wide range of individual characteristics, such as temperature and density.

Beginning at about 10^4 K it appears that the radial temperature gradient, dT/dz, may be positive only in the statistical sense and that, in fact, it is frequently negative in localized regions. Associated with this convoluted temperature profile is a somewhat chaotic mixture of radiation associated with relatively high and low temperatures. In some sense, of course, the temperature structure is as much the result of the radiative properties as it is the cause of them. To what extent the radiation determines the detailed structure of the corona is still unclear, however.

Problems of radiation–hydrodynamic coupling need much more investigation. Considerable work has been done with respect to the effects of shifting and broadening the line absorption profiles, which are related to macro- and microturbulence. A little has been done relative to the effects of streaming motion on ionization and excitation equilibria in optically thin layers in which strong temperature gradients are present. However, little or nothing has been done for the optically thick case. The other half of the radiation–hydrodynamic coupling problem – viz. the effect of the radiation on the motion itself – has been limited largely to radiation damping of photospheric waves. The deeper problems concerning the forces exerted by the radiation on individual fluid elements and the effects of such forces on the internal field of motion within the atmosphere have hardly been touched.

7.3. OBSERVATIONS

The benefit of improved observations is well illustrated by the evolution of magnetic field observations and the attendant changes in concepts brought about by the improved observations. The first magnetographs capable of measuring low-intensity magnetic fields with modest spatial resolution displayed complex patterns of opposite magnetic polarities near active regions, large migratory unipolar areas at somewhat higher latitudes, and weak polar fields of opposite polarity at the two poles. These observations immediately dispelled the popular notion that the Sun's magnetic field was predominantly bipolar on a global scale.

Subsequent refinements in magnetographs, coupled with efforts to resolve discrepancies between measurements made in different spectral lines and with different instruments, led to the discovery that the picture was still more complex. Opposite magnetic polarities were discovered to occur on small spatial scales at all solar latitudes and the fields were found to occur predominantly in small flux tubes less than 0.5 arcsec in diameter. Thus, the unipolar areas previously found at middle and high latitudes were now shown to be unipolar only in the statistical sense that one polarity dominated the other. It was also, discovered, that the magnetic flux emerging as small-scale, quiet Sun dipoles probably exceeded the magnetic flux emerging in active regions.

The small photospheric flux tubes were discovered to lie predominantly at the borders of supergranule cells and underneath the chromospheric network. These discoveries established a close link between the magnetic flux tubes and the fluid circulation in the

supergranule cells and, in addition, correlations between magnetic fields and chromospheric heating and between magnetic fields and the relatively massive chromospheric
outflow is spicules.

Later discoveries that the corona seemed to be almost universally controlled by the
magnetic geometry amplified the association between the magnetic field and atmospheric
heating. These associations, in turn, essentially destroyed and long-held belief that the
corona is heated by pressure waves before other confirming evidence forced the same
conclusion.

Many solar physicists now believe that a large portion of the atmospheric heating is
due to magnetic field confined to flux tubes too small to be resolved; that most of the
vertical mass flow in the chromosphere and transition region is confined to spicules too
small to be studies in detail; and that most of the energy represented by fluid motions
resides in 'microturbulent' eddies too small to be resolved. This is an awkward and
unsatisfactory situation that needs to be rectified. There is no technical reason, other
than limited resources, why such phenomena cannot be resolved and studied in appropriate detail.

Progress in solar physics in the past has been closely coupled to expanded access to
new spectral regions and to improved spatial, and spectral resolution. Of the three areas,
spatial resolution has improved the least over the last three decades. Also, in most of the
major areas where data quality is the limiting factor, the foremost limitation is spatial
resolution. It is abundantly clear, therefore, that the major new thrust in observational
solar physics must be in the direction of improved spatial resolution.

The Sun, in all probability and from all indications, is not a peculiar star. Solar-type
phenomena abound in the other stars, and what happens in the Sun probably happens
in most stars — more subdued in some cases and more extreme in others. It is only
through our ability to partially resolve the structural features of the solar atmosphere
that we have been able to progress to our present incomplete, but, nevertheless, much
improved, understanding of the basic physical processes of the solar atmosphere. This
understanding, in turn, has provided the key to the present understanding of stellar
atmospheres.

References

Athays, R. G.: 1970a, *Solar Phys.* **11**, 347.
Athay, R. G.: 1970b, *Solar Phys.* **12**, 1975.
Athay, R. G.: 1970c, *Astrophys. J.* **161**, 713.
Athay, R. G.: 1972, *Radiation Transport in Spectral Lines*, D. Reidel, Dordrecht.
Athay, R. G.: 1976, *The Solar Chromosphere and Corona: Quiet Sun*, D. Reidel, Dordrecht.
Athay, R. G.: 1981a, in S. Jordan (ed.), *The Sun as a Star*, NASA SP-450, Washington, D. C.
Athay, R. G.: 1981b, *Astrophys, J.,* **249**, 340.
Athay, R. G. 1981c, *Astrophy. J.* **250**, 709.
Athay, R. G. and T. Holzer: 1982, *Astrophys. J.* **255**, 743.
Athay, R. G., Mihalas, D., and Shine, R. A.: 1975, *Solar Phys.* **45**, 15.
Avery, L. W. and House, L. L.: 1968, *Astrophys. J.* **152**, 493.
Avrett, E. H.: 1981, in R. M. Bonnet and A. K. Dupree (eds), *Solar Phenomena in Stars and Stellar
    Systems*, D. Reidel, Dordrecht.
Beckers, J. M.: 1972, *Ann. Rev. Astron. Astrophys.* **10**, 73.

Bohlin, J. D.: 1977, in J. B. Zirker (ed.), *Coronal Holes and High Speed Wind Streams*, Colorado Assoc. Univ. Press, Boulder.

Bonnet, R. M., Bruner, E. C. Jr, Acton, L. W., Brown, W. A., and Decaudin, M.: 1980, *Astrophys. J.* **237**, L47.

Bonnet, R. M. *et al.*, 1982, *Astron. Astrophys.* **111**, 125.

Bray, R. J. and Loughhead R. E.: 1974, *The Solar Chromosphere*, Chapman and Hall, London.

Cannon, C. J.: 1973, *J. Quan. Spec. Rad. Trans.* **18**, 1011.

Cayrel, R.: 1963, *Compt. Rend. Astron. Soc. Paris* **257**, 3309.

Cuny, Y.: 1971, *Solar Phys.* **16**, 293.

Dere, K. P. and Mason, H.: 1981, in F. Q. Orrall (ed.), *Skylab Workshop on Active Regions*, Colorado Assoc. Univ. Press, Boulder, p. 129.

Dere, K. P., Manson, H., Widing, K., and Bhatia, A. K.: 1979, *Astrophys J. Supp.* **40**, 341.

Deubner, F. 1976, in R. M. Bonnet and Ph. Delache (eds), *The Energy Balance and Hydrodynamics of the Solar Chromosphere and Corona*, G. de Bussac, Clermont Ferrand.

Doschek, G. A. and Feldman, U.: 1978, *Astron. Astrophys.* **67**, 11.

Doschek, G. A., Feldman, U., and Bohlin, J. D.: 1976, *Astrophys. J.* **205**, L177.

Dupree, A. K. and Goldberg, L.: 1967, *Solar Phys.*, **1**, 229.

Durrant, C. J.: 1979, *Astron. Astrophys.* **73**, 137.

Elwert, G. and Raju, P. K.: 1972, *Solar Phys.* **25**, 314.

Endler, F., Hammer, R. and Ulmschneider, P.: 1979, *Astron. Astrophys.* **73**, 190.

Feldman, U., Doschek, G. A., and Mariska, J. T.: 1979, *Astrophys. J.* **229**, 369.

Foukal, P. and Vernazza, J.: 1979, *Astrophys. J.* **234**, 707.

Frisch, H.: 1975, *Astron. Astrophys.* **40**, 267.

Frisch, U. and Frisch, H.: 1975, *Monthly Notices Roy. Astron. Soc.* **173**, 167.

Froeschle, Ch. and Frisch, H.: 1980, *Astron. Astrophys.* **91**, 202.

Gabriel, A. H.: 1976, *Phil. Trans. Roy. Soc. Lond.* **A281**, 339.

Gabriel, A. H. and Jordan, C.: 1969, *Monthly Notices Roy. Astron. Soc.* **145**, 241.

Gail, H. P., Hundt, E., Kegel, W. H., Schmidt-Burgk, J., and Traving, G.: 1974, *Astron. Astrophys.* **32**, 65.

Gebbie, K. B. and Thomas, R. N.: 1970, *Astrophys. J.* **161**, 229.

Golub, L., Krieger, A. S., Silk, J. K., Timothy, A. F. and Vaiana, G. S.: 1974, *Astrophys J.* **189**, L93.

Gouttebroze, P.: 1977, *Astron. Astrophys.* **54**, 203.

Gouttebroze, P. and Leibacher, J. W.: 1980, *Astrophys. J.* **238**, 1134.

Hearn, A. G.: 1975, *Astron. Astrophys.* **40**, 355.

Heasley, J. N. Jr, Landi'Degl'Innocenti, E., and Querfeld, C.: 1981 (unpublished).

Jeffries, J. T. and Thomsa, T. N.: 1958, *Astrophys. J.* **127**, 667.

Jeffries, J. T.: 1968, *Spectral Line Formation*, Blaisdell, Waltham, Mass.

Jones, H.-P. and Skumanich, A.: 1973, *Astrophys. J.* **185**, 167.

Jones, H.-P. and Skumanich, A.: 1980, *Astrophys. J. Supp.* **42**, 221.

Jordan, C.: 1980, *Astron. Astrophys.* **86**, 355.

Jordan, S. (ed.): 1981, *The Sun as a Star*, NASA SP-450; Washington, D.C.

Kopp, R. A. and Kuperus, M.: 1968, *Solar Phys.* **4**, 212.

Kourganoff, V.: 1963, *Basic Methods in Transfer Problems*, Dover, New York.

Kreiger, A. S.: 1977, in J. B. Zirker (ed.), *Coronal Holes and High Speed Wind Streams*, Colorado Assoc. Univ. Press, Boulder.

Kreplin, R. W., Dere, K. P., Horan, D. M., and Meekins, J. F.: 1977, in O. R. White *The Solar Output and its Variation*, Colorado Assoc. Univ. Press, Boulder.

Kuperus, M. and Athay, R. G.: 1967, *Solar Phys.* **1**, 361.

Kurucz, R.: 1974, *Solar Phys.* **34**, 17.

Kulander, J. L.: 1968, *J. Quant. Spec. Rad. Trans.* **8**, 273.

Livingston, W., Holweger H. and White, O. R.: 1981, *Workshop on Solar Constant Variations*, GSFC.

Managency, A. and Souffrin, P.: 1979, *Astron. Astrophys.* **78**, 36.

Mibalas, D.: 1978, *Stellar Atmospheres*, Freeman, San Francisco.

Milkey, R. W. and Mihalas, D.: 1974, *Astrophys. J.* **192**, 769.

Orrall, F. Q.: 1981, *Solar Active Regions*, Colorado Assoc. Univ. Press, Boulder.

Osterbrock, D. E.: 1961, *Astrophys. J.* **184**, 347.
Pneuman, G. W. and Kopp, R. A.: 1978, *Solar Phys.* **57**, 49.
Pottasch, S. R.: 1964, *Space Sci. Rev.* **3**, 816.
Praderie, F. and Thomas, R. N.: 1972, *Astrophys. J.* **172**, 485.
Reeves, E. M.: 1976, *Solar Phys.* **46**, 53.
Rybicki, G. B.: 1967, *IAU Symp.* 28, 481.
Scharmer, G. B.: 1981, *Astrophys. J.* **249**, 720.
Schmahl, E. J. and Orrall, F. Q.: 1979, *Astrophys. J.* **231**, L41.
Sedlmayer, E.: 1976 in R. Cayrel and M. Steinberg (eds), *Physique des mouvements dans les atmo-sphere stellaire*, CNRS, Paris.
Shine, R. A., Milkey, R. and Mihalas, D.: 1975, *Astrophys. J.* **199**, 724.
Stahlberg, J.: 1979, *Astron. Nach.* **300**, 261.
Stenhölm, L. G.: 1977, *Astron. Astrophys.* **54**, 577.
Sturrock, P. A.: 1980, *Solar Flares*, Colorado Assoc. Univ. Press, Boulder.
Summers, H. P. and McWhirter, R. W. P.: 1979, *J. Phys. B.* **14**, 2387.
Thomas, R. N.: 1965, *Non-Equilibrium Thermodynamics*, Colorado Assoc. Univ. Press, Boulder.
Thomas, R. N.: 1960, *Astrophys. J.* **131**, 429.
Thomas, R. N. and Athay, R. G.: 1961, *Physics of the Solar Chromosphere*, Interscience, New York.
Tripp, D. A., Athay, R. G., and Peterson, V. L.: 1978, *Astrophys J.* **220**, 314.
Vernazza, J., Avrett, E. H., and Loeser, R.: 1981, *Astrophys J. Suppl.* **45**, 635.
Vidal-Madjar, A.: 1977, in O. R. White (ed.), *The Solar Output and its Variations*, Colorado Assoc. Univ. Press, Boulder.
Webb, D. F.: 1981, in F. Q. Orrall (ed.), *Skylab Workshop on Active Regions* Colorado Assoc. Univ. Press, Boulder.
White, O. R. (ed.): 1977, *The Solar Output and its Variations*, Colorado Assoc. Univ. Press, Boulder.
White, O. R. and Livingston, W. C.: 1981, *Astrophys. J.* **249**, 798.
Withbroe, G. L.: 1970, *Solar Phys.* **11** 208.
Wilson, R. C., Gulkis, S., Janssen, M., Hudson, H. S., and Chapman, G. A.: 1981, *Science,* **211**, 700.

High Altitude Observatory,
National Centre for Atmospheric Research,
Boulder, CO 80307,
U.S.A.

CHAPTER 9

CHROMOSPHERIC FINE STRUCTURE

R. G. ATHAY

1. Introduction

1.1. ROLE OF FINE STRUCTURE

The mechanical (nonradiative) energy that is responsible for the solar chromosphere and corona is generated in the subphotospheric layers and propagates into the outer layers where it is dissipated. The propagating energy probably carries momentum as well as energy. Also, because the energy source is mechanical it is expected to be associated with temporal and spatial fluctuations of physical variables. It is logical to assume, therefore, that the transport of mechanical energy and momentum will produce local fluctuations in intensity and velocity and possibly in magnetic field strength. Solar fine structure, at least in part, is the observational manifestation of such processes. The dissipation process, of course, may involve intermediate phenomena that influence the fine structure, and, in turn, the fine structure undoubtedly influences the energy propagation. In any case, the nature of the propagating energy and its dissipation is undoubtedly imprinted in the fine structure.

Once we are able to describe the fine structure in terms of fluctuations in physical variables, we should have the clues necessary to identify the energy and momentum processes producing the fluctuations, and ultimately, to identify the source of heating. Although we are still far from this goal, several notable milestones have been reached.

Studies of short period oscillations, for example, have shown that the energy flux in compressive waves in the upper chromosphere and transition region is insufficient to produce the observed coronal heating (Bruner, 1978; Athay and White, 1978, 1979). On the other hand, the energy flux in compressive waves in the photosphere may be sufficient to heat the low and middle chromosphere (Deubner, 1976; Ulmschneider, 1979), but even here there are serious reservations (Cram, 1977). A second striking example is the discovery of the selective heating of individual coronal loops (Webb, 1981), which points strongly towards a heating mechanism closely coupled to the magnetic field.

In a somewhat different vein, studies of network and spicules indicate that vertical motions are widespread and sufficient to play an important role in the energy balance. The strong upflow observed in spicules (cf. Beckers, 1972; Athay, 1976) is apparently balanced by a return downflow in the network at transition region temperatures (Doschek et al., 1976; Gebbie et al., 1981).

A more indirect, but vitally important, role played by solar fine structure arises in connection with spectral diagnostics. In order to interpret spectroscopic data in terms

Peter A. Sturrock (ed.), Physics of the Sun, Vol. II, pp. 51–69.

of physical variables, it is necessary to know the geometrical structure of the region producing the observed radiation. The customary assumption of a plane-parallel atmosphere may suffice for some purposes, whereas for other purposes it is totally inadequate. The clustering of photospheric magnetic fields into small knots covering some 0.1% of the surface with field strengths of nearly 2000 gauss (Livingston and Harvey, 1969; Stenflo, 1973) is poorly represented by an average field strength of 2 gauss even though the average may be accurate.

A good example of how far astray we can be led by misinterpretation of atmospheric geometry is provided by the prolonged debate about chromospheric support. Eclipse observations of the chromosphere were erroneously interpreted early in this century as meaning that the chromosphere extended more or less spherically symmetrically to a height of 12 000 km beyond the solar limb. From then until the 1950s, the commonly stated central problem of chromospheric physics was how to provide such an extended region at chromospheric temperatures. The serious proposals included radiation support by E. A. Milne, turbulent pressure by W. H. McCrea and a super hot chromosphere by R. O. Redman. Although the problem posed generated a lively and extended debate, the problem, in fact, was baseless.

Later eclipse observations in 1952 combined with observations of spicules at the limb revealed that the chromosphere could be reasonably approximated as spherically symmetric for about the first 2000 km only. Beyond that, the chromosphere consists almost entirely of long, vertical spicules covering no more than 1% of the solar surface. At the limb, the spicules merge to form an apparently solid mass because of their elongated shape combined with the large solar radius and the resultant long path length over which the spicules are visible.

1.2. DEFINITION OF FINE STRUCTURE AND THE OBSERVATIONAL CHALLENGE

Solar fine structure is usually defined in observational terms as local fluctuations in the specific intensity of radiation. This is convenient for classifying and for studying morphological properties such as lifetimes, dimensions, growth rates, and geometrical patterns. The ultimate goal of observations, however, is to describe the fine structure in terms of the physical variables, pressure, temperature, density, velocity, and magnetic field strength as well as their morphological properties. The task of determining the full set of physical variables appropriate to a given class of structural features requires a rather extensive set of carefully designed observational data coupled with equally carefully designed diagnostic methods. In practice, observational and instrumental constraints have severely limited the available data, which, in turn, has limited the objectives.

Because of the relative ease with which observations can be made in a single spectral band, most studies of fine structure fall within this category. Studies of this type provide valuable morphological data. Also, growth rates provide some indication of velocities and the geometrical patterns provide clues to the magnetic field geometry. At a somewhat higher level of sophistication, observations are made in opposite sides of a spectral line. The difference between the two images is then used as a measure of the Doppler shift in the line, which is interpreted as a streaming velocity. Such observations have added a great deal to the study of fine structure and have led to the discovery of important phenomena that were undetected in single band, monochromatic data.

Efforts to measure magnetic field strengths in the chromosphere have been frustrated by the limited choice of spectral lines and by the complexity of the line structure itself. No reliable quantitative data have been regularly available.

The main limitation in attempts to obtain sufficient data to determine the fluctuations in physical variables associated with fine structure is atmospheric seeing. Given the best in achievable instrumentation, an observer would still require extended periods of seeing at a fraction of an arcsec in order to obtain sufficient data. With the realization that such observing conditions are not likely to occur, there has been little incentive to develop the necessary instrumentation.

Studies of chromospheric fine structure began in the mid-nineteenth century with the perfection of spectroscopes. These instruments made possible the study of monochromatic images at the extreme limb of the Sun both at times of total eclipse and, with somewhat less detail, outside of eclipse. Such studies led to the description of the chromosphere at the limb by A. Secchi as resembling a 'burning prairie'. The invention and perfection of the spectroheliograph in the 1890s by G. E. Hale and H. Delandres opened a fruitful new era in which monochromatic observations of the disk revealed a wealth of structural detail. In the 90 years, or so, that has followed, many important phenomena have been discussed and carefully described in their changing center-to-limb aspects.

In spite of the long history of observations and detailed studies, we are still unable to provide a completely satisfactory description of all chromospheric fine structure in terms of a set of physical features. The problems of arriving at such a description are manifold. Structure seen in one spectral line looks quite different from that seen in another, and in fact, changes markedly in appearance with wavelength within a single spectral line. In addition, much of the fine structure has geometrical scale lengths of less than 1 arcsec. Only rarely does a ground-based observer find good enough seeing to obtain a 'good' picture of the fine structure, and, even then much of the structure is still too small to be resolved. Time-scales present still another challenge. Many of the fine structure features have lifetimes of only a few minutes, during which they may undergo marked evolution.

The combined need to attain high spatial and time resolution in enough simultaneous monochromatic images to provide an integrated description of chromospheric structure is a formidable challenge, occasionally realized in part but never in total. As a compromise, we have been forced to rely on statistical properties of features seen at different wavelengths and times, and with different spatial resolution, to attempt to identify common features at different wavelengths. Such an approach, at best, is only partially successful. For example, a demonstration that bright grains observed in Ca II have similar statistical properties to dark grains seen in Hα suggests, but does not prove, that the two sets of grains are cospatial and cotemporal.

Because the same solar feature can have quite a different appearance in different observing circumstances, it often takes some time to establish that two different views of a single set of objects, in fact, have a common identity. This creates an environment in which 'new discoveries' sometimes turn out to be simply new views of known phenomena. Also, it creates a situation where it is probable that a given physical phenomenon may have different names depending on how it is observed.

The nature of the solar atmosphere imposes limits on the smallest scale of fine structure that can be observed. The factor that ultimately limits achievable resolution in most circumstances is the distance over which photon properties can respond to changes in atmospheric variables. The minimum distance for such response in an optically thick medium is the photon mean free path. Near optical depth unity, which is the mean source for the observed radiation, the mean free path is of the order of the density scale height, H. In the middle chromosphere, $H \approx 200$ km. Features smaller than this can be resolved if they are in optically thin layers, but one should not expect to observe features smaller than about 200 km in the layers near optical depth unity and larger. In any given layer of the chromosphere and upper photosphere, of course, there are specific spectral bands, such as the wings of strong lines, for which the layer in question is optically thin but still has enough opacity to be observed. By judicious choice of such spectral bands, the resolution of fine structure can be maximized.

A further limitation on resolution is imposed by line-of-sight integration. A typical effective contribution length for line-of-sight integration near optical depth unity is approximately $2H/\cos \theta$, where θ is the angle between the line-of-sight and the solar radius vector. The line-of-sight integration tends to blend features smaller than $2H/\cos \theta$ with the background, which adds to the difficulty of observing small features.

From a practical standpoint, therefore, the observable fine structure with limited wavelength coverage is limited in the middle chromosphere to features of approximately 200 km, i.e. ~ 0.3 arcsec. Near the base of the chromosphere the density scale height is near 100 km, and there is more likelihood for observing smaller structures at these depths than in the middle and upper chromosphere. Given sufficient choice of wavelengths for observations, these limits can be reduced somewhat.

Although we have not limited the definition of fine structure to intensity fluctuations, it is nearly always the case that fluctuations in any of the physical variables are correlated with intensity fluctuations. Fluctuations in either velocity or magnetic field intensity, for example, are likely associated with fluctuations in gas pressure, which in turn are expected to result in intensity fluctuations.

2. Properties of Chromospheric Structure

2.1. NETWORK AND SUPERGRANULE CELLS

Typically, chromospheric emission shows only weak center-to-limb changes in intensity except in the case of optically thin radiation. The absence of a marked center-to-limb effect at wavelengths for which the chromosphere is optically thick is a direct result of the fine structure, which roughens the surface and removes spherical symmetry. On intermediate scales, there is a coarse mottling of the disk associated with supergranule cells and their bordering network. Supergranule cells are observed in the high photosphere as large convection eddies with horizontal flow diverging from the cell center and subsiding flow at the cell borders. The cell size defined by the network system appears to coincide with the supergranule cell size observed in the photosphere, i.e. approximately 40–50 arcsec diameter. The network, which borders the cells, first becomes visible in the upper photosphere where it covers some 10% of the surface and becomes increasingly prominent with increasing height, mostly as a result of the increasing width of the network.

The network has a different appearance in different spectral bands. In most spectral lines, the network appears bright against a darker background. However, in some subordinate spectral lines such as hydrogen Balmer lines (excluding the center of Hα) and helium lines the network appears dark against a brighter background. The changing character of the network with observations in different spectral lines is a radiative transfer effect resulting from non-LTE. The network is more distinct when observed with moderate spatial resolution and near active regions. Illustrations of network structure are shown in Figures 1 and 2

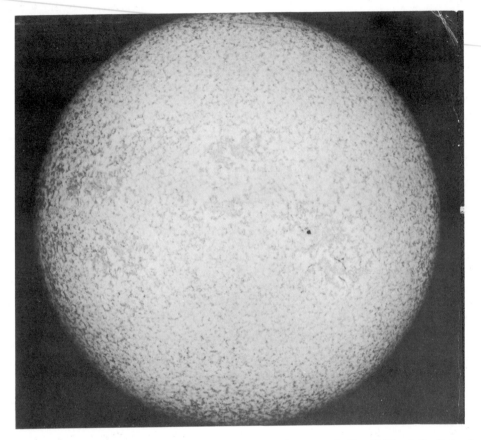

Fig. 1. Dark network structure observed at + 0.7 Å from line center in Hα. (California Institute of Technology.)

From the middle chromosphere upward, the width of the network remains relatively constant (Reeves, 1976). However, in the low chromosphere and upper photosphere the network appears somewhat narrower. In the upper chromosphere, the network contributes approximately 60% of the observed radiation in emission lines in the extreme ultraviolet but occupies only about 35–40% of the surface area (Reeves, 1976). The only comparable data available in the visual spectral regions are for the emission components of the Ca II K line formed in the middle chromosphere. In this line, the network covers 39% of the surface and contributes 45% of the radiation flux in the K_2 and K_3 components of the line. (Skumanich *et al.*, 1975)

Fig. 2. Bright network structure observed at the center of the Ca II K line. (California Institute of Technology.)

The lifetime of the network associated with a given supergranule cell is about 20 h (Simon and Leighton, 1964) which is similar to the lifetime of the cell. However, much of the network is made up of a variety of smaller features with lifetimes ranging from about 1 h to a few minutes. Also, the network is the locus for most of the nonactive region magnetic flux as well as for most of the vertical component of fluid velocity.

Observations of the network at moderate spatial resolution statistically show a downflow with average velocities of 40 m s^{-1} in a photospheric iron line (λ5233) (Skumanich *et al.*, 1975), 800 m s^{-1} in a middle chromospheric Si II line (λ1817) (November *et al.*, 1979), and 4 km s^{-1} in a lower transition region line of C IV (λ1548) (Gebbie *et al.*, 1981; Athay *et al.*, 1982). If the flow is continuous at the three levels, the quantity ρVa should be constant, where ρ is density, V is velocity and a is the fraction of the surface covered by network. However, estimates of ρVa at the levels where the three velocities are measured show no such tendency. The downflow of mass flux in the transition region is only 1% of that in the chromosphere, which in turn, is only 1% of that in the photosphere. Evidently the downflow is supplied almost entirely at photospheric and chromospheric levels by the lateral outflow from the supergranule cell center to the network.

Although with moderate resolution the observed network flow is downward, the same is not necessarily true with high spatial resolution. Spicules rise vertically out of the network beginning in the upper chromosphere and extend through the transition region into the low corona (see illustration in Figure 3). The estimated upward mass flux in

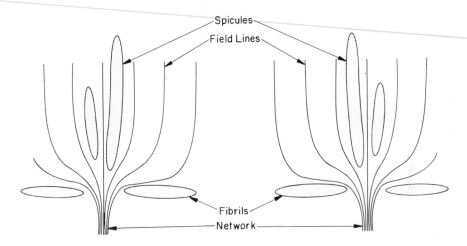

Fig. 3. A schematic illustration of the association of spicules and fibrils with the magnetic field emerging from the network.

spicules (cf. Beckers, 1968; Athay, 1976) is of the same order as the downward mass flux observed in C IV which, averaged over the surface of the Sun, corresponds to a flux of 10^{15} hydrogen atoms cm^{-2} s^{-1}. This is some two orders of magnitude greater than the mass flux needed for the solar wind. It is plausible to conclude, therefore, that the network downflow in the transition region is primarily the return flow of the spicule upheaval.

Although the picture of the average downflow in the network as the return flow of spicule material is plausible, the caution noted in Section 1.1 needs to be kept in mind. The evidence is entirely circumstantial.

An additional phenomenon observed at high spatial resolution is the explosive, high-speed ejections occurring at transition region temperatures near 10^5 K (Brueckner, 1980). The explosive events are less numerous, shorter lived, and have higher velocities than spicules. Also, the total mass flux in the explosive events is much less than in spicules.

In the middle and high chromosphere, the network radiates fine, hairlike fibrils (Figure 4) seen prominently in narrow band Hα pictures near line center. The fibrils diverge from the network into the supergranule cells and have been interpreted as tracers of low-lying magnetic flux tubes. Attempts to determine flow velocities in fibrils, none of which is entirely conclusive (cf. Bray and Loughhead, 1974), show some evidence of flow directed towards the network from the supergranule cell (Bhavilai, 1965), i.e. in the same direction as the overall flow.

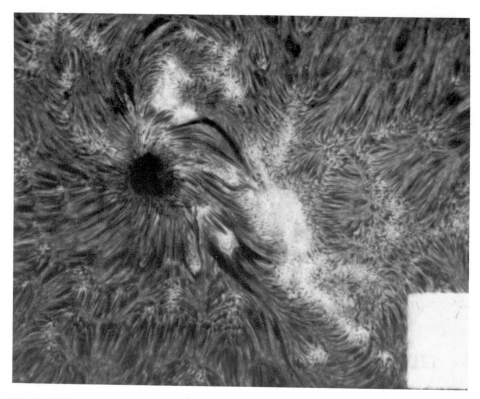

Fig. 4. Fibril structure observed in Hα. Network fibrils are seen away from the dark sunspot and bright plage regions. (Big Bear Observatory, California Institute of Technology.)

2.2. NETWORK COARSE AND FINE STRUCTURE

The recognition that chromospheric structure can be related to a more or less global mosaic of supergranule cells and network chains is a relatively recent concept. Part of the early failure to emphasize the network structure is due to the complex microstructure within the network. The coarser elements of the microstructure are easily resolved and the natural tendency has been to attempt to resolve the fine structure within these coarse features. Table I summarizes the basic features of network microstructure. For a more comprehensive review, the reader is referred to Bray and Loughhead (1974).

The coarse features of the network observed in Hα near line center are the rosettes consisting of clusters of dark fine mottles (Figure 4). Near the limb, the rosettes take on a somewhat different appearance referred to by some authors as bushes. When viewed off-band in Hα, − that is, in the line wings, − the bushes become more prominent and reveal the network as a collection of fine vertical columns rising above the general background (Figure 5). The individual vertical columns appear to be identical with spicules seen at the exteme limb.

Clustered around the centers of the rosettes and bushes and filling the gaps between the dark mottles of the rosette are the bright mottles. It is the collective appearance of the bright mottles that forms much of the network chain.

TABLE I

Structure and feature	Diameter (km)	Length (km)	Lifetime (min)	Aliases	Suggested identity	Closely associated features	Remarks
NETWORK							
Bright mottles	1000–8000	Similar	12	Bright flocculi	Network coarse	Dark mottles	Irregular shapes
Dark mottles	700–2000	8000–12 000	5–15	Rosettes, bushes dark flocculi	Spicules	Bright mottles	Clusters of dark mottles form rosettes or bushes
Rosettes	5000	Similar	1000	Bushes, dark flocculi	—	Bright mottles	Part of network coarse structure
Spicules	700–2000	10 000	10–15	—	Dark mottles, burning prairie	Bright mottles	Primarily vertical orientation
Fibrils	1000–2000	Similar	6–12	Dark mottles	Spicules	Bright mottles	Seen mainly in broadband Hα observation
SUPERGRANULE CELLS							
Bright grains	1000	Similar	3–4	—	Dark grains in Hα	—	Visible mainly in violet side of K-line, K_2V.
Fibrils	1000–1000	5000–10000	10–30	—	—	—	Seen in broadband Hα radiating from network
ACTIVE REGIONS							
Fibrils	700–2000	7000–15 000	3–50	—	—	Sunspot super-penumbra	Longer and more organized than other fibrils.
Arch filament systems	1000–3000	30 000	30	—	—	Sunspot regions	Connect regions of opposite polarity

Fig. 5. Spicules observed in the wing of Hα overlying the network near the solar limb. (Big Bear Observatory, California Institute of Technology.)

A class of features known as network fibrils (Marsh, 1976) appears to be dark mottles (spicules) viewed more nearly along their axes. A second class of features known as cell fibrils (Foukal, 1971a, b; Marsh, 1976; Bray and Loughhead, 1974) diverge from the network and overlie the supergranule cell. These are low-lying relatively horizontal features of considerable elongation and enhanced lifetime relative to the fine dark mottles.

It is important to note that fibrils do not have the appearance of complete loops with opposite ends anchored in magnetic fields of opposite polarity. A more accurate description associates the fibrils with a segment of a magnetic flux tube emanating from the network but whose other terminus is unknown.

The picture of the network fine structure that emerge from these features is a mixture of the more nearly vertical spicules and more nearly horizontal fibrils (Figure 3) emanating from a somewhat continuous chain of brighter features. Whether these brighter features have a wall-like component that constitutes the basic network structure or are simply collections of well separated individual bright fine structures is not clear at this time.

2.3. SUPERGRANULE CELL STRUCTURE

The primary features of the supergranule cells other than the canopy of fibrils overlying the cells are the bright grains seen (Table I) in the Ca II lines. Dark grains seen in Hα are probably identical with the bright Ca II grains (Beckers, 1968).

Liu (1974) identified the bright grains as vertically propagating features appearing first in the far wings of the Ca II K line and moving progressively towards line center. At the violet emission peak, the grains flash into view for 3–4 min. They are much less conspicuous in the red emission peak. Beckers (1968) reports that grains tend to recur on a quasiperiodic basis.

2.4. ACTIVE REGION STRUCTURE

It is not the intent, here, to describe all of the fine structure of active regions. Instead, we limit the discussion to active region fibrils and arch filament systems (Table I), which appear to have a close resemblance with quiet Sun fibrils and spicules.

Active region fibrils, illustrated in Figure 6, tend to define well-organized patterns extending well beyond the photospheric boundaries of sunspot penumbra. This has given rise to the phrase 'superpenumbral fibrils' used by some authors. Aside from their greater length and closer organization, the active region fibrils have many similarities to cell fibrils. One notable difference is that the active region fibrils more nearly resemble magnetic loops in that they tend to show a consistent center of curvature throughout their length. One has the impression that the active region fibrils are magnetic loops that are visible over most of their length.

Although individual fibrils may last only minutes, the fibril patterns in active regions persist for long periods of time. The diameter of the superpenumbral fibril pattern (Figure 6) around large sunspots grows continually as the active region develops (Bumba and Howard, 1965) with a typical expansion velocity of 200 m s^{-1}. Adjacent fibril patterns show evidence of orderly interaction, suggesting an even larger overall pattern than that associated with individual sunspots.

Arch filament systems (Bruzek, 1967) are distinct from fibrils in that they clearly arch over magnetic netural lines and connect to regions of opposite polarity. They tend to be notably darker than fibrils when observed near the center of Hα and often more sharply curved than fibrils. Also, they are elevated above the normal levels of fibrils. Examples are shown in Figure 6.

3. Velocity and Magnetic Structure

3.1. MAGNETIC STRUCTURE

Quiet Sun magnetic fields in the deep photosphere are confined largely to small intense flux tubes less than 400 km in diameter. These flux tubes are scattered along the borders of supergranule cells and underlie the chromospheric network. Phenomena of the network, therefore, occur in association with magnetic fields, and it is logical to identify such features as fibrils and spicules with individual strands of magnetic field.

Areas in which one magnetic polarity dominates on the Sun (unipolar regions) are typically much larger than the supergranule cells as shown in Figure 7. Hence, the flux tubes associated with neighboring supergranules tend to be of the same polarity. Because

Fig. 6. Fibril structure in Hα in an active region. (Sacramento Peak Observatory, Association of
Universities for Research in Astronomy, Inc.)

of this, it is often assumed that the field lines in photospheric flux tubes expand to more
or less fill the chromospheric network but to remain essentially vertical into the transition
region. Thus, the network is supposed to map the area of the magnetic flux tube as it
expands into the corona as illustrated in Figure 3.

Several lines of evidence argue against such a field topology: (1) the fibrils over the
supergranule cells are more or less parallel over the entire cell rather than converging near
the center of the cell, (2) there is no particular reason to assume an absence of magnetic
field outside the network and (3) observations show strong evidence of appreciable
horizontal components to network fields even in the low chromosphere and temperature
minimum region.

In the models proposed for network magnetic fields, as illustrated in Figure 3, the
field does not develop a strong horizontal component until the temperature approaches
10^6 K (cf. Gabriel, 1976; Athay, 1981). Also, the horizontal components of the field
converge towards the center of the supergranule cells in a pattern similar to spokes

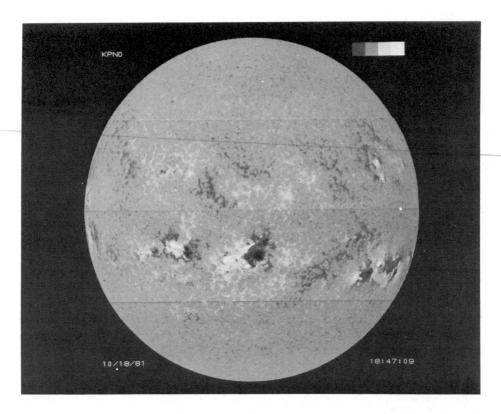

Fig. 7. A photospheric magnetogram showing network and active region structure. White and black are opposite polarities. (Kitt Peak National Observatory, Association of Universities for Research in Astronomy, Inc.)

converging at the hub of a wheel. None of this is consistent with the observed pattern of fibrils over the supergranule cells. The fibrils are seen in Hα, which requires that the field lines have strong horizontal components at chromospheric temperatures ($< \sim 10^4$ K), and the fibrils do not converge towards the cell center.

Within the chromosphere the network has essentially constant width. If the photospheric fields simply expanded to fill the network, the expected field strength in the chromospheric network would be essentially constant with height. On the other hand, the gas pressure in the network decreases by some three orders of magnitude through the chromosphere. In the deep photosphere, the ratio, β, of gas pressure, P_g, outside the flux tube to magnetic pressure, $B^2/8\pi$, is of order unity. If the field were confined to the network, the resultant value of β would rise to about 10^3 in the low chromosphere and return to order unity in the upper chromosphere. It is difficult to understand why the field strength should remain constant in such circumstances.

Recent observations of longitudinal photospheric magnetic field strength by Giovanelli (1980) show strong evidence for horizontal field components observed in strong Fraunhofer lines near the solar limb. This suggests that in the extreme upper photosphere and low chromosphere the field is neither confined to the network nor entirely verical.

It is possible, therefore, that the magnetic fields confined to the small flux tubes in the photosphere behave similarly to sunspot fields. The latter are known from observations to exhibit rapid spreading and strong horizontal components in the upper photosphere.

Direct measurements of the magnetic fields at middle and upper chromospheric levels are difficult to achieve, and it is doubtful that reliable observations have been made. The problem is complicated by the richness and pervasiveness of the fine structure coupled with strong departures from local thermodynamic equilibrium in the few spectral lines formed at such levels in visual spectral regions.

3.2. VELOCITY STRUCTURE

Measurements of chromospheric velocities are attempted by utilizing Doppler shifts of spectral lines and displacements of individual features. The former gives the line-of-sight component and the latter gives the transverse component. Both methods are subject to large errors. Displacements of features are uncertain because of the difficulty of identifying the same feature at different times. For example, a fibril that is growing in length is very likely changing simultaneously in overall visibility. Thus, the apparent 'end' of the fibril may not move with the same velocity as the material within the fibril.

Doppler shifts of spectral lines are quite readily observed but highly ambiguous in interpretation. The problem here is that the line is formed over a scale length that is large compared to the scale length for velocity. Vivid evidence that this is the case is shown by the nonthermal component of line width in the chromosphere. One rarely, if ever, observes Doppler shifts of lines on the disk that are as large as the nonthermal broadening of the lines. As a result, velocity amplitudes derived from Doppler shifts of individual features observed on the disk are only crudely known. In some cases, even the sign of the velocity is uncertain. These reservations should be kept in mind through the following summary.

At chromospheric levels, material flows from the supergranule cell to the network then flows downward in the network. The influx of material into the net work leads to a downward increase of the mass flux in the network, as noted earlier. These, however, are statistical trends and do not imply that every feature in the supergranule cell flows towards the network and that downward motion occurs everywhere in the network.

Motion around sunspots at chromospheric levels is similar to that around the network. The dominant tendency is for inflow towards the spot (reverse Evershed flow) and downflow within the spot. Again, the effects are statistical rather than universal. The same flow patterns for the network and spot are also found in the lower transition region.

From a rather extensive study of fibrils, Marsh (1976) proposes a model in which a fibril grows by injection of material at its base and contracts again as the material falls back under gravity. If such a model were sufficient to explain all fibrils, the net flow would be zero and would show no preferred direction. On the other hand, Marsh's own data, as well as data from other sources, show clear statistical trends. For sunspot fibrils, for example, Marsh (1976) finds 80% showing inward flow and 20% showing outward flow. This is consistent with the reverse Evershed flow. Most summaries of flow in network and cell fibrils favor inflow towards the network as the dominant trend (cf. Beckers, 1968; Bray and Loughhead, 1974; Athay, 1976). However, conclusive evidence is somewhat lacking, and not all authors agree with the majority opinion (cf.

Bruzek, 1967). Even though the case is hard to prove for individual fibrils, the existence of a general flow from cell to network is clearly established (Novermber *et al.*, 1979). Also, the downwardly increasing mass flux in the network demands a compensating inflow through the network borders.

Velocity amplitudes are difficult to determine with much confidence for fine structure on the solar disk. Grossmann-Doerth and von Uexküll (1973) found the most common velocities to be 4–8 km s^{-1} for a number of small Hα features observed near the network. Both upward and downward velocities were observed with the average being 3.9 km s^{-1} downward. November *et al.* (1979) found a cell to network flow velocity in the middle chromosphere of 3 km s^{-1}.

Flow velocities are much less ambiguous in spicules than in fibrils due to the ability to observe individual spicules at the limb. Both the rate of elongation of spicules and their Doppler shifts give a strong predominance of upflow at a mean velocity of 20–25 km s^{-1}. Few, if any, spicules reverse the direction of their Doppler shift (Mouradian, 1965), and most show either constant velocity or velocity increasing with time (Pasachoff *et al.*, 1969, Kulidzanishvili, 1980; Nikolsky, 1970).

It is clear that similarities between spicules and fibrils do not extend to their dominant motions. Also, the upflow in spicules runs counter to the general downflow in the network even though the spicules are themselves part of the network.

4. Magnetohydrodynamic Structure

4.1. A NEW LOOK AT FINE STRUCTURE

Studies of the solar and stellar atmospheres were dominated for several decades by the naïve assumptions of radiative and hydrostatic equilibrium. Soon after the discovery of the high coronal temperature it was recognized that such assumptions could not account for the basic properties of the chromosphere and corona. The first assumption to be dropped was radiative equilibrium. One then envisioned the solar atmosphere as being heated by some aerodynamic phenomenon but still maintaining a clear approximation to hydrostatic equilibrium both globally and locally. Although it has been recognized for some time that the heating mechanism very likely transfers momentum as well as heat to the atmosphere, the consequences of the momentum addition have been largely ignored. Such attitudes are still perhaps dominant at the present time. However, studies of solar fine structure with increasingly sophisticated instruments are forcing the recognition that we are dealing with magnetohydrodynamic phenomena rather than simply aerodynamic phenomena.

The two most striking concepts emerging from fine structure studies in recent years are: (1) magnetic fields are widespread and play a major role in fine structure; and (2) fluid motions are of large amplitudes and clearly coupled with magnetic field structure. Furthermore, it is now clear that the logarithmic gradients of magnetic field strength, B, and velocity, V, are often large, i.e. the scale length, l_B and l_V, defined by d ln $B/ds = l_B^{-1}$ and d ln $V/ds = l_V^{-1}$ are of the same order as the scale of the fine structure.

In the photosphere, for example, where magnetic field strength has been measured with some confidence (Livingston and Harvey, 1969; Stenflo, 1973) the magnetic flux is concentrated in small flux tubes of less than 400 km diameter and field strengths of approximately 1500 gauss. It is believed that the downflow at the borders of supergranule cells is largely confined to the flux tubes. In the chromosphere, the magnetic fields are no longer confined to small flux tubes of high field strength to the same degree as in the photosphere. The evidence suggests that the fields are spread throughout much of super-granule cell as well as the network. However, the rather well-defined features of the fine structure and the highly structured velocity field associated with the fine structure suggests the possibility that the magnetic field is finely structured even though it is widespread.

Throughout the upper chromosphere and low corona in the quiet Sun, the average gas pressure is of the order of 10^{-1} dyne cm^{-2}. If the magnetic fields were spread uniformly in the chromospheric network, and throughout the corona, the expected field strength would be 2–4 gauss in the upper chromosphere and 1–2 gauss in the low corona. The corresponding magnetic pressures, $B^2/8\pi$, are within a small factor of the average gas pressure. Also, the macroscopic velocities associated with fine structure are often a substantial fraction of the local sound speed. Under these conditions, one expects the magnetic field and velocity field to be mutually interactive. In a structured velocity field, such a situation is likely to result in a structured magnetic field. The evidence that chromospheric magnetic fields are still partially confined to network features and that there is still a supergranule flow pattern with flow towards the network at chromospheric levels clearly indicates that the magnetic field is structured by the flow. On the other hand, the thread-like appearance of fibrils and spicules suggests that the flow associated with these features is guided by magnetic lines of force.

The seemingly inescapable conclusion from recent studies of solar fine structure is that we are dealing with a true magnetohydrodynamic situation in which the fluid motions help to shape the magnetic fields, which, in turn, help to determine the fluid motion. If one adopts such a picture, then it becomes mandatory to approach the study of the magnetic and velocity structure as an integral problem rather than as two separate problems. Some such studies have already been attempted. However, the challenge presents major difficulties and we have made only a modest beginning.

4.2. LIFTING FORCES

One of the striking properties of chromospheric fine structure is the relatively common occurrence of large amplitude upward motions. Spicule upheavals are commonplace events. Their number on the Sun at any one time is about 4×10^5, which corresponds to about 40 per supergranule. The birthrate for spicules within the network is about 0.5 per square megameter per hour, and the average upward particle flux within the network is 3×10^{15} protons cm^{-2} s^{-1}.

If spicules were pushed upward by a localized force at the spicule base, the pressure gradient within the spicule would be expected to exceed the ambient hydrostatic pressure gradient. Just the opposite is observed, viz. the pressure gradient along the axis of a spicule is smaller by a considerable factor than the hydrostatic pressure gradient for the temperature within the spicule. Pikelner (1971) has argued on this basis that the accelerating force for spicules must be a body force distributed along the spicule axis rather than a localized force near the spicule base.

A number of mechanisms for spicule acceleration have been proposed. None of these is without serious reservations, but the models by Unno *et al.*, (1974) and Parker (1974) appear to be the most promising.

The spicule mass flux is two orders of magnitude larger than the observed solar wind flux. Thus, nearly all of the matter lifted in spicules apparently falls back. The conjectured downfalling spicule material can account for the observed downflow in the network at transition region temperatures. However, this requires that the spicule matter be heated to temperatures of at least 2×10^5 K.

Both the lifting and heating of spicule material represent energy input to the atmosphere. Potential energy acquired by lifting the spicule material eventually reappears as heat energy when the spicule materials falls back to its rest levels (Athay and Holzer, 1981). Such energy appears to be an important contributor to the total energy input to the transition region and possibly the low corona.

5. Major Problems

5.1. ENERGY BALANCE

Before we can claim to have an acceptable physical model of the solar atmosphere we must identify the major phenomena involved in the energy and momentum balances and gain an understanding of how the balances are established and maintained. These challenges are closely linked to our understanding of solar fine structure.

Energy balance in the solar atmosphere, in broad terms, involves the dissipation of mechanical energy as heat and the removal of the heat input by radiation or by the solar wind. The immediate suspects as the source of mechanical energy are the subphotosphere convection zone and rotation. However, the chain of events connecting these sources of mechanical energy to the ultimate production of radiation and solar wind clearly involves a variety of physical processes. For example, mechanical energy may be transferred from the subphotospheric layers to the chromosphere and corona by propagating waves or by magnetic fields. The waves and magnetic fields may do work against the atmosphere, such as lifting, or they may generate electric currents prior to dissipating as heat. Once the energy is dissipated as heat it may be transported elsewhere by thermal conduction or enthalpy flow before it is carried away either as radiation or as solar wind expansion.

Each of these phenomena is present with varying degrees of importance at different locations in the solar atmosphere. Furthermore, most of them are interactive. The magnetic fields, for example, strongly influence wave generation and propagation, thermal and electric conduction and enthalpy flow.

Since the fine structure in the solar atmosphere is coupled with major fluctuations in the radiation flux as well as in both the magnetic flux and velocity field, it is logical to identify the fine structure as a major manifestation of the physical processes associated with energy and momentum balance. By the same reasoning, it is logical to conclude that we shall not arrive at an understanding of the energy and momentum balance without also understanding the fine structure.

5.2. MOMENTUM BALANCE

The complex combination of flows associated with supergranulation and network in the chromosphere presents a number of challenging problems. What is the structure of the flow in fibrils and how is the fibril flow related, if at all, to the spicule flow? What happens to the spicule flow after the spicules fade from visibility in the visual spectrum? How high is the spicule matter lifted and how much is it heated? How is continuity achieved between outflow in supergranule cells, downflow in the network, and upflow in spicules? What forces drive the flows? These are but some of the more obvious questions, none of which is answerable at this time.

The preceding questions asked about the flows could be rephrased, of course, as questions about the magnetic fields, and we emphasize again that the magnetic structure and velocity structure are inseparably linked together.

It is both troubling and challenging that the unresolved velocity amplitudes evidenced by nonthermal line widths are usually larger than the residual velocity amplitudes. Understanding the nature of the motions associated with nonthermal line width represents one of the major challenges of high-resolution solar studies.

5.3. THE SUN AS A STAR

Studies of solar fine structure are both helped by and indispensable to studies of atmospheres of other stars. Fine structure is no more and no less important towards an understanding of the solar atmosphere than it is towards an understanding of stars of other spectral types. Similarly, observations of other stars provide clues as to how atmospheric structure relates to such properties as rotation rate, surface gravity, spectral type, etc. Such knowledge may not solve the riddles of solar fine structure, but it provides useful boundary conditions that may ultimately prove to be indispensable.

References

Athay, R. G.: 1976, *The Solar Chromosphere and Corona: Quiet Sun*, D. Reidel, Dordrecht, Chap. 3.
Athay, R. G.: 1981, *Astrophys. J.* **249**, 340.
Athay, R. G., Gurman, J. B., Henze, W. A. and Shine, R. A.: 1982, *Astrophys. J.,* **261**, 684.
Athay, R. G. and Holzer, T.: 1981, *Astrophys. J.*
Athay, R. G. and White, O. R.: 1978, *Astrophys. J.* **226**, 1135.
Athay, R. G. and White, O. R.: 1979, *Astrophys. J.* **229**, 1147.
Beckers, J. M.: 1968, *Solar Phys.* **3**, 367.
Beckers, S. M.: 1972, *Ann. Rev. Astron. Astrophys.* **10**, 73.
Bhavilai, R.: 1965, *Monthly Notices Roy. Astron. Soc.* **130**, 411.
Bray, R. J. and Loughhead, R. E.: 1974, *The Solar Chromosphere*, Chapman and Hall, London, Chap. 3.
Brueckner, G. E.: 1980, in P. A. Wayman (ed.), *Highlights of Astronomy*, Vol. 5, D. Reidel, Dordrecht, p. 557.
Bruner, E. C. Sr.: 1978, *Astrophys. J.* **226**, 1140.
Bruzek, A.: 1967, *Solar Phys.* **2**, 451.
Bumba, V. and Howard, R.: 1965, *Astrophys. J.* **141**, 1492.
Cram, L. E.: 1977, *Astron. Astrophys.* **59**, 151.
Deubner, F.: 1976, *IAU Coll.* **36**.

Doschek, G. A., Feldman, U. and Bohlin, J. D.: 1976, *Astrophys. J.* **205**, L177.

Foukal, P.: 1971a, *Solar Phys.* **19**, 59.

Foukal, P.: 1971b, *Solar Phys.* **20**, 298.

Gabriel, A. H.: 1976, *Phil. Trans, Roy. Soc. London* **A281**, 339.

Gebbie, K. B., Hill, F., Toomre, J., Novermber, L. J., Simon, G. W., Athay, R. G., Bruner, E. C., Rehse, R. A., Gurman, J. B., Shine, R. A., Woodgate, B. E. and Tandberg-Hanssen, E. A.: 1981, *Astrophys. J. (Letters)*, **251**, L115.

Giovanelli, R. G.: 1980, *Solar Phys.* **68**, 49.

Grossmann-Doerth, U. and von Uexküll, M.: 1973, *Solar Phys.* **28**, 319.

Kulidzanishvili, V. I.: 1980, *Solar Phys.* **66**, 251.

Liu, S. Y.: 1974, *Astrophys. J.* **189**, 359.

Livingston, W. and Harvey, J.: 1969, *Solar Phys.* **10**, 294.

Marsh, K. A.: 1976, *Solar Phys.* **50**, 37.

Mouradian, Z.: 1965, *Ann. Astrophys.* **28**, 805.

Nikolsky, G. M.: 1970, *Solar Phys.* **12**, 379.

November, L. J., Toomre, J., Gebbie, K. B. and Simon, G. W.: 1979, *Astrophys. J.* **227**, 600.

Parker, E. N.: 1974, *Astrophys. J.* **190**, 429.

Pasachoff, J. M., Noyes, R. M., and Beckers, J. M.: 1969, *Solar Phys.* **5**, 131.

Pikelner, S. B.: 1971, *Solar Phys.* **20**, 236.

Reeves, E. M.: 1976, *Solar Phys.* **46**, 53.

Simon, G. W. and Leighton, R. B.: 1964, *Astrophys. J.* **140**, 1120.

Skumanich, A., Smythe, C., and Frazier, E. N.: 1975, *Astrophys. J.* **200**, 747.

Stenflo, J. O.: 1973, *Solar Phys.* **32**, 41.

Ulmschneider, P.: 1979, *Space Sci. Rev.* **24**, 71.

Unno, W., Ribes E., and Appenzeller, I.: 1974, *Solar Phys.* **22**, 358.

Webb, D. F. 1981, in F. Q. Orrall (ed.), *Solar Active Regions*, Colorado Assoc. Univ. Press, Boulder, p. 165.

High Altitude Observatory,
National Center for Atmospheric Research,
Boulder, CO 80307,
U.S.A.

CHAPTER 10

STRUCTURE, DYNAMICS, AND HEATING OF THE SOLAR ATMOSPHERE

GERALD W. PNEUMAN AND FRANK Q. ORRALL

1. Introduction

There is no observable level of the solar atmosphere where the magnetic field does not interact with the solar plasma to produce inhomogeneities, to accelerate flows, and to channel the flow of heat and momentum. At the level of the photosphere the motions are still largely hydrodynamic except where these motions have assembled the magnetic field into intense flux tubes, or where intense flux tubes have newly intruded into the atmosphere from below the surface. In these low layers, plane-parallel or mean models of the atmosphere are still very useful and capable of reproducing the center-to-limb behavior of most of the solar radiation. Inhomogeneities and the magnetic field can for many purposes be considered important, but local, intrusions into an otherwise homogeneous atmosphere.

With increasing height, the magnetic field expands to become important everywhere in the atmosphere. This transition occurs at some uncertain and variable height in the chromosphere. Certainly from the upper chromosphere outward into interplanetary space the magnetic field cannot be neglected, and mean models of the atmosphere become increasingly inadequate, even as a first approximation.

It is in the solar corona, however, that magnetohydrodynamic effects are most obvious and where they have been most studied. Here, too, the need for nonradiative heating is most clearly related to the magnetic field. The corona is constantly evolving on both small and large scale. The 'quiet' corona is really slow-varying. The emphasis of this chapter is therefore on coronal activity.

2. Basic Theory and Processes

Observations of the upper solar atmosphere in recent years have pointed increasingly to the dominance of magnetic fields in the energy balance, force balance, and dynamics of almost every phenomenon we see. For this reason, the physics of the corona, transition region, and, perhaps, chromosphere lies within the realm of magnetohydrodynamics. In this section we shall outline the basic equations relevant to the study of these processes, evaluate the transport properties, and examine how effectively and in what manner this rather complex interaction between gas and magnetic fields takes place. We begin by evaluating the importance of thermal conduction, magnetic diffusion, and viscosity. In

Peter A. Sturrock (ed.), Physics of the Sun, Vol. II, pp. 71–134.
© 1986 *by D. Reidel Publishing Company.*

the presence of a magnetic field, the coefficients for these three processes are anisotropic: transport is perpendicular to the field rather than along it. We shall postpone consideration of these effects here except to note that this reduction is very important in the chromosphere and corona, so that the flow of heat is channelled along the field.

The importance of thermal conduction in determining the large-scale temperature structure of the solar corona was first recognized by Chapman (1957) who pointed out that thermal conduction alone could extend the solar corona to great distances into interplanetary space. For example, in the absence of all other energy processes, the divergence of the conductive flux must be zero, i.e.

$$\nabla \cdot [\kappa(T) \nabla T] = 0, \qquad (2.1)$$

where $\kappa(T)$ is the coefficient of thermal conductivity given (Chapman, 1954; Spitzer, 1956) by

$$\kappa(T) = 6 \times 10^{-7} \, T^{5/2} \, \text{erg cm}^{-1} \, \text{s}^{-1} \, \text{K}^{-1}.$$

Integrating Equation (2.1) with respect to the radial distance r, we have

$$T = T_0 \left(\frac{r_0}{r}\right)^{2/7},$$

where T_0 is the temperature at the coronal base, r_0. This is an extremely slow decline in temperature giving a temperature at 1 AU that is only a factor of 4.6 lower than that at the coronal base. The fact that this slow decline in temperature is inconsistent with hydrostatic equilibrium, combined with the requirement that the gas pressure vanish at large distances, helped lead Parker (1958) to postulate that the corona must expand, producing the now well-known solar wind.

The relative importance of convection of magnetic fields by plasma motions and magnetic diffusion is demonstrated by the magnetic Reynolds number R_m, given by

$$R_m = \frac{LV\sigma}{c^2},$$

where $\sigma = 2 \times 10^7 \, T^{3/2} \, \text{s}^{-1}$ (Cowling, 1953; Spitzer, 1956). Again, for $L = 10^{10}$ cm, $V = 10$–100 km s^{-1}, and $T = 1$–2×10^6 K, $R_m = 2 \times 10^{11}$–6×10^{12}. At first glance, it would seem as if resistivity would be totally negligible, which is true on the large scale. Near neutral sheets, however, the characteristic lengths are much smaller, and the classical concept of the electrical conductivity may not hold. Here, magnetic merging and reconnection can take place and the physics is quite different. Reconnection with rapid diffusion of magnetic fields can occur in a variety of ways, such as through standing Alfvén waves (Petschek, 1964; Sonnerup, 1970). Also instabilities such as Bohm diffusion (Bohm, 1949), the tearing mode instability (Furth et al., 1963; Jaggi, 1963; Coppi and Friedland, 1971; Biskamp and Schindler, 1971), anomalous resistivity (Linhart, 1960; Alfvén and Carlqvist, 1967; Hamberger and Friedman, 1968; Sagdeev and Galeev, 1969; Coppi and Mazzucato, 1971; Papadopoulos, 1976), and the interchange instability

(Bernstein *et al.*, 1958; Rosenbluth, 1960; Artsimovich, 1964; Boyd and Sanderson, 1969) can reduce the effective magnetic Reynolds number substantially and permit reconnection at a rate even up to a significant fraction of the Alfvén speed regardless of the magnitude of the electrical conductivity. The reader is referred to other chapters in this book for a more detailed discussion of the very important phenomenon of magnetic reconnection as applied to all levels of the solar atmosphere.

The coefficient of viscosity for fully ionized hydrogen (Chapman, 1954) is

$$\mu = 1.2 \times 10^{-16} \, T^{5/2} \text{ g cm}^{-1} \text{ s}^{-1},$$

where T is the temperature in degrees Kelvin. The relative importance of inertial forces to viscous forces is measured by the nondimensional Reynolds number, R, given by

$$R = \frac{\rho V L}{\mu} \, .$$

where ρ is the density, V the velocity, and L a characteristic length over which the speed varies. For the solar corona, $T \cong 10^6$ K, $V \cong 100$ km s^{-1}, and $L \cong 10^{10}$ cm. This gives $R = 14$ for a particle density of 10^7 cm^{-3}. For these conditions, then, viscous forces should not be very important. However, for the same density a temperature of 2×10^6 K and a velocity of 10 km s^{-1} gives $R = 0.24$. In this case, viscous forces should be important. Although viscous effects are usually neglected when applying the MHD equations to the solar atmosphere, it should be kept in mind that, under certain conditions, these forces could play a significant role.

The equation of force balance including inertial, pressure, gravitational, magnetic, and viscous forces is

$$\rho \left[\frac{\partial \mathbf{V}}{\partial t} + (\mathbf{V} \cdot \nabla)\mathbf{V} \right] = -\nabla P - \frac{GM_\odot}{r^2} \, \hat{e}_r + \frac{1}{4\pi} \, (\nabla \times \mathbf{B}) \times \mathbf{B} +$$

$$+ \mu [\nabla^2 \mathbf{V} + \tfrac{1}{3} \nabla(\nabla \cdot \mathbf{V})], \tag{2.2}$$

where P is the gas pressure, $\mathbf{B}$ the magnetic field vector, and $\hat{e}_r$ a unit vector in the radial direction. Here, we have neglected forces due to electrical fields, since charge neutrality is usually closely maintained, and radiation pressure which is not important in the solar atmosphere.[1] Conservation of mass requires that

$$\frac{\partial \rho}{\partial t} + \nabla \cdot (\rho \mathbf{V}) = 0. \tag{2.3}$$

In Equation (2.2) an important nondimensional number, β, can be obtained by taking the approximate ratio of the pressure forces to magnetic forces, i.e

$$\beta = \frac{8\pi P}{B^2} \, .$$

[1] Here, we are not referring to the F-corona or dust component for which radiation pressure is crucial.

If $\beta \gg 1$, pressure forces dominate and the magnetic field can perhaps be neglected. If, on the other hand, $\beta \ll 1$, the magnetic forces dominate over the pressure. For this case, we have $(\nabla \times \mathbf{B}) \times \mathbf{B} \approx 0$ or $\nabla \times \mathbf{B} = \alpha \mathbf{B}$ with $\nabla \alpha \cdot \mathbf{B} = 0$. Fields of this type are called 'force-free' magnetic fields with the simplest type ($\alpha = 0$) being a potential field. Of all the force-free field configurations for given photospheric boundary conditions, the potential field has the lowest energy and, therefore, is of considerable interest in describing the magnetic structure of the corona. For a temperature of 1.5×10^6 K and a particle density of 10^8 cm^{-3}, $\beta = 1$ indicates a magnetic field strength of about 1 gauss. Field strengths at the coronal base in quiet regions are generally believed to be larger than this so that potential fields may be a valid first approximation to the overall magnetic structure of the inner corona.

Another important nondimensional number can be derived from Equation (2.2) by taking the ratio of the inertial to magnetic forces. This ratio is the so-called Alfvénic Mach number given by

$$M_a = \frac{\sqrt{4\pi\rho}\, V}{B}$$

For the solar wind, this number is less than one close to the Sun but greater than one at 1 AU. Hence, a transition from sub-Alfvénic to super-Alfvénic flow takes place during the travel of the wind from the Sun to the Earth. Similarly, the ratio of inertial to pressure forces is given by the ordinary Mach number $M = (\sqrt{m_p/kT})V$. This number also changes from less than one to greater than one in the outer corona although, in most models, inside the Alfvénic transition point. Finally, we have already discussed the ratio of inertial to viscous terms, R. Even though this number is not exceptionally large, viscosity is usually neglected in atmospheric flow problems.

Maxwell's equations can be reduced to two equations — one for the conservation of magnetic flux and the other, the so-called induction equation, i.e.

$$\nabla \cdot \mathbf{B} = 0 \tag{2.4}$$

$$\frac{\partial \mathbf{B}}{\partial t} = \nabla \times (\mathbf{V} \times \mathbf{B}) - \nabla \times \left[\frac{c^2}{\sigma}(\nabla \times \mathbf{B})\right] \tag{2.5}$$

We have already mentioned that the ratio of the first term on the right-hand side of Equation (2.5) (convective) to the second (diffusive) is given by the magnetic Reynolds number R_m. Since, away form neutral points and sheets, R_m is quite large, magnetic diffusion is usually unimportant. If this is the case, then Equation (2.5) becomes a mathematical statement of a fundamental concept basic to all MHD processes in the solar atmosphere — the 'frozen-field' concept (see Ferraro and Plumption (1961) for a formal derivation from Equation (2.5)). This concept means that, *if at any time a magnetic field line threads through a given set of fluid elements, it must thread through those same elements at all later times regardless of the relative displacements of those fluid elements.* In other words, there can be no relative motion between the magnetic field and the fluid.

The equation for conservation of energy can be put in the form

$$\frac{\partial e}{\partial t} + \nabla \cdot \mathbf{q} = Q_R ,$$

(2.6)

where

$$e = \frac{1}{2}\rho V^2 + \frac{3kT\rho}{2m_p} + \frac{B^2}{8\pi}$$

and $\mathbf{q}$, the energy flux, is given by

$$\mathbf{q} = \rho \mathbf{V}\left(\frac{1}{2}V^2 + \frac{5kT}{2m_p} - \frac{GM_\odot}{r}\right) +$$

$$+ \frac{1}{4\pi}\,\mathbf{B} \times (\mathbf{V} \times \mathbf{B}) - \frac{c^2}{4\pi\sigma}\,\mathbf{B} \times (\nabla \times \mathbf{B}) - K\,\nabla T + \mathbf{V} \cdot \tau,$$

where τ is the viscous stress tensor. Q_R is the radiation loss term: for an optically thin gas it represents a loss, and can be approximated by

$$Q_R = n^2 Q(T).$$

(2.7)

The form of $Q(T)$ depends on the chemical composition of the plasma as well as on the atomic processes included. $Q(T)$ has been calculated by several authors on the basis of slightly different assumptions (see Rosner *et al.* (1976a, b) for a comparison). It has a broad maximum near $T = 10^5$ K which has important implications for thermal stability of the corona. At coronal temperatures.

$$Q_R \approx 9 \times 10^{35}\,\rho^2 \text{ erg cm}^{-3}\text{ s}^{-1}.$$

In the above, k is Boltzmann's constant and m_p is the proton mass. In Equation (2.6), all terms can be important depending upon the circumstance. In the corona, however, the radiative loss term is often only significant within about the first scale height. Also, the magnetic terms are generally neglected for the large-scale energy balance but are important if hydrodynamic waves, ejected magnetic flux, or reconnection are present.

3. Overall Atmospheric Structure

Almost all the Sun's radiation, extending in wavelength from the extreme ultraviolet (EUV) to the centimetre radio emission, comes from a geometrically-thin layer of the solar atmosphere at most a few thousand kilometres in thickness. In Chapter 8 of this volume, R. G. Athay discusses the physical processes that produce the spectrum and cause the peculiar temperature distribution in this layer, which includes the photosphere, the chromosphere, and the base of the transition region. In that chapter, Figure 4 illustrates such a temperature distribution using a semi-empirical model for the average quiet Sun from a recent elaborate study by Vernazza, Avrett, and Loeser (1981) known as VAL-III.

Given such a semi-empirical model, the net radiative loss from the atmosphere (that is, the divergence of the radiative flux vector) can be calculated as a function of height and used to constrain models of nonradiative heating, and the overall problem of energy balance.

A major difficulty is that this thin layer is obviously not plane-parallel but in fact possesses an exquisite dynamic fine structure. This may be seen on high-resolution spectroheliograms taken in radiations arising in this layer, such as the photographs in Lyman-α (Lα) and at 1600 Å taken by Bonnet *et al.* (1980) that appear as Figures 6, 7, and 8 of Chapter 8. Those making theoretical models are thus faced with a dilemma that is characteristic of solar physics, but that is especially acute when building models of the chromosphere and prominences: namely, that a uniform plane-parallel geometry, assumed in order to simplify the radiative transfer calculations, seems unrealistic in view of the obvious fine structure, while most attempts to include the flows and fine structure, compromise the radiative transfer calculations. This dilemma has been faced in a number of different ways, and in fact, our present knowledge of the thermodynamic structure of the photosphere and chromosphere is the result of a synthesis of many different empirical, semi-empirical, and theoretical studies. In the future, advances in computer technology should allow multidimensional radiative transfer calculations to play a larger role in this modeling. The reader is referred to Athay (1976) for a detailed discussion and review of photospheric and chromospheric modeling and for a comprehensive description of complex phenomenology of the chromospheric fine structure. Here we can consider only the most obvious ways in which the magnetic field influences the structure of this region.

The influence of the magnetic field is most clearly visualized at the level of the photosphere. There the field is observed to be concentrated into isolated intense magnetic flux tubes (Stenflo, 1976; Harvey, 1977). The smaller flux tubes are brighter than the surrounding photosphere and correspond to the faculae, while the larger ones are darker than the surrounding photosphere and correspond to pores and sunspots. Spruit and Zwaan (1981) find an apparently continuous distribution in size and contrast. The smaller facular points may be <0.1 arcsec in diameter. The transition from bright faculae to dark pore occurs at a diameter of about 1 arcsec. The magnetic field strength is believed to be between 1 and 2 kilogauss in the facular elements and 2–3 kilogauss in pores and spots. The problem of constructing an inhomogeneous model of the photosphere is thus reduced to constructing models for individual flux tubes, and for the surrounding field-free region that can be considered homogeneous and plane parallel (see, for example, Chapman, 1981, and references therein).

Since the pressure falls off with height, these flux tubes must expand with height in order to remain in horizontal pressure balance with the surrounding atmosphere. The magnetic field lines, assumed vertical and parallel low in the photosphere, will fan out with height and acquire an increasing horizontal component. This suggests that there is a horizontal *canopy* of magnetic field at some height above the field-free region of the photosphere. Any model of this horizontal field must be highly idealized. Neighboring magnetic flux tubes may be of the same or opposite polarities. Gabriel (1976b) assumed the height of this canopy to lie 1500 km above the photosphere in his inhomogeneous model of the transition region. Recently Giovanelli (1980) and Giovanelli and Jones (1981) found direct observational evidence for this canopy in the vicinity of active regions. It appears as diffuse structure on Kitt Peak magnetograms in such lines as $\lambda 8542$ Ca II, Mg I b_2 and Hα. They infer the height of the canopy base to be often less than 400–500 km near the active region, and even lower within it. This is an unexpected result which suggests that our understanding of chromospheric magnetic fields is not very complete.

4. The Chromosphere–Corona Transition Region and the Base of the Inner Corona

4.1. INTRODUCTION

The mean temperature of the solar atmosphere rises abruptly above the top of the chromosphere, and then more slowly to reach 10^6 K within 5–10 Mm of the visible surface of the Sun. Above this transition region, it continues to rise in the inner corona to reach some temperature maximum whose value depends on the local structure of the corona. Here it is useful to distinguish broadly between regions of the corona that are 'open' and 'closed'. Open regions are threaded by magnetic field lines that extend out into interplanetary space. Here mass and energy can flow out in the solar wind, and the temperature maximum almost certainly occurs within one solar radius of the surface, and probably much closer. In closed regions the magnetic field lines are loops or arches whose vertex is no more than about a solar radius above the solar surface and usually much less. The plasma confined to these loops has its temperature maximum near the loop vertex. Both mass and energy may be transferred through the corona from one foot point to the other. In both the open and closed regions, some large fraction of the energy deposited in the corona as heat by the dissipation of waves, electric currents, or other processes, is lost to the corona by thermal conduction to the underlying transition region and chromosphere. Short-lived jets of chromospheric material in the form of spicules penetrate the transition region and carry a large flux of mass into the corona, most of which must return to the surface. Thus, energy and mass are exchanged between the photosphere and inner corona, and some fraction of them escapes from the Sun in the form of radiation and the solar wind. The upper chromosphere, transition region, and inner corona thus form a closely interacting and interpenetrating system of great complexity and astrophysical interest.

A number of reviews treat the transition region and the coronal base from different points of view (e.g. Athay, 1971, 1976; Jordan and Wilson, 1971; Gabriel, 1976a; Kopp and Orrall, 1977; Noyes, 1971; Vaiana and Rosner, 1978; Withbroe, 1976, 1977a, b; Withbroe and Noyes, 1977). These document a rapid increase in the quantity and quality of pertinent observational data, and also an evolving appreciation for the physical complexity of the solar transition region. With the launch of the International Ultraviolet Explorer (IUE) in 1978 it became possible to observe emission lines formed at temperatures between $10^{3.8}$ to $10^{5.4}$ K in large number of stars. These observations suggest that chromospheric and transition region phenomena are common and varied in cool stars (see reviews by Ayres, 1981; Linsky, 1980). Observations with the Einstein Orbiting Observatory indicate that stars are also commonly emitters of soft X-rays. For cool dwarfs the ratio of X-ray to total flux tends to increase down the main sequence (Vaiana *et al.*, 1981). Perhaps all cool stars have chromospheres, transition regions and coronae that correspond in some way to their solar counterparts, although their observed properties vary widely even among stars of the same spectral class. Knowledge of how the hot outer atmospheres of stars vary with rotation, mass, surface gravity, age, composition, effective temperature, and other gross properties is expected to contribute greatly to our understanding of the physics of the Sun's outer atmosphere (cf. Bonnet and Dupree, 1981; Cassinelli and MacGregor in Chapter 18 of this work). Until now, however, the Sun, which can be observed in exquisite detail, has provided most of our physical insight into the nature of nonradiatively heated stellar atmospheres and the role of the magnetic field in their formation.

In the limited space of this section we consider first the determination of the empirical mean (horizontally averaged) temperature structure of the transition region and coronal base, and its implications for the energy budget. (The mean structure of the chromosphere and its significance is discussed by Athay in Chapter 8 of this volume.) We then consider the inhomogeneous and dynamic structure of the region, which is implicated in most current outstanding problems.

4.2. EMPIRICAL MODELS OF THE MEAN TEMPERATURE STRUCTURE

About 20 years ago spectrophotometric observations of the Sun in the XUV became available and provided the basic data needed to study solar plasma in the temperature range $10^{4.5}$–10^6 K. This opened the transition region to detailed exploration.

A set of total intensities of resonance emission lines from a number of ionic species formed at a range of different temperatures provide the most useful observational data for deriving the temperature structure. Most of these lines lie in the XUV and are optically thin. A surprisingly simple approximate expression can be written for the emissivity ϵ(erg cm^{-3} s^{-1} ster^{-1}) of solar plasma for such lines; namely

$$\epsilon = CA_i f_i G_i n^2 . \tag{4.1}$$

Here $C \approx 1.7 \times 10^{-16}$ (c.g.s.); A_i is the chemical abundance relative to hydrogen by number; f_i the oscillator strength; n the electron density (cm^{-3}); and $G_i(T)$ contains all of the temperature dependent terms including a gaunt factor. However, by the fractional abundance of the ion giving rise to the line provides the dominant variation in $G_i(T)$. This causes $G_i(T)$ to be quite sharply peaked with a maximum at T_i, the optimum temperature for the formation of the line. The subscript i indicates the particular line observed. The primary assumptions or approximations used in deriving Equation (4.1) are: (a) that the line is collisionally excited from the ground level and radiatively de-excited; (b) that most ions are in the ground level; and (c) that the fractional ionization depends only on temperature. These approximations have been used in most studies since the first detailed analysis of solar XUV data by Pottasch (1964).

The total specific intensity I_i(erg cm^{-2} s^{-1} ster^{-1}) in a particular emission line i observed from the vicinity of the Earth will be

$$I_i = (c/a)A_i f_i \int n^2 G_i(T)\, dV, \tag{4.2}$$

where the integral is over the entire volume of the observing column. Here a is the cross-sectional area (cm^{-2}) of the observing column at the Sun and I_i the intensity as measured with the limited spatial resolution of the spectrometer. Making the transformation

$$n^2 \, dV = \xi(T) \, dT \qquad (4.3)$$

we can write

$$\frac{I_i}{(c/a)A_i f_i} = \int_0^\infty \xi(T)G_i(T) \, dT, \qquad (4.4)$$

where the function $\xi(T)$, called the differential emission measure (DEM), evidently contains all the information about the temperature structure of the atmosphere available in the data. Given a set of values for the left-hand side of Equation (4.4) and the functions $G_i(T)$ for ions with a range of characteristic temperatures T_i, one can invert the system of equations to derive $\xi(T)$. (Craig and Brown (1976) discuss the inherent instability and nonuniqueness of this inversion process in the presence of errors and also give a more rigorous definition of the DEM and clarify its formal meaning. Further discussion is given by Dere and Mason (1981). Jeffries *et al.* (1972) have treated the problem of interpreting a set of total line intensities from an optically thin gas more generally in terms of a bivariate distribution function $\mu(T,n) \, dT \, dn$ which is also appropriate for lines whose emission does not vary simply as n^2. But the recovery of this function suffers from the same instability as does the emission measure.)

If the atmosphere can be assumed plane-parallel within an observing column at the center of the Sun's disk, and if $T(h)$ is a monotonic function of height h, then from (4.3)

$$\frac{dT}{dh} = \frac{P_e^2}{(kT)^2 \, \xi(T)} \qquad (4.5)$$

Here $P_e = nkT$ is the electron pressure, which in hydrostatic equilibrium will be roughly constant in the transition region since its geometrical thickness is small compared to a pressure scale height. Given $\xi(T)$ and P_e, Equation (4.5) determines the temperature distribution $T(h)$ above an arbitrary initial height h_0, and the thermal conductive flux which is proportional to $T^{5/2} \, (dT/dh)$. The same analysis yields the relative chemical abundances of the elements represented in the observed ions although determining the abundances relative to hydrogen (A_i) is more difficult (see the review by Athay, 1976).

Estimates of P_e can be derived from analysis of spectral lines whose emission does not depend on density as n^2. Such lines are usually either forbidden lines or lines excited from metastable levels. Thus, for example, the intensity ratio of the lines $\lambda 1176$ to $\lambda 977$, both of C III $(T_i \approx 10^{4.95}$ K), depends sensitively on n at transition region densities. Here $\lambda 977$ is a resonance line and $\lambda 1176$ is excited from the metastable levels $2s \, 2p^3 \, P$. Dere and Mason (1981) review a number of such sensitive line ratios. Withbroe (1977a) summarizes pressure determinations in the chromosphere, transition region, and corona by several workers and by a number of techniques. P_e is uncertain by at least a factor of 2 and hence the thermal flux is uncertain by a factor of 4 or more. Withbroe and Noyes (1977) suggest typical total gas pressures in the transition region of 0.07, 0.2, and 2 dyne respectively in open regions (coronal holes), quiet closed regions, and active regions.

Giovanelli (1949) first recognized that thermal conductive flux flowing back from the corona through the transition region must play an important role in the energy balance of the underlying atmosphere. For example, a typical estimate of the thermal conductive flux F_c from the corona is $\sim 10^{5.3}$ erg cm^{-2} s^{-1} (Withbroe and Noyes, 1977). This is large enough (within the uncertainties) to supply the mean net radiative losses of the transition region and upper chromosphere. But although F_c is large in the upper transition region ($T > 10^{5.3}$ K) it is nearly constant there. Its divergence is too small to compensate for local radiative losses, at least, in a spherically symmetric model. On the other hand, in the lower transition region ($T < 10^5$ K), F_c is relatively small so that $\nabla \cdot F_c$ must be large near $T = 10^5$ K; that is, most of the coronal conductive flux must be deposited in the middle transition region. Thus, for example, Gabriel (1976b) finds that a spherically symmetric model requires an additional source of 4×10^6 erg cm^{-2} s^{-1} between $T = 10^{5.5}$ and $10^{5.8}$ K, and an additional sink of 3×10^6 erg cm^{-2} s^{-1} between $T = 10^{4.8}$ and $10^{5.5}$ K.

As Gabriel pointed out, these large energy sources and sinks must be artifacts of the assumption of static spherical symmetry, at least in part. Kuperus and Athay (1967) suggested that the implied large deposition of conducted heat in the middle transition region would be unstable and generate motions — possibly spicules — so that a kinetic energy term is essential. More generally we can now say that convective energy transport is important in both the transition region and inner corona. Kopp and Kuperus (1968) pointed out that the coronal magnetic fields would channel the downward thermal flux into the supergranule network boundaries where the photospheric magnetic fields are observed to be concentrated. They stress that the geometry imposed by this channeling cannot be neglected in the energy balance equation. Unfortunately, the geometry is still uncertain. There is now abundant direct observational evidence for both inhomogeneities and flows in the transition region, and recent models increasingly attempt to take them into account.

4.3. INHOMOGENEOUS AND DYNAMIC STRUCTURE OF THE TRANSITION SHELL

As soon as the spherically symmetric approximation for the atmosphere is abandoned, the terms 'upper chromosphere', 'transition region', and 'coronal base' lose their meaning as the names of distinct overlying physical layers, and are instead commonly used to designate temperature regimes. There is thus no longer a consistently used name for the spherical shell 5–15 Mm in thickness that separates the low chromosphere from the corona. For convenience in this section we shall hereafter refer to it as the 'transition shell'. Within this complex and dynamic shell, material at chromospheric and coronal temperatures interpenetrate, but are separated evidently by a thin interface at transition region temperatures.

The simplest type of inhomogeneity of the transition shell is that imposed by the supergranular network magnetic fields. If there is any model that may be presently called the 'standard model' of the transition shell, it is certainly the 'network' or 'magnetic' model hich takes the network fields into account in an idealized way. A number of semi-empirical versions of this model have appeared in recent years (cf. Gabriel, 1976a, b; Elzer and Elwert, 1980a, b; Athay, 1981b, 1982). All impose energy balance, but with quite different assumptions. The basic new data that constrain these and models are EUV spectroheliograms and spectra with sufficient resolution to distinguish between the network and cell interiors.

The first EUV spectroheliograms showed that the network was brighter than the cell interiors in emission lines arising in the transition region (cf. Feldman *et al.*, 1976; Reeves, 1976). Reeves (1976), using spectroheliograms of 5 arcsec resolution, found that the quiet Sun network has about the same width ($\approx$10 arcsec) and, has the same form when observed in ions formed over the range in temperature $10^{3.8}$ K $< T < 10^{5.8}$ K. It covers $\approx$46% of the solar surface, which is not significantly different from the 40% found on Ca II K-line spectroheliograms observed from the ground (Skumanich *et al.*, 1975). The brightness contrast between cell and network reaches a maximum at $T \approx 10^{5.2}$ K, where the network produces $\approx$75% of the emission, and decreases at higher temperatures. Thus in Mg X ($T \approx 10^{6.15}$ K) only remnants of the network remain visible. Within coronal holes, the contrast between network and cells is less pronounced than for the quiet Sun in transition region emissions although the cells in holes and the quiet Sun seem identical (Reeves, 1976). At higher temperature, the emission within holes drops in both cell and network so that the Mg X soft X-ray emission is fainter than in quiet regions by a factor of 5 to 10.

The first magnetic network model to incorporate these new observations was that of Gabriel (1976b). In this two-dimensional model in hydrostatic equilibrium, Gabriel idealizes that magnetic field as a potential field generated by magnetic line sources that extend in the third boundaries. The lines represent supergranular diameter. Energy is supplied by thermal conduction from the corona and is lost by radiation. No other terms of the energy equation (Equation (2.6)) are included. The thermal conductive flux entering the top of the atmosphere is treated as a parameter chosen for best fit to observed data. Gabriel's model is quite a good reproduction of the width of the network, the observed emission above the limb, and the emission measure curve in the range $10^{5.3}$ K $\leq T \leq 10^6$ K. Elzer and Elwert (1980a, b) have constructed similar models for both the quiet Sun and coronal holes, in a somewhat more realistic three-dimensional geometry. The coronal hole models were made to match critical solutions of the solar wind equations and the extended thickness above the limb observed in Ne VII. To achieve this agreement it was necessary to invoke a nonradial coronal geometry (which the observations suggest) and also a secondary heating peak low in the transition region (which is purely speculative).

These models assume that coronal heat conduction supplies enough to the transition zone. Pneuman and Kopp (1977, 1978), however, have pointed out that the enthalpy flux associated with the downflows observed in the supergranular network (White, 1976; Gebbie *et al.*, 1980; Brueckner *et al.*, 1980) may exceed the thermal conductive flux. They suggest that this downflow is spicular material that has been heated to coronal temperatures and is returning to the chromosphere. Of course, their suggestion imposes even greater demands on the coronal heating mechanism. Athay (1981b, 1982) has included both this enthalpy and the thermal conductive flux in a series of magnetic network models that satisfy the observed EUV emission measure for $T \geq 10^{5.3}$ K. He finds the calculated emission measure is quite insensitive to either the magnetic geometry, or the dependence of downflow velocity with height.

None of the above network models reproduces or even approximates the observed emission measure curve for $T < 10^5$ K. Neither do the spherically symmetric models (cf. Jordan, 1980). Pneuman and Kopp (1978) suggest that this deficiency arises from neglecting the variation of the downflow with height. Athay (1981b) remarks that it is probably related to transition region dynamics rather than to an external source of heat. However, quiescent prominences observed well above the Sun's limb show an emission measure curve similar in shape to that of the corona at the center of the quiet disk (Orrall and Schmahl, 1976). This result strongly suggests that the same process reverses the emission measure curve near $T \sim 10^{5.3}$ K in both the transition shell and in prominences.

Additional observational data that can constrain empirical models include millimetric and centimetric radio measurements (reviewed by Athay, 1976; and Furst, 1980); and XUV intensity measurements center-to-limb and as a function of height above the limb (e.g. Withbroe, 1970; Withbroe and Mariska, 1976; Brueckner and Nicolas, 1973; Doschek *et al.*, 1976; Feldman *et al.*, 1980a, b). These show clearly that the transition region is highly structured and suggest that spicules, fibrils, loops, and other structures not only absorb XUV emission in the Lyman continuum, but may also emit an uncertain but possibly large fraction of the line emission. Schmahl and Orrall (1979) present evidence for a Lyman continuum absorption of transition region lines that is too large to explain by spicules or fibrils, and that suggests a more intimate mixing of neutral hydrogen with the transition region's ions.

5. Closed Coronal Regions

Above the transition region lies the solar corona which represents the outermost region of the solar atmosphere. As discussed in Section 4, the temperature here rises to over a million degrees Kelvin and then declines relatively slowly outward, falling to about 2×10^5 K at the orbit of Earth. This initial rise in temperature is now thought to result from the dissipation of mechanical or magnetic energy generated by the convection zone.

As was argued by Parker (1958), and subsequently verified by direct observation (Shklovskii *et al.*, 1960; Bridge *et al.*, 1961), this high coronal temperature and the high gas pressure associated with it results in a general expansion of the corona into interplanetary space. However, this expansion does not occur everywhere but only where the gas pressure is sufficient to distend the large-scale magnetic fields outward into an 'open' configuration. In regions of strong field, the closed field lines can prohibit expansion and produce in a quite different physical environment from that in the open field regions. The physics of the expanding open field regions will be discussed in Section 6. Here we concentrate on the closed field configurations.

5.1. BASIC STRUCTURE

The inner corona is most dense and radiates most strongly in regions where the magnetic field is 'closed' locally — that is, in those regions where the magnetic field is sufficiently strong to resist the dynamical forces of the coronal expansion. Closed regions are highly structured and consist almost entirely of plasma-filled loops or systems of loops, whose topology is consistent (to first order) with that of the expected coronal magnetic field. Thus, for example, systems or arcades of loops often bridge magnetic neutral lines to connect regions of opposite polarity in the underlying photosphere. Almost all the solar soft X-ray emission, the coronal EUV emission, and visible coronal forbidden line emission arises in these closed plasma loops. The results from ATM experiments on Skylab stimulated many recent observational and theoretical studies of coronal loops. Many of these were presented at the Skylab Solar Workshop on Active Regions and are discussed in the workshop monograph (edited by Orrall, 1981). More recent impetus has come from the Einstein Orbiting Observatory which finds that soft X-ray emission from stars is typical rather than exceptional (Vaiana *et al.*, 1981, and Chapter 18 of this work by Cassinelli and MacGregor), suggesting that plasma loops may dominate the outer atmosphere of many stars and are thus of broad astrophysical importance.

Because most of the observed coronal emission comes from loops, it is not surprising that we presently know more about them than about any other coronal structure. Indeed, at present we know very little about the coronal plasma in which these loops might be imbedded. Thus most of this section on closed coronal regions will be devoted to the physics and phenomenology of loops. First, we shall classify three types of region of the corona that are magnetically closed and describe briefly how they evolve.

Magnetic fields enter the Sun's atmosphere from below the photosphere in the form of ropes or bundles of magnetic flux. They first appear in the photosphere as compact bipolar magnetic regions (BMRs) of zero net flux (see, e.g., reviews by Sheeley (1981) and by Gilman in Chapter 5 of this work). These strong compact magnetic loops are often filled with brightly emitting coronal plasma. The largest and longest lived BMRs surface in the latitudes of the sunspot zone, and produce sunspot groups and all the attendant phenomena that collectively comprise a 'solar active region'.

Small, short-lived BMRs may surface at any latitude, and a few hundred of these 'ephemeral' regions, each having a typical lifetime of one-third of a day, are distributed over the solar surface at any one time. These also intrude into the corona and produce a small compact cluster of bright coronal loops known as 'X-ray bright points' (e.g. Golub *et al.*, 1974; Harvey and Martin, 1973).

The third type of closed region is loosely called 'the quiet corona' which usually covers most of the solar surface. As active region BMRs are slowly dispersed by subphotospheric motions, their flux spreads to produce large, roughly unipolar regions separated by magnetic neutral lines. The neutral lines are spanned by magnetic loops of large scale, which may become filled with corona plasma and are clearly visible on soft X-ray images (see Figure 4).

5.2. PHENOMENOLOGY OF CORONAL LOOPS

The loop-like structure of the inner corona has long been known from ground-based observations at eclipse and with the coronagraph (see, e.g., Dollfus, 1971; Dunn, 1971; and reviews by Kiepenheuer, 1953; Van de Hulst, 1953). More recently, the numerous soft X-ray and EUV images from Skylab have shown systems of loops directly against the disk in varying projections and in relation to the underlying chromospheric and photospheric structures. Moreover, EUV emission from numerous ionic species provides detailed temperature diagnostics. As a consequence of these new Skylab data, a clear and consistent picture of the morphology and evolution of coronal loops is beginning to emerge (see, e.g., the review by Webb, 1981).

Active region loops have been studied more intensively than any other type, and much of what we know of the general properties of coronal loops is based on active region studies. In a typical nonflaring active region, the differential emission measure as a function of temperature for ($T > 10^5$ K), has a maximum between $T \approx 10^{6.0}$ and $10^{6.4}$ K, with very little emission above $10^{6.6}$ K (see, e.g., Pye et al., 1978; Levine and Pye, 1980). Emission lines characteristic of $T < 10^6$ K are formed both in the underlying atmosphere (especially from the network) and in coronal loops or portions of loops. Emission characteristic of $T > 10^6$ K comes almost entirely from coronal loops (Dere, 1982). A comparison of monochromatic images of a typical active region, taken in emission lines with increasing formation temperatures, shows that loops tend to become longer lived, more numerous, and more diffuse as their temperature approaches $T \cong 10^{6.4}$ K.

This effect is most clearly seen on the NRL XUV spectroheliograms from Skylab (e.g. Sheeley, 1980, 1981). Sheeley contrasted images of a mature active region observed in $\lambda 465$ Ne VII ($T \approx 10^{5.7}$ K), $\lambda 368$ Mg IX ($T \approx 10^6$ K), and $\lambda 284$ Fe XV ($T \approx 10^{6.4}$ K). In Ne VII the most striking features are spikes extending into the corona, most prominent in the outer regions of BMRs where they appear to radiate from the regions of strongest field. In Mg IX the spikes are more often seen as complete loops and are longer lived, typically ≈ 1.5 h. In Fe XV the plasma is concentrated in systems of diffuse loops that bridge the neutral line of the BMR. A typical loop system lives for several days, and an individual loop about 6 h, although individual loops are often not clearly distinguishable. Thus, the morphology and evolution of a loop depends on its temperature and it has proved useful to distinguish between 'hot loops' and 'cool loops'.

Hot loops are the diffuse loops formed at $T > 10^6$ K and seen, for example, in $\lambda 284$ Fe XV, soft X-ray, or in the forbidden line $\lambda 5303$ Fe XIV with the coronagraph (Dunn, 1971). These large systems of hot loops fade out towards their footpoints (in contrast to the cool loops) so that Fe XV and soft X-ray emission is often missing directly above large sunspot umbrae. The hot loops evidently correspond to the maximum in the active region emission measure curve between $T \approx 10^{6.0}$ and $10^{6.4}$ K. The cool loops are those with $T \leq 10^6$ K. In contrast to the hot loops, cool loops are rooted in the underlying transition region and fade upward, with the hotter ones extending to greater heights. Hot loops and cool loops are not ordinarily cospatial — that is, quiescent hot loops do not have cool cores (Cheng, 1980; Cheng et al., 1980; Dere, 1980, 1982; Mariska et al., 1980; Sheeley, 1980, 1981; Webb, 1981). Indeed, observations suggest that quiescent loops in active regions are isothermal and isobaric, except near their footpoints.

On the other hand, there is evidence that developing loops may have a radial temperature and pressure gradient, with a 'cool core', and some coaxial loop models have been constructed (Foukal, 1975, 1976, 1978; Levine and Withbroe, 1977; Van Hoven et al., 1977). These models have been applied especially to the EUV 'plumes' sometimes seen above sunspot umbrae that emit very strongly in emission lines characteristic of the transition region (Foukal et al., 1974).

5.3. QUASISTATIC LOOP MODELS IN ENERGY BALANCE

Since many hot coronal loops are observed to change only slowly over many hours, it is reasonable to construct static models in energy and pressure balance in order to represent their basic observed properties. The assumption of static equilibrium has been the starting point for a number of theoretical investigations (referenced in part by Rosner *et al.*, 1978a; and in reviews by Vaiana and Rosner, 1978; Withbroe, 1981).

Briefly, it is assumed that the magnetic field provides a symmetric loop geometry that channels the flow of heat and matter. Energy deposited along the loop by some process given by a heating term H (erg cm^{-3} s^{-1}), is redistributed by thermal conduction, some fraction of it flowing out through the footpoints into the underlying transition region or chromosphere as a basal heat flux, F_c (erg cm^{-2} s^{-1}). The remainder is lost by radiation with local emission $n^2 Q(T)$(erg cm^{-3} s^{-1}) as in Equation (2.7).

In these one-dimensional static models the equation of static energy balance can be written as

$$\frac{1}{A(s)} \cdot \frac{d}{ds}\left[A(s)\kappa T^{5/2}\frac{dT}{ds}\right] + H - n^2 Q(T) = 0, \tag{5.1}$$

where s is the distance measured from one footpoint along the loop, $\kappa T^{5/2}$ is the coefficient of thermal conductivity parallel to the field, and $A(s)$ is the cross-sectional area. In hydrostatic equilibrium.

$$\frac{dP}{ds} + \frac{GM_\odot}{r^2}\rho\hat{e}_r\hat{e}_s = 0, \tag{5.2}$$

where $\hat{e}_r$ is a unit vector in the radial direction, $\hat{e}_s$ a unit vector directed along the field, and P the total pressure including the gas pressure, and also perhaps the wave or turbulent pressure (e.g. McWhirter *et al.*, 1975). The boundary conditions are defined by the symmetry, and by specifying the basal heat flux $(F_c)_0$ at the basal temperature T_0. For simplicity, this base is usually taken above the top of the chromosphere $(T > 10^{4.5}$ K) where the plasma can be assumed effectively fully ionized gas and optically thin. The basal heat flux for quiescent loops lies in the range $0 \leq (F_c)_0 \leq 4 \times 10^6$ erg cm^{-2} s^{-1}, the upper limit being set by the ability of the underlying chromosphere to radiate this heat flux principally in Lα (cf. Athay, Chapter 8 of this volume).

A number of plausible mechanisms for depositing heat in coronal loops have been suggested and explored in varying detail. Some are discussed below (see Section 5.6). But to unique or definitive specification of the heating term H or associated wave pressure term is yet possible. Indeed, a major incentive for coronal loop studies is that they may provide insight into these processes. One difficulty is that we cannot expect coronal heating to depend only on the local temperature and pressure. Nevertheless, approximate parametrized expressions for H can be assumed for a variety of processes, and these have been the basis of a number of studies.

As noted in the previous section. observations suggests that hot slowly evolving loops are isobaric and isothermal, except near their footpoints. Rosner *et al.* (1978a) show that stable loops in hydrostatic equilibrium must reach their maximum temperature T_m at their vertices of the loop. The pressure scale height corresponding to T_m for hot loops $(\approx 10^{10}$ cm) exceeds the height of most loops, so that the pressure P is nearly a constant. The loop's size can be described by a length L (cm) measured from one footpoint to the other along the loop. A loop can thus be roughly characterized by a set of these fundamental observable parameters (P, T_m, L) even though physical conditions must change rapidly at the footpoints. Rosner *et al.* (1978a) point out that these parameters are related by simple 'scaling laws'. If $(F_c)_0 = 0$, for example, they find that

$$T_m = 1.4 \times 10^3 \, (PL)^{1/3},$$

(5.3)

a relation with no disposable constants. Static loop models calculated by a number of authors with increasing sophistication confirm this relationship to first order and indicate that the constant of proportionality is not strongly dependent on the exact form of the radiative loss term the temperature dependence of the heating function, the basal heat flux $(F_c)_0$, or the rate of expansion of the loop cross-section with height.

This scaling law has been the principal point of comparison between theory and observation. The relations seem to hold quite well (to within a factor of 2 in predicted T_m) for hot loops even though flare loops, loops in mature active regions, and loops of large scale are included (cf. the review of Withbroe, 1981). This agreement suggests that these models describe the gross physics of the quiescent closed corona, although it does not prove that the quiescent loops are in strict pressure and energy balance. On the other hand, the weak dependence of the scaling law on the details of the heating function is disappointing; these first-order studies have set no strong constraints on the heating process. Similarly, the shape of the predicted differential emission measure (DEM) curve projected by these theoretical models is not very sensitive to the temperature dependence of the heating function. It is, however, sensitive to the basal heat flux and, to some extent, loop geometry. Comparison with the observed integrated DEM for a typical active region implies that $(F_c)_0 < 10^6$, − a value that the upper chromosphere can radiate.

5.4. Evolution and stability

Static loop models suggest at least qualitatively why some closed magnetic field lines become filled with plasma to form visible quiescent loops, and how these loops respond to a change in heat input. Rosner *et al.* (1978a) find that the heat input term varies with loop pressure and length

$$H \approx p^{7/6} L^{-5/6}.$$

(5.4)

Or, together with Equation (5.3) and the perfect gas laws, we can write

$$n_v \approx L^{1/7} H^{4/7},$$

(5.5)

where n_v is the electron density at the loop vertex. Thus, to remain in energy balance, the density and pressure of a loop of fixed length must increase in response to an increased heat input, and *vice versa*. The scaling laws thus provide a heuristic approach to understanding the evolution and existence of discrete loops. Increasing the nonradiative heating within a given bundle of field lines will cause the plasma on them to increase in pressure and density, and they will become manifest as a discrete coronal loop. Other nearby field lines might have less energy, less plasma deposited on them and, hence, would not produce visible loops. Equation (5.4) is also consistent with the observation that small low loops (such as those associated with newly emerged flux in active regions or in ephemeral regions) show a larger rate of local energy deposition than large loops found in mature active regions or in quiet regions of the corona.

We can also see how this adjustment of a loop to a change in heating (or length) might come about. An increase in energy input to a loop of fixed geometry initially in energy balance will cause the temperature to rise, since the radiative loss factor $Q(T)$ decreases with temperature for $T \approx 2 \times 10^6$ K. Only thermal conduction can dispose of this added energy, and the basal heat flux $(F_c)_0$ will increase. This may result in the 'evaporation' of chromospheric material into the loop causing an increase in P and n. The resulting increase in radiative losses can reduce $(F_c)_0$ and permit the establishment of a new equilibrium state at higher values of T_m and P, consistent with Equation (5.3). Alternatively, a decrease in heating will result in radiative cooling with subsequent condensation and 'draining' of material from the loop. Similar readjustments follow an increase or decrease in L.

This response of a static loop to a change in heating has been explored in numerical experiments by Krall and Antiochos (1980) and Craig and McClymont (1980). They solved the one-dimensional, time-dependent hydrodynamic equations in initially static loops subject to different initial and boundary conditions and different perturbations in heating. These experiments (discussed by Priest, 1981) support the heuristic model described above in that they show that energy and mass may be exchanged between the loop and the chromosphere at the footpoints in response to a perturbation and heating, and that the plasma in the loop may relax to a static configuration on a time-scale roughly expected on the basis of the radiative and conductive relaxation times. One new interesting result of the above studies is that the approach to equilibrium is a slowly damped oscillation. It would certainly not be surprising if these quasistatic loops were in continual oscillation triggered from time to time by changes in geometry or heating. These calculations are exploratory and much work remains to be done. Finding realistic yet tractable boundary conditions near the footpoints is especially difficult, and the steep temperature gradients at the base call for high numerical resolution.

Since a finite decrease in heating can cause material to drain from a loop, it is natural to ask if static loops in energy balance are thermally stable. Parker (1953) formulated the problem of thermal stability, applied it to the corona, and concluded that, since radiative losses probably increase with n^2 but decrease with T, the corona would be unstable 'unless stabilized by a strongly temperature dependent energy supply'.

Recently, Antiochos (1979), Habbal and Rosner (1979), and Hood and Priest (1980) have considered the thermal stability of model static loops in energy balance. Each study has used somewhat different approximations, especially for the crucial lower boundary conditions, and each reaches somewhat different conclusions (cf. the discussion by Priest, 1981). Antiochos (1979) and Hood and Priest (1980) seem to agree that isolated loops (those with $(F_c)_0 = 0$) are unstable. However, a more recent study by Chiuderi et al. (1981) finds that stability depends most critically on the basal temperature T_0 and $(F_c)_0$, and that even thermally isolated loops with $T_0 < 1.3 \times 10^4$ K are thermally stable. Further, Antiochos (1979) notes that the instability is associated with the cool material at the base of the loop. He makes the conjecture that the coronal portions may be in static equilibrium while the lower, cooler portions may be in an oscillatory or dynamic equilibrium. Unfortunately, these studies seem to agree that thermal instability is not sufficiently sensitive to the parameters of the heating function to place any presently useful constraints on it.

Some loops disrupt violently, evidently because of large-scale MHD instabilities (e.g. Gibson, 1977). Small-scale MHD instabilities must also be important, but the apparent MHD stability of most observed coronal loop systems seems striking. This has been the subject of several theoretical investigations (cf. Van Hoven et al., 1981, and references therein) which confirm that loops are stable to many types of perturbations.

5.5. SYSTEMATIC FLOW

In addition to transient flows, such as those associated with evaporation and draining discussed above, there is ample direct evidence for sustained flow in the corona, especially within plasma loops (see, e.g., the review by Webb 1981). Moreover, the chromosphere and transition region where loops are rooted, are almost everywhere observed to be in oscillatory, transient, or systematic motion (see, e.g., reviews by Athay, 1981a; Brueckner, 1981). Small flows at the footpoints may be magnified in the lower density corona and even small pressure differences between the endpoints of the loop can either decelerate or drive plasma in siphonic flow.

5.6. HEATING

The heating of the solar corona remains one of the most challenging and fascinating problems of theoretical and observational solar physics. At the present time it is fair to say that we do not understand coronal heating. We cannot now, with confidence, calculate the local rate of energy deposition or its dependence on local physical parameters. A number of plausible mechanisms have been suggested and explored in increasing detail. Few of the detailed studies seem to have escaped some serious criticism, yet most of the suggested mechanisms remain viable candidates. It is generally agreed that fluctuations in the convective zone are the major source of nonradiative energy in the Sun's atmosphere. These fluctuations must first be generated, then propagated upward, and finally dissipated. Major difficulties are that the pertinent processes are often nonlinear; that they are more or less globally controlled and hence model dependent; and that they usually have no unique, strong, observable signature that would make it possible to assess the relative importance of different processes or otherwise constrain a particular theory. Thus the difficulties are both theoretical and observational. On the other hand, recent rapid progress towards understanding coronal structure and dynamics has greatly clarified the overall problem of coronal heating and stimulated theoretical interest in specific mechanisms. It is now possible to discuss some of the conflicting demands that coronal phenomenology and modeling impose on a theory of coronal heating, and some mechanisms that might satisfy them.

It seems very likely that the magnetic field is somehow implicated in many heating mechanisms, since almost everywhere on the Sun, regions of strong magnetic field are regions of enhanced heating (except in and above sunspots). Nonmagnetic mechanisms cannot be ruled out, but they seem less likely in the low beta corona. Nonmagnetic acoustic waves (Biermann, 1948; Schwarzschild, 1948) dissipated by periodic shock waves (Schatzman, 1949; Osterbrock, 1961; Uchida, 1963; Kuperus, 1965, 1969; Kopp,

1968) was long considered important, but now seem a less likely mechanism for the highly structured upper chromosphere, transition region and corona (see, e.g., the review by Athay, 1976). Moreover, the needed acoustic wave flux does not appear to be present in the upper chromosphere (Athay and White, 1978, 1979a, b; White and Athay, 1979). Gravity modes, probably important lower in the chromosphere (e.g. Stein and Leibacher, 1974), are not expected to be important in the corona because of the high thermal conductivity, and suppression by magnetic fields (Hollweg, 1981a, b).

In closed magnetic regions, energy deposited in the corona is lost in part by radiation. The remainder flows by conduction back downward to be radiated in part by the transition region, and in part by Lyman α from the 'second temperature plateau' at the top of the chromosphere where Lyman α is effectively thin (see Athay, Chapter 8, this volume). In open coronal regions, less energy is lost by coronal radiation than in quiet regions because of the lower density, but solar wind losses to the open corona more than compensate (Pneuman, 1973; Noci, 1973). Withbroe and Noyes (1977) estimate that the total coronal energy flux that must be supplied by some mechanism to replace conductive, radiative, and solar wind losses as $10^{5.5}$ in the quiet Sun, $10^{5.9}$ in the coronal holes, and 10^7 in active regions (see also Athay, this volume). Since the corona is highly structured, these are of course average estimates.

Observed coronal loops can often be fitted quite well to potential fields with $\nabla \times \mathbf{B} = 0$, (Levine and Altschuler, 1974; Poletto $et\ al.$, 1975), and the potential continuation of the longitudinal component of the photospheric magnetic field distribution over the solar surface reproduces quiet well the gross properties of the large-scale coronal topology. On the other hand, the topology is sometimes more consistent with force-free fields (e.g. Nakagawa and Raadu, 1972; Rust $et\ al.$, 1975; Levine, 1976). Evidently currents do flow in the corona and their dissipation may be an important source of heating. Even where the topology of the corona appears potential, small departures from the potential form would be undetectable since individual loops are often diffuse and only barely resolved.

Gold (1964) pointed out that in the turbulent high beta plasma below the photosphere, the magnetic field has a large curl. He contrasted this with the relatively small or zero curl present in the coronal magnetic field and argued that subphotospheric fields, upon intruding into the upper atmosphere, must evidently 'shed their curls' by dissipating the associated current in the chromosphere and inner corona. That relaxation of stressed magnetic fields might produce impulsive flare heating and perhaps also other heating. This process seems likely to be important for X-ray bright points in ephemeral regions, or newly emerged flux within active regions. These small loops require a high rate of energy deposition. X-ray bright points often show rapid fluctuations in brightness (Sheeley and Golub, 1979; Habbal and Withbroe, 1981). They may flare (e.g. Nolte $et\ al.$, 1979) and eject matter (Moore $et\ al.$, 1977). Parker (1975) has explored the emergence of twisted flux tubes and shows that they are likely to form unstable kinks with subsequent reconnection. Thus the relaxation of stresses in newly emerged magnetic flux tubes seems one plausible component of plasma loop heating.

Tucker (1973) suggested that heating by the relaxation of magnetic stresses might also take place as a steady process in slowly evolving active regions. The dissipation rate of parallel currents would be balanced by continued twisting. This suggestion has been explored in greater detail by Rosner $et\ al.$ (1978b) who consider the anomalous dissipation of the parallel currents to take place in a thin sheath surrounding the plasma loop. More recently, this work has been extended by several studies. A critical and detailed review of this work is given by Hollweg (1981a) who points out that, if twists maintain the parallel currents, these twists propagate as torsional Alfvén waves. Hence, sustained heating by parallel currents is closely related to wave heating to which we now turn.

Numerous studies of Alfvén waves in the context of coronal heating have appeared over the past 35 years. Van de Hulst (1950) pointed out that Alfvén waves are not easily damped in the corona by linear processes, while Osterbrock (1961) found that Alfvén waves generated in the convective zone would be heavily damped in passing through the photosphere and low chromosphere. This latter difficulty is greatly reduced by the more recent recognition that magnetic fields are concentrated in small flux tubes of high field strength (1−1.5 kilogauss) at the levels of the low chromosphere and photosphere. The wave velocity amplitude δv needed to carry a given energy flux can then be small relative to the Alfvén speed (since $\delta v / V_A = \sqrt{4} \, \pi \rho B^{-2} \, \delta v$) thus greatly reducing the damping at low levels (Uchida and Kaburaki, 1974). The problem of dissipating Alfvén waves directly in the corona remains, but wave-mode coupling to other MHD modes can occur in the structured corona (e.g. Osterbrock, 1961; Kaburaki and Uchida, 1971).

Long period Alfvén waves observed in the solar wind at 1 AU evidently come from within the Alfvénic critical point (that is, where the flow speed reaches the Alfvénic velocity at $\approx 30 \, R_\odot$) and possibly directly from the inner corona (see, e.g., reviews by Hollweg, 1975, 1978a). They show evidence of nonlinear damping. Because they can deposit both head and momentum, they have been incorporated in detailed models of the solar wind (e.g. Hollweg, 1973, 1978a; Jacques, 1977, 1978). Alfvén waves also have attractive properties for heating quiescent coronal loops. They travel only along the magnetic field lines suggesting that nonradiative energy enters loops through their footpoints (e.g. Webb, 1981). Because they are noncompressive, their modeling is more tractable. Thus Hollweg (1981b) has studied the propagation of Alfvénic twists on both open and closed flux tubes, and finds that strong resonances can exist within closed loops due to reflections at the transition region in the footpoints. The forces exerted by high-frequency waves in these resonances on both the chromosphere and corona is found to be large. Neither the effects of these forces nor the problem of wave damping has been addressed in detail. Alfvénic surface waves can exist at a discontinuity in Alfvén speed or in a region where the Alfvén speed has a gradient. They have been recently studied in detail (Wentzel, 1978, 1979). Ionson (1978) has suggested that coronal loops can be heated by surface waves dissipated in a very thin sheath surrounding the loop.

Fast mode MHD waves can carry energy across the magnetic field and thus offer the possibility that waves from a large area at the base of the corona can deposit energy locally within coronal structures. This possibility has been explored by Habbal et al. (1979) who applied it to the heating of coronal loops. Fast waves propagating upward can be refracted into regions with low Alfvén speed where collisionless Landau/transit-time damping occurs. Thus, energy deposition tends to be concentrated in particular loops. Moreover, this collisionless damping has the property that it increases with plasma beta in the corona, so that the heating rate increases with heating in positive feedback, further favoring energy deposition within certain loops. Zweibel (1980) has studied the thermal stability of an atmosphere heated by fast mode waves. The fastest growing perturbations are found to be perpendicular to the field lines, which may contribute to coronal fine structure. Fast waves have been detected in the solar wind (Hollweg, 1978a) but not yet identified in the corona. It is not certain what fraction of fast waves generated in the photosphere would reach the corona (e.g. Osterbrock, 1961), and the expected flux there is not known.

6. Open Coronal Regions and the Coronal Expansion into Interplanetary Space

6.1. INTRODUCTION

Regions of the inner corona where the magnetic field is predominantly 'open' — that is, regions threaded by magnetic field lines that have been drawn out into interplanetary space by the dynamical forces of the coronal plasma — contrast sharply with the closed field regions discussed in Section 5 above. First, the electron density in open regions is lower — on average by roughly a factor 1/3. This has long been known for the open regions near the poles from white light eclipse observations. The more recent discovery that open regions in low latitude are also regions of low density has led to the unifying concept of 'coronal holes' discussed below. Because coronal emission depends on the square of the electron density, EUV and soft X-ray emission from open coronal regions is very low compared to the 'quiet' closed corona that surrounds them. Hence, open regions or 'coronal holes' can be readily recognized on soft X-ray photographs of the Sun (see Figure 4).

Closed bipolar magnetic regions, especially ephemeral regions, are observed to intrude as islands within open regions or coronal holes. As expected, they give rise to small loops, bright in soft X-rays, characteristic of the closed corona. The only known open field lines that appear to emerge from predominantly closed regions of the corona are those above neutral lines in coronal streamers (see Section 7). As a consequence, almost all the solar wind must originate in the compact open regions of the inner corona, which may cover less than 20% of the solar surface. Within about two solar radii of the Sun's center, however, the open regions spread to fill the total solid angle. Because the inner corona in open regions loses mass, heat, and convective energy to interplanetary space, and because the observable radiation from it is small, the physics and the diagnostic tools to study it are quite different from the closed corona.

In this section we consider first the solar wind, then the open regions of the inner corona, and finally, the problem of corona-interplanetary modeling.

6.2. THE SOLAR WIND

Theoretical and observational studies of the solar wind have abounded and the reader should refer to several excellent reviews of solar wind research over the past 23 years (Parker, 1963; Brandt, 1970; Hundhausen, 1972, 1977; Holzer, 1979). We shall not attempt to review all this work but, instead, shall concentrate on several important but still unresolved problems related to this fundamental interplanetary phenomenon.

The average 'quiet' solar wind as observed at 1 AU has a speed of approximately

350 km s^{-1} with a density of about 10 cm^{-3}. This implies a particle flux of about 10^{36} particles s^{-1}. In coronal holes, however, where the high-speed streams originate (cf. Hundhausen, 1977), the observed speeds are much higher, being about 700–800 km s^{-1}, but such streams have lower densities of about 5 cm^{-3}. Hence, the particle flux is about the same for both high and low-speed streams. These numbers must be considered as only crude averages since there is considerable structuring of the wind between the Sun and Earth both by coronal magnetic fields (cf. Durney and Pneuman, 1975; Pneuman, 1976; Richter and Suess, 1977; Ripkin, 1977; Riesebieter, 1977a, b) and stream interactions produced by solar rotation (Carovillano and Siscoe, 1969; Goldstein, 1971; Siscoe and Finley, 1972; Matsuda and Sakurai, 1972; Lewis and Siscoe, 1973; Hundhausen, 1972; Pizzo, 1978).

One major difficulty with present solar wind theories has become apparent with the discovery of the high-speed streams during the Skylab period. These streams exhibit speeds at 1 AU of 700–800 km s^{-1} – much higher than had previously been observed. Ordinary pressure driven solar wind theory assumes that thermal conduction from a hot corona provides the energy supply required to predict successfully the observed speeds of the low-speed solar wind. The deficiency of thermal conduction models, however, lies in the fact that the predicted energy flux at 1 AU lies almost entirely in the thermal conduction, whereas observations show that about 99% of the flux is in kinetic energy. Evidently actual conversion of energy from conduction into kinetic energy is far more efficient than predicted. The main problem, however, is that *none* of the purely pressure driven wind models can predict the 700–800 km s^{-1} observed in the high-speed streams. Several modifications of the classic theory have been tried. Mechanical heating low in the corona either has little effect or actually produces a decrease in the wind speed (Pneuman, 1980c; Leer and Holzer, 1980). Solar heating increases the mass flux, thus reducing the energy per unit mass. Extended heating beyond the critical point by hydromagnetic waves has been proposed (Barnes, 1969; Hollweg, 1973). However, the observed wind temperature at 1 AU imposes rather delicate conditions on the heating functions. In order to circumvent these difficulties, theoretical work has also centered upon the direct addition of momentum to the wind by MHD waves of various sorts (Belcher, 1971; Barnes and Hung, 1972; Hollweg, 1973; Belcher and Olbert, 1975; Jacques, 1977).

High wind speeds are inferred close to the Sun and provide another motive for studying momentum deposition by waves. Consider the following: the observed particle flux at 1 AU is about 4.4 × 10^6 cm^{-2} s^{-1} (Hundhausen, 1977). This flux extrapolates back to 2 × 10^{13} cm^{-2} s^{-1} at the solar surface, when averaged over the whole Sun. Since coronal hole regions occupy only about 20% of the solar surface, the flux at the Sun necessary to supply the wind is about 10^{14} cm^{-2} s^{-1}. Hence, if the particle density at 1.2 $R_\odot$ is 2 × 10^7 (van de Hulst, 1950), the velocity there must be 50 km s^{-1}. This conclusion was underlined by the work of Munro and Jackson (1977) in their analysis of the polar coronal hole observed in white light by Skylab in 1973. Using white-light data above 2 $R_\odot$, data at intermediate heights from HAO's K-coronameter at Mauna Loa, Hawaii, and data near the surface from the soft X-ray photographs from AS&E's experiment on Skylab (Nolte *et al.*, 1976), they were able to estimate the cross-section of the coronal hole boundary as a function of radial distance from the Sun (see Figure 1). They also were able to obtain the electron density as a function of height from the white-light

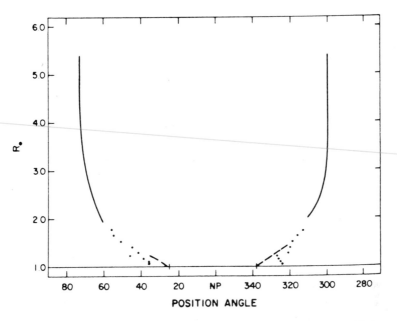

Fig. 1. Boundary of a polar coronal hole observed by Skylab in 1973 from Munro and Jackson (1977). The solid curves are from the High Altitude Observatory white-light coronagraph; the dashed curves from the AS&E soft X-ray equipment (Nolte *et al.*, 1976); and the dotted curves from the ground-based K-coronameter at Mauna Loa, Hawaii. (Reprinted by courtesy of R. Munro and *The Astrophysical Journal*, published by the University of Chicago Press; © 1977, The American Astronomical Society.)

data. Then, they assumed the observed particle flux at 1 AU, and determined the velocity as a function of height, i.e.

$$V = \frac{m}{N(r)A(r)}$$

and the velocity gradient:

$$\frac{\mathrm{d}V}{\mathrm{d}r} = -\frac{m}{n^2 A^2}\left[n\,\frac{\mathrm{d}A}{\mathrm{d}r} + A\,\frac{\mathrm{d}n}{\mathrm{d}r}\right],$$

where m is the particle flux, n the particle density, and A the cross-sectional area. Now, insertion of these terms into the momentum equation determines the pressure gradient as a function of radial distance under the assumption that no forces other than inertial, pressure, and gravitational exist. The pressure gradient, when combined with the gas law and the observed density, yields a temperature distribution. The temperatures Munro and Jackson obtained were improbably high, which indicated that forces other than pressure must exist.

In view of the above arguments, it is now generally agreed that there must exist accelerating forces other than gas pressure to produce the high speeds observed at the orbit of Earth and the high speeds in the low corona that are required by the conservation of mass. Although wave momentum is presently the favored concept, it is disconcerting that *waves of the magnitude needed to accelerate the solar wind have not as yet been observed in the low corona*. However, there is evidence for magnetic disturbances which could be interpreted as waves at 1 AU (cf. Barnes, 1979). Such waves could be generated at the Sun but could also be produced during the transit of the solar wind between the surface and the orbit of Earth. Using OSO-8 time series of profiles of Si II (Athay and White, 1979a) and C IV (Athay and White, 1979b), an upper limit of 1×10^4 erg cm^{-3} for the acoustic flux in the chromosphere was deduced. This is far below the flux necessary to supply the radiation and solar wind losses in the low corona. Also, Athay and White (1978) concluded that the situation is no better for Alfvén waves. Similar conclusions were reached by Cheng *et al.* (1979) from the line intensities and widths of EUV forbidden lines of Si VIII, Fe XI, and Fe XIII. Furthermore, high-resolution observations of the transition region and low corona suggest explosive-like radial motions, both upward and downward, rather than wave-like motions (cf. Brueckner, 1980). In view of these difficulties, further exploration into the theoretically interesting problems of wave momentum addition to the solar wind should perhaps be strongly supplemented by a serious observational attempt to either verify or eliminate the existence of these waves which form the fundamental basis of the theories.

If wave momentum proves ultimately to fail on observational grounds to provide the acceleration for the solar wind, then other mechanisms for providing the necessary force must be sought. One such mechanism might be found involving the active participation of magnetic fields which emerge into the corona in the form of active regions and X-ray bright points. The bright points, for example, represent most of the flux emergence on the Sun except during periods of maximum solar activity (Golub *et al.*, 1977; Davis *et al.*, 1977), contributing about 5×10^{22} Mx day^{-1} (Golub, 1980). Since the estimated total flux in the Sun's polar field is only about 8×10^{21} Mx during sunspot minimum, the bright points could replenish the polar field in less than 4 h. In order to maintain a more-or-less steady state coronal field, this flux must either be dissipated locally or expelled. Both processes could be important for the solar wind since these fields represent a large energy supply to the corona. Reconnection of this emerging flux in the corona could provide a coronal heating mechanism (Vaiana and Rosner, 1978) and the ejection of the reconnected fields could accelerate the solar wind to high velocities (Mulland and Ahmad, 1982; Pneuman, 1983). Spicules have been interpreted as evidence for magnetic reconnection (Pikel'ner, 1969; Uchida, 1969). Also, EUV observations in the transition region and low corona show similar explosive events (Brueckner, 1980). All these ideas, however, are presently in a primitive state of development and much more study and observational work is needed to fully assess the effectiveness of these electromagnetic mechanisms.

6.3. MAGNETIC FIELDS IN THE SOLAR WIND

As been pointed out earlier, the coronal magnetic field has a profound influence on the structure of the solar wind, both on large and small scales. The overall magnetic field can be best understood by considering it to be composed of two parts: first, a large-scale field determined by the photospheric field pattern and modified as one proceeds outward from the Sun by the solar wind forces and solar rotation; and, second, small-scale fluctuations on this large-scale field produced by discontinuities, shocks, waves, etc. The large-scale fields vary with time scales representative of the evolving photospheric field patterns during the course of the solar cycle although fast changes can take place during period of transient activity. The origin of magnetic fluctuations which occur on a much shorter time-scale is more unclear.

To a first approximation, the large-scale field in the inner corona seems to be reasonably well approximated by a potential field out to a distance where the solar wind dominates (Newkirk *et al.*, 1968, 1972; Schatten, 1971; Altschuler *et al.*, 1977; Levine *et al.*, 1977; Pneuman *et al.*, 1978). This is not to say the inner corona is current-free but only that the overall coronal structure — such as the location of coronal holes, streamers, etc. — can be reasonably specified by an extrapolated potential field using observed photospheric line-of-sight fields. Beyond about $2\,R_\odot$ the field is distended into a more or less radial direction by the solar wind and modified into a spiral by rotation.

The effect of rotation on the field structure can be seen by examining the azimuthal component of Equation (2.5) with $\sigma = \infty$ and $\partial/\partial t = 0$ and Equation (4.1) (Parker, 1958) which gives

$$B_\Phi \propto \frac{1}{rV}\,; \qquad B_r \propto \frac{1}{r^2}$$

where we have assumed the flow is purely radial with velocity V. If α is the angle between the field vector and the radius vector, then we have

$$\tan \alpha = \frac{B_\Phi}{B_r} \propto \frac{r}{V} \tag{6.1}$$

Equation (6.1) is the formula for an Archimedes spiral for V = const and shows that the field becomes increasingly twisted into the azimuthal direction as the solar wind proceeds outward from the Sun (see Figure 2). This predicted spiral or 'hose' angle in the interplanetary field has been confirmed both by spacecraft at 1 AU (cf. Hundhausen, 1972) and by Helios 1 and 2 measurements between 0.3 and 1 AU (Mariani *et al.*, 1979).

Magnetic discontinuities in the solar wind are present in two forms; rotational discontinuities interpreted as transverse Alfvén waves and tangential discontinuities in which equality of magnetic plus gas pressure is maintained across the discontinuity. Surveys of observational and theoretical considerations of these structures can be found in Burlaga (1972) and Siscoe (1974).

As might be expected, hydromagnetic waves are constantly present in the solar wind. Fluctuations on virtually all time-scales were first observed by Mariner 2 (Davis *et al.*, 1966; Neugebauer and Snyder, 1966) and first interpreted as hydromagnetic waves by Coleman (1966, 1967) by comparing velocity and magnetic field spectra from Mariner 2 data. The most fundamental advance in this area, however, was due to Belcher and Davis (1971) who established, through a correlation study of the radial components of velocity and magnetic field fluctuations, that most interplanetary fluctuations were, indeed, Alfvén waves and that they dominated the small-scale structure of the solar wind more than 50% of the time. The waves always propagate outward from the Sun rather than inward, suggesting they are of solar origin.

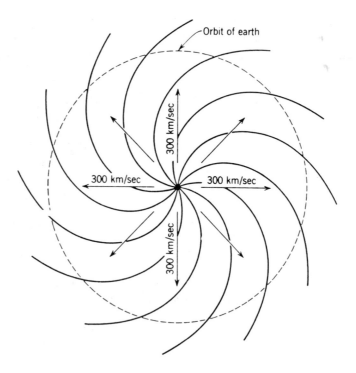

Fig. 2. Extension of a radial solar field out into interplanetary space produced by a radial solar wind of 300 km s^{-1} combined with solar rotation, taken from Parker (1963). (Reprinted by courtesy of John Wiley & Sons, © 1963.)

6.4. CORONAL HOLES

The most visible signature of open field regions in the solar corona are the so-called coronal holes' which appear in white light as dark regions (see lower right of Figure 3). Although coronal holes became a subject of widespread interest as a result of Skylab, their existence has been known for some time, beginning probably with the observations of Waldmeier (1957).

The most extensive studies of these structures, however, were provided by Skylab. Figure 4 shows a typical soft X-ray photograph clearly showing the large dark coronal hole extending downward from the pole, surrounded by bright regions presumably occupied by closed magnetic fields. The contention that holes are bounded by closed fields was discussed by McIntosh et al. (1976) who showed that holes were bounded by magnetic arches bridging neutral lines in the photospheric magnetic field usually lined with prominences.

In general, XUV emission is sharply depressed in coronal holes, with Mg IX being entirely absent there and the depression in He II revealing sharp boundaries. An indication that the coronal temperature above the transition region in coronal holes may be lower than in other regions is suggested by the observations that the limb-brightened ring is significantly elevated with respect to other lines (Tousey et al., 1973; Huber et al., 1974;

Fig. 3. High Altitude Observatory photograph of the eclipse of 1970 showing a large coronal hole in the lower right-hand quadrant. The solar poles are oriented along a line extending from lower left to upper right. This photograph was obtained by using a radially graded filter which enhances the brightness far out relative to that closer in. (Reprinted by courtesy of D. Reidel, Dordrecht, Holland, © 1971.)

Bohlin *et al.*, 1975). This could be interpreted to indicate that the temperature is reduced there (Bohlin *et al.*, 1975).

The nature of the chromospheric network in holes has been investigated in the XUV by Huber *et al.* (1974), Reeves (1976), and Mango *et al.* (1978). In general, the hole becomes less visible with decreasing temperature and height indicating that the holes are truly coronal.

Coronal holes can also be distinguished from ground-based observations — by the modulation of the overall density structure of the K-corona and by radio observations. During periods of minimum activity, a typical map of the K-corona obtained by the K-coronameter instrument at Mauna Loa, Hawaii, shows a rather broad band of enhanced coronal brightness weaving across the Sun above and below the equator with low brightness regions outside. This pattern is consistent with potential field calculations which show a single 'warped' neutral line weaving across the solar equator (Pneuman, 1976). This pattern was quite stable during the latter part of the Skylab mission and suggests a 'warped' dipole with coronal holes on the poleward side of the closed field regions (Hundhausen, 1977). Two other properties of coronal holes worth noting are their lifetimes and rotational characteristics. Holes are fairly long-lived even away from the

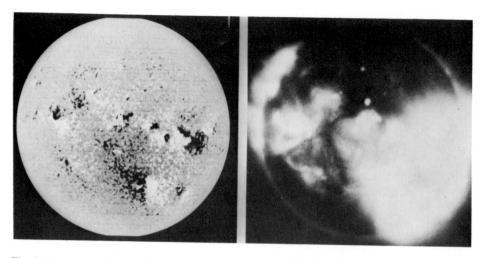

Fig. 4. Concurrent photographs of the Sun for the period covering the March 7, 1970 eclipse showing, on the right, the coronal X-ray emission and, on the left, the polarity pattern of the photospheric magnetic field (taken from Krieger *et al.*, 1971). Note the one-to-one coincidence of bright X-ray sources and bipolar magnetic regions (adjacent light and dark regions of the left). These regions are areas of high field strength and are probably also closed. (Reprinted by courtesy of D. Reidel, Dordrecht, Holland, © 1971.)

poles with the lifetime averaging about six solar rotations. The growth and decay rate of holes is about $(1.5 \pm 0.5) \times 10^4$ km^2 s^{-1} (Bohlin, 1977). Coronal holes seem to show little or no differential rotation (Timothy *et al.*, 1975; Wagner, 1975; Adams, 1976; Bohlin, 1977). If this is indeed true, an interesting theoretical problem presents itself as to how these structures can maintain such a rigid rotation pattern in the presence of the observed differential rotation of the photosphere.

The good correlation of recurrent high-speed streams with low-density regions in the corona has been known for some time (Wilcox, 1968; Hundhausen, 1972). The first identification of coronal holes with open magnetic field lines in the corona was made by Altschuler *et al.* (1972) who compared coronal electron densities derived from K-coronameter observations with potential magnetic field maps showning that coronal holes coincided with regions of weak diverging field lines. Then, the suggestion that the unusually low densities observed in coronal holes were somehow related to energy losses from the corona incurred by the solar wind was put forward independently by Pneuman (1973) and Noci (1973) based upon different but equivalent arguments Pneuman calculated the expected difference in density and temperature at the same height for both open and closed magnetic regions based upon a thermally conductive model of the solar wind in the open regions and an isothermal temperature in the closed regions, both regions having the same base conditions. He found that the open regions could have a temperature of a factor of 2 lower than the closed regions and densities of a factor of 10 lower. Thus, reduced heating in coronal hole regions is not required to produce the low densities, only the increased energy losses by thermal conduction and the solar wind. Noci (1973), assuming the corona is heated by waves, argued that the outward flux of mechanical energy was proportional to the pressure, whereas the inward conductive flux varies as the pressure squared (Munro and Withbroe, 1972). Hence, as one goes from quiet regions to coronal holes, the decrease in the energy input is more than compensated for by the increase in losses with the difference going into the solar wind.

This first positive identification of coronal holes as the source of high-velocity solar wind streams was made by Krieger *et al.* (1973). Since that time, the association of high-speed streams with coronal holes has been firmly established, mostly due to the work done during the Skylab Workshop on Coronal Holes (Zirker, 1977). The reader is referred to Hundhausen (1977) for a comprehensive summary of that work. During the Skylab period, about 20% of the solar surface was occupied by coronal holes. However, 75% of this area was covered by the two polar holes indicating that, by far, the poles are the biggest contributor of open field lines to the interplanetary medium.

6.5. CORONAL-INTERPLANETARY MODELING

It is possible, in principle to model the solar corona between the Sun and Earth in three dimensions, given measured quantities such as magnetic field strength, density, and temperature as a function of latitude and longitude at the coronal base provided the correct flow and energy equations are known. The results of such a calculation could then be compared to observed conditions at 1 AU to provide a definitive test of the solar observations and the theory. Attempts at doing this have been successful in part but have also revealed some glaring problems in our understanding of both the solar observations and the solar wind processes which operate in the interplanetary medium.

The key to coronal-interplanetary modeling lies in the specification of the role of the coronal magnetic fields which directly modulate the spatial distribution of the flow and the density structure in the inner corona. Although a full magnetohydrodynamic description is possible (Pneuman and Kopp, 1971; Endler, 1971), it would require an enormous amount of computational complexity and computer time and would only be feasible if solar boundary conditions and the solar wind physics were precisely known. Therefore, as a compromise to a full MHD treatment, a magnetic geometry is calculated, based upon physical assumptions, and the solar wind equations integrated along individual flux tubes in order to obtain a global description of the wind at 1 AU. The field in the inner corona is assumed to be potential (current-free) based on the observed line-of-sight photospheric fields and is constrained to become radial at a prescribed 'source surface' in order to simulate the domination of the solar wind at larger radial distances. Most descriptions have employed a spherical source surface (Altschuler and Newkirk, 1969; Schatten and Wilcox, 1969; Levine and Altschuler, 1974; Durney and Pneuman, 1975; Altschuler *et al.*, 1975; Adams and Pneuman, 1976; Riesebieter, 1977a, b; Pneuman *et al.*, 1978; Riesebieter and Neugebauer, 1979). Recently, however, nonspherical source surfaces have been employed (Schulz *et al.*, 1978; Levine *et al.*, 1979). Most of the above calculations of field geometry, however, did not include any solar wind calculations. These solutions do predict the location of coronal holes (Levine *et al.*, 1977; Pneuman *et al.*, 1978) although the calculated areas of open field lines are somewhat smaller than the areas of the X-ray holes. They also show reasonable correspondence with the K-coronameter brightness contours (Pneuman *et al.*, 1978).

Global solutions incorporating the solar wind have been attempted (Durney and Pneuman, 1975; Pneuman, 1976; Riesebieter, 1977a, b). Although these solutions are reasonably successful in predicting the location of high-speed streams and the polarity of the interplanetary fields associated with them (see Figure 5), they also predict quantities

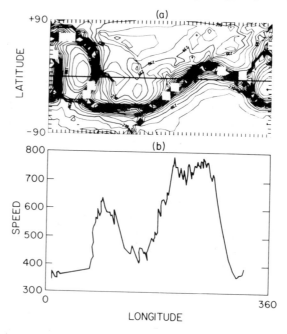

Fig. 5. Comparison of (a) computed velocity contours at 1 AU with (b) *in situ* measurements by Imp spacecraft (furnished by S. Bame and colleagues) for the same period. The horizontal line in (a) denotes the path of the spacecraft. The region of closely packed contours is the neutral sheet or sector boundary across which the magnetic field changes polairty. Note the presence of high-speed streams in locations where the computed sector boundary departs from the spacecraft path. The departure at the extreme left was not covered by observations as indicated by the straight line in (b) that location. (Reprinted by courtesy of the American Geophysical Union, © 1976.)

that are irreconcilable with observations. For example, the predicted speeds are too low, being about 300 km s^{-1} rather than the 700 cm s^{-1} required in high-speed streams. Also the solutions are almost totally dominated by heat conduction at 1 AU as opposed to the observed fact that 99% of the energy there is kinetic. Obviously, our understanding of thermal conduction in the solar wind is lacking (see, e.g., Hollweg, 1978a). Another serious problem is that all potential field models incorporating a source surface at a reasonable height yield magnetic field strengths at the orbit of Earth that are a factor of 5–20 too low. Thus, either the photospheric measurements are too low or there are many open field regions not accounted for in potential magnetic field models. Notwithstanding these difficulties, coronal-interplanetary modeling can provide a useful tool for testing our coronal observations and solar wind theory.

7. Coronal Streamers: Gas–Magnetic Field Interactions in the Solar Corona

To conclude our discussion of the steady-state properties of the solar atmosphere, we shall discuss coronal streamers which cannot be properly classified as either closed or open coronal regions, since they contain elements of both. They can be seen with the naked eye during eclipse as bright structures extending outward from the coronal base to a considerable distance.

Streamers are unusually bright structures extending far out into the corona as seen in Figure 6. Michard (1954) divided a typical streamer into two distinct parts — a base and

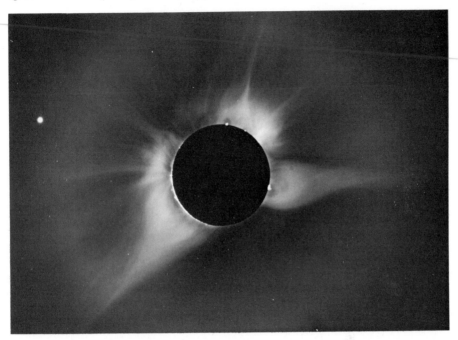

Fig. 6. Photograph of the 1966 solar eclipse taken by the High Altitude Observatory using a radially graded filter which enhances the brightness far out relative to that closer in. Note the large conspicuous coronal streamers with prominences (bright points) at their base. The coronal cavities surrounding these prominences can be detected especially for the streamers situated at the lop and to the right.

a stem. This distinction is real with the division produced, as we shall see, by the fact that a streamer is magnetically dominated near the base and flow dominated at higher altitudes. The density enhancement in a streamer over the adjacent regions is low near the streamer base but increases with increasing distance from the Sun to a factor of 3–10 (Schmidt, 1953; Michard, 1954; Hepburn, 1955; Saito, 1959; Saito and Billings, 1964; Saito and Owaki, 1967; Leblanc, 1970; Newkirk *et al.*, 1970; Koutchmy, 1971). Normally, large quiescent streamers overlie a quiescent prominence with a 'coronal cavity' surrounding the prominence and imbedded in the streamer (see Figure 6). Little is known about the temperature structure in streamers directly, although the observed increase in density enhancement with height has been interpreted to be due to the increase in scale height provided by a more isothermal temperature distribution in the streamer core than outside (Pneuman and Kopp. 1970; Pneuman, 1973).

Basically, a helmet streamer can be understood as a system of closed field lines rooted to the Sun with regions to the outside and above being stripped open by the expanding solar wind, as shown in Figure 7. In this picture, there is no flow in the closed core but expansion everywhere else. Since the streamer overlies a bipolar magnetic region in the photosphere, a neutral sheet in the magnetic field, which extends outward into interplanetary space, is produced at the top of the core. It is these neutral sheets which may well form the well-known 'sector boundaries' observed by spacecraft at the orbit of earth.

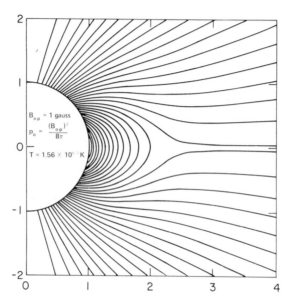

Fig. 7. Computed magnetic field topology for an idealized coronal streamer with axial symmetry (from Pneuman and Kopp, 1971). Base conditions and coronal temperature used are shown inside the figure of the Sun. Here, the transition from closed to open field lines at the equator occurs at $r = 2.5\, R_\odot$ where the domination of the solar wind becomes most apparent. (Reprinted by courtesy of D. Reidel, Dordrecht, Holland, © 1971.)

Helmet streamers last for several rotations even though the magnetic field patterns beneath them may change from rotation to rotation. Bohlin (1969) found a lifetime for a large high-latitude streamer of 4.5–5.5 solar rotations. Little is known about the birth of helmet streamers. Based upon physical considerations, however, at least two possibilities seem plausible. One is that they form by the emergence of new flux which rises up into the corona until the solar wind pulls out the uppermost field lines to infinity, forming the streamer structure. Another is that they can form from the reconnection of initially open field lines of a coronal hole (Rust, 1981).

The physics of a helmet streamer is quite complex since the structure encompasses both open field regions with solar wind flow and adjacent closed regions where the gas is at rest. Thus, the physical processes taking place in the two regions can be quite different. For example, on the boundary between the two regions, the gas pressures will be different. Since the quantity $P + B^2/8\pi$ must be continuous across the boundary, the magnetic field will change rapidly there, across a current sheet. Also, a neutral sheet will extend from the top of the closed region out to infinity, separating the regions of opposite magnetic polarity.

In the closed region, sufficient thermal conduction forces the temperature to be isothermal provided the temperature is uniform along the coronal base. Since the velocity there is zero and the magnetic force has no component along field lines, Equation (2.2) can be integrated along field lines to yield the pressure distribution, i.e.

$$P = P_0 \exp\left(-\frac{GM_\odot m_p}{r_0 kT}\left(1 - \frac{r_0}{r}\right)\right), \tag{7.1}$$

where P_0 is the pressure at a reference level r_0 at the base of the closed region and m_p is the mean particle mass. In the adjacent open field region, the pressure, again from Equation (2.2) with $V \neq 0$, is given by

$$P = P_0 \exp \frac{m_p}{k} \int_{s_0}^{s} \frac{1}{T} \frac{\partial}{\partial s}\left[\frac{GM_\odot}{r} - \frac{V^2}{2}\right] ds, \tag{7.2}$$

where s is the curvilinear coordinate along a field line. A comparison of Equations (7.1) and (7.2) shows that a discontinuity in pressure will exist at the boundary unless the temperatures in the two regions are equal and $V = 0$ or constant. This discontinuity in pressure requires a corresponding discontinuity in field strength of

$$[B^2] = 8\pi[P], \tag{7.3}$$

where the brackets denote the jump across the boundary. From Ampere's law, a current sheet of magnitude j^* running perpendicular to the magnetic field is then required, where

$$\mathbf{j}^* = \frac{1}{4\pi} \hat{n} \times [\mathbf{B}],$$

where $\hat{n}$ is a unit vector oriented perpendicular to the magnetic field. The structure and thickness of this current sheet in coronal streamers has not been studied. Its structure could be quite complex since both resistivity (due to the magnetic field discontinuity) and viscosity (due to the velocity discontinuity) become important in the same region.

At the top of the helmet, the magnetic field must be zero at a neutral point (neutral line in three dimensions). Sturrock and Smith (1968) have discussed the three possible types of neutral points in such a configuration. These are the Y-type where the field approaches zero from all directions, the T-type where the field vanishes as the neutral point is approached from the outside of the helmet but is finite inside, and the 'cusp' type where the field vanishes on the inside of the helmet but is finite outside. The three configurations are shown in Figure 8. *Only the cusp-type neutral point is permissible for helmet streamers* where the gas flows smoothly past the neutral point (Sturrock and Smith, 1968; Pneuman and Kopp, 1971).

The first attempts to describe coronal streamers mathematically were quite crude and consisted of averaging the conservation equations across two regions — a streamer core and an interstreamer region. Fully self-consistent MHD models of a coronal streamer were later obtained by Pneuman and Kopp (1971) see Figure 7) and Endler (1971). Both these models employed identical boundary conditions but used drastically different

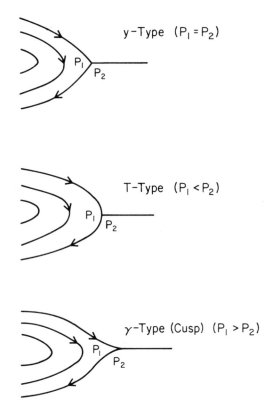

Fig. 8. Three possible types of neutral point configurations showing the required relationship between internal and external gas pressures required for each. Since observations indicate that the pressure in closed magnetic regions is higher than in open regions, the cusp-type neutral point would seem to be the favored topology. (Reprinted by courtesy of D. Reidel, Dordrecht, Holland, © 1974.)

methods of solution. As should be expected, both solutions were identical. These models assumed a totally isothermal corona, a normal component of the field at the coronal base equal to that of a dipole, and a uniform base pressure of magnitude such that the gas pressure was equal to the surface magnetic pressure at the poles.

Another novel way at looking at streamers was provided by Weber (1977). He presumed the streamer to be in a constant evolutionary state with fields being stripped off the top of the helmet as new ones replace them from below giving a steady-state appearance of the streamer. The major difference between this model and those of Pneuman and Kopp and of Endler is that flux must be brought up through the solar surface continuously during the lifetime of the streamer. It would seem that, to maintain a steady-state appearance with this process, a lateral motion of footpoints through photospheric material would be required.

In conclusion, it can be said that many aspects of the physics of streamers are now understood although a nonisothermal streamer model has never been attempted. A complete understanding of their physics is essential for understanding the corona as a whole. For example, during solar minimum, the corona resembles a warped dipole with large coronal holes at the poles. In this case, the major portion of the lower corona consists of one coronal streamer winding around the Sun. Hence, the above solutions actually represent an isothermal axisymmetric solution for the whole corona rather than a single structure. During periods when the photospheric field patterns are more complex the coronal structure will be correspondingly more complex. Nevertheless, the basic gas—magnetic field interaction process remains the same.

8. Coronal Activity

Although we have devoted considerable time in the previous sections discussing steady-state properties of the upper solar atmosphere, the reader should not be misled into thinking that it is totally quiescent. Although coronal activity is most pronounced during the maximum phase of the sunspot cycle, changes on a rapid time-scale are occurring virtually all the time. The most noticeable rapid changes occur with prominences, flares, impulsive flows such as surges, coronal whips, etc., and coronal transients. Although the precise physical relationships between these events is still uncertain, it is clear that an intimate relationship does exist and that they all involve magnetic fields in a major way.

8.1. THE PROMINENCE PHENOMENON

Prominences remain a subject of continued interest in solar physics, since they are involved in almost all aspects of solar activity and display a wide range of astrophysical processes. They exhibit such hydromagnetic effects as magnetic containment, support, and reconnection. They can be observed as isolated objects in emission against the sky with high contrast, and their internal structure or geometry is sometimes simple. They are thus well suited for testing models of energy balance and radiative transfer in an inhomogeneous atmosphere. Prominences take part in a complex interchange of mass between the photosphere, chromosphere and corona. They are markers of the large-scale solar magnetic fields and their evolution, and may serve to outline large-scale internal circulation in the Sun. It seems likely that prominences or prominence-like phenomena exist in all stellar atmospheres with magnetic fields and extended atmospheres.

The past few years has seen a rapid increase in the understanding of the physics of prominences. Spatially resolved EUV and radio observations have become available for the first time, and there has been an important increase in the quantity and accuracy of ground-based spectrophotometric and polarimetric data. This has been matched by increasingly sophisticated models of the radiative transfer, especially in quiescent prominences. White light and EUV observations from space, and ground-based coronagraph and radioheliograph observations have shown that coronal transient events are more frequent than formerly supposed. Prominences are implicated in many of these events, and the need for improved understanding of prominence support, stability, and dynamics is now apparent.

The following definition is consistent with common usage of the term in the literature: *A prominence is any structure existing in the Sun's atmosphere that can be observed above the limb in the H-alpha line.* In other words, prominences are any relatively cool $(5-10 \times 10^3$ K) dense structure imbedded within the hot $(10^6$ K) corona. A number of quite different and almost wholly unrelated phenomena, some transient and some long-lived, will satisfy the above definition. Nevertheless, so many physical processes and diagnostic techniques are common to all of them that the term remains useful.

Prominence material is inferred to be roughly 100-fold cooler than the corona, and about 100-fold more dense. This would seem to suggest that it may be isobaric and in gas-pressure equilibrium with the surrounding corona. However, prominences always contain magnetic fields as part of their structure and the magnetic pressure associated with these fields (1—200 gauss) could compete with or dominate the gas pressure. Prominence material is so dense that its continued presence in the corona implies either a degree of static or dynamic support that can only be provided by the magnetic field; or that there is a source of downflowing material continuously resupplied. Both situations are observed, sometimes within the same prominence.

A number of schemes of prominence classification have been proposed and are in more or less general use. These depend on such properties as morphology, motions, spectra, or association with other activity, and have been summarized and compared by Tandberg-Hanssen (1974). The list of types given below is not proposed as a new classification scheme, but is intended to show the rather different phenomena that are collectively called 'prominences'. The fine structures of the chromosphere and transition region discussed earlier are prominence-like in many respects. *Spicules* extend into the corona and are considered prominences by some authors. Other prominence-like fine structures observed against the disk just above the chromosphere, but not certainly above the limb (and so not strictly prominences by our definition) include *fibrils, field transition arches*, and *arch filament systems*. The latter are believed to be a direct manifestation of the emergence of new magnetic flux from below (see, e.g., Athay, 1976). A class of spicules so much larger than ordinary spicules are called *macrospicules* and are associated with flaring X-ray bright points (Moore *et al.*, 1977).

Prominence material ejected from flaring active regions, which includes *surges* and *sprays*, are the principle manifestations of mass motions produced by solar flares. Surges are jets of material that are obviously constrained and guided by the magnetic field, and which fall back into the Sun along the same path. Sprays are more explosive, fragmented ejections that escape from the active region and often from the Sun.

Downflowing material observed in active regions includes *loop prominences* and *coronal rain*. In both, the material for them appears to condense from the corona and fall under gravity into the Sun along magnetic field lines. Loop prominence systems or post flare loops consist of a system of loops that grow in height in a few hours from about 10^4 to 10^5 km. These prominences are an integral part of certain major energetic flares, and are discussed further in Section 8.2. Coronal rain consists of isolated arcs or threads of material that appear suddenly and fall along the magnetic field lines.

Structures that form along the boundaries of regions of opposite photospheric magnetic polarity are most aptly described by the term *hedgerow*. These long, blade-like palisades are called filaments when seen in absorption against the disk in the cores of strong Fraunhofer lines (such as $H\alpha$ or Ca II H and K). They are almost always present on the Sun, both on the disk and at the limb, and it is to these structures that the term 'prominence' is most often applied. The dimensions of a well-developed filament in megameters are typically 5 wide by 30 high by 200 long. They are relatively stable and long-lived in the sense that the same filament can often be identified for many solar rotations, even though the size, position, and internal arrangement may be constantly changing. Filaments are born within active regions and move poleward when the active region breaks up (d'Azambuja and d'Azambuja, 1948; Kiepenheuer, 1953). When within active regions, they often show rapid development and are usually referred to as 'active' prominences. The stable, slowly evolving hedgerows or filaments are often called 'quiescent' prominences. Most filaments (and perhaps all) pass through a stage of disruption or disparation brusque in which they erupt or ascent from the Sun's surface, often reforming a few days later.

8.1.1. *Quiescent Prominences*

Most of our knowledge of the detailed physical nature of prominences has come from the study of quiescents. As described in Section 7, they lie along magnetic neutral lines overarched by arcades of coronal plasma loops that form the base of coronal streamers. The low-density coronal cavity that underlies this arcade, long known from eclipse and coronagraph studies at the limb, can now be studied against the Sun's disk in soft X-rays and radio emission (see reviews by Kundu, 1979; Schmahl, 1979). The quiescent prominence fills only a small volume of this cavity. The temperature and density of the corona in the immediate vicinity of the prominence is not well known.

(a) FINE STRUCTURE. While the large-scale gross structure and evolution of quiescent hedgerows is well known from classical studies of filaments (e.g. d'Azambuja and d'Azambuja, 1948), the internal fine structure is much more difficult to observe, describe, classify, or even visualize in three dimensions. The best observations have been made in Hα with the coronagraph, when prominences were crossing the limb broadside. The largest collection of large-scale, high-quality ciné observations of this type has been made by Dunn (1960, 1971) at Sacramento Peak. Other studies based on his observations include those of Menzel and Wolbach (1960) and Engvold (1976). Often, the internal structure consists of wavy, well-defined vertical threads which terminate abruptly at their lower ends, sometimes on sharply defined arches, but other more chaotic thread-like structures are also common.

Engvold (1976) found thicknesses for the threads ranging from 0.4 to 1.5 Mm with the thickness often increasing with height, although some prominences have much coarser threads and some are almost amorphous. Prominence threads vary in brightness and thickness along their length and often contain bright knots of such short lifetimes that prominence material may not be in statistical equilibrium (Engvold, 1980). In any event, the fine structure obvious in most prominences complicates any attempt to interpret their spectrum or detailed physics.

The question of how the magnetic field is oriented with respect to the threads is unanswered. For some highly ordered threads (often called loops or arches) it seems obvious that the magnetic field lies along the loop. Consistent with this, material on these threads usually falls downward. On the other hand, it is not at all obvious that the magnetic field lies along the more numerous vertical threads. On these, material does not appear to be free-falling. The systematic downward velocities observed in filaments are only ≈ 1 km s^{-1} (Maltby, 1976; Kubota, 1981), although apparent downward motions observed in limb prominences are higher (Engvold, 1976). When prominences erupt we can see an unfolding of their internal structure as they rise and expand (see Figure 10 below). Often this structure is helical and strongly ordered. A strong suggestion of helical structure is also often present in nonerupting quiescents. The systems of wavy vertical threads could be simply a compacted helix, except that then the support of material within them must be explained (Malville, 1976). The question of overall prominence support is discussed below.

Spectropolarimetric observations show promise to provide the crucially needed determination of the local magnetic field strength and direction. This information is needed not only to understand the channeling of heat and matter within prominence fine structures, but also to understand the overall magnetic configuration that supports and stabilizes the prominence. Both the measurements and their interpretation and difficult, but they are of fundamental importance for solar MHD (see Leroy, 1979, for a review). There is some spectropolarimetric evidence that the prominence fine structure is indeed directed along the field (Engvold and Leroy, 1979).

(b) THERMODYNAMIC STRUCTURE. Many interpretations of spectra which may derive basic physical parameters within the cool $H\alpha$ emitting portion of quiescent prominences have been published (see, e.g., Hirayama, 1979). Some basic parameters include the atom or ion kinetic temperatures T_i; the electron temperature T_e; the electron density n_e; the H and He ionization; the chemical composition, especially the He/H abundance ratio; the nonthermal random ('turbulent') velocity ξ; the systematic line-of-sight velocity; the total mass density ρ; the optical thickness of certain critical radiations such as the Lyman continuum; the magnetic field; and geometric parameters such as the effective thickness in the line-of-sight and the fractional emitting volume. None of these quantities can be derived independently, directly, and uniquely from the data. Some theoretical modeling is often required to guide the interpretation.

The most direct determination of T_i comes from the widths of the lower H I Balmer lines (corrected for self-absorption and nonthermal broadening) and the widths of lines of He I and metals. A value of T_i and of ξ is chosen to give the best fit to all of the widths. He I is evidently excited by coronal radiation. Hirayama (1979) summarizes the results of a number of authors and concludes that T_i in the central region of quiescents ranges from 4500 to 8500 K, and suggests 6500 K as an average value. At the edges of quiescents T_i is higher, 8000–12000 K. The corresponding value of ξ and 3–8 km s^{-1} in the center and 10–20 km s^{-1} at the edges.

The Lyman continuum (H I LyC) yields the most direct and unambiguous determination of the electron temperature (Schmahl et al., 1974). Orrall and Schmahl (1980) found a mean color temperature of 7520±740 K for the nine quiescent prominences well observed by Skylab. Since they find the LyC to be optically thick at $\lambda912$, the color temperature must equal the electron temperature T_e. These values of T_e lie within the observed range of T_i but are on the average higher than Hirayama's (1979) estimated average of 6500 K. This difference may occur because T_i is derived from optically thin emissions which sample the entire prominence in the line-of-sight while the H I LyC is thicker and gives less weight to the inner cooler regions of the prominence.

Stark broadening of the high Balmer lines provides the most direct determination of the electron density and has been used by a number of observers. Hirayama's (1979) summary suggests a range of $10^{10.0}$–$10^{11.4}$ cm^{-3} with $10^{10.5}$ cm^{-3} a typical value in quiescents.

A number of other observations set important but less direct constraints on temperature and electron density. These include observations of the Balmer, Paschen, and other continua, the Balmer decrement, and the ratios of a number of selected line emission ratios. The interpretation of these data requires geometrical information such as the effective length of the observing column and a fractional emitting volume, or depends on detailed non-LTE calculations that are in some way model dependent.

In recent years two interacting parallel developments have stimulated much renewed interest in the interpretation of prominence spectra. One is the development of improved photometric techniques. In particular, consistent sets of selected line intensity ratios have been measured with high precision at many places in a number of prominences of different brightness (see, e.g., Landman, 1980, and references therein). The other development has been in the construction of increasingly sophisticated and realistic numerical models of radiative transfer within prominences (e.g. Heasley and Milkey, 1978; Milkey, 1979). The improved photometric data have provided the observational constraints needed to test these theoretical models.

8.1.2. *Prominence Support*

Since the relatively heavy prominence material is suspended above the solar surface and enveloped by tenuous coronal gas, it seems clear that the support of such a structure must rely on some force other than gas pressure. Magnetic fields are presumably involved. A crude estimate of the pressure, gravitational, and magnetic forces substantiates this conjecture.

Magnetic field strengths are generally reported to lie in the range 0–30 gauss (see Rust, 1972; Smolkov and Bashkirtsev, 1973; Tandberg-Hanssen, 1974; Jensen *et al.*, 1979, for comprehensive reviews). Taken together with the estimates of temperature and density given above, we find a gas pressure of 6.2×10^{-3} to 2.9×10^{-1} dyne, a gravitational pressure $(GM_\odot \rho/R_\odot)$ of 3.2×10^1 to 8×10^2 dyne, and a magnetic pressure up to 3.6×10^1 dyne. Thus, it appears that gravitational and magnetic forces are dominant in prominence equilibrium. However, it must be kept in mind that magnetic forces have no component along field lines, and where the field is not horizontal the gas pressure must oppose gravity. Most theories of prominence support involving magnetic fields center around three themes — suspension in a sagging hammock type of field configuration, confinement of a current sheet, and helical models.

Models of the first type employ magnetic fields which lie in a plane perpendicular to the long axis of the prominence where upward $\mathbf{J} \times \mathbf{B}$ forces support the prominence material against gravity (Bhatnager *et al.*, 1951; Dungey, 1953; Kippenhahn and Schlüter, 1957; Priest *et al.*, 1979; Milne *et al.*, 1979). The best known of these models is that of Kippenhahn and Schlüter, a schematic of which is shown in Figure 9(a). Although this model has remained attractive for many years, it does have some difficulties. For example, recent observations of magnetic fields in prominences suggest that the field strength increases with height (Rust, 1966; Leroy, 1977; House and Smartt, 1979). If this is the case, the magnetic pressure gradient is downward in the same direction as gravity. If gas pressures are truly small, this means that both gravity and the magnetic pressure force must be balance by an upward magnetic tension. As can be seen from Figure 9(a), the tension force certainly is in the right direction. However, during the formation of such a structure, one would expect the sinking material to compress field

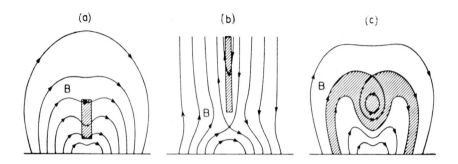

Fig. 9. Three suggested prominence configurations (from Anzer, 1979): (a) a sagging hammock topology as suggested by Kippenhahn and Schluter (1957); (b) a current sheet proposed by Kuperus and Tandberg-Hanssen (1967), Raadu and Kuperus (1973), Kuperus and Raadu (1974), and Raadu (1979); and (c) an inverted tear-drop model from Malville (1979). Reprinted by courtesy of the Institute of Theoretical Astrophysics, Oslo, Norway, © 1989).

lines near the bottom of the prominence and produce a higher field strength there where both the magnetic pressure and tension oppose gravity. This is not observed. Also, if the prominence were formed by condensation of material out of the corona onto a loop-like structure, it is difficult to see how the inverted magnetic geometry shown in Figure (9a) would form in the first place; i.e. why the material would not merely slide down the legs of the loops towards the chromosphere.

Models employing current sheets have been attempted by Kuperus and Tandberg-Hanssen (1967). Raadu and Kuperus (1973), Kuperus and Raadu (1974), and Raadu (1979). Kuperus and Tandberg-Hanssen (1967), suggest that the condensation of material begins in a neutral sheet separating regions of opposite magnetic polarity. Then, due to the tearing mode instability, field lines begin to reconnect forming isolated bubbles in the sheet which sink until they are supported by the horizontal field beneath. This model naturally explains the knots of falling material seen in quiescent prominences. Raadu and Kuperus (1973) argued that, at a neutral line between distinct bipolar regions, a current sheet should be present that separates the magnetic fields and which should be suitable for prominence formation. Supporting forces result from internal restructuring of the currents and their interaction with the photosphere (Kuperus and Raadu, 1974). The basic structure for all these models is shown schematically in Figure 9(b).

Two objections to these models can be raised. First, observations of prominence magnetic fields using the Zeeman effect have shown that the prominence field seems to thread through the prominence in the direction predicted by the Kippenhahn–Schlüter model rather than in the opposite direction as in the current sheet models. Second, the current sheet models predict an 'open' field configuration above the prominence. This is hard to reconcile with the closed coronal loops observed above filaments.

The possibility of helical structure in prominences has intrigued solar physicists probably since the great prominence eruption of 1946 when particularly good Hα observations showed suggestive evidence of helical structure in the expanding material (see Figure 10). Also, Malville (1969), from an extensive study in search of helical structure in quiescent prominences, concluded that it was not an uncommon characteristic (see also Rompolt,

1971; Ohman, 1972). A helical field configuration can be visualized as a superposition of a two-dimensional field in a plane perpendicular to the prominence axis combined with a field oriented along the axis of the prominence. Iospha (1968), Nakagawa and Malville (1969), and Anzer and Tandberg-Hanssen (1970) have included such a field in their prominence models, particularly for the purpose of stability considerations. These authors have assumed a circular field perpendicular to the filament — clearly in contradiction with the observation that prominences are thin vertical structures. However, one could easily modify the cross-section of the helix into a tall, thin oval shape which could meet the observational requirements. Although not expressly shown in Figure 9(c), the suggested inverted tear-drop model of Malville (1979) could also be visualized as a helix in three dimensions.

It is difficult to assess the relative merits of the three basic ideas expressed above since they all possess some good and bad features. It seems that better observations are needed to clarify our ideas. Clearly, since a prominence is a magnetic structure, high-resolution observations of prominence magnetic fields, particularly the vector field, are of paramount importance.

8.1.3. *Prominence Stability: Disparition Brusques*

Quiescent prominences occasionally activate, erupt, and disappear. These events often precede large two-ribbon flares (see Section 8.3) and are closely associated with coronal transients (Section 8.5). Therefore, they are of paramount importance to the whole solar activity problem. Figure 10 shows a classic example of a disparition brusque, perhaps the largest ever observed.

A typical disparition brusque event is preceded by an increase in the random velocities in the prominence to about $30-50$ km s^{-1}. Then, the whole prominence starts to rise, and accelerate to velocities of several hundred kilometres per second. The continual acceleration of the material implies a driving force. However, the origin of this force is, as yet, unknown. The prominence is often seen to exhibit a spiraling motion as it rises (Severny and Khoklova, 1953; Zirin, 1968; Malville, 1979; Slomin, 1969; Anzer and Tandberg-Hanssen, 1970; Jockers and Engvold, 1975), suggestive of the unwinding of a helix. Usually a new filament with very nearly the same shape reforms following the eruption, although this is not always the case, particularly with high-latitude filaments. Often, prominences undergo violent internal activations without actually erupting. For example, internal motions have been observed by Valnicek (1964) to increase to up to 300 km s^{-1} in an activated filament and then die down again.

Chiefly as a result of Skylab, the association of filament activity and soft X-ray emission has come to be appreciated. During the period of activation, which may precede the eruption by a fraction of an hour to several hours, X-ray emission is observed coming from a mass of material equal to about 10% of the filaments mass at a temperature of $\approx 2.5 \times 10^6$ K (Webb *et al.*, 1976). The emission measure in X-ray continues to rise, however, even after the filament disappears, indicating that perhaps the heated material lies beneath the filament rather than in the prominence material itself. This emission leads to the well-known 'long-decay enhancements' or 'gradual-rise-and-fall' events characteristically associated with erupting filaments (Sheeley *et al.*, 1975; Kahler, 1977). This emission is very similar to, but less energetic than, that seen above two-ribbon

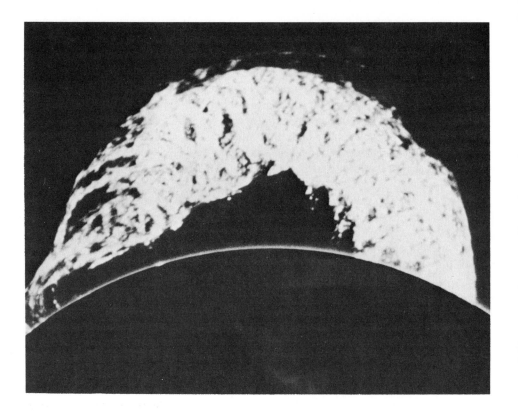

Fig. 10. Disparition brusque of June 4, 1946, one of the largest ever seen. Note the suggestion of an
unwinding helical structure as the prominence rises.

solar flares. This suggests that we are seeing a common process associated with eruptive
prominences and flares.

The spectacular nature of eruptive prominences has inspired many studies of prominence stability (Kippenhahn and Schlüter, 1957; Coppi, 1964; Anzer, 1969, Nakagawa
and Malville, 1969; Nakagawa, 1970, Sakurai 1976; Milne *et al.*, 1979). We cannot discuss
each of these models separately. However, the whole question of prominence stability is
complicated by the fact that even the equilibrium structure of a prominence is presently
unknown. Thus, all the stability analyses are extremely model dependent. Since prominence eruptions often precede solar flares they may be caused by a large instability
elsewhere.

8.2. SOLAR FLARES

Perhaps the most spectacular and energetic form of solar activity is the solar flare, both
because of its dominance as a primary energy release mechanism in the solar atmosphere
and of its effect upon the upper Earth atmosphere. Recently, it has been suggested that
flares can be divided into two quite unique, although not necessarily physically different,
classes (Pallavicini *et al.*, 1977; Pallavicini and Vaiana, 1980). One comprises small,

compact loop-like structures with short event lifetimes (1 h) and only a modest energy release but with large energy densities usually termed 'compact flares', 'subflares', or 'single-loop flares'. They are not usually associated with coronal transients. The other class occurs in large diffuse rising loop systems, usually brighter at the top, visible in Hα, the EUV, and soft X-rays. They are must larger and more energetic than the compact flares ($\sim 10^{30-32}$ erg), have lifetimes > 3 h, are almost always preceded by prominence eruptions, and are associated with coronal transients. Because of their association with spreading Hα flare ribbons in the chromosphere, they are termed 'two-ribbon' flares. Because of space limitations, and as flares are covered in detail in other chapters of this book, we shall discuss here only the most important large-scale aspects of these two classes of flares as well as the physical implications of the observations.

8.2.1. *Two-Ribbon Flares*

The dominating characteristic of two-ribbon flares is the large flare-loop systems which appear at the very onset of the flare and rise upward slowly into the corona. The term 'post-flare' loops was formerly applied to these structures. It is quite misleading and probably originated historically because the loops were first clearly seen on the limb in Hα only after the system had risen quite high in the corona since they were occulted by the disk until then. Also, as we shall see, the loops appear initially at temperatures much higher than Hα temperatures and appear in Hα only after cooling for about 20 min, which is quite late in the flare. As a result of EUV and soft X-ray observations from Skylab and concurrent Hα observations taken from the ground, a much more unified picture of these events is emerging. We now see the flare loops as the primary aspect of the flare with the Hα flare ribbons at the base of these loops being produced by energy propagating down from the loop tops rather than the other way around. It now appears that the primary energy release in these flares occurs at the tops of the flare loops.

A classic description of flare loops as seen in Hα was given by Bruzek (1964a). He noted that the most conspicuous feature of four large flares observed in July 1961 were two long flare ribbons which expanded laterally away from a neutral line in the magnetic field. Loops connecting the flare ribbons became visible with an apparent outward expansion of 5–10 km s^{-1}. However, there was not much real expansion of the loops but, rather, the appearance of new loops at successively greater heights in the corona. It is now quite clear, from Hα, EUV, and soft X-ray observations (cf. Bruzek, 1964a, b; Neupert *et al.*, 1974; Kahler *et al.*, 1975; Cheng and Widing, 1975; Vorpahl *et al.*, 1975; Pallavicini *et al.*, 1975; Rust and Webb, 1977; MacCombie and Rust, 1979; Nolte *et al.*, 1979; Martin 1979; Pallavicini and Vaiana, 1980) that the basic structure is an arcade of arches which successively appear at greater heights in the corona. The brightest X-ray emission occurs at the top (Pallavicini *et al.*, 1975; Rust, 1976; Vorphal *et al.*, 1977; Nolte *et al.*, 1979; Petrasso *et al.*, 1979; MacCombie and Rust, 1979).

The heights of the loops depend strongly upon the type of emission observed. The lowest loops appear in Hα, whereas the highest are seen in soft X-rays rising to heights of over 100 000 km (Nolte *et al.*, 1979; MacCombie and Rust, 1979). Loops at intermediate temperatures are seen in the EUV lines (Withbroe, 1978). Figure 11 shows the loop height as a function of time for both Hα and X-ray observations for the classic loop system of the July 29, 1973 two-ribbon flare. The X-ray loops were not seen early in

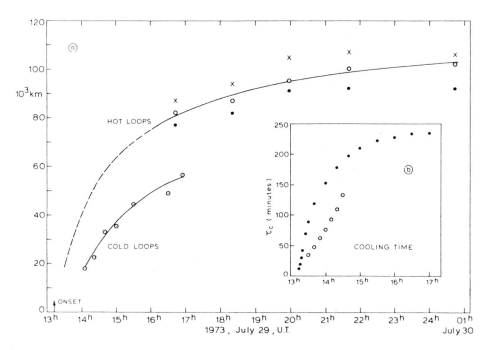

Fig. 11. Heights versus time of the cool (Hα) and hot (X-ray) flare loops during the July 29, 1973 flare. The arrow denotes the Hα flare onset time. The symbols ○, ⊙, ●, and X refer to various geometrical assumptions used to deduce the loop heights and are described in Moore *et al.* (1980). The box in the lower right-hand corner shows the cooling time of individual loops derived from the loop growth (○), and ribbon separation (●). (From Moore *et al.*, 1980.) (Reprinted by courtesy of the Colorado Associated University Press, Boulder, Colorado, © 1980.)

this flare since the satellite was initially in the Earth's shadow. However, earlier coverage was obtained for the September 7, 1973 flare which showed that the X-ray loops began very early in the flare. The reader is referred to MacCombie and Rust (1979) for a discussion of loop growth for other flares observed by Skylab.

The velocity of loop growth for the July 29, 1973 flare is shown in Figure 13. Note that the Hα and X-ray observations tend to fall on the same general velocity curve although the two sets of observations for this particular flare do not overlap. We might speculate that the observations might follow the curve at all heights. If this is true, it means that the velocity of ascent of the system is independent of the physical properties of the individual loops (e.g. temperature) and depends only upon the height in the corona at which the system is observed. This is a strong argument in favor of the viewpoint that we are observing a rising system of discrete, stationary loops.

The overall evolution of the flare ribbons in the chromosphere follows the flare loop evolution very closely. Figure 13, for example, shows that the velocity of separation of the ribbons and the rise velocity of the loop system at a given time are similar — as they should be if, indeed, the ribbons are the loci of the footpoints on the loops. Recent observations indicate that the Hα loops are rooted on the inside edge of the ribbons (Roy, 1972; Rust and Bar, 1973; Beckers, 1977) whereas the X-ray loops originate in

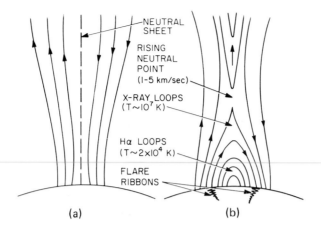

Fig. 12. Schematic of flare loop evolution taken from Pneuman (1981) based upon the model of Kopp and Pneuman (1976). (a) Open field configuration immediately after the eruptive phase, and (b) the field topology at some arbitrary time during the flare. Here, the open field is reconnected to form closed loops with the hottest loops at the top where the magnetic energy is being released, and the cool (Hα) loops lower down. These cool loops were formerly hot but have lost most of their thermal energy content due to radiation and thermal conduction. The chromospheric flare ribbons are presumably produced by energy traveling downward via either thermal conduction or energetic particles from the reconnecting region. (Reprinted by courtesy of Gordon and Breach, New York, London, Paris, © 1981.)

the middle and outer portions of the ribbons (Moore *et al.*, 1980). This is consistent with the observation that the hot loops are the highest with the cool loops nested beneath.

A conspicuous feature of the flare loop system is the downward flow observed in the legs. This motion is seen as soon as the loops appear in Hα. Kleczek (1964) has estimated that the total mass delivered to the chromosphere by this downflow is about $10^{15}-10^{16}$ g. Since this mass is a significant percentage of the total mass of the corona, it is difficult to account for it by condensation out of coronal material (Jeffries and Orrall, 1963a, b, 1965; Kleczek, 1963, 1964). To make matters worse, these mass estimates were based upon Hα observations alone. Recently, Pneuman (1982) integrated the derived density in the flare loop system for the July 29, 1973 flare using X-ray observations. He obtained a total mass for that event of 7.5×10^{16} g. This much greater mass implies that the higher loops are never seen in Hα because they never cool below coronal temperatures. If coronal condensation is ruled out as the mass supply, the material must be supplied from the chromosphere. Since the primary energy supply is probably at the tops of the loops, the energy needed to lift the mass does not originate locally in the chromosphere but propagates downward from above. This process of chromospheric 'evaporation' has been suggested by many authors (Hudson and Ohki, 1972; Sturrock, 1973; Hirayama, 1974; Lin and Hudson, 1976; Antiochos and Sturrock, 1978; Colgate, 1978; Withbroe, 1978). The chromospheric material can be heated, either by energetic particles propagating downward (Sturrock, 1973; Lin and Hudson, 1976) or by thermal conduction (Hirayama, 1974; Antiochos and Sturrock, 1978; Colgate, 1978; Withbroe, 1978). Since most of the energy resides in the thermal plasma we must question whether nonthermal processes are efficient enough.

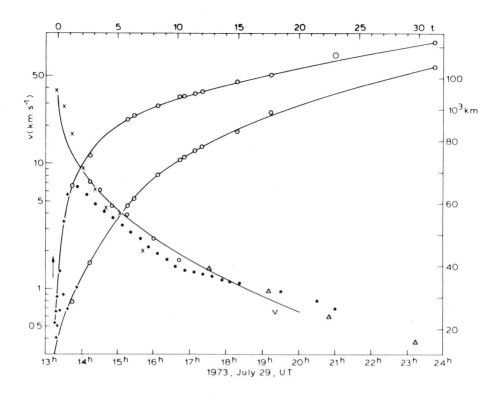

Fig. 13. Velocities (left scale) of loop growth (○, cool loops; △, hot loops) and ribbon separation (×, ●) for the July 29, 1973 flare. The solid curve is the empirical fit obtained by Svestka *et al.* Also shown in the figure is the ribbon separation distance (⊙, right scale) as measured from the outer (O) and inner (I) edges. (From Moore *et al.*, 1980.) (Reprinted by courtesy of the Colorado Associated University Press, Boulder, Colorado, © 1980.)

The physical structure of loop prominence systems is quite complex since the system is evidently composed of individual loops in different stages of evolution. Hence, the spatial and temporal distributions of any given physical property are intimately linked. Nevertheless, the structure of the whole system can be understood by combining two concepts: (1) the temporal variation of the physical properties of a single flux tube, once it is formed, depends upon the characteristics of that flux tube only and is independent of variations in flux tubes around it; (2) the composite structure of the whole system evolves in time as the individual loops evolve, with the youngest at the top and the oldest lower down.

The hottest part of the loop prominence system is at the top where temperatures of over 10^6 K are attained (Hudson and Ohki, 1972; Cheng and Widing, 1975; Widing and Dere, 1977; Vorpahl *et al.*, 1977; MacCombie and Rust, 1979; Petrasso *et al.*, 1979; Moore *et al.*, 1980). The Hα loops at the bottom of the system have temperatures of only about 2×10^4 K (Jeffries and Orrall, 1964). This temperature structure, combined with the observed rise of the loop system through the formation of discrete new loops, suggests that the Hα loops were previously X-ray loops that cooled down as new X-ray loops were formed above. Since the cooling time for an individual loop is only about one-half hour, whereas the lifetime of the system is many hours (Culhane *et al.*, 1970; Moore and Datlowe, 1975; Kopp and Pneuman, 1976; Antiochos and Sturrock, 1976; Krieger, 1977; Withbroe, 1978; Moore *et al.*, 1980), some form of continuous heating must take place during the entire event. Presumably, energy is added to the new loops (at their tops) just as they are formed. Hence, the observed decay with time of the thermal energy density (Moore *et al.*, 1980; MacCombie and Rust, 1979) should not be interpreted as the cooling of individual loops. Rather, the decay follows that rate at which energy is being added to the most newly formed loops.

All the observed properties discussed above can be plausibly explained as the consequence of the *magnetic reconnection of lines of force which were distended outward outward by some eruptive process* such as a filament eruption (Carmichael, 1964; Bruzek, 1969; Sturrock, 1972; Roy, 1972; Hirayama, 1974; Kopp and Pneuman, 1976). Normally, the eruption of a pre-flare filament precedes a two-ribbon flare. Of 297 importance 1 or larger flares studies by Martin and Ramsey (1972), 53% were preceded by rising prominence activity. percentage of these that were accompanied by two-ribbon flares is unknown. The pre-flare role of the filament in specific events has been discussed by many authors (cf. Foukal, 1970; Zirin and Tanaka, 1973; Pallavicini *et al.*, 1975; Thomas, 1975; Widing, 1975; Rust, 1976; Svestka, 1976; Martin, 1980). According to this model, the opening of field lines during the prominence eruption produces the flare loops along with the two flare ribbons in the chromosphere. The rise of the loop system is explained by the rising neutral line as the reconnection proceeds upward. This picture accounts for the outward expansion of the flare ribbons without requiring motion of magnetic field lines through the photosphere, and is consistent with the observed fixed and discrete nature of the loops (see Figure 12). The observations that imply the loops are hottest at the tops are also nicely explained by the reconnection process, since one expects most of the magnetic energy to be released in the vicinity of the neutral line which lies at or above the tops of the X-ray loops. This energy could be added in the form of shock waves (Kopp and Pneuman, 1976; Heyvaerts *et al.*, 1977; Cargill and Priest, 1981) or by Ohmic dissipation (Pneuman, 1982).

In conclusion, we feel that the process of magnetic reconnection following a filament eruption is compatible with both Hα and X-ray observations and is to be expected from MHD theory. The basic unanswered question in this whole process is: How is the energy put into the magnetic field in the first place? Since reconnection only removes energy from the magnetic field, we must find a mechanism for storing energy in magnetic fields. This mechanism is probably related to subphotospheric convective motions and their influence upon emergent flux through twisting and shear. New observations from the Solar Maximum Mission may help to clarify this point.

8.2.2. *Compact Flares*

Compact flares or subflares are far more frequent than the larger two-ribbon flares and about 10–100 times smaller in area. Whereas the large flares occupy all or a large part of an active region, these flares are usually limited to certain regions away from the main neutral line. They appear as small, bright loop structures nested beneath the larger loops of the active region. This is to be contrasted with two-ribbon flares where the loops form a long arcade over a main neutral line and are rooted in two long chromospheric flare ribbons. With sufficient resolution, one sees that the thermal X-ray plasma in compact flares, as in two-ribbon flares, consists of a continuum of loops with hotter gas occupying the outermost loops and cooler gas underneath. During the decay phase the outermost loops cool more slowly than the inner ones, similar to two-ribbon flares. The most striking difference between these and the larger flares, however, is the apparent absence of any separation of the footpoints during the process of the flare, or expansion of the loop system as a whole. Perhaps, this is a true physical difference. However, the much shorter lifetimes, the smaller size, and the possibility that a smaller magnetic field geometry may be involved in the loop formation process, may also mask important similarities.

The electron density in the compact flare loops lies in range $10^{11}-10^{12}$ cm^{-3}. These densities are about an order of magnitude larger than those observed in the large flares. Since flares of intermediate scale correspondingly have intermediate scale density values (cf. Moore et al., 1980), it appears that as flare volume increases, the plasma density decreases. Part of the reason for this relationship is probably that electron densities in large flares are usually reported for the tops of the flare loops which lie much higher in the corona than compact flares. The temperatures in compact flare loops do not seem to differ greatly from those measured in two-ribbon flares. Typically the temperature of the X-ray plasma lies in the range $1-2 \times 10^7$ K (Moore and Datlowe, 1975; Moore, 1976; Doschek et al., 1979; 1980; Feldman et al., 1980a, b) similar to that found for two-ribbon flares. From the above estimates of temperature and density, and with the observed flare volume, the thermal energy content of subflares is estimated to be about 10^{29} erg.

There is no strong evidence for continued heating during the decay phase of compact flames, as opposed to two-ribbon flares. Radiative cooling dominates during the decay phase, although conductive cooling may be important early in the flare. However, for temperatures below 10^7 K, the gradual fall-off of emission measure observed is incompatible with the large radiative decays expected (Moore et al., 1980) which indicates, perhaps, some heating. The mass supply to the loops may be chromospheric evaporation driven by thermal conduction (Krieger, 1977; Withbroe, 1978).

All in all, the differences between compact flares and two-ribbon flares do not appear to be as great as initially suggested. The suggested heating and cooling processes, and mass supply mechanism are similar. The main differences are in the scale and duration, and in the loop growth which seems to be absent in compact flares. If compact flares are produced by magnetic reconnection, then the lack of growth of the compact loop systems could be attributed to discrete impulsive reconnection events that do not involve the relaxation of a large-scale magnetic field.

8.3. IMPULSIVE FLOWS

8.3.1. *Surges*

Although EUV data suggests that all flares are accompanied by some sort of surge activity, Hα observations show that they occur with only about 20% of flares. Typically, surges grow rapidly out of a small bright amount of material at or near the flare site, reach velocities of about 100–200 km s^{-1}, and last 10–20 min. As seen in Hα, the surge material sometimes falls back to the solar surface after reaching its maximum height while at other times merely fades from view. This might be for thermodynamic reasons or because of the limited passband (i.e. velocity range) of the observations. As contrasted to sprays and eruptive prominences, surges seem to follow a predefined trajectory along an existing large-scale magnetic field.

Surges originate deep in the chromosphere as evidenced by their association with 'Ellerman bombs' (Ellerman, 1917) from which all nonflare associated surges originate (Roy, 1973). These bombs lie near the temperature minimum. They appear as a rapid brightening (on a time-scale of a few minutes), followed by a period of about constant brightness lasting for 10–20 min, and then by a rapid decay (Bruzek, 1972; Roy and Laparskas, 1973). There is a strong tendency for surges to recur at a rate of greater than one per hour at a given location (Tandberg-Hanssen, 1967). Surges are commonly associated with Type-III radio bursts, which usually appear much later (about 10 min) than the surge itself (Westin, 1969).

The mechanism for accelerating surges has received very little attention, mainly due the general lack of good observational material. The observations of Schmahl (see Rust *et al.*, 1980) suggests that pressure gradients are sufficient to drive the material to the observed speeds. Theoretical models, prompted chiefly by Skylab observations, have employed gas pressure as the driving force for surges (Steinolfson *et al.*, 1979; McClymont, 1980). Carlqvist (1979), on the other hand, has proposed that surges are produced by current interruption in much the same manner as in flares (Alfvén and Carlqvist, 1967; Carlqvist, 1969, 1973). Whatever their cause, surges do appear to be controlled and guided by the overlying magnetic field so that they probably do not result directly from a general reorientation of the field as in large flares.

8.3.2. *Sprays*

In the past, the observed clumpiness of spray material led observers to conjecture that the ejected material could not be contained by a magnetic field (cf. Tandberg-Hanssen, 1974). Newer techniques in optical observing (high spatial resolution cinephotography, tunable passband filters, multislit spectroscopy, and extended angular field coronagraphs) show, however, that sprays are indeed confined within magnetic loops with material draining down one or both sides of the legs (Tandberg-Hanssen *et al.*, 1980). One important difference between sprays and 'disparition brusques', however, is in their acceleration profiles. Whereas eruptive prominences tend to accelerate slowly at first, gaining speed with height, sprays show a very rapid acceleration to speeds of over 500 km s^{-1} within a few minutes.

Sprays are intimately connected with flares. Tandberg-Hanssen *et al.* (1980), from a study of 13 well-observed sprays during the period 1969–1974, concluded that half originated from observable flares. For the other half, the flares were not well defined mostly because they were behind the limb.

Sprays can also be observed in soft X-rays. Rust and Webb (1977), studied 156 soft X-ray enhancements during the Skylab period and found 126 to be associated with Hα filament activity.

8.3.3. *Coronal Whips*

Another interesting type of coronal event is the so-called 'coronal whip'. These events have been thus far observed only in the coronal green line. They appear as rapid transverse motions of thin magnetic structures. In this case, mass does not appear to be flowing along field lines, but instead, is pulled rapidly through the corona in a whipping fashion by a readjustment of the whole field structure. The entire process begins gradually and accelerates to velocities of about 100 km s^{-1} (Evans, 1957; Kleczek 1963; Bruzek and De Mastus, 1970; Dunn, 1970). This is clearly involved with a rapid large-scale change in the coronal field structure, and probably is the result of a reconnection process. One possible configuration that could lead to a 'whip' is shown in Figure 14 (Pneuman, 1974).

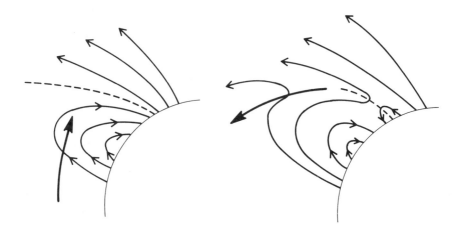

Fig. 14. Possible magnetic evolution leading to a coronal whip. At the left, a closed configuration merges with an open one shown by the broad arrow (or vice versa) leading to reconnection at the boundary shown at the right. The released field lines then whip outward and sideways. (Reprinted by courtesy of D. Reidel, Dordrecht, Holland, © 1974.)

Here a loop structure moves towards open field lines (or vice versa) of opposite polarity. Reconnection then occurs, freeing the photospheric connection of one side of the loops. Once released, these now open field lines will whip out into the corona at some fraction of the Alfvén speed, carrying their load of coronal material along. If this picture is correct then whips could preferentially occur near the edges of coronal holes. However, such an identification has not been attempted.

8.4. CORONAL TRANSIENTS

We may define a 'coronal transient' as a large-scale disturbance in the corona which propagates continuously outward from the Sun. Such a disturbance is ordinarily recognized as an enhancement in electron density although, recently, outward moving depletions have also been identified as transients (Fisher *et al.*, 1981; Fisher and Poland, 1981)..However, even for these, density enhancements appear as the event progresses. Due to many obscuring effects (such as line-of-sight smoothing, perspective differences, etc.), transients defy a precise geometrical definition to a majority of cases. Some appear as magnetic loops or arcades, while others resemble amorphous clouds without a clearly recognizable structure. This ambiguity has posed a problem for theoreticians since no assumed geometry can claim universality.

The White Light Coronagraph on Skylab established for the first time that large-scale coronal transient events are common. Observations of white-light coronal transients have also been made by NRL's coronagraph aboard OSO-7, the HAO Coronal Activity Monitor at Mauna Loa, Hawaii, NRL's Solwind coronagraph (Michels *et al.*, 1979a, b; Sheeley *et al.*, 1979, 1980a, b; Poland *et al.*, 1980), and recently with the HAO Coronagraph/Polarimeter aboard the Solar Maximum Mission Satellite (Wagner *et al.*, 1981; House *et al.*, 1981). Koomen *et al.* (1974) combined observations from OSO-7 and the Mauna Loa coronagraph of a huge coronal cloud moving outward from the Sun on June 16, 1972. Abrupt deletions of material from the inner corona are commonly observed during such events (Hansen *et al.*, 1974). In some particularly dramatic cases, an estimated $10^{39}-10^{40}$ electrons are expelled into interplanetary space.

Coronal transients are also observed infrequently in the green coronal emission line, $\lambda 5303$ (De Mastus *et al.*, 1973; Wagner *et al.*, 1974; Wagner and De Mastus, 1977).

At radio wavelengths, coronal transients can leave their signatures in moving Type IV radio bursts and in the Type II bursts that are presumably associated with a shock traveling ahead of the transient.

The radiation from the moving Type IV burst is generally attributed to gyrosynchotron emission from relativistic electrons (cf. Dulk, 1973). An alternative process involving coherent plasma oscillations has been proposed by Stepanov (1974). Whether these bursts occur in flux tubes rooted in the Sun or in isolated field structures which travel along with the transient is not completely clear. Evidence for the former hypothesis is given by Krivsky and Kruger (1973), Sakurai (1973), and Sakurai and Chao (1974) while that for the latter was argued by Riddle (1970), Smerd and Dulk (1971), Dulk and Altschuler (1971), McLean (1973), and Stewart *et al.* (1974a, b).

One strong argument for the latter hypothesis is that the moving Type IV bursts are contained. One needs a closed magnetic geometry to keep the relativistic electrons from rapidly spreading. One very important feature of the radio observations is that the magnetic field strength in transients can be estimated. These estimates indicate that the magnetic energy density exceeds the thermal energy density by a large factor. For example, for the September 15, 1973 event, Dulk *et al.* (1976) estimate the ratio of the thermal energy density to magnetic energy density to be 0.007. Furthermore, they suggest that the magnetic energy density also exceeds the kinetic energy density.

The white-light observations taken from Skylab and the Solar Maximum Mission provide a wealth of information about coronal transients (see Figure 15). In 227 days of observation in 1973 and early 1974, 115 transient brightenings and 77 definite mass ejections were seen by Skaylab (Munro *et al.*, 1979). Most of these were low-latitude events and encompassed an angular extent of usually less than 68° (Hildner, 1977). The observed speeds of these transients varied considerably but usually ranged between about 300 and 1000 km s^{-1}. Typically, the velocity profiles show either a constant speed or a slight acceleration (Munro *et al.*, 1979). Gosling *et al.* (1976) have shown that no events show strong decelerations. The ejected mass is usually between 10^{15} and 10^{16} g (Hildner *et al.*, 1975; Jackson and Hildner, 1978).

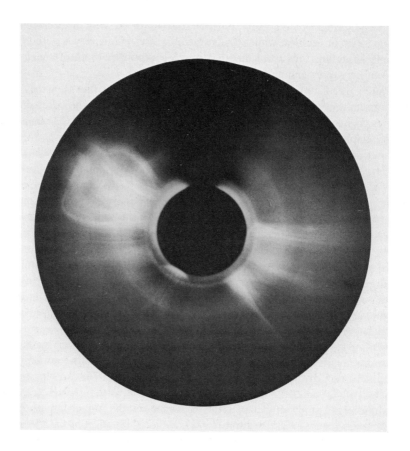

Fig. 15. White-light photograph of a huge coronal transient which occurred on June 10, 1973 (upper left). The photograph was taken by the High Altitude Observatory's ATM white-light coronagraph and clearly shows the outward moving magnetic structure comprising the transient.

The total energy involved in transients (exclusive of magnetic energy) is in the range $10^{30}-10^{32}$ erg and is therefore comparable to the energy released in large flares. The gravitational potential energy usually exceeds the kinetic energy by a small amount. The thermal and magnetic energies injected into the interplanetary medium by the flare of September 5, 1973 were estimated by Dulk and McLean (1978) to be, respectively, a few times 10^{30} and 10^{31} erg. This is consistent with the findings of Dulk *et al*. (1976) and Gergely *et al*. (1979) who report that the magnetic energy in a transient exceeds the thermal energy by a factor of 10 or more.[2] It should be noted that the thermal and magnetic energy resides in the corona before the transient (Rust *et al*., 1980) and are converted into potential and kinetic energy during the event. In any case, we can surmise from the observations that a coronal transient is probably magnetically controlled and results from the relaxation of a higher magnetic energy state to a lower one.

---

[2] It must be pointed out here that the measurements applied to dense plasmoids *within* the transient and not to the entire large-scale structure as seen in white light.

Jackson and Hildner (1978) have pointed out the existence of broad regions in the corona ahead of the transient, called 'forerunners', where the density is slightly enhanced over the background. The forerunners carry typically about 20% of the excess mass associated with the transient.

The material forming the transient appears to come from the lower corona (Koomen et al., 1974; Stewart et al., 1974 a, b; Rust and Hildner, 1976; Webb and Jackson, 1981) where coronal depletions are commonly observed with the K-coronameter in Mauna Loa, Hawaii. Standard coronal models (cf. Saito, 1970; Saito et al., 1977) show that more than enough material is available in the corona below 1.3 $R_\odot$ to account for the mass of the transient.

The most common event associated with coronal transients is an eruptive prominence. About 70% of all transients accompany these eruptions (Munro et al., 1979). Many, but not all, of these eruptive prominence events also precede two-ribbon flares. In fact, every two-ribbon flare observed during Skylab produced a coronal transient provided it was close enough to the limb to be observed (Munro et al., 1979).

Recently, the new High Altitude Observatory's Coronal Dynamics experiment at Mauna Loa, Hawaii has shown some new and interesting aspects of coronal transients at low heights not attainable by satellite-borne coronagraphs (Fisher et al., 1981; Fisher and Poland, 1981). Using subtraction techniques, this instrument has revealed the presence of dark voids moving upward with the prominence and surrounding it. Gradually, enhancements formed around and above this void. The physical interpretation of these depletions is not clear. It must be remembered that voids normally surround prominences underlying coronal streamers in equilibrium (Wesley, 1927; Waldmeier, 1941, 1970; Van de Hulst, 1953; Kleczek and Hansen, 1962; Saito and Hyder, 1968) and would be expected to expand with the transient. Also, coronal depletions are commonly produced by transients. This is to be expected since the $10^{15}-10^{16}$ g of ejected material must be supplied from the inner corona. On the other hand, these rising depletions could be showing some new physical process of importance to the initiation of transients.

Theoretical efforts towards modeling coronal transients have fallen into two categories: global models and single-structure models. In the global models, the transient is initiated by a prescribed pressure pulse at the coronal base and its effect upon the large-scale magnetic field geometry is calculated. Of these models, some are purely hydrodynamic (Nakagawa et al., 1975; Wu et al., 1975) while later models have incorporated magneto-hydrodynamic processes (Nakagawa et al., 1978; Wu et al., 1978; Steinolfson et al., 1978; Dryer et al., 1979; Nakagawa et al., 1981; Wu et al., 1981). In these later models, the thermal pressure provides the energy for the transient while the magnetic field inhibits and channels the resulting expansion. The thermal pressure pulse that initiates the transient is assumed to be caused by a flare-induced temperature increase at the base of the corona from $\approx 10^6$ to $\approx 10^7$ K. This high-temperature plasma is confined to localized magnetic loop systems. Although these models do reproduce some interesting observed properties of transients, the observation that the magnetic energy dominates the thermal energy by more than a factor of 10 would seem to make models with a thermal origin questionable. Furthermore, the observed high-temperature plasma seen in two-ribbon flares appears to be confined to localized magnetic loop systems evolving slowly in time. It is difficult to see how thermal forces associated with this plasma could do work in producing a coronal transient.

All the single-structure models of coronal transients to date produce the active acceleration of the transient by magnetic fields. If the gas pressure is indeed unimportant as compared to magnetic and gravitational forces, then only three forces come into play in determining the resulting velocity profile, i.e. gravity, magnetic pressure, and magnetic tension. Of these, the gravity force and the tension force (at least in the front part of the transient) are directed downward. Hence, outward acceleration can be achieved only by providing excess magnetic pressure the transient, or a decrease above it. A reduction in the strength of the overlying field has not yet been incorporated into a transient theory although, in principle, it could work. All other theories have invoked an increased magnetic pressure at the bottom.

Theories by Mouschovias and Poland (1978), Anser (1978), and Van Tend (1979) propose to accelerate a transient with a twisted magnetic configuration. The common occurrence of transients with eruptive prominence events led Pneuman (1980a) to construct a theory incorporating both events. The weakness in this model lies in the fact, so that the problem of the initiation of the transient is not addressed. The fact that every two-ribbon flare observed close enough to the limb was accompanied by a coronal transient (Munro et al., 1979) has led to theories by Pneuman (1980b) and Anzer and Pneuman (1982) in which the eruptive prominence, transient, and flare were incorporated into a single model. A completely new approach of the transient problem, involving magnetic buoyancy, has been suggested by Low (1981).

Most of the theories discussed above do predict velocity profiles which agree qualitatively with observation. However, a simple argument shows that such a profile is to be expected for a wide variety of forcing functions. For example, let an applied force per unit mass F exist that declines radially as r^{-n}, i.e.

$$F = F_0 \left(\frac{r_0}{r}\right)^n .$$

Then, neglecting gas pressure, the equation of motion for the transient in Lagrangian form is

$$V \frac{dV}{dr} = F_0 \left(\frac{r_0}{r}\right)^n - \frac{GM_\odot}{r^2} .$$

Thus the transient will accelerate provided that $F_0 > GM_\odot/r_0^2$. Integrating and letting $V = 0$ at $r = r_0$, we have

$$V = \left\{ \frac{2F_0 r_0}{n-1} \left[1 - \left(\frac{r_0}{r}\right)^{n-1} \right] - \frac{2GM_\odot}{r_0} \left[1 - \left(\frac{r_0}{r}\right) \right] \right\}^{1/2}$$

We see that, for any $n > 1$, the transient will accelerate initially and approach a constant asymptotic speed given by

$$V_\infty = \left\{ \frac{2F_0 r_0}{n-1} - \frac{2GM_\odot}{r_0} \right\}^{1/2} .$$

The requirement for a positive constant speed at large distances shows that

$$\frac{F_0}{n-1} > \frac{GM_\odot}{r_0^2} \, .$$

Thus, the observed speed chracteristics of a transient can be achieved for a sufficiently large force (large F_0) which does not fall too rapidly with radial distance (n not too much greater than 1).

All the above theories have their advantages and drawbacks and we must regard the coronal transient problem as presently unsolved. The intimate relationships among prominences, flares, and transients must be understood as a whole in order to arrive at the final answer.

References

Adams, W.: 1976, *Solar Phys.* **47**, 601.

Adams, J. and Pneuman, G. W.: 1976, *Solar Phys.* **46**, 185.

Alfvén, H. and Carlqvist, P.: 1967, *Solar Phys.* **1**, 220.

Altschuler, M. D., Levine, R. J., Stix, M., and Harvey, J. W.: 1977, *Solar Phys.* **51**, 345.

Altschuler, M. D. and Newkirk, G., Jr: 1969, *Solar Phys.* **9**, 131.

Altschuler, M. D., Trotter, D. E., Newkirk, G., Jr, and Howard, R.: 1975, *Solar Phys.* **41**, 225.

Altschuler, M. D., Trotter, D. E., and Orrall, F. Q.: 1972, *Solar Phys.* **26**, 354.

Antiochos, S. K.: 1979, *Astrophys. J.* **232**, L125.

Antiochos, S. K. and Sturrock, P. A.: 1978, *Astrophys. J.* **220**, 1137.

Anzer, U.: 1969, *Solar Phys.* **8**, 37.

Anzer, U.: 1978, *Solar Phys.* **57**, 111.

Anzer, U.: 1979, *IAU Coll.* **44**, p. 322.

Anzer, U. and Pneuman, G. W.: 1982, *Solar Phys.* **79**, 129.

Anzer, U. and Tandberg-Hanssen, E.: 1970, *Solar Phys.* **11**, 61.

Artsimovich, L. A.: 1964, *Controlled Thermonuclear Reactions*, Gordon and Breach, New York.

Athay, R. G.: 1971, in C. J. Macris (ed.), *Physics of the Solar Corona*, D. Reidel, Dordrecht, p. 36.

Athay, R. G.: 1976, *The Solar Chromosphere and Corona: Quiet Sun*, D. Reidel, Dordrecht.

Athay, R. G.: 1981a, in F. Q. Orrall (ed.), *Solar Active Regions*, Colorado Assoc. Univ. Press, Boulder, Colo., p. 83.

Athay, R. G.: 1981b, *Astrophys. J.* **249**, 340.

Athay, R. G.: 1982, *Astrophys. J.* **263**, 982.

Athay, R. G. and White, O. R.: 1978, *Astrophys. J.* **226**, 1135.

Athay, R. G. and White, O. R.: 1979a, *Astrophys. J. Suppl.* **39**, 333.

Athay, R. G. and White, O. R.: 1979b, *Astrophys. J.* **229**, 1147.

Ayres, T.: 1981 in R. D. Chapman (ed.), *The Universe at Ultraviolet Wavelengths: The First Two Years of IUE*, NASA Conf. Pub. 2171, p. 237.

Barnes, A.: 1969, *Astrophys. J.* **155**, 311.

Barnes, A.: 1979, in E. N. Parker, C. R. Kennel, and L. J. Lanzerotti (eds), *Solar System Plasma Physics*, North-Holland, Amsterdam, p. 249.

Barnes, A. and Hung, R. J.: 1972, *J. Plasma Phys.* **8**, 197.

Beckers, J.: 1977, Private communication to R. Moore.

Belcher, J. W.: 1971, *Astrophys. J.* **168**, 509.

Belcher, J. W. and Davis, L., Jr: 1971, *J. Geophys. Res.* **79**, 3534.

Belcher, J. W. and Olbert, S.: 1975, *Astrophys. J.* **200**, 369.

Bernstein, I. B., Friedman, E. A., Kruskal, M. D., and Kulsrud, R. M.: 1958, *Proc. Roy. Soc.* **A244**, 17.

Bhatnager, P. L., Krook, M., and Menzel, D. H.: 1951, *Dynamics of Ionized Media*, Rept. Conf. On Dynamics of Ionized Media, Univ. College, London.

Biermann, L.: 1948, *Z. Astrophys.* **25**, 161.
Biskamp, D. and Schindler, K.: 1971, *Plasma Phys.* **13**, 1013.
Bohlin, J. D.: 1969, Ph. D. Thesis, Dept of Astro-Geophysics, Univ. of Colorado, Boulder.
Bohlin, J. D.: 1977, *Solar Phys.* **51**, 377.
Bohlin, J. D., Sheeley, N. R., Jr, and Tousey, R.: 1975, *Space Res.* **15**, 651.
Bohm, D.: 1949, in A. Gutherie and R. Wakerling (eds), *The Characteristics of Electrical Discharges in Magnetic Fields*, McGraw-Hill, New York, p. 5.
Bonnet, R. M., Bruner, E. C., Jr, Acton, L. W., Brown, W. A., and Decaudin, M.: 1980, *Astrophys. J.* **237**, L47.
Bonnet, R. M. and Dupree, A. K.: 1981, *Solar Phenomena in Stars and Stellar Systems*, D. Reidel, Dordrecht.
Boyd, J. T. M. and Sanderson, J. J.: 1969, *Plasma Dynamics*, Barnes and Noble, New York.
Brandt, J. C.: 1970, *Introduction to the Solar Wind*, Freeman, San Francisco.
Bridge, H., Dilworth, C., Lazrus, A. J., Lyon, E. F., Rossi, B., and Scherb, F.: 1961, *Int. Conf. Cosmic Rays and Earth Storm, Kyoto, Japan; J. Phys. Soc. Japan*, **17**, Suppl. A-11.
Brueckner, G. E.: 1980, in P. A. Weyman (ed.), *Highlights of Astronomy*, **5**, 557.
Brueckner, G. E.: 1981, in F. Q. Orrall (ed.), *Solar Active Regions*, Colorado Assoc. Univ. Press. Boulder, p. 113.
Brueckner, G. E., Bartoe, J.-D. F., and Dykton, M.: 1980: *Bull. Amer. Astron. Soc.* **12**, 907.
Brueckner, G. E. and Nicolas, K. R.: 1973, *Solar Phys.* **29**, 301.
Bruzek, A.: 1964a, in W. N. Hess (ed.), *AAS-NASA Symp. on physics of Solar Flares*, NASA SP-50, Washington, D.C., p. 301.
Bruzek, A.: 1964b, *Astrophys. J.* **140**, 2.
Bruzek, A.: 1969, *Solar Flares and Space Research*, North-Holland, Amsterdam.
Bruzek, A.: 1969, *Plasma Instabilities in Astrophysics*, Gordon and Breach, New York.
Bruzek, A.: 1972, *Solar Phys.* **26**, 94.
Bruzek, A. and De Mastus, H.: 1970, *Solar Phys.* **12**, 447.
Burlaga, L. F.: 1972, *Astrophys. Space Sci. Library*, **29**, 135.
Cargill, P. J. and Priest, E. R.: 1981, *Solar Phys.* **65**, 251.
Cargill, P. and Priest, E. R.: 1983, *Astrophys. J.* **266**, 383.
Carlqvist, P.: 1969, *Solar Phys.* **7**, 220.
Carlqvist, P.: 1973, Tech, Rept. TRITA-EPP-05, Dept of Plasma Phys., Roy. Inst. Tech., Stockholm.
Carlqvist, P.: 1979, *Solar Phys.* **63**, 353.
Carmichael, H.: 1964, in W. N. Hess (ed.), *AAS-NASA Symp. on Physics of Solar Flares*, NASA SP-50, Washington, D.C., p. 451.
Carovillano, R. L. and Siscoe, G. L.: 1969, *Solar Phys.* **8**, 401.
Chapman, G.: 1981, in F. Q. Orrall (ed.), *Solar Active Regions*, Colorado Assoc. Univ. Press, Boulder, p. 43.
Chapman, S.: 1954, *Astrophys. J.* **120**, 151.
Chapman, S.: 1957, *Smithsonian Contrib. Astrophys.* **2**, 1.
Cheng, E. C.: 1980, *Astrophys. J.* **238**, 743.
Cheng, C., Doschek, G. A., and Feldman, U.: 1979, *Astrophys. J.* **227**, 1037.
Cheng, C. C., Smith, J. B., Jr, and Tandberg-Hanssen, E.: 1980, *Solar Phys.* **67**, 259.
Cheng, C. and Widing, K. G.: 1975, *Astrophys. J.* **201**, 735.
Chiuderi, C., Einaudi, G., and Torricelli-Ciamponi, G.: 1981, *Astron. Astrophys.* **97**, 27.
Colgate, S. A.: 1978, *Astrophys. J.* **221**, 1068.
Coleman, P. J. Jr: 1966, *Phys. Rev. Lett.* **17**, 207.
Coleman, P. J., Jr: 1967, *Planet, Space Sci.* **15**, 953.
Coppi, B.: 1964, *Ann. Phys.* **30**, 178.
Coppi, B. and Friedland, A. B.: 1971, *Astrophys. J.* **169**, 379.
Coppi, B. and Mazzucato, E.: 1971, *Phys. Fluids* **64**, 134.
Cowling, T. G.: 1953, in C. P. Kuiper (ed.), *The Sun,* Univ. of Chicago Press, Chicago, p. 532.
Craig, I. J. D. and Brown, J. C.: 1976, *Astron. Astrophys.* **49**, 239.
Craig, I. J. D. and McClymont, A. N.: 1980, Presented to the Skylab Workshop on Solar Active Regions, Boulder, Colorado.

Culhane, J. L., Vesecky, J. F., and Phillips, K. J. H.: 1970, *Solar Phys.* **15**, 394.

Davis, J. M., Golub, L., and Krieger, A. S.: 1977, *Astrophys. J.* **214**, L141.

Davis, L. Jr, Smith, E. J., Coleman, P. J., Jr, and Sonnet, C. P.: 1966, in R. J. Mackin Jr and M. Neugebauer (eds), *The Solar Wind*, Pergamon Press, New York, p. 35.

d'Azambuja, L. and d'Azambuja, M.: 1948, *Ann. Obs. Paris, Meudon* **6**, 7.

De Mastus, H. L., Wagner, W. J., and Robinson, R. D.: 1973, *Solar Phys.* **31**, 449.

Dere, K. P.: 1980, Thesis, the Catholic Univ. of America (Univ. Microfilms International No. CGM80–17720).

Dere, K. P.: 1982, *Solar Phys.* **75**, 189.

Dere, K. P. and Mason, H. E.: 1981, in F. Orrall (ed.), *Solar Active Regions*, Colorado Assoc. Univ. Press, Boulder, p. 129.

Dollfus, A.: 1971, in C. J. Macris (ed.), *Physics of the Solar Corona*, D. Reidel, Dordrecht, p. 97.

Doschek, G. A., Feldman, U., Van Hoosier, M. E., and Bartoe, J. D.: 1976, *Ap. J. Suppl.* **31**, 417.

Doschek, G. A., Kreplin, R. W., and Feldman, U.: 1979, *Ap. J.,* **233**, list.

Doschek, G. A., Feldman, U., Kreplin, R. W., and Cohen, L.: 1980, *Astrophys. J.* **239**, 725.

Dryer, M., Wu, S. T., Steinolfson, R. S., and Wilson, R. M.: 1979, *Astrophys. J.* **227**, 1059.

Dulk, G. A.: 1973, *Solar Phys.* **32**, 491.

Dulk, G. A. and Altschuler, M. A.: 1971, *Solar Phys.* **20**, 438.

Dulk, G. A. and McLean, D. J.: 1978, *Solar Phys.* **57**, 279.

Dulk, G. A., Smerd, S. F., MacQueen, R. M., Gosling, J. R., Magun, A., Stewart, R. T., Sheridan, K. V., Robinson, R. D., and Jacques, S.: 1976, *Solar Phys.* **49**, 369.

Dungey, J. W.: 1953, *Mon. Not. Roy. Astron. Soc.,* **113**, 180.

Dunn, R.: 1960, Thesis, Harvard Univ.

Dunn, R.: 1970, *AIAA Bull.* **7**, 564.

Dunn, R.: 1971, *The Menzel Symp. on Solar Phys.,* NBS Special Pub. **353**, 71.

Durney, B. R. and Pneuman, G. W.: 1975, *Solar Phys.* **40**, 461.

Ellerman, F.: 1917, *Astrophys. J.* **45**, 298.

Elzer, L. R. and Elwert, G.: 1980a, *Astron. Astrophys.* **86**, 181.

Elzer, L. R. and Elwert, G.: 1980b, *Astron. Astrophys.* **86**, 188.

Endler, F.: 1971, PhD. Thesis, Göttingen Univ.

Engvold, O.: 1976, *Solar Phys.* **49**, 283.

Engvold, O.: 1980, *Solar Phys.* **67**, 1980.

Engvold, O. and Leroy, J. L.: 1979, *IAU Coll.* **44**, 97.

Evans, J. W.: 1957, Publ. *Astron. Soc. Pacific* **69**, 421.

Feldman, U., Doschek, G. A., and Kreplin, R. W.: 1980a, *Astrophys. J.* **238**, 365.

Feldman, U., Doschek, G. A., Kreplin, R. W., and Mariska, J. T.: 1980b, *Astrophys. J.* **241**, 1175.

Feldman, U., Doschek, G. A., and Patterson, N. P.: 1976, *Astrophys. J.* **209**, 270.

Ferraro, V. C. A. and Plumpton, C.: 1961, *An Introduction to Magneto-Fluid Mechanics*, Oxford Univ. Press, Oxford.

Fisher, R., Garcia, C. J., and Seagraves, P.: 1981, *Astrophys. J.* **246**, L161.

Fisher, R. and Poland, A. I.: 1981, *Astrophys. J.* **246**, 1004.

Foukal, P.: 1970, *Solar Phys.* **13**, 330.

Foukal, P.: 1975, *Solar Phys.* **43**, 327.

Foukal, P.: 1976, *Astrophys. J.* **210**, 575.

Foukal, P.: 1978, *Astrophys. J.* **223**, 1046.

Foukal, P., Huber, M. C. E., Noyes, R. W., Reeves, E. M., Schmahl, E. J., Timothy, J. G., Vernazza, J. E., and Withbroe, G. L.: 1974, *Astrophys. J.* **193**, L143.

Furst, E.: 1980, *IAU Symp.* **86**, p. 25.

Furth, H. P., Killeen, J., and Rosenbluth, M. N.: 1963, *Phys. Fluids* **6**, 459.

Gabriel, A. H.: 1976a, in R.-M. Bonnet and P. Delache (eds), *Energy Balance and Hydrodynamics of the Solar Chromosphere and Corona*, G. de Bussac, Clermont Ferrand, p. 375.

Gabriel, A. H.: 1976b, *Phil. Trans. Roy. Soc. London* **A281**, 339.

Gebbie, K. B., Hill, F., Toomre, J., November, L. J., Simon, G. W., Athay, R. G., Bruner, E. C., Rense, R., Gurman, J. B., Shine, R. A., Woodgate, B., and Tandberg-Hanssen, E.: 1980, *Bull. Amer. Aston. Soc.* **12**, 907.

Gergely, T. E., Kundu, M. R., Munro, R. H., and Poland, A. I.: 1979, *Astrophys. J.* **230**, 57.
Gibson, E. G.: 1977, *Solar Phys.* **53**, 127.
Giovanelli, R. G.: 1949, *Monthly Notices Roy. Astron. Soc.* **109**, 372.
Giovanelli, R. G.: 1980, *Solar Phys.* **68**, 49.
Giovanelli, R. G. and Jones, H. P.: 1981, *Solar Phys.* **70**, 207.
Gold, T.: 1964, *AAS-NASA Symp. on Physics of Solar Flares*, NASA SP-50, Washington, D.C., p. 389.
Goldstein, B.: 1971, unpublished rept.
Golub, L.: 1980, *Phil Trans. Roy. Soc. London* **297**, 595.
Golub, L., Krieger, A. S., Harvey, J. W., and Vaiana, G. S.: 1977, *Solar Phys.* **53**, 111.
Golub, L., Krieger, A. S., Silk, J. K., Timothy, A. F., and Vaiana, G. S.: 1974, *Astrophys. J.* **189**, L93.
Gosling, J. T., Hildner, E., MacQueen, R. M., Munro, R. H., Poland, A. I., and Ross, C. L.: 1976, *Solar Phys.* **48**, 389.
Habbal, S. R., Leer, E., and Holzer, T. E.: 1979, *Solar Phys.* **64**, 287.
Habbal, S. R. and Rosner, R.: 1979, *Astrophys. J.* **234**, 1113.
Habbal, S. R. and Withbroe, G. L.: 1981, *Solar Phys.* **69**, 77.
Hamberger, S. M. and Friedman, M.: 1968, *Phys. Rev. Lett.* **21**, 674.
Hansen, R. T., Garcia, C. J., Hansen, S. F., and Yasukawa, E.: 1974, *Publ. Astron. Soc.* **86**, 500.
Harvey, J.: 1977, in *Highlights of Astronomy* Vol. 4, Pt II, p. 223.
Harvey, K. L. and Martin, S. F.: 1973, *Solar Phys.* **32**, 389.
Heasley, J. H. and Milkey, R. W.: 1978, *Astrophys. J.* **221**, 677.
Hepburn, N.: 1955, *Astrophys. J.* **122**, 445.
Heyvaerts, J., Priest, E. R., and Rust, D. M.: 1977, *Astrophys. J.* **216**, 123.
Hildner, E.: 1977, *Astrophys. Space Sci. Library* **71**.
Hildner, E., Gosling, J. T., Hansen, R. T., and Bohlin, J. D.: 1975, *Solar Phys.* **45**, 363.
Hirayama, T.: 1974, *Solar Phys.* **34**, 323.
Hirayama, T.: 1979, *IAU Coll.* **44**, 4.
Hollweg, J. V.: 1973a, *Astrophys. J.* **181**, 547.
Hollweg, J. V.: 1973b, *J. Geophys. Res.* **78**, 3643.
Hollweg, J. V.: 1975, *Rev. Geophys. Space Phys.* **13**, 263.
Hollweg, J. V.: 1978a, *Rev. Geophys. Space Res.* **16**, 689.
Hollweg, J. V.: 1978b, *Solar Phys.* **56**, 305.
Hollweg, J. V.: 1981a in F. Q. Orrall (ed.), *Solar Active Regions* Colorado Assoc. Univ. Press, Boulder, p. 227.
Hollweg, J. V.: 1981b, *Solar Phys.* **70**, 25.
Hood, G. W. and Priest, E. R.: 1980, *Astron. Astrophys.* **87**, 126.
Holzer, T. E.: 1979, in E. N. Parker, C. R. Kennel, L. J. Lanzerotti (eds), *Solar System Plasma Physics* North Holland, Amsterdam, p. 101.
House, L. L. and Smartt, R. N.: 1979, *IAU Coll.* **44**, p. 81.
House, L. L., Wagner, W. J., Hildner, E., Sawyer, C., and Schmidt, H. V.: 1981, *Astrophys. J.* **244**, L117.
Huber, M. C. E., Foukal, P. V., Noyes, R. W., Reeves, E. M., Schmahl, E. J., Timothy, J. G., Vernazza, J. E., and Withbroe, G. L.: 1974, *Astrophys. J.* **194**, L151.
Hudson, H. S. and Ohki, K.: 1972, *Solar Phys.* **23**, 155.
Hundhausen, A. J.: 1972, *Coronal Expansion and the Solar Wind*, Springer-Verlag, New York.
Hundhausen, A. J.: 1977, in J. B. Zirker (ed.), *Coronal Holes and High Speed Wind Streams*, Colorado Assoc. Univ. Press, Boulder, p. 225.
Ionson, J. A.: 1978, *Astrophys. J.* **226**, 650.
Iospha, B. A.: 1968, *Results of Research on the Intl. Geophys. Projects*, Solar Activity No. 3 Nauka, Moscow, p. 44.
Jackson, B. V. and Hildner, E.: 1978, *Solar Phys.* **60**, 155.
Jacques, S. A.: 1977, *Astrophys. J.* **215**, 942.
Jacques, S. A.: 1978, *Astrophys. J.* **226**, 632.
Jaggi, R. K.: 1963, *J. Geophys. Res.* **68**, 4429.
Jeffries, J. T. and Orrall, F. Q.: 1963a, *The Solar Spectrum*, Utrecht. p. 131.
Jeffries, J. T. and Orrall, F. Q.: 1963b, *Astrophys. J.* **137**, 1232.

Jeffries, J. T. and Orrall, F. Q.: 1965, *Astrophys. J.* **141**, 519.

Jeffries, J. T., Orrall, F. Q., and Zirker, J. B.: 1972, *Solar Phys.* **22**, 307.

Jensen, E., Maltby, P., and Orrall, F. Q.: 1979, *IAU Coll.* **44**, p. 1.

Jockers, K. and Engvold, O.: 1975, *Solar Phys.* **44**, 429.

Jordan, C.: 1980, *Astron. Astrophys.* **86**, 355.

Jordan, C, and Wilson, R.: 1971, in C. J. Macris (ed.), *Physics of the Solar Corona*, D. Reidel., Dordrecht, p. 219.

Kaburaki, O. and Uchida, Y.: 1971, *Publ. Astron. Soc. Japan* **23**, 405.

Kahler, S.: 1977, *Astrophys. J.* **214**, 891.

Kahler, S. W., Krieger, A. S., and Vaiana, G. S.: 1975, *Astrophys. J.* **199**, L57.

Kiepenheuer, K. O.: 1953 in G. P. Kuiper (ed.), *The Sun*, Univ. of Chicago Press, Chicago, p. 322.

Kippenhahn, R. and Schlüter, A.: 1957, *Z. Astrophys.* **43**, 36.

Kleczek, J.: 1963, *Bull. Astron. Inst. Czech.* **14**, 167.

Kleczek, J.: 1964, in W. Hess (ed.), *AAS-NASA Symp. on Physics of Solar Flares*, NASA SP-50, Washington D.C., p. 77.

Kleczek, J. and Hansen, R. T.: 1962, *Publ. Astron. Soc. Pacific* **74**, 507.

Koomen, M. J., Howard, R. A., Hansen, R. T., and Hansen, S. F.: 1974, *Solar Phys.* **34**, 447.

Kopp, R.: 1968, Thesis, Harvard Univ.

Kopp, R. A. and Kuperus, M.: 1968, *Solar Phys.* **4**, 212.

Kopp, R. A. and Orrall, F. Q.: 1977, in J. B. Zirker (ed.), *Coronal Holes and High Speed Wind Streams*, Colorado Assoc. Univ. Press, Boulder, p. 179.

Kopp, R. A. and Pneuman, G. W.: 1976, *Solar Phys.* **50**, 85.

Koutchmy, S.: 1971, *Astron. Astrophys.* **13**, 79.

Krall, K. R. and Antiochos, S. K.: 1980, *Astrophys. J.* **242**, 374.

Krieger, A. S.: 1977, *Solar Phys.* **56**, 107.

Krieger, A. S., Timothy, A. F., and Roelof, E. C.: 1973, *Solar Phys.* **29**, 505.

Krieger, A. S., Vaiana, G. S., and VanSpeybroeck, L. P.: 1971, in R. Howard (ed.), *IAU Coll.* **43**, 397.

Krivsky, L. and Kruger, A.: 1973, *Bull. Astron. Inst. Czech.* **24**, 291.

Kubota, J.: 1981, in F. Mariyana and J. C. Henoux, (eds), *Proc. Japan-France Seminar on Solar Physics*, p. 178.

Kundu, M. R.: 1979, *IAU Coll.* **44**, 122.

Kuperus, M.: 1965, *Rech. Astron. Obs. Utrecht* **17**, 1.

Kuperus, M.: 1969, *Space Sci. Rev.* **9**, 713.

Kuperus, M. and Athay, R. G.: 1967, *Solar Phys.* **1**, 361.

Kuperus, M. and Raadu, M. A.: 1974, *Astron. Astrophys.* **31**, 189.

Kuperus, M. and Tandberg-Hanssen, E.: 1967, *Solar Phys.* **2**, 39.

Landman, D. A.: 1980, *Astrophys. J.* **237**, 988.

Leblanc, Y.: 1970, *Astron. Astrophys.* **4**, 315.

Leer, E. and Holzer, T. E.: 1979, *Solar Phys.* **63**, 143.

Leer, E. and Holzer, T. E.: 1980, *J. Geophys. Res.* **85**, 4681.

Leroy, J. L.: 1977, *Astron. Astrophys.* **60**, 79.

Leroy, J. L.: 1979, *IAU Coll.* **44**, 56.

Levine, R. H.: 1976, *Solar Phys.* **46**, 159.

Levine, R. H. and Altschuler, M. D.: 1974, *Solar Phys.* **36**, 345.

Levine, R. H., Altshculer, M. D., Harvey, J. W., and Jackson, B. V.: 1977, *Astrophys. J.* **215**, 636.

Levine, R. H. and Pye, J. P.: 1980, *Solar Phys.* **66**, 39.

Levine, R. H., Schulz, M., and Frazier, E. N.: 1979, *Bull. Amer. Astron. Soc.* **11**, 697.

Levine, R. H. and Withbroe, G. L.: 1977, *Solar Phys.* **51**, 83.

Lewis, R. R. and Siscoe, G. L.: 1973, *J. Geophys. Res.* **78**, 6443.

Lin, R. P. and Hudson, H. S.: 1976, *Solar Phys.* **50**, 153.

Linsky, J. L.: 1980, *Ann. Rev. Astron. Astrophys.* **18**, 439.

Linhart, J. G.: 1960, *Plasma Physics*, North Holland, Amsterdam.

Low, B. C.: 1981, *Astrophys. J.* **251**, 352.

MacCombie, W. J. and Rust, D. M.: 1979, *Solar Phys.* **61**, 69.

MacQueen, R. M., Gosling, J. T., Hildner, E., Munro, R. H., Poland, A. I., and Ross, C. L.: 1976, *Phil. Trans. Roy. Soc. London* **281**, 405.

Maltby, P.: 1976, *Solar Phys.* **46**, 149.

Malville, J. M.: 1969, Private communication to E. Tandberg-Hanssen.

Malville, J. M.: 1976, *Solar Phys.* **50**, 79.

Malville, J. M.: 1979, *IAU Coll.* **44**, p. 192.

Mango, S. A., Bohlin, J. D., Glackin, D. L., and Linsky, J. L.: 1978, *Astrophys. J.* **220**, 68.

Mariani, F., Villante, U., Bruno, R., Bavassano, B., and Ness, N. F.: 1979, *Solar Phys.* **63**, 411.

Mariska, J. T., Feldman, U., and Doschek, G. A.: 1980, *Astrophys. J.* **240**, 300.

Martin, S. F.: 1979, *Solar Phys.* **64**, 165.

Martin, S. F.: 1980, *Solar Phys.* **68**, 217.

Martin, S. F. and Ramsey, H. E.: 1972, in P. S. McIntosh and M. Dryer (eds.), *Solar Activity, Observations and Predictions*, MIT Press, Cambridge, Mass, p. 370.

Matsuda, T. and Sakurai, T.: 1972, *Cosmic Electrodynamics* **3**, 97.

McClymont, A. N.: 1980 (see Rust *et al.*, 1980).

McIntosh, P. S., Krieger, A. S., Nolte, J. T., and Vaiana, G.: 1976, *Solar Phys.* **49**, 57.

McLean, D. J.: 1973, *Proc. Astron. Soc. Austral.* **2**, 222.

McWhirter, R. W. P., Thonemann, P. C., and Wilson, R.: 1975, *Astron. Astrophys.* **40**, 63.

Menzel, D. H. and Wolbach, J. G.: 1960, *Sky and Tel.* **20**, 252 and 330.

Michard, R.: 1954, *Ann. d'Astrophys.* **17**, 429.

Michels, D. J., Howard R. A., Koomen, M. J., and Sheeley, N. R., Jr: 1979a, *IAU Symp.* **86**, 439.

Michels, D. J., Howard, R. A., Koomen, M. J., Sheeley, N. R., Jr, and Rompolt, B.: *IAU Symp.* **91** p. 387.

Milkey, R. W.: 1979, *IAU Coll.* **44**, 349.

Milne, A. M., Priest, E. R., and Roberts, B.: 1979, *Astrophys. J.* **232**, 304.

Moore, R. L.: 1976, Big Bear Solar Obs. Rept No. 0158.

Moore, R. L. and Datlowe, D. W.: 1975, *Solar Phys.* **43**, 189.

Moore, R. L., McKenzie, D. L., Svestka, Z., Widing, K. G., Antiochos, S. K., Dere, K. P., Dodson-Prince, H. W., Hiei, E., Krall, K. R., Krieger, A. S., Mason, H. E., Petrasso, R. D., Pneuman, G. W., Silk, J. K., Vorpahl, J. A., and Withbroe, G. L.: 1980, in P. A. Sturrock (ed.), *Solar Flares*, Colorado Assoc. Univ. Press, Boulder, p. 341.

Moore, R. L., Tang, F., Bohlin, J. D., and Golob, L.: 1977, *Astrophys. J.* **218**, 286.

Mouradian, A. and Leroy, J. L.: 1977, *Solar Phys.* **51**, 103.

Mouschovias, T. Ch. and Poland, A. I.: 1978, *Astrophys. J.* **220**, 675.

Mullan, D. J. and Ahmad, I. A.: 1982, *Solar Phys.* **75**, 347.

Munro, R. H., Gosling, J. T., Hildner, E., MacQueen, R. M., Poland, A. I. and Ross, C. L.: 1979, *Solar Phys.* **61**, 201.

Munro, R. H. and Jackson, B. V.: 1977, *Astrophys. J.* **213**, 874.

Munro, R. H. and Withbroe, G. L.: 1972, *Astrophys. J.* **176**, 511.

Nakagawa, Y.: 1970, *Solar Phys.* **12**, 419.

Nakagawa, Y. and Malville, J. M.: 1969, *Solar Phys.* **9**, 102.

Nakagawa, Y. and Raadu, M. A.: 1972, *Solar Phys.* **25**, 127.

Nakagawa, Y., Wu, S. T., and Han, S. M.: 1978, *Astrophys. J.* **219**, 314.

Nakagawa, Y., Wu, S. T., and Han, S. M.: 1981, *Astrophys. J.* **244**, 331.

Nakagawa, Y., Wu, S. T., and Tandberg-Hanssen, E.: 1975, *Astrophys. J.* **41**, 387.

Neugebauer, M. and Snyder, C. W.: 1966, in R. J. Mackin Jr and M. Neugebauer (eds), *The Solar Wind*, Pergamon Press, New York, p. 3.

Neupert, W. M. and Pizzo, V.: 1974, *J. Geophys. Res.* **79**, 3701.

Neupert, W. M., Thomas, R. J., and Chapman, R. D.: 1974, *Solar Phys.* **34**, 349.

Newkirk, G. A., Altschuler, M. D., and Harvey, J.: 1968, in O. Kiepenheuer (ed.), *Structure and Development of Solar Active Regions*, D. Reidel, Dordrecht, Holland, Vol. 35, p. 379.

Newkirk, G. A., Jr, Dupree, R. G., and Schmahl, E. J.: 1970, *Solar Phys.* **15**, 15.

Newkirk, G. A. Jr., Trotter, D. E., and Altschuler, M. D.: 1972, *Solar Phys.* **75**, 347.

Noci, G.: 1973, *Solar Phys.* **28**, 403.

Nolte, J. T., Gerassimenko, M., Krieger, A. S., and Petrasso, R. D.: 1979, *Solar Phys.* **62**, 123.

Nolte, J. T., Krieger, A. S., Timothy, A. F., Vaiana, G. S., and Zombeck, M. V.: 1976, *Solar Phys.* **46**, 291.
Nolte, J. T., Solodyna, C. V., and Gerassimenko, M.: 1979, *Solar Phys.* **63**, 113.
Noyes, R. W.: 1972, *Ann. Rev. Astron. Astrophys.* **9**, 209.
Ohman, Y.: 1972, *Solar Phys.* **23**, 134.
Orrall, F. Q. (ed.): 1981, *Solar Active Regions*, Colorado Assoc. Univ. Press, Boulder.
Orrall, F. Q. and Schmahl, E. J.: 1976, *Solar Phys.* **50**, 365.
Orrall, F. Q. and Schmahl, E. J.: 1980, *Astrophys. J.* **240**, 908.
Osterbrock, D. E.: 1961, *Astrophys. J.* **134**, 347.
Pallavicini, R., Serio, S., and Vaiana, G. S.: 1977, *Astrophys. J.* **216**, 108.
Pallavicini, R., Vaiana, G. S., Kahler, S. W., and Krieger, A. S.: 1975, *Solar Phys.* **45**, 411.
Pallavicini, R. and Vaiana, G. S.: 1980, *Solar Phys.* **67**, 127.
Papadopoulos, K.: 1976, NRL Memo, Rept 3275, Naval Research Laboratories, Washington D. C.
Parker, E. N.: 1953, *Astrophys. J.* **117**, 431.
Parker, E. N.: 1958, *Astrophys. J.* **128**, 664.
Parker, E. N.: 1963, *Interplanetary Dynamical Processes*, Interscience, New York.
Parker, E. N.: 1975, *Astrophys. J.* **201**, 494.
Petrasso, R. D., Nolte, J. T., Gerassimenko, M., Krieger, A. S. Krogstad, R., and Seguin, F. H.: 1979, *Solar Phys.* **62**, 133.
Petschek, H. E.: 1964, in W. N. Hess (ed.), *AAS-NASA Symp. on Physics of Solar Flares*, NASA SP-50, Washington, D.C., p. 425.
Pikel'ner, S. B.: 1969, *Soviet Astron.* **13**, 259.
Pizzo, V.: 1978, *J. Geophys. Res.* **83**, 5563.
Pneuman, G. W.: 1973, *Solar Phys.* **28**, 247.
Pneuman, G. W.: 1974, in G. Newkirk (ed.), *Coronal Disturbances,* D. Reidel, Dordrecht, p. 35.
Pneuman, G. W.: 1976, *Proc. Int. Symp. on Terrestrial Physics, June 7–18*, Boulder, Colorado, p. 428.
Pneuman, G. W.: 1980a, *Solar Phys.* **65**, 369.
Pneuman, G. W.: 1980b, *IAU Symp.* **91**, 317.
Pneuman, G. W.: 1980c, *Astron. Astrophys.* **81**, 161.
Pneuman, G. W.: 1981, in E. Priest (ed.), *Solar Flare Magnetohydrodynamics*, Gordon and Breach, New York, London, Paris, p. 379.
Pneuman, G. W.: 1982, *Solar Phys.* **78**, 229.
Pneuman, G. W.: 1983, *Astrophys. J.* **265**, 468.
Pneuman, G. W., Hansen, S. F., and Hansen, R. T.: 1978, *Solar Phys.* **59**, 313.
Pneuman, G. W. and Kopp, R. A.: 1970, *Solar Phys.* **13**, 176.
Pneuman, G. W. and Kopp, R. A.: 1971, *Solar Phys.* **18**, 258.
Pneuman, G. W. and Kopp, R. A.: 1977, *Astron. Astrophys.* **55**, 305.
Pneuman, G. W. and Kopp, R. A.: 1978, *Solar Phys.* **57**, 49.
Poland, A. I., Howard, R. A., Kooman, M. J., Michels, D. J., and Sheeley, N. R. Jr: 1981, *Solar Phys.* **69**, 169.
Poletto, G., Vaiana, G. S., Zombeck, M. V., Krieger, A. A., and Timothy, A. F.: 1975, *Solar Phys.* **44**, 83.
Pottasch, S. R.: 1964, *Space Sci. Rev.* **3**, 816.
Priest, E. R.: 1981, in F. Q. Orrall (ed.), *Solar Active Regions*, Colorado Assoc. Univ. Press, Boulder, p. 213.
Priest, E. R., Milne, A. M., and Roberts, B.: 1979, *IAU Coll.* **44**, p. 184.
Pye, J. P., Evans, K. D., Hutchson, R. J., Gerassimenko, M., Davis, J. M., Krieger, A. S., and Vesecky, J. F.: 1978, *Astron. Astrophys.* **65**, 123.
Raadu, M. A.: 1979, *IAU Coll.* **44**.
Raadu, M. A. and Kuperus, M.: 1973, *Solar Phys.* **28**, 77.
Reeves, E. M.: 1976, *Solar Phys.* **46**, 53.
Richter, A. K. and Suess, S. T.: 1977, *J. Geophys. Res.* **82**, 593.
Riddle, A. C.: 1970, *Solar Phys.* **13**, 448.
Riesebieter, W.: 1977a, Ph. D. Thesis, Tech. Univ. at Braunschweig.
Riesebieter, W.: 1977b, *Trans. AGU* **58**, 485.

Riesebieter, W. and Neugebauer, F. M.: 1979, *Solar Phys.* **63**, 127.

Ripkin, H. W.: 1977, *Trans. AGU* **58**, 1225.

Rompolt, B.: 1971, *Physics of Prominences*, Coll. Anacapri, German Solar Obs.

Rosenbluth, M. N.: 1960, in F. Clauser (ed.), *Symp. on Plasma*, Addison Wesley, Reading, p. 19.

Rosner, R., Golub, L., Coppi, B., and Vaiana, G. S., 1978a, *Astrophys. J.* **222**, 317.

Rosner, R., Tucker, W. H., and Vaiana, G. S.: 1978b, *Astrophys. J.* **222**, 317.

Roy, J. R.: 1972, *Solar Phys.* **26**, 418.

Roy, J. R.: 1973, *Solar Phys.* **28**, 95.

Roy, J. R. and Leparskas, H.: 1973, *Solar Phys.* **30**, 449.

Rust, D. M.: 1966, NCAR Coop. Thesis No. 6, Univ. of Colorado Boulder.

Rust, D. M.: 1972, *Air Force Survey in Geophysics*, p. 237.

Rust, D. M.: 1976, *Solar Phys.* **47**, 21.

Rust, D. M.: 1981, *Solar Phys.* (submitted).

Rust, D. M. and Bar, V.: 1973, *Solar Phys.* **33**, 445.

Rust, D. M. and Hildner, E.: 1976, *Solar Phys.* **48**, 381.

Rust, D. M., Hildner, E., Dryer, M., Hansen, R. T., McClymont, A. N., McKenna Lawlor, S. M. P., McLean, D. J., Schmahl, E. J., Steinolfson, R. S., Tandberg-Hanssen, E., Tousey, R., Webb, D. F., and Wu, S. T.: 1980, in P. A. Sturrock (ed.), *Solar Flares*, Colorado Assoc. Univ. Press, Boulder, p. 273

Rust, D. M., Nakagawa, Y., and Neupert, W. M.: 1975, *Solar Phys.* **41**, 397.

Rust, D. M. and Webb, D. F.: 1977, *Solar Phys.* **54**, 53.

Sagdeev, R. Z. and Galeev, A. A.: 1969, *Nonlinear Plasma Theory*, Benjamin, New York.

Saito, K.: 1959, *Publ. Astron. Soc. Japan* **11**, 234.

Saito, K.: 1970, *Ann. Tokyo Astron. Obs.*, 2nd Series, XII, 53.

Saito, K. and Billings, D. E.: 1964, *Astrophys. J.* **140**, 760.

Saito, K. and Hyder, C.: 1968, *Solar Phys.* **5**, 61.

Saito, K. and Owaki, N.: 1967, *Publ. Astron. Soc. Japan* **19**, 535.

Saito, K., Poland, A. I., and Munro, R. H.: 1977, *Solar Phys.* **55**, 121.

Sakurai, K.: 1973, *Solar Phys.* **31**, 483.

Sakurai, K.: 1976, *Publ. Astron. Soc. Japan* **28**, 177.

Sakurai, K. and Chao, J. K.: 1974, *J. Geophys. Res.* **79**, 661.

Schatzman, E.: 1949, *Ann. Astrophys.* **12**, 203.

Schatten, K. H.: 1971, in R. Howard (ed.), *Solar Magnetic Fields* **43**, 569.

Schatten, K. H. and Wilcox, J. M.: 1969, *Solar Phys.* **6**, 442.

Schmahl, E. J.: 1979, *IAU Coll.* **44**, 102.

Schmahl, E. J., Foukal, P. V., Huber, M. C. E., Noyes, R. W., Reeves, E. M., Timothy, J. G., Vernazza, J. E., and Withbroe, G. L.: 1974, *Solar Phys.* **39**, 337.

Schmahl, E. J. and Orrall, F. Q.: 1979, *Astrophys. J.* **231**, L41.

Schmidt, M.: 1953, *B.A.N.* **12**, 61.

Schulz, M., Frazier, E. N., and Boucher, D. J.: 1978, *Solar Phys.* **60**, 83.

Schwarzschild, M.: 1948, *Astrophys. J.* **107**, 1.

Severny, A. B. and Khoklova, V. L.: 1953, *Isv. Krymsk. Astrofiz. Observ.* **10**, 9.

Shelley, N. R.: 1980, *Solar Phys.* **66**, 79.

Sheeley, N. R.: 1981, in F. Q. Orrall, (ed.), *Solar Active Regions*, Colorado Assoc. Univ. Press, Boulder, p. 17.

Sheeley, N. R., Jr, Bohlin, J. D., Brueckner, G. E., Purcell, J. D., Scherrer, V. E., Tousey, R., Smith, J. B. Jr, Speigh, D. M., Tandberg-Hanssen, E., Wilson, R. M., Deloach, A. C., Hoover, R. B., and McGuire, J. P.: 1975, *Solar Phys.* **45**, 377.

Sheeley, N. R. and Golob, L.: 1979, *Solar Phys.* **63**, 119.

Sheeley, N. R., Jr, Howard, R. A., Michels, D. J., and Koomen, M. J.: 1979a, *IAU Symp.* **91**, p. 37.

Sheeley, N. R., Jr, Michels, D. J., Howard, R. A., and Koomen, M. J.: 1980a, *Astrophys. J.* **237**, L99.

Sheeley, N. R., Jr, Howard, R. A., Koomen, M. J., and Poland, A. I.: 1980b, *Astrophys. J.* **238**, L161.

Shklovskii, I. S., Moroz, V. I., and Kurt, V. G.: 1960, *Astron. Zh.* **37**, 931.

Siscoe, G. L.: 1974, in C. T. Russell (ed.), *Solar Wind Three*, Pub. Inst. of Geophys. and Planetary Phys., Univ. of Calif., Los Angeles, Calif., p. 151.

Siscoe, G. L. and Finley, L. T.: 1972, *J. Geophys. Res.* 77, 35.

Skumanich, A., Smythe, C., and Frazier, E. H., 1975, *Astrophys. J.* 200, 747.

Slomin, Yu. M.: 1969, *Soviet Astron.* 13, 450.

Smerd, S. F. and Dulk, G. A.: 1971, *IAU Symp.* 43, 616.

Smolkov, G. Y. and Bashkirtsev, V. S.: 1973, in J. Sykora (ed.), *7th Regional Consultation on Solar Phys.*, Slovak Acad. of Science, p. 175.

Sonnerup, B. U. O.: 1970, *J. Plasma Phys.* 14, 283.

Spitzer, L.: 1956, *The Physics of Fully Ionized Gases*, Interscience, New York.

Spruit, H. C. and Zwaan, C.: 1981, *Solar Phys.* 70, 207.

Stein, R. F. and Leibacher, J.: 1974, *Ann. Rev. Astron. Astrophys.* 12, 407.

Steinolfson, R. S., Wu, S. T., Dryer, M., and Tandberg-Hanssen, E.: 1978, *Astrophys. J.* 225, 259.

Steinolfson, R. S., Schmahl, E. J., and Wu, S. T.: 1979, *Solar Phys.* 63, 187.

Stenflo, J. O.: 1976, in R.-M. Bonnet and P. Delache (eds), *The Energy Balance and Hydrodynamics of the Solar Chromosphere and Corona*, G. de Bussac, Clermont Ferrand, p. 143.

Stepanov, A. V.: 1974, *Soviet Astron.* 17, 781.

Stewart, R. T., McCabe, M. K., Koomen, M. J., Hansen, R. T., and Dulk, G. A.: 1974a, *Solar Phys.* 36, 219.

Stewart, R. T., Howard, R. A., Hansen, F., Gergely, T., and Kundu, M.: 1974b, *Solar Phys.* 36, 219.

Sturrock, P. A.: 1972, *Solar Phys.* 23, 438.

Sturrock, P. A.: 1973, *Symp. on High Energy Phenomena on the Sun*, NASA GSFC Preprint, SP-342, Washington, D. C.

Sturrock, P. A. and Smith, S. M.: 1968, *Solar Phys.* 5, 87.

Suess, S. T.: 1979, *Space Sci. Rev.* 23, 159.

Svestka, Z.: 1976, *Solar Flares*, D. Reidel, Dordrecht.

Tandberg-Hanssen, E.: 1967, *Solar Activity*, Blaisdell, Waltham, Mass.

Tandberg-Hanssen, E.: 1974, *Solar Prominences*, D. Reidel, Dordrecht.

Tandberg-Hanssen, E., Martin, S. F., and Hansen, R. T.: 1980, *Solar Phys.* 65, 357.

Thomas, R. J.: 1975, *IAU Symp.* 68, p. 25.

Timothy, R. J., Krieger, A. S., and Vaiana, G. S.: 1975, *Solar Phys.* 42, 135.

Tousey, R., Bartoe, J. D. F., Bohlin, J. D., Brueckner, G. E., Purcell, J. D., Scherrer, V. E., Sheeley, N. R., Jr, Schumacher, R. J., and Van Hoosier, M. E.: 1973, *Solar Phys.* 33, 265.

Tucker, W. H.: 1973, *Astrophys. J.* 186, 285.

Uchida, Y.: 1969, *Publ. Astron. Soc. Japan* 21, 128.

Uchida, Y.: 1963, *Publ. Astron. Soc. Japan* 15, 376.

Uchida, Y. and Kaburaki, O.: 1974, *IAU Symp.* 56, p. 299.

Vaiana, G. S., Cassinelli, J. P., Fabbiano, G., Gracconi, R., Golob, L., Gorenstein, P., Haisch, B. M., Harnder, F. R., Jr, Johnson, H. M., Linsky, J. L., Maxon, C. W., Mewe, R., Rosner, R., Seward, F., Topka, K., and Zwaan, C.: 1981, *Astrophys. J.* 244, 163.

Vaiana, G. S. and Rosner, R.: 1978, *Ann. Rev. Astron. Astrophys.* 16, 393.

Valnicek, B.: 1964, *Bull. Astron. Inst. Czech.* 15, 207.

Van de Hulst, H. C.: 1950, *B.A.N.* 11, 135.

Van de Hulst, H. C.: 1953, in G. K. Kuiper (ed.), *The Sun*, Univ. of Chicago Press, p. 296.

Van Hoven, G., Chiuderi, C., and Giachetti, R.: 1977, *Astrophys. J.* 213, 869.

Van Hoven, R. M., Ma, S. S., and Einaudi, G.: 1981, *Astron. Astrophys.* 97, 232.

Van Tend, W.: 1979, *Solar Phys.* 61, 89.

Vernazza, J. E., Avrett, E. H., and Loeser, R.: 1981, *Astrophys. J. Supp.* 45, 635.

Vorpahl, J. A., Gibson, E. G., Landecker, P. B., McKenzie, D. L., and Underwood, J. H.: 1975, *Solar Phys.* 45, 199.

Vorpahl, J. A., Tandberg-Hanssen, E., and Smith, J. R.: 1977, *Astrophys. J.* 212, 550.

Wagner, W. J.: 1975, *Astrophys. J.* 198, L141.

Wagner, W. J. and DeMastus, H. L.: 1977, *Bull. Amer. Astron. Soc.* 9, 369.

Wagner, W. J., Hansen, R. T., and Hansen, S. F.: 1974, *Solar Phys.* 34, 453.

Wagner, W. J., Hildner, E., House, L. L., and Sawyer, C.: 1981, *Astrophys. J.* 244, L123.

Waldmeier, M.: 1941, *Ergbn. Probleme der Sonnen-Forsch.* Leipzig, p. 234.

Waldmeier, M.: 1957, *Die Sonnenkorona*, Vol. 2, Verlag Birkhauser, Basel.

Waldmeier, M.: 1970, *Solar Phys.* **15**, 167.
Webb, D. F.: 1981, in F. Q., Orrall, (ed.), *Solar Active Regions*, Colorado Assoc. Univ. Press, Boulder, p. 165.
Webb, D. F. and Jackson, B. V.: 1981, *Solar Phys.* **73**, 341.
Webb, D. F., Krieger, A. S., and Rust, D. M.: 1976, *Solar Phys.* **48**, 159.
Weber, W. J.: 1977, *Solar Phys.* **53**, 59.
Wentzel, D. G.: 1978, *Rev. Geophys. Space Phys.* **16**, 757.
Wentzel, D. G.: 1979, *Astrophys. J.* **229**, 319.
Wesley, W. H.: 1927, *Mem. Roy. Astron. Soc.* **64**, Appendix.
Westin, H.: 1969, *Solar Phys.* **7**, 393.
White, O. R.: 1976, in R.-M. Bonnet and P. Delache (eds), *Energy Balance and Hydrodynamics of the Solar Chromosphere and Corona*, G. de Bussac, Clermont Ferrand, p. 75.
White, O. R. and Athay, R. G.: 1979, *Astrophys. J. Supp.* **39**, 317.
Widing, K. G.: 1975, *IAU Symp.* **68**, p. 153.
Widing, K. G. and Dere, K. P.: 1977, *Solar Phys.* **55**, 431.
Wilcox, J. M.: 1968, *Space Sci. Rev.* **8**, 258.
Withbroe, G. L.: 1970, *Solar Phys.* **11**, 208.
Withbroe, G. L.: 1976, in R.-M. Bonnet and P. Delache (eds), *Energy Balance and Hydrodynamics of the Solar Chromasphere and Corona*, G. de Bussac, Clermont Ferrand, p. 263.
Withbroe, G. L.: 1977a in E. Hansen and S. Schaffner (eds), *Proc. of the November 7–10, 1977 OSO-8 Workshop*, Univ. of Colorado, Boulder, p. 2.
Withbroe, G. L.: 1977b, in J. B. Zirker (ed.), *Coronal Holes and High Speed Wind Streams*, Colorado Assoc. Univ. Press, Boulder, p. 145.
Withbroe G. L.: 1978, *Astrophys. J.* **225**, 641.
Withbroe, G. L.: 1981, in F. Q. Orrall, (ed.), *Solar Active Regions* Colorado Assoc. Univ. Press, Boulder, p. 199.
Withbroe, G. L. and Mariska, J. T.: 1976, *Solar Phys.* **48**, 21.
Withbroe, G. L. and Noyes, R. W.: 1977, *Ann. Rev. Astron. Astrophys.* **15**, 363.
Wu, S. T., Dryer, M., McIntosh, P. S., and Reichmann, E. J.: 1975, *Solar Phys.* **44**, 117.
Wu, S. T., Steinolfson, R. S., Dryer, M., and Tandberg-Hanssen, E.: 1981, *Astrophys. J.* **243**, 641.
Wu, S. T., Dryer, M., Nakagawa, Y., and Han, S. M.: 1978, *Astrophys. J.* **219**, 324.
Zirin, H.: 1968, *Nobel Symp.* **9**, 131.
Zirin, H. and Tanaka, K.: 1973, *Solar Phys.* **32**, 173.
Zirker, J. B.: 1977, in J. B. Zirker (ed.), *Coronal Holes and High-Speed Wind Streams*, Colorado Assoc. Univ. Press, Boulder, p. 1.
Zweibel, E.: 1980, *Solar Phys.* **66**, 305.

Frank Q. Orrall,
Institute for Astronomy,
University of Hawaii,
2680 Woodlawn Dr.
Honolulu, HI 96822,
U.S.A.

Gerald W. Pneuman,
High Altitude Observatory,
National Center for Atmospheric Research,
Boulder, CO 80307,
U.S.A.

CHAPTER 11

PHYSICAL PROCESSES IN THE SOLAR CORONA

R. ROSNER, B. C. LOW, and T. E. HOLZER

1. Introduction

Since the early 1970s, it has become evident that the entire range of solar surface activity is governed by the interaction between the surface magnetic fields of the Sun and the ambient plasma: the rich phenomenology discussed by others in this work indicates that magnetic fields play a central role in the energetics of both transient events (including flares and magnetic flux emergence associated with solar active regions), as well as in the far more quiescent 'quiet' Sun. Thus, solar magnetic fields not only organize the spatial structure of photospheric, chromospheric, and coronal plasma, but also largely govern the physical processes which control local conditions — heat conduction, viscous and joule heating, particle acceleration, radiative emission. These various processes involve the complications of microscopic plasma effects well known to laboratory and space plasma physicists, albeit under vastly different conditions; and, most crucially from the astrophysical point of view, these processes probably also play a central role in the physics of other astrophysical systems, such as other stars, the interstellar medium, supernova remnants, and accretion disks.

Thus, the general problem we are faced with has three complementary aspects, (1) To what extent can the theoretical results of laboratory and space plasma physics be translated (and perhaps extended) to the solar domain? (2) From the more general astrophysical perspective, can our knowledge of solar plasma physics be usefully extended to more general astrophysical circumstances? (3) Can the results of solar and other astrophysical studies provide meaningful new constraints on plasma physics developed in the terrestrial domain, but now tested in vastly different physical regimes? In any case, we are faced with the difficulty that limitations on spatial resolution and *in-situ* measurements mitigate against the kind of detailed confrontation of theory with observations that laboratory physicists are accustomed to: instead, we must be content with available 'macroscopic' observational hints at the microphysics; and, as our theoretical models gain in complexity, we can only hope that theoretical predictions can become sufficiently detailed so that observations can meaningfully constrain the models. Because of the Sun's proximity, solar physics has, in this respect, a great advantage over much of astrophysics; thus, much as has occurred in spectroscopy for the past three decades, solar physics may play a central role in the interpretation of plasma phenomena in distant astronomical objects.

In this chapter, we shall take the point of view that the phenomenology is as given,

Peter A. Sturrock (ed.), Physics of the Sun, Vol. II, pp. 135–180.
© 1986 *by D. Reidel Publishing Company.*

and explore what is currently known about some of the principal physical processes thought to lead to activity on the solar surface. We shall focus on three issues in particular: the microphysics governing local transport processes (Section 2); the magnetohydrodynamics governing the global (equilibrium or dynamic) state of the outer solar atmosphere (Section 3); and the mechanisms underlying mass loss from the solar surface (Section 4). In putting these topics together, we hope to illustrate the different facets of consideration at which we pose, and attempt to answer, the physical questions. Our aim is not to be encyclopedic; we hope instead to call attention to those issues which seem to us to define the present limits of our theoretical understanding and computational ability.

2. Transport Theory

Virtually all current efforts to model the fluid and magnetohydrodynamic (MHD) properties of the solar outer atmosphere are based upon the use of the standard hydrodynamic equations which follow from taking the first three moments of the Boltzmann equation, together with the macroscopic Maxwell equations, an appropriate equation of state, and the set of linear constitutive relations connecting the heat flux with the temperature gradient, the net current density with the effective electric field, and the Reynolds stress to the velocity gradient (which serve to close the set of moment equations). In the limit of vanishingly small gradients in the fluid variables, the constitutive relations can be derived without difficulty and the atmosphere may be properly regarded as a well-behaved multicomponent fluid; these by now classical results can, for example, be found summarized in Spitzer's (1962) monograph.

If gradients in the fluid variables are not small, then the choice of appropriate constitutive relations and, indeed, the very concept of a *local* description of the equations of motion, becomes problematical. It is important to realize that although the astrophysical literature has largely adopted *de facto* standards in response to this problem (such as, for example, expressions for the thermal heat flux under various extreme limiting conditions on the temperature gradient scales and particle mean free paths (see Cowie and McKee, 1977)), recent experimental and theoretical studies of transport processes in the laboratory have raised some doubts about the validity of these *de facto* standards. It is the purpose of this section to examine this problem and to focus on the corresponding impact on our understanding of physical processes in the solar outer atmosphere.

2.1. FUNDAMENTAL PARAMETER REGIMES

To begin with, it is useful to describe the basic physical regime encountered in the plasma overlying the solar surface; in the following, variables have their customary meaning, and subscripts are used to rescale physical quantities (thus, $T_6 \equiv T/10^6$ K). Because the number of particles within a Debye sphere, N_D, is large

$$N_D = \frac{4\pi}{3} n_e \lambda_d^3 \cong 1.4 \times 10^8 \, n_8^{-1/2} \, T_6^{3/2}, \qquad (2.1)$$

(λ_d the Debye length), and because the electron–ion mean free path is typically (but not always!) small compared to canonical density and temperature scale lengths

$$\lambda_{mfp} \cong 1.1 \times 10^8 \, n_8^{-1} \, T_6^2 (\ln \Lambda/20)^{-1} \text{ cm} \tag{2.2}$$

(ln Λ the standard Coulomb logarithm), classical plasma kinetic and transport theory (cf. Braginskii, 1965) is an appropriate description of most of the solar outer atmosphere. The principal exception arises in the solar transition region, the thermal boundary layer separating the solar chromosphere and corona; this important exceptional case will be discussed further below. Furthermore, because the plasma β

$$\beta \equiv p_{gas}/(B^2/8\pi) \cong 6.9 \times 10^{-3} \, n_8 \, T_6 \, B_1^{-2} \ll 1, \tag{2.3}$$

is small, and because the electron and proton gyroradii

$$\lambda_{ge} \cong 2.21 \, T_6^{1/2} \, B_1^{-1} \text{ cm} \ll \lambda_{mfp} \tag{2.4}$$

are small, much of the solar chromospheric and inner coronal plasma is dynamically dominated by magnetic fields; in particular, the latter inequality (2.4) implies that cross-field transport of mass and heat is severely curtailed because the effective mean random displacement leading to diffusive transport perpendicular to B is just λ_{ge} (or λ_{gi}), rather than λ_{mfp}. We shall return to the issue of cross-field transport below.

Although the foregoing represents a fair description of 'standard' conditions in the solar outer atmosphere, it is important to realize that much of the dynamics relevant to the mass, energy, and momentum balance of the atmosphere as a whole may occur in (possibly isolated) regions which can sharply depart from these conditions. For example, the local value of β may become large in regions of rapid magnetic flux reconnection (because B/n may vanish in singular layers), and may be $O(1)$ as the solar corona merges into the interplanetary medium; in the photosphere, $\beta \approx O(1)$ is obtained (for typical values of $B \approx 1500$ G) at a temperature of approximately 5800 K, corresponding to an optical depth $\tau(5000 \text{ Å})$ of approximately 0.4 (Spruit, 1977). In this context it is useful to note that equipartition between the magnetic field and the kinetic energy density for a field strength of 1500 G does not occur until $3 \sim 1 \times 10^8$ cm below the surface, where the gas pressure has risen to the point that $\beta \cong 50$; this is to remind us that β is not a universally meaningful measure of the relative importance of the magnetic field in determining the dynamical behavior of the fluid. For example, although $\beta > 1$ in the interplanetary medium, local momentum balance is in fact dominated by flow accelera-tion (because the flow is both supersonic and super-Alfvénic); thus, estimates of β say nothing about the magnitude of the ratio of total kinetic to magnetic energy density. It is furthermore noteworthy that appeal to equipartition between, for example, kinetic and magnetic energy densities (as is at times done in the context of the interaction between magnetic fields and turbulent fluids) is not a reliable guide to estimates of local magnetic field strengths; even in the simplest case, the extent of local field amplification by fluid motions depends upon the balance between the dissipation of energy by field diffusion and the rate of work performed by the fluid in concentrating the fields (for details in the context of solar and stellar photospheres, see Galloway and Weiss (1981) and references therein).

2.2. THE 'IDEAL' PROBLEM

Because the classical fluid and magnetic Reynolds numbers in the solar atmosphere are extremely large, it has been common practice to regard the limit in which the resistivity and viscosity are set to zero ('ideal' MHD) as an appropriate lowest order description of the atmosphere; in the case of the corona, the large thermal conductivity at high ($\gtrsim 10^6$ K) temperatures has been furthermore invoked in order to assume (to first approximation) isothermality. A natural question is then whether the results of the 'ideal' problem can be regarded as reliable guides to an understanding of the (more realistic) nonideal equations, and thence of the actual physical system.

Two very distinct issues are involved. On the one hand, the power (and attraction) of the dissipationless (vanishing diffusive transport) limit derives from the fact that analytic calculations can be carried through, and therefore detailed analytic study of highly nonlinear phenomena in an area in which our physical intuition is very poorly developed can be carried through (see Section 3; also Moffatt, 1978, and Parker, 1979). Indeed, much of our current understanding of, for example, magnetostatic equilibria (Section 3) and MHD wave propagation (Section 4) is based on studies carried out in the ideal MHD limit; and the development of reliable numerical schemes for simulations could hardly have been carried out in the absence of benchmark analytic solutions to well-defined test problems with which simulations can be compared.

On the other hand, the crucial limitation of the diffusionless regime as a reliable guide to observed phenomenology lies in the poor definition of the limits of applicability of the theory itself. Because of its explanatory power, the practitioner of, for example, ideal MHD may be readily tempted to generalize results to domains in which nonideal effects can enter; is it then possible to estimate the importance of such departures from ideal behavior? There is at present no general answer to this question, but some formal studies have been carried out to resolve the issue, in particular in the context of turbulent diffusion and the behavior of singular layers.

One aspect of the problem can be appreciated by considering the simplest case, in which we begin with the classic induction equation

$$\left[\frac{\partial}{\partial t} - \eta \nabla^2 \right] \mathbf{B}(\mathbf{r},\, t) = \text{curl } \mathbf{v} \times \mathbf{B} \tag{2.5}$$

in the kinematic limit, in which the flow $\mathbf{v}(\mathbf{r},\, t)$ is specified *a priori*; η is the (assumed to be spatially constant) magnetic diffusivity. If the flow $\mathbf{v}$ is turbulent, and if its statistical properties are specified (for example, by observations), then one may ask whether the solution of Equation (2.5) depends to any significant extent upon η; in particular, are the solutions to the simpler ideal equation

$$\frac{\partial}{\partial t} \mathbf{B}(\mathbf{r},\, t) = \text{curl } \mathbf{v} \times \mathbf{B} \tag{2.6}$$

qualitatively different from those obtained from (2.5) in the limit $\eta \to 0$? The remarkable fact is that there is no straightforward answer to this question. The formal theoretical techniques developed for the solution of Equation (2.6) [see review by Van Kampen,

1976; also Knobloch, 1980] cannot be applied to Equation (2.5); and, indeed the asymptotic behavior ($t \to \infty$) obtaining by applying the limit $\eta \to 0$ depends critically on the ordering of the applied limits (cf. Moffatt, 1978, §7.7). The situation, which obtains in the kinematic limit, is not likely to be ameliorated if one relaxes the kinematic assumption, so that one also solves dynamically for $\mathbf{v}$.

Suppose then that we regard Equation (2.5) as a *deterministic* problem; that is, let us consider the possibility that the above difficulty is due solely to the stochastic nature of the problem as posed above, and therefore let us view $\mathbf{v}(\mathbf{r}, t)$ and $\mathbf{B}(\mathbf{r}, t)$ as functions of space and time to be specified (or calculated) without spatial or ensemble averaging. For example, we might consider the evolution of photospheric magnetic fields and their extension into the chromosphere and corona when subjected to photospheric flow pattern which may be specified *a priori* or calculated self-consistently. This problem has been extensively studied in both regimes; both the kinematic calculation in which the back-reaction of the field on the perturbing flow is not taken into account, and the fields 'deform' passively (Heyvaerts, 1974; Jockers, 1978; Sakurai and Levine, 1981). and the fully dynamic problem, in which full account is taken of the Lorentz force back-reaction (Weiss, 1966; Parker, 1979; Proctor and Weiss, 1982; Nordlund, 1982) suggest that if the flow correlation time-scale is short when compared with the (initial) diffusion time-scale, then the solutions of Equations (2.5) and (2.6) are qualitatively similar almost everywhere (this fact indeed provides some of the motivation for ideal studies of the sort discussed in Section 3 below). Differences between these solutions arise primarily because of the generation of singular layers, that is, regions where gradients in $\mathbf{B}$ may become locally large; in such regions of large field gradients, $\eta \nabla^2 \mathbf{B}$ can be large no matter how small η.

In essence, (2.5) can then be regarded as a singular perturbation problem, in which the order of the governing differential equation changes in certain boundary (= singular) layers. As a consequence, the global response of the system may be significantly affected, as was first recognized by Furth *et al.* (1963) and Rosenbluth *et al.* (1966): for example, the very character of instabilities in such boundary layers can depend crucially on whether or not η vanishes *identically*; thus, certain classes of instabilities occur precisely *because* the magnetic diffusivity does *not* vanish identically. For these reasons, proper inclusion of diffusive effects in MHD problems such as numerical simulations in which singular regions may appear is a challenging task; to be convincing, such simulations must be able to demonstrate than the $\eta = 0$ solutions are in fact the proper limiting solutions for finite η, as $\eta \to 0$.

2.3. VISCOSITY

For reasons similar to those which motivate studies of ideal MHD (Section 2.2 above), it is common to assume in solar fluid dynamics that viscous forces are unimportant. Using the standard expression for the dynamic viscosity $\mu_\| \cong 2 \times 10^{-15} T^{5/2} A_i^{1/2}/Z^4$ $\ln \Lambda$ (cf. Lang, 1980), it is easy to show that viscous force/Lorentz force $\approx 1.3 \times 10^{-6} B_2^{-2} T_6^{5/2} v_5 L_7^{-1} (\ln \Lambda/20)^{-1}$, and viscous force/pressure gradient $\approx 3.6 \times 10^{-3}$ $n_9^{-1} T_6^{3/2} v_5 L_7^{-1} (\ln \Lambda/20)^{-1}$, where L is a typical scale length and the other plasma parameters take on their usual meanings. (Note that $\mu_\|$ refers to viscous momentum transport parallel to the magnetic field.) Thus, for motions on scales equal to or larger than can be presently resolved from ground-based observations (i.e. $\gtrsim 1$ arcsec), viscous forces are relatively unimportant (cf. Tandberg-Hanssen, 1967; his equations 1.31 and 1.33a). However, viscosity not only acts to transfer momentum (and so plays a role in the Navier–Stokes equation), but also serves to thermalize macroscopic motions (and so appears as a heating term in the energy conservation equation); in this latter role, viscosity may play a very important role; for example, it is commonly invoked for damping of various wave modes (see below).

Viscosity can also play an important role in momentum transport if the velocity gradient scale is sufficiently small. This can commonly occur when $\mathbf{B} \neq 0$; in that case, since the electron gyroradius λ_{ge} can be (and usually is) substantially smaller than the mean free path λ_{mfp}, fluid behavior on spatial scales well below the mean free path can occur. In particular, consider a magnetic configuration in which magnetic surfaces are well defined: local perturbations which lie in these magnetic surfaces can then result in shear Alfvén waves which may propagate along 'adjacent' flux surfaces. Such waves can interact when the flux surface separation is of the order of a few ion gyroradii, hence, one can obtain extremely large velocity gradients

$$|\nabla v| \sim [B/(4\pi n m)^{1/2}]/\lambda_{gi} \approx 2.31 \times 10^6 \, n_8^{-1/2} \, T_6^{-1/2} \, B_1^2 , \qquad (2.7)$$

so that viscous effects can become extremely important (Ionson, 1978; Hollweg, 1981; Heyvaerts and Priest, 1983), and may require – because of the smallness of scale – a kinetic rather than a fluid treatment. Thus, heating due to viscous damping can indeed compete with other thermalization processes for organized motions in the solar outer atmosphere; for example, in the recent study of shear Alfvén wave propagation by Heyvaerts and Priest (1983), wave dissipation occurs both because of shear (Kelvin–Helmholtz) instabilities as well as tearing mode instability.

2.4. PARALLEL THERMAL ENERGY TRANSPORT

Thermal conduction plays a central role in our understanding of the structure of the solar outer atmosphere. Because of the relatively high temperatures encountered, thermal conduction is an efficient energy transport mechanism, and because the ratio of the electron gyroradius to the mean free path is very small, such thermal energy transport largely occurs along magnetic field lines (Spitzer, 1962). In the fully collisional regime ($\nu_{ei} \gg$ other frequencies; $\lambda_{mfp} \ll L_T \equiv T/|\nabla T|$, $L_n \equiv n/|\nabla n|$), one writes

$$F_c = -\kappa_{\parallel} \nabla_{\parallel} T \, \text{erg cm}^{-2} \, \text{s}^{-1} , \qquad (2.8)$$

where $\kappa_{\parallel}$ is the Spitzer–Harm conductivity. A simple order-of-magnitude estimate of the conductive flux is given by $F_c \sim 10^7 \, (\lambda_{mfp}/L_T) n_8 \, T_6^{3/2} \, \text{erg cm}^{-2} \, \text{s}^{-1}$, essentially carried entirely by electrons because of their high mobility. For a fully-ionized plasma, the exact expression then depends weakly on the density (through the Coulomb logarithm) and the mean Z of the plasma, and in simplified form is tabulated by Spitzer (1962); a more recent tabulation, which furthermore includes effects of partial ionization and allows for various possible elemental abundance conditions, is given by Nowak and Ulmschneider (1977).

Now, as discussed in the classic papers by Spitzer and Harm (1953) and Braginskii (1965), the above results depend crucially on the assumption that the electron velocity distribution is locally only slightly skewed from Maxwellian; in particular, the strong inequalities $\lambda_{mfp} \ll L_T, L_n$ must be obeyed. The obvious question is then how large λ_{mfp} can be without leading to a significant departure from the Spitzer–Harm results: one would like to know, first, whether the thermal conductive flux is bounded from above as $\lambda_{mfp}/\lambda_T \to 1$; second, what the functional form of the modified thermal conductivity is, so that the temperature variation in space can be calculated; and third, what the nature of the departure from Maxwellian form is, so as to be able to assess the importance of enhanced 'tails' of the electron velocity distribution function.

In the astrophysical domain, such questions have generally been answered by appealing to microinstabilities. That is, one asks whether, upon evaluating the Spitzer conductive flux F_c, the relative drift speed of the electrons

$$v_D \cong |F_c|/n_e m_e v_{te}^2 \tag{2.9}$$

exceeds some instability threshold (see, e.g., Coppi, 1976); if so, one then determines the character of the relevant instability and, most crucially, the resulting effective scattering rate (or effective diffusion rate) of thermal electrons resulting from the microinstability (cf. Forslund, 1970; Manheimer, 1977, and references therein). An important aspect of this approach is that the enhanced effective scattering rate for electrons is usually calculated at the stage of marginal stability, so that the departures of local velocity distributions from Maxwellian form can be determined in a straightforward way. This approach has in particular characterized studies of heat flux saturation during solar flares and related transients (cf. Sturrock, 1980; Spicer, 1979), and has been particularly studied in the interstellar medium context (cf. Cowie and McKee, 1977). Are these calculations relevant to the 'quiescent' solar atmosphere?

The first question is whether the heat flux calculated on the basis of the Spitzer–Harm conductivity ever approaches the presumptive upper bound on the heat flux, the so-called free-streaming value (Parker, 1964a, b; Cowie and McKee, 1977, and references therein),

$$F_{\text{sat}} \cong 6.2 \times 10^4 \, pT^{1/2} \text{ erg cm}^{-2} \text{ s}^{-1}, \tag{2.10}$$

obtained by assuming that the heat flux-carrying electrons all stream down the temperature gradient at the local electron thermal speed. Using simple analytic energy balance models for the transition region, in which the (observed) local radiative losses are entirely balanced by the divergence of the conductive flux (this places an extreme upper bound on the temperature gradient), one can show that the maximum heat flux is given by

$$F_{\text{max}} \cong 5 \times 10^2 \, pT^{1/2} \text{ erg cm}^{-2} \text{ s}^{-1}, \tag{2.11}$$

so that $F_{\text{max}} \ll F_{\text{sat}}$ for typical transition region temperatures (cf. Rosner et al., 1984). However, one can easily show from Equation (2.2) that the mean free path for *thermal* electrons is not much smaller than the temperature scale length L_T. Thus, for a typical heat flux F_c of 10^7 erg cm^{-2} s^{-1} (cf. Withbroe and Noyes, 1977), we have

$$\lambda_{mfp}/L_T \propto n^{-1} \, T^{3/2} F_c \approx 0.1 n_9^{-1} \, T_6^{-3/2} \, (F_c/10^7), \tag{2.12}$$

where L_T is the temperature gradient scale $T/|\nabla T|$; thus, at active region temperatures of 2.5×10^6 K or larger, thermal coronal electrons may penetrate fairly deeply into lower atmospheric layers (see Roussel-Dupre, 1980; Shoub, 1983b). Hence, whereas heat flux saturation by microinstabilities is *not* likely to be relevant to studies of the quiescent solar atmosphere, one ought to be somewhat concerned by the relatively large electron penetration depth, and its implications for locally determined thermal equilibrium and ionization balance (which is sensitive to precisely that part of the electron velocity distribution − the high energy tail − which can penetrate furthest into the cooler, denser lower atmosphere).

This latter problem has recently received much attention in the laboratory. In particular, studies of thermal conduction in ablation experiments using particle or photon beams (as, for example, used in laser fusion research; see review by Max, 1981) have raised the strong possibility that thermal conduction may depart strongly from 'classical' Spitzer form even within the context of standard transport theory (i.e. without the necessity of invoking excitation of microinstabilities); this may occur because the local electron distribution function can depart significantly from a local displaced Maxwellian (Parker, 1964b; a good recent review is given in Max, 1981). That such difficulties can arise is seen immediately if one recalls that the heat flux is obtained from the third moment of the electron velocity distribution, so that somewhat suprathermal electrons carry most of the heat flux; and that because of the very sensitive dependence of the ion−electron collisional cross-section on velocity for suprathermal electrons, the effective mean free path of such electrons is far larger than the thermal mean free path (and hence, by comparing with Equation (2.12), may easily become larger than the temperature scale length L_T in the solar atmosphere). These laboratory studies show that the Spitzer−Harm results do not give the proper temperature dependence of the heat flux for even fairly modest temperature gradients (in the sense that the Spitzer−Harm conductivity somewhat overestimates the heat flux (see below; also Max, 1981)). We are therefore led to ask: (1) At what point do the Spitzer−Harm (or Braginskii) asymptotic expansions in λ_{mfp}/L_T break down? (2) What is the appropriate conductivity in the regime of 'large' λ_{mfp}/L_T? (3) To what extent is the heat flux carried by very hot electrons, for which the thermal boundary layer (transition region) may be entirely collisionless? In this context, it should be noted that the accuracy of the 'classical' heat transport calculations depends entirely on what one wishes to calculate: for example, in the transition region below the corona, the errors incurred in calculating the divergence of the heat flux may be relatively unimportant when compared with the errors in data interpretation which ignores the shifts in ionizational equilibrium due to enhanced high-energy electron velocity distribution tails (Shoub, 1983a). This important point is discussed separately below.

Most of the available theoretical studies which address these questions have been carried out in the context of laboratory observations, although recent studies by Shoub (1983a) and Scudder and collaborators (Scudder and Olbert, 1979a, b) have now also raised similar issues in the context of the solar transition region and solar wind. The principal results may be summarized as follows: the standard Spitzer−Harm prescription begins to fail when λ_{mfp}/L_T is of order 0.015 or larger (Gray and Kilkenny, 1980). The reason is simply that these 'classical' conduction models are based on a linearization of the velocity distribution function $f(x, v, \theta)$

$$f(x, v, \theta) \approx f_0(x, v) + f_1(x, v) \cos \theta$$

(with f_0 the unperturbed Maxwellian, $f_1 \cos \theta$ the perturbation, and θ the angle between the particle motion and the direction of the temperature gradient), such that when $\lambda_{mfp}/L_T \gtrsim 0.015$, $f_1 \sim f_0$, i.e. the linearization fails. For example, we find for typical active region structures of scale length $\sim 10^9$ cm

$$\lambda_{mfp}/L_T \cong 0.044(n/10^9 \text{ cm}^{-3})^{-1} (T/2 \times 10^6 \text{ K})^2 (\ln \Lambda/20)^{-1} \, l_9^{-1} > 0.015, \quad (2.13)$$

where l_9 is the typical loop length in units of 10^9 cm. Hence, the 'classical' expression for $\kappa_\parallel$ given by Spitzer (1962) or Braginskii (1965) fails to accurately describe thermal conduction *anywhere* within such loops (much less within their transition regions); in 'quiet' Sun structures, for which $l_9 > 10$, this difficulty probably does not arise. This discrepancy is generally significant, but not extreme: for example, the results of analyses of laboratory experiments by Matte and Virmont (1982) may be fitted fairly well ($r^2 = 0.91$) by a simple power law expression of the form

$$F_c/F_{\text{classical}} \approx 0.11 \left(\frac{\lambda_{mfp}}{L_T} \right)^{-0.36} ; \quad (2.14)$$

similar corrections to the classical heat flux have also recently been calculated by Bell *et al.* (1981). Evidently, for thermal mean free paths of order 1/30 of the temperature gradient scale length or larger, the classical heat flux expression overestimates the true heat flux by a least a factor of 2, reaching a discrepancy of a factor of 10 when the mean free path and temperature scale length are equal. It should be noted that laboratory data suggest that the actual heat flux is depressed further than Equation (2.14) would predict (by a factor of up to 30, rather than 10); presumably, the implication is that at some point, increased electron scattering by microinstabilities must also be taken into account in the 'corrected' classical calculation.

A further issue in understanding the conductive contribution to local energy balance is to what extent the conduction problem is truly local. The energy balance relation is usually written in the form of a local conservation law, i.e. as a differential equation in which divergences of the various energy fluxes appear; this can be done as long as distant points (i.e. points separated by distances large when compared to the typical scale lengths characterizing the thermodynamic (or fluid) variables such as the temperature, density, or velocity) cannot communicate *directly*. The classic example where this local approximation fails is, of course, in radiative transfer theory, where indeed integrodifferential equations must be used to describe the propagation of photons in a medium of moderate optical depth. In the present case, the difficulty arises because of the presence of the 'tail' electrons, whose relatively large mean free paths can turn the thermal conduction problem into a nonlocal process. In this case, laboratory studies are not very relevant to the solar problem because the overall geometric relation between hot and cold plasma now becomes important (thus, quite unlike the solar case, the mean free path of hot electrons in laser fusion studies may become large when compared to the dimensions of the cold target itself). The most recent calculations relevant to the Sun are those of Shoub (1983a), which indeed show that (depending upon the actual value of λ_{mfp}/L_T) the tails of local electron distribution functions at the base of transition regions can depart strongly from Maxwellian form for electrons with energies 10 times thermal or more; for example, for a transition region pressure of 0.167 dyn cm^{-2} and a coronal temperature of 2×10^6 K, Shoub finds that the relative enhancement of the tail population (at, say, an electron energy of 20 times thermal) can vary from a factor of 10 at

10^6 K to over 100 at 10^5 K. Some of the implications for modeling the temperature and density structure of the solar outer atmosphere are discussed below; for present purposes the key point is that the energy balance equation is no longer local. That is, much as in radiative transfer theory, local energy gains and losses (viz. conduction) depend upon the plasma conditions elsewhere, so that the formal problem to be solved is an integrodifferential equation; the solution to the energy balance equation thus depends on the global conditions of the atmosphere. This very difficult problem has, to our knowledge, yet to be attacked self-consistently in the astrophysical context.

2.5. PERPENDICULAR TRANSPORT

A general discussion of mass and thermal energy transport across magnetic field lines in the high-conductivity limit is well beyond our scope; we refer the reader to the thorough review by Hinton and Hazeltine (1976), which deals with the general problem of transport processes in inhomogeneous, magnetized plasmas (albeit in the laboratory plasma context). Instead, we shall focus on the more restrictive problem of accounting for the observational suggestions that considerable cross-field transport occurs in the solar outer atmosphere (cf. Foukal, 1978); the issue is to reconcile such transport with the fact that $\lambda_{gi}/\lambda_{mfp} \ll 1$ (so that 'classical' diffusive transport across the magnetic field ought to be negligible).

Consider first the problem of transferring mass across magnetic field lines. The relative motion of matter and magnetic fields is, in the simplest case, governed by a diffusive slip equation (Parker, 1979); clearly, in the absence of externally — applied electric fields, relative diffusion of matter and fields can only occur *because* of some effective scattering process. Since $\omega_g/\nu_{ei} \gg 1$, classical transport (i.e. electron–ion collsions) cannot be called upon; and so one commonly invokes some 'anomalous' collisional process, e.g. scattering by electrostatic modes (see Tucker, 1973; Ionson, 1978 and references therin). As pointed out by Rosner *et al.* (1978), this leads to the difficulty that significant levels of plasma microturbulence must occur throughout the domain of enhanced cross-field mass transport, so that the driving mechanism underlying the microinstability (viz. almost field-aligned current flow) must act everywhere as well and, similarly, the invariably associated gradient in some physical attribute (viz. magnetic field) must be large everywhere in the region of anomalous transport.

Similar considerations apply to heat transport perpendicular to the magnetic field: in contrast to parallel heat transport, in which case scattering of the heat flux-carrying particles *inhibits* transport, perpendicular heat transport occurs precisely *because* of the scattering of heat flux-carrying particles. Significant perpendicular heat transport thus requires some means of enhancing the effective scattering rate of the heat-carrying particles throughout the region of enhanced transport; and, as just pointed out, it is not obvious how one is to arrange such broad spatial extent of the microturbulent region.

This difficulty raises the intriguing question of whether diffusive transport is really required by the data. In the laboratory domain, similar questions regarding diffusive (radial) transport in toroidal geometry have arisen, along with similar difficulties: it does not seem possible to provide sufficiently enhanced scattering levels throughout the apparent 'diffusion' region. One possibility, which is explored in great depth in Hinton and Hazeltine (1976), is that magnetic field inhomogeneities allow for convective cross-field transport; relatively little has been done in this regard in the solar context (see, however, Ionson, 1978), and we shall not discuss this possibility any further. Instead, we wish to consider the basis for assuming that cross-field transport is occurring: we recall that the very notion of cross-field transport carries with it the implicit assumption that the magnetic field is almost everywhere locally well-ordered; that is, one assumes that, locally, one can sensibly define appropriate level surfaces (which may be identified, for example, with the level surfaces of the Euler potentials) within which field lines lie and across which some heat transport can occur. This assumption may, however, be unwarranted. One possibility which has been raised is that the notion of well-defined magnetic field flux surfaces is restricted to only very special cases of exceptional symmetry (cf. Rechester and Rosenbluth, 1978; see Cary and Littlejohn, 1983, and Tsinganos et al., 1984, for recent discussions and references, as well as Section 3 below). In this case, one argues that arbitrary magnetic field configurations (including those represented by the actual, constantly — perturbed configurations observed at the solar surface) are characterized by field lines which are ergodic within the structure in question, so that well-defined 'flux surfaces' may not exist, and 'cross-field transport' may not be meaningful. In that case, heat transport in directions orthogonal to the local mean orientation of the magnetic field (which may be inferred from observations) may be understood as the consequence of the conjunction of efficient transport along the magnetic field, and a random displacement of the field lines orthogonal to their mean orientation (cf. Colgate, 1978). Essential to this process is the fact that, in the stochastic regime, field lines separate exponentially on the scale of the electron gyroradius (Rechester and Rosenbluth, 1978). That is, consider the motion of a set of particle guiding centers in phase space; we assume that these start with approximately identical initial conditions, and so occupy a compact neighborhood in phase space. In the collisionless regime, the evolution of these orbits is volume-preserving (because as long as the drift effects discussed by Hinton and Hazeltine (1976) are neglected, orbits follow the magnetic field, and div $\mathbf{B} = 0$); however, in the stochastic regime, orbits with almost identical initial conditions rapidly separate. The result is that the compact initial phase space volume is rapidly sheared out: it diffuses into phase space, but since the evolution is volume-preserving, the initial compact set becomes 'stringy' (for some simple examples, including the classic 'Arnold's cat' transformation, see Arnold and Avez, 1968); this is the essence of the irreversibility of the evolution on the macroscopic (gyroradius and larger) scale — we obtain 'diffusion'. Thus, the difficulty of accounting for enhanced collisional transport throughout a loop volume is eliminated, but in its stead we must now confront the processes which underlie such ergodic field line behavior. Future very high resolution imaging of solar coronal structures might be able to shed some light on the problem: at the very least, we would like to address the observational question of whether 'cross-field' transport is a diffusive or convective process.

2.6. SOME COMMENTS ON MODEL BUILDING

In addition to the obvious impact the preceding has on discussions of the physics of transport processes in the solar outer atmosphere, the recent results of 'classical' thermal transport theory significantly affect the interpretation of commonly used solar diagnostics. Best known (and perhaps most commonly used) are emission measure analyses of the transition region based upon UV emission lines (cf. Withbroe, 1975, and references therein). This diagnostic tool is fundamentally based on the fact that in local thermodynamic equilibrium, and in the fully collisional regime, the line intensity I_i scales simply with the square of the local density; for present purposes, it suffices to write

$$I_i \sim \int dV g G_i(T) \cdot n^2 \approx \langle g G_i(T) \rangle_T \cdot \int dV n^2 (T_*),\qquad(2.15)$$

where $g G_i(T)$ contains the temperature-dependent atomic physics for the transition in question, and is averaged over the range of temperature of line formation; the density is evaluated at the temperature T_* of maximum abundance of the ion responsible for the line in question; and the volume integral is carried out over the entire (possibly unresolved) source region in which the ion in question is formed. The differential emission measure (which is a measure of the amount of material lying within the temperature range of line formation) can be rewritten simply by noting that, in the limit of a monotonic dependence of the plasma temperature on the line-of-sight coordinate,

$$n^2 \, dV \approx [n^2/(dT/ds)] \, dT \, dA,\qquad(2.16)$$

where s is the line-of-sight coordinate, and dA is the resolution element area. Using the expression for the heat flux given above, we obtain

$$n^2 (dT/ds)^{-1} \propto p^2 \, T^{1/2} F_c^{-1},\qquad(2.17)$$

so that Equation (2.15) now reads

$$I_i \approx \langle g G_i(T_*) \rangle \cdot p^2 \cdot \kappa T_*^{1/2}/F_c.\qquad(2.18)$$

Thus, both the line intensity and the differential emission measure are strongly affected by the functional form of the heat flux. This is indeed a difficulty because the results of the numerical simulations and laboratory observations discussed above show that the 'classical' conductivity tends to overestimate the actual heat flux in transition region boundary layers by factors of 2–10 (that is, for fixed temperature and temperature gradient, we can use Equation (2.14) to estimate F_{actual}); and, perhaps even more important, that the local electron distribution is not fully determined by local conditions, particularly in the 'tail' of the distribution.

There are two immediate consequences. The first (more obvious) difficulty is that the derived semi-empirical temperature stratification cannot easily be used to determine the local contribution of the thermal conductive flux to local energy balance; that is, the local conductive flux, and its divergence, cannot be derived from local measurements of the temperature gradient alone, but instead depend on the temperature profile as a whole. More insidiously, the enhanced electron tails can significantly affect ionization balance in the transition region (Shoub, 1983a): in particular, the lower bound on temperatures of formation of specific ions can be shifted to much lower values, so that the associated emission lines are no longer good diagnostics for plasma in narrow temperature ranges, and the classic inversion techniques for deriving differential emission measures from spectral data become very ill-conditioned (see Craig and Brown, 1976). This problem is particularly acute for low-pressure regions, i.e. the quiet Sun, because the penetration of suprathermal electrons is very sensitive to the mean free path, and hence to the density. Unfortunately, these regions have been traditionally considered least affected by spatial resolution effects (as occur in modeling transition layers of active region loops), and hence most reliable from the point of view of constructing semi-empirical atmospheres. Thus, it seems that the construction of semi-empirical model atmospheres may not be separated from the problem of deriving the thermal conductive flux: one requires estimates of the departure from Maxwellian behavior in order to calculate the ionization structure, and thence the differential emission measure profile with temperature.

3. Magnetohydrodynamic Processes in the Corona

Magnetohydrodynamics describes a plasma in terms of its fluid-like properties. It is more a physical model than a fundamental theory in that it describes an idealized plasma. Not all properties of realistic plasmas, both in the astrophysical environment and in the laboratory, can be explained in terms of this model. Indeed, the study of 'anomalous' effects, associated with departure from fluid behavior, is a major concern of plasma physics. Notwithstanding this limitation, MHD is an important, useful tool. It is a reliable theory if we are dealing with phenomena with large spatial scales and low frequencies. Such phenomena are often the ones of immediate first interest in the solar corona. The set of MHD equations forms a self-contained self-consistent system, leaving no ambiguity as to what are the MHD properties. This is a useful quality for a lowest order description.

In this section, our current theoretical understanding of coronal MHD processes is reviewed. Emphasis will be given on understanding the basic physics, as motivated and illustrated by phenomena observed in the corona, and we will have occasion to point out the limitations of the MHD description. The theoretical developments surveyed will be related to those of fusion plasmas (see the recent review of Freidberg, 1982), pointing out the similarities and differences of the physical problems encountered. To keep the physical discussion simple, only the ideal, or nondissipative, one-fluid model (see Section 2.2 above) will be considered. The ideal MHD equations are:

$$\rho \, \frac{d\mathbf{v}}{dt} = \frac{1}{c} \, \mathbf{J} \times \mathbf{B} - \nabla p - \rho \nabla \phi, \tag{3.1}$$

$$\nabla \times \mathbf{B} = \frac{4\pi}{c} \, \mathbf{J}, \tag{3.2}$$

$$\nabla \cdot \mathbf{B} = 0, \tag{3.3}$$

$$\nabla \times \mathbf{E} = -\frac{1}{c} \, \frac{\partial \mathbf{B}}{\partial t}, \tag{3.4}$$

$$\mathbf{E} + \frac{1}{c} \, \mathbf{v} \times \mathbf{B} = 0, \tag{3.5}$$

$$\frac{\partial \rho}{\partial t} + \nabla \cdot (\rho \, \mathbf{v}) = 0, \tag{3.6}$$

$$\frac{\mathrm{d}}{\mathrm{d}t} (p \, \rho^{-\gamma}) = 0, \tag{3.7}$$

where p, ρ, $\mathbf{v}$, $\mathbf{B}$, $\mathbf{E}$, $\mathbf{J}$, and ϕ are, respectively, pressure, density, velocity, magnetic field electric field, electric current density, and the externally imposed gravitational potential. The constants γ and c are the adiabatic index and the speed of light, respectively. We have also introduced the Lagrangian derivative $\mathrm{d}/\mathrm{d}t = \partial/\partial t + (\mathbf{v} \cdot \nabla)$. We begin by discussing static ($\partial/\partial t = 0$, $\mathbf{v} = 0$) and stationary ($\partial/\partial t = 0$, $\mathbf{v} \neq 0$) models for quasisteady structures in the corona (Section 3.1). In Section 3.2, we discuss the question of linear stability of equilibrium magnetic fields. What constitutes the necessary conditions for an equilibrium state to exist? This question is quite separate from whether a given, existing, equilibrium state is stable, and we also take it up in Section 3.2. Time-dependent phenomena are discussed in Section 3.3. We conclude this section with an overview in Section 3.4.

3.1. EQUILIBRIUM MAGNETIC FIELDS

Plasma structures of dimensions $10^3 - 10^6$ km are often observed to be quasisteady in the corona, with lifetimes of hours to weeks. These structures are quasisteady in the sense that their lifetimes are much larger than their characteristic time-scales for fully developed dynamical evolution. Quiescent prominences and their associated helmet-streamer structures are well-known examples (Newkirk, 1967; Tandberg-Hanssen, 1974). X-ray loops, first observed during the OSO-4 and Skylab missions, are another class of examples (Vaiana et al., 1968, 1976). The predominance of loops in the X-ray corona has led to the point of view that they are the constituent elements making up a highly inhomogeneous component of the corona (Vaiana and Rosner, 1978). The simplest theoretical model for these long-lived objects is based on the assumption of static equi-librium. This is only a first approximation for the spatial and time-scale of interest. For instance, on a small enough spatial scale, motion can be detected in a quiescent prominence although the large-scale structure remains static in appearance. Setting the time derivatives and the velocity to zero in the ideal MHD equations, static equilibrium is described by

$$\frac{1}{4\pi} (\nabla \times \mathbf{B}) \times \mathbf{B} - \nabla p - \rho g \hat{z} = 0, \tag{3.8}$$

where the Lorentz force has been written in terms of the magnetic field alone and, for simplicity, we take a uniform gravity of acceleration g in the $-\hat{z}$ direction. We shall review recent progress made in investigating Equation (3.8) quantitatively, to suggest on theoretical grounds the sort of magnetostatic states that are admissible in the corona.

The magnetostatic problems for the corona differ in two ways from those encountered with laboratory plasmas. Coronal structures often can be of such large dimensions that gravity plays an important role in determining the equilibrium plasma structure. If we resolve the force-balance equation (Equation (3.8)) along the magnetic field line, the Lorentz force has no component and we obtain

$$\mathbf{B} \cdot (\nabla p + \rho g \hat{z}) = 0, \tag{3.9}$$

showing the hydrostatic support of the plasma weight by the pressure gradient in this direction. Solutions of Equation (3.9) show the hydrostatic scale height $h = kT/mg$ to be the crucial parameter: if $h/L \gg 1$, L being a typical scale along $\mathbf{B}$, the pressure is approximately constant and gravity is unimportant. Hence, gravity may be neglected for a compact hot coronal structure to first order as in the description of laboratory equilibrium states. Setting $g = 0$ leads to the well-known condition that the equilibrium pressure is constant on each magnetic field line. Coronal and laboratory plasmas also differ in the kind of boundary conditions applicable to their respective environments. Whereas laboratory plasmas are controlled, isolated systems created and engineered towards designated goals, naturally occurring coronal plasmas have no rigid boundaries. Much of the interesting physics of a coronal plasma arises from its interaction with the natural environment which includes the cool dense photosphere, and part of the difficulty is in properly translating these physical interactions into appropriate boundary conditions (see Section 3.2).

The magnetic field plays a central role in coronal physics. Unfortunately, it cannot be readily measured, except in the case of the field in the photosphere. The line-of-sight component of the photospheric field is routinely measured by the use of the Zeeman effect. The transverse component of the photospheric field can also be measured, but with less confidence (Harvey, 1977; Stenflo, 1978; Hagyard et al., 1982). It was thought that one could meaningfully extrapolate the measured photospheric field to the corona from these limited observational data, using suitable theoretical models. The simplest procedure was to assume no coronal current, so that the corona was filled with the potential field generated by the photosphere (e.g. Schmidt, 1964; Altschuler and Newkirk, 1969; Levine, 1975). This model was extended to allow for coronal currents that flow parallel to the magnetic field such that $\nabla \times \mathbf{B} = \alpha \mathbf{B}$ for a constant α, i.e. the case of the linear force-free field (e.g. Nakagawa and Raadu, 1972; Seehafer, 1978). The computed magnetic field configurations of these models have been compared with observed plasma structures based on the assumption that the latter outline the actual coronal magnetic field lines. Cases of both good and bad agreements between model and observed plasma structures have been reported. Two difficulties prevent clear physical conclusions to be drawn from these models. First, the comparison between model and data is based on an assumption that cannot be justified physically. Second, there is little physical ground to expect that a given magnetic field should be force free (or potential) of the linear type. The intense magnetic field in the active region is probably approximately force-free (Gold, 1964). However, physical considerations show that any realistic twisting of a magnetic field is likely to produce a nonlinear force-free field with a nonconstant α (Sakurai, 1979; Low, 1982a). Once the simplicity of linearity is lost, it is not known how the magnetic field is to be determined in a general way from knowledge of its boundary value (Barnes and Sturrock, 1972; Low, 1982a). The problem becomes even more complex if we take it a step closer to reality by allowing the Lorentz force in Equation (3.8) not to be zero. The magnetic field and plasma are now coupled nonlinearly and have to be treated self-consistently. The formulation and solution of these problems for realistic situations is not feasible at present. A more modest goal than that envisioned by the early linear models takes priority. Effort has been concentrated on studying idealized magnetostatic states, both to learn the basic properties contained in Equation (3.8) and to develop physical intuition for understanding the more complicated structures one expects to find in the corona.

If the magnetic field is idealized to be invariant in the x-direction, it can be expressed in terms of two scalars, A and B_x:

$$\mathbf{B} = \nabla \times (A\hat{x}) + B_x \hat{x} \tag{3.10}$$

To keep the discussion simple, let us consider the case $B_x = 0$. The electric current density is then

$$\mathbf{J} = -\frac{c}{4\pi} \nabla^2 A\hat{x}. \tag{3.11}$$

Equation (3.8) reduces to

$$p(y, z) = P(A) \exp\left[-\int_0^z \frac{mg\, dz'}{kT(A, z')}\right], \tag{3.12}$$

$$\nabla^2 A + 4\pi \frac{\partial p(z, A)}{\partial A} = 0, \tag{3.13}$$

where we have used the ideal gas law,

$$p = \rho k T/m, \tag{3.14}$$

with T, k, and m being the temperature, the Boltzmann constant, and mean particle mass, respectively. These equations, specialized to the case of an isothermal atmosphere, were first obtained by Dungey (1953) and later extended by Low (1975a) to the form given here for a free-floating temperature distribution. The physical meaning is as follows. Magnetic field lines are curves of constant A in the $y-z$ plane. Along each field line, the Lorentz force has no component and the purely hydrostatic equilibrium described by Equation (3.9) results in the barometric relation (3.12). On each of the yet unspecified field lines, the pressure profile is fixed by the temperature distribution on that field line. Force balance perpendicular to the field, as described by Equation (3.13), then determines the field and pressure self-consistently. Although there are two independent directions perpendicular to a given field direction, only one equation of perpendicular force-balance arises because of the assumed invariance of the system in the x-direction. Note that Equation (3.13) is in fact Ampere's law (equation (3.2)) put in a form that specifies the electric current distribution (Equation (3.11)) as required for force balance. Another point to note is that if gravity is neglected, with $g = 0$, pressure is constant along field lines. In this case, Equation (3.13) is the Cartesian equivalence of the Grad–Shafranov equation which describes toroidal equilibrium laboratory plasmas (Grad and Rubin, 1958; Shafranov, 1960).

As a theoretical problem, Equations (3.12) and (3.13) attracted much interest with regard to the magnetostatics of quiescent prominences. Various explicit solutions are known, ranging from models of vertical infinite sheets to single and multiple horizontal filaments in a unbounded medium (Dungey, 1953; Kippenhahn and Schlüter, 1957; Lerche and Low, 1980a; Low *et al.*, 1983). Other models include a lower boundary to bring in the effect of the photosphere (Lerche and Low, 1980b; Low, 1981a; Van Tend and Kuperus, 1978; Zweibel and Hundhausen, 1982). These studies illustrate how pockets of plasma can be trapped and supported against gravity in local valleys of winding magnetic field lines. There is no simple rule about how magnetic field geometry and the structure of the embedding plasma are related. In general, density contours do not outline the magnetic field lines because of the presence of gravity. The same basic clump of plasma can be supported in different field geometries, as distinguished topologically by the presence of magnetic neutral points and where these neutral points are positioned (Low, 1981a; Anzer, 1979; Low *et al.*, 1984). These different topologies have dynamical implications that we have yet to explore. It is fortunate that prominence plasmas are dense and cool enough to have the right type of emissions which can be detected and analyzed for magnetic signatures. There are ongoing programs to measure the field strength and orientation in prominences, using Stokes parameter analysis and the Hanle effect (House and Smartt, 1982; Leroy *et al.*, 1984). Observational data of this type promise to be instructive in discriminating between the different theoretical model developed so far, but the observational programs are in their infancy.

The following two directions for future work would seem to be worthwhile. The first concerns the various modes of energy transport in the corona that are ignored in ideal MHD. Thermal conduction is efficient along magnetic field lines. Depending on the plasma conditions, a range of radiative mechanisms operate, ranging from the optically thin to thick regimes (Withbroe and Noyes, 1977). Finally, there is the ubiquitous heating of the plasma (see, e.g., Rosner *et al.*, 1978; Kuperus *et al.*, 1981; Parker, 1983a, b). Equations (3.12) and (3.13) carry a free-floating temperature. This degree of freedom allows the imposition of an energy balance equation of the form

$$\nabla \cdot [\kappa_\| (\mathbf{B} \cdot \nabla T)\mathbf{B}/B^2] = R - H, \tag{3.15}$$

where the thermal flux is assumed to be channeled along field lines with conductivity $\kappa_\|$ and the radiative loss R and heating function H are taken, depending on the specific microscopic mechanisms at work, to be expressible in terms of the thermodynamic field. Equations (3.12)–(3.15) determine a static equilibrium where force balance and energy balance are coupled nonlinearly. In this complex situation, the magnetic field has a dual role to play — to exert a Lorentz force and to influence energy transport by introducing anisotropic thermal conduction — as well as various plasma effects that can contribute to heating (e.g. Rosner *et al.*, 1978).

The complete problem has a simple tractable example, reducible to a set of ordinary differential equations. This is the Kippenhahn–Schlüter vertical infinite prominence sheet, extended with a free-floating temperature profile (Poland and Anzer, 1971; Low, 1975b; Lerche and Low, 1977; Heasley and Mihalas, 1976). Milne *et al.* (1979) presented a numerical study, extending the model of Lerche and Low (1977), to allow for a radiative loss based on the standard semiempirical model for an optically thin gas at 10^5-10^6 K (e.g. Cox and Tucker, 1969). This radiative loss is balanced locally by classical thermal conduction and a heating rate set heuristically proportional to the local density. More recently, a similar model was found by Low and Wu (1981) which can be integrated analytically in closed form. In these studies, the interplay between the demands of force and energy balance in determining the static distributions of plasma and magnetic field is illustrated. Priest (1978) has made the interesting point that incompatibility between force-equilibrium and energy balance may lead to impossibility of equilibrium in some situations. One such situation was suggested by Milne et al. (1979), whose numerical results on the complete problem showed no equilibrium solutions for large values of the plasma beta. This conclusion was found not to be tenable by the analytic work of Low and Wu (1981), the artificial result being due to unphysical boundary conditions used in the numerical model. For this essentially one-dimensional model, the plasma along individual field lines can adjust to meet both demands of force equilibrium and energy balance for any value of the plasma beta. The problem should be investigated further, and in particular, more sophisticated magnetic geometries than that of the sheet model should be considered. Numerical techniques are necessary for these problems and have to be developed.

With nonlinear equations, both the existence and uniqueness of solutions are not guaranteed. The hazard of judging whether a numerical result is physically meaningful cannot be underestimated. Numerical iteration is a popular procedure for treating nonlinear problems. Since the solution, assuming that it exists, is sometimes not unique, different forms of iteration can converge to different solutions (see, e.g., Low and Nakagawa, 1975). It should be emphasized that this nonuniqueness is not due to an incomplete specification of mathematical conditions. For example, a Dirichlet boundary condition is mathematically complete for determining a solution of the elliptic Equation (3.13) if solutions do exist. However, for nonlinear forms of $p(A, z)$ there may be no solution or, if solutions exist, there may be more than one, for a given Dirichlet boundary condition. In this connection, the procedure of Mouschovias (1974) is physically appealing. With an interest in the magnetostatic equilibrium of interstellar clouds, Mouschovias treated the isothermal equation

$$\nabla^2 A + 4\pi \frac{dP_0(A)}{dA} \exp(-mgz/kT_0) = 0, \tag{3.16}$$

where $T = T_0$, a constant. As a Dirichlet problem it would be tempting to fix the functional form $P_0(A)$ and perform a standard iteration, say

$$\nabla^2 A_n + 4\pi \frac{dP_0(A_{n-1})}{dA_{n-1}} \exp(-mgz/kT_0) = 0, \tag{3.17}$$

where $n = 1, 2, 3, \ldots$, assuming that solutions exist. Mouschovias augments such a procedure by reconstructing the functional form of $P_0(A)$ at each iteration to preserve a given fixed amount of mass in each flux tube. The rationale is that such an iteration mimicks a dynamically admissible evolution consistent with the 'frozen-in' condition applicable to a medium with an infinite electrical conductivity. The iteration with this dynamical constraint singles out one of many possible iteration paths towards a final state. This technique should prove useful in treating the problems posed by solar plasmas. Other dynamical invariants should be considered; for example, one which has been used in theoretical construction of laboratory plasmas is the magnetic helicity (Taylor, 1974). In the solar atmosphere, the field lines are tied to the dense photosphere. This leads to the invariance of the footpoints, which should also be imposed (see, e.g., Kruskal and Kulsrud, 1958; Barnes and Sturrock, 1972; Jockers, 1978; Low, 1982a).

The second direction for future work is to extend the present study to three-dimensional structures with no obvious symmetry. This raises the prospect of comparing theory with realistic coronal structures (Low, 1982b). A great deal of interest centers on X-ray coronal loops, which are, by nature, three dimensional. Work is in progress and a first, analytic, magnetostatic model for an isothermal inverted U-shaped plasma loop has been found (Low, 1982c).

Another important class of equilibrium magnetic field is that of steady magneto-hydrodynamic flows, described by the MHD equations ((3.1)–(3.7)) by setting only the time derivatives to zero. For a system with an ignorable coordinate, Equations (3.12) and (3.13) are, respectively, generalized to a Bernoulli integral along velocity stream lines and a transverse force-balance equation. Such a reduction has been carried out in a general analysis by Tsinganos (1981, 1982a) and Morozov and Solov'ev (1980). It should be pointed out that induction Equations ((3.4) and (3.5)) do not necessarily imply that the steady flow is everywhere parallel to the magnetic field, although such flows obviously satisfy the equations. In general, the flow lines and magnetic lines lie in surfaces defined by the electrostatic potential. The analysis of Tsinganos unifies and generalizes various particular results obtained by various workers through the years (see, e.g., Unno et al, 1974). It would seem that a neat reduction of complicated equations to simpler ones would pave way for new development. Work is in progress to treat the reduced equations (Tsinganos, 1984).

An important application of steady MHD flows concerns the solar wind flow around a helmet-streamer structure first treated numerically by Pneuman and Kopp (1971). The solar wind dominates in the upper corona, drawing bipolar field lines into a radial configuration. Where the field polarities of opposite signs meet, an electric current sheet forms. Lower down, the magnetic tension force dominates in a region of closed bipolar loops and this localized region relaxes into a static state. Pneuman and Kopp (1971) obtained a global solution by numerical iteration, treating the steady MHD equations directly. Since this work, little progress has been made to study the basic physics of this important structure of the corona.

3.2. LINEAR STABILITY AND NONEQUILIBRIUM

A most important property of an equilibrium state is whether it is stable to infinitesimal perturbations. It is sometimes said that if an equilibrium is not stable, it is of no interest because it cannot be realized physically. Actually, the equilibrium states one considers in theoretical calculations are so idealized that they can hardly be purported to represent reality. It seems that the real primary interest of stability analysis is to learn the basic physics of why certain simple states are stable and others not. Armed with this knowledge, one attempts to understand the more complicated realistic situations. To construct a model that matches all the details of a real phenomenon is not feasible; nor is it a desirable thing to do, since the model would be equally difficult to understand in terms of essential physics. This point of view is reflected in the present state of development in MHD stability theory. Although the known solutions of magnetostatic equilibrium have highly idealized geometries, they are nonetheless sufficiently complex so that the first-order time-dependent MHD equations linearized about these equilibrium states, can lead to eigenvalue problems that are generally intractable. Consequently, stability analyses can be carried out only for the simplest types of equilibrium states. Not only is the problem more tractable, but the basic physics is more transparent to understand. A considerable body of knowledge has thus been built up (e.g. Chandrasekhar, 1961; Freidberg, 1982).

The observation of long-lived plasma structures, such as X-ray loops, implies the existence of stable equilibrium in the corona. One often distinguishes between stabilities, or instabilities, of mechanical or thermal origin, although the two are realistically not unrelated. We shall be concerned with only mechanical stability based on the ideal MHD Equations ((3.1)–(3.7)), neglecting various dissipative processes such as those described by Equation (3.15). Thermal instability is an important subject in its own right and recently has attracted much attention in connection with active region plasma loops. We refer the reader to the monograph edited by Orrall (1981) for recent reviews.

A spatially isolated magnetized plasma in the absence of gravity cannot contain itself against indefinite expansion. This is a well-known consequence of the virial theorem. This effect is overcome in the laboratory by the use of rigid boundaries. In the solar corona, two physical effects allows for confinement and stability of a localized structure, when the circumstance is right. The first is often referred to as line-tying (Kulsrud, 1967). The other is, of course, solar gravity. Coronal magnetic fields emerge as bipolar arches with feet 'tied' in the dense photosphere below. There is always a competition between the resulting downward-directed magnetic tension force, combined with gravity, and the natural tendency of the magnetized plasma to expand.

A cylindrical force-free magnetic field is intrinsically unstable if not suitably anchored to a boundary. Parker (1966) showed, by the use of the virial theorem, that a twisted field, if sufficiently tightly wound, will develop internal magnetic stress amounting to longitudinal compression and is thus unstable to buckling. Anzer (1968) performed a detailed stability analysis to show that an infinitely long flux rope is kink-unstable for any degree of twist. That line-tying at rigid boundaries can stabilize the force-free field was shown by Raadu (1972), who found that kink instability can be avoided if the flux rope is of finite length, anchored rigidly at both ends and if the total twist along its length is moderate, typically not more than a complete twist along the length. This model has been investigated further by Hood and Priest (1979) to consider equilibrium states with the Lorentz force balanced by the pressure gradient, leading to the same basic result.

A simple tractable model that simulates the magnetic arcade structure (Newkirk, 1967) assumes a magnetic field that is invariant in the x-direction with field lines rooted at the plane $z = 0$ taken to be the coronal base (e.g. Low and Nakagawa, 1975). The field is of the form given in Equation (3.10). For a force-free field, we require $B_x(y, z) = B_x[A(y, z)]$, and

$$\nabla^2 A + B_x(A) \, \frac{dB_x(A)}{dA} = 0. \tag{3.18}$$

Although the magnetic field is independent of x, it has a three-dimensional geometry if $B_x \neq 0$. One may visualize $B_x \neq 0$ to be the result of displacing bipolar pairs of footpoints of arching field lines in the x-direction. The model has been studied extensively, with the principal result that a sequence of stable equilibria, through which the magnetic field evolve as its footpoints are displaced quasisteadily, can terminate abruptly at the first appearance of a marginally stable state (Low, 1977a, b; Birn et al., 1978; Jockers, 1978). As the terminal state is approached, the small gas pressure that has been ignored in the force-free approximation becomes important in spite of its smallness in magnitude, and equilibrium breaks down (Low, 1980a). The terminal state is characterized by a footpoint displacement in the x-direction which is of comparable magnitude to the initial footpoint separation defined by the state with $B_x = 0$. This result is evidently of the same physical nature as that found by Raadu (1972) for a finite-length flux rope. It appears that a magnetic field cannot be twisted or deformed indefinitely. As the field topology, such as indicated by the degree of twist, becomes too complex, instability sets in. Parker (1972, 1975, 1979a) suggests that instability, once it has set in, leads to a buckling of flux rope, which brings oppositely oriented fields together and leads to resistive reconnection (Vasyliunas, 1975; Sonnerup, 1979). Such a process liberates energy and may be the origin of flares, X-ray bright points, as well as the more quiet heating of the corona (Parker, 1979b, c).

Let us now turn to the question of boundary conditions. In the standard stability analysis, the photosphere is regarded as extremely massive relative to the coronal gas. This leads to the idealization that the photosphere is effectively a rigid boundary. Furthermore, for an infinite electrical conductivity, the normal photospheric magnetic flux distribution cannot change with time. Therefore, no slippage of the plasma along the photosphere is allowed; this boundary condition is additional to requiring zero normal velocity at the rigid photosphere. In the standard notation, the linearized MHD equations for an infinitesimal plasma displacement $\xi(\mathbf{r}, t)$ can be reduced to

$$-\omega^2 \rho_0 \xi_\omega = F(\xi_\omega), \tag{3.19}$$

where ρ_0 is the equilibrium density, F is a linear second-order operator independent of time and expressible in terms of the equilibrium state, and we have Fourier analyzed $\xi(\mathbf{r}, t) = \xi_\omega(\mathbf{r}) \exp(-i\omega t)$ (e.g. Bernstein et al., 1958). The rigid boundary model subjects Equation (3.19) to the homogeneous boundary condition $\xi_\omega = 0$, leading to an eigenvalue problem for ω^2 and ξ_ω. If an eigenmode with $\mathrm{Im}(\omega) > 0$ exists, the system is unstable.

If we have an energetically isolated system, the conservation of total energy leads to the principle of Bernstein et al. (1958) which has been widely used in stability analyses. This principle allows us to show the existence of an unstable mode without having to solve the usually intractable eigenvalue problem posed by Equation (3.19). For an unstable eigenmode to exist, it is sufficient to find a static plasma displacement $\xi_{\text{test}}(\mathbf{x})$ that satisfies the same boundary conditions for $\xi_\omega(\mathbf{x})$ such that the potential energy change $\delta W = -1/2 \int_V \xi_{\text{test}} F(\xi_{\text{test}}) \, dV$ is negative.

It has been suggested that more realistic boundary conditions for the coronal base should allow for free plasma flows across it, parallel to the boundary magnetic field (e.g. Van Hoven et al., 1981). Several calculations have appeared in the literature which pursue this idea but their physical meaning is unclear. For an energetically conservative system, the operator F in Equation (3.19) is self-adjoint relative to the specific boundary conditions applied to ξ_ω. It then follows that the eigenvalue ω^2 is a real number. The mathematical statement for a conservative system involves the surface integral, in standard notation,

$$\delta W_s = -1/2 \int_s \{(\xi \times \mathbf{B}_0) \times [\nabla \times (\xi \times \mathbf{B}_0)] + \xi(\gamma p_0/\rho_0)\nabla \cdot (\rho_0 \xi)\} \cdot \mathrm{d}S, \qquad (3.20)$$

which represents the energy exchange between the system and its exterior, when the system is subject to a displacement ξ. For a conservative system, the boundary conditions on ξ must be such that δW_s vanishes for all time. Einaudi and Van Hoven (1981) suggested the boundary condition

$$\xi_\perp = (\xi \times \mathbf{B}_0) = 0, \qquad (3.21)$$

$$\xi_z(z = -L) - \xi_z(z = L) = \frac{\partial}{\partial z} \, \xi_z(z = -L) - \frac{\partial}{\partial z} \, \xi_z(z = L) = 0 \qquad (3.22)$$

for a finite length of cylindrically symmetric magnetic flux rope anchored at the end-planes $z = \pm L$. With these boundary conditions, the first term in the integrand of δW_s vanishes identically, whereas the second term gives integrals at $z = \pm L$ that sum to zero. The system conserves energy and can be subjected to an energy principle analysis. Attention is drawn to the fact that the equilibrium quantities $\mathbf{B}_0$, p_0, ρ_0 at $z = \pm L$ must be the same in order that these boundary conditions imply $\delta W_s = 0$. We note that a simple modification of these boundary conditions that also leads to $\delta W_s = 0$ is to replace Equation (3.22) with

$$\xi_z(z = -L) + \xi_z(z = L) = \frac{\partial}{\partial z} \, \xi_z(z = -L) + \frac{\partial}{\partial z} \, \xi_z(z = L) = 0. \qquad (3.23)$$

This set of boundary conditions was not treated by these previous authors, but will be relevant to our discussion below. In subsequent work, only boundary condition (3.21) was used explicitly, leaving ambiguous the boundary conditions to be applied to $\xi_\parallel$, the displacement parallel to the magnetic field (Van Hoven et al., 1981; Ray and Van Hoven, 1982). Consider the example treated by Van Hoven et al. (1981). They introduced the general Fourier expansion

$$\xi_\parallel = \sum_{m,n} \{\xi_s(r, m, n) \cos (m\theta + \delta_s) \sin n\pi z/L +$$

$$+ \xi_c(r, m, n) \sin (m\theta + \delta_c) \cos n\pi z/L\} , \qquad (3.24)$$

where m and n are integers and the phase angles depend on r, m, and n. A finite-length flux rope is again treated, but anchored at $z = 0, L$. This expansion was regarded to be of special interest because it includes as a particular case, the expansion for a rigid boundary system with $\xi_\parallel = 0$ at $z = 0, L$. The latter is obtained by setting $\xi_c = 0$ for all m and n in Equation (3.24). An energy principle analysis for stability was carried out based on the general expansion. A careful examination shows that Equation (3.24) contains two distinct sets of Fourier modes. For simplicity, consider the case where the equilibrium magnetic field is everywhere aligned in the z-direction. Then, the parallel displacement $\xi_{\parallel}$ is identical to ξ_z and we can discuss ξ_z without bothering about $\xi_{\pm}$. The even n modes in Equation (3.24) satisfy boundary conditions (3.22) whereas the odd n modes satisfy boundary conditions (3.23). Each of these two sets is complete in itself, describing a distinct conservative system. Mixing the two sets is not meaningful — that is, the combined set does not describe a system with well-defined boundary conditions — and, in particular, $\delta W_s \neq 0$ so that no energy principle applies. Once it is recognized that the even and odd modes correspond to two distinct systems, setting $\xi_c = 0$ in either subset does not yield the rigid boundary system. The rigid boundary system is, instead, yet another distinct physical system made of only the sine modes of both odd and even n. The results of the above stability studies based on the mixed modes therefore do not have clear physical meaning.

It is an important physical point that the rigid boundary conditions are an approximation that leaves out a great part of the physics of the transition layer at the coronal base. This approximation is valid in the limit of infinite inertia for the coronal base. If this limit is to be relaxed for $\xi_{\parallel} \neq 0$, then $\xi \neq 0$ altogether at the boundary. The corona freely interacts with the transition layer and the region below, and this interaction is the essential physics in this more realistic situation. Any constraint to make the corona energetically isolated in spite of $\xi_{\parallel} \neq 0$ is artificial and without a physical purpose. Boundary conditions (3.22) and (3.23) are examples of such artificial constraints. They require both the equilibrium states and the linear perturbations to have special symmetry over large spatial separations in order that the system is energetically isolated. Moreover, the coronal base so defined is no longer a material boundary when there is a mass flux across it. The coronal base must then be a form of wave surface fixed in space. What process, then, can produce such a standing wave? We conclude that the recent work attempting to relax the rigid boundary conditions to allow for $\xi_{\parallel} \neq 0$ is not physical. A proper treatment should regard the coronal base to be a freely adjusting material or contact surface. An explicit model for the region below this contact surface needs to be incorporated explicitly as part of a larger conserved system. A similar problem has been encountered in fusion research, where the confined plasma is separated from the rigid wall of the container by a volume of vacuum magnetic field (Bernstein et al., 1958). Ultimately, the physics of the transition layer must be introduced to properly understand coronal structures. The above contact surface separating plasmas of two diverse thermodynamic states — the cool, dense state below and the hot, tenuous state above — owes its existence to solar gravity and the various processes of heating, cooling, and energy transport.

Plasma stability without gravity has been studied extensively in fusion research (see the review by Freidberg, 1982). Gravity introduces the effect of buoyancy, which can have dramatic results if coupled with the effects of a magnetic field. A magnetic field behaves like an inertialess fluid that exerts a pressure, if one overlooks the magnetic tension force. For example, a heavy fluid overlying a magnetic field is Rayleigh–Taylor unstable (Chandrasekhar, 1961). An isolated magnetic flux tube tends to be lighter than its surrounding and becomes buoyant (Parker, 1955). These simple effects can be quite complicated when the magnetic tension force is involved in a field with complex geometry. In the study of interstellar cloud formation, Parker (1966) suggested that clouds form because of the tendency for heavy matter to slide down along magnetic field lines due to the gravity of the galactic disk. The instability runs away with the sinking of the heavily loaded valleys of undulating field lines as the upper density-depleted portions of the field rise upward because of buoyancy. Similar processes of condensation should be explored for relevance to the solar corona (Sweet, 1971).

The interplay between buoyancy and magnetic field is illustrated by a simple model of Low (1981b). In two-dimensional geometry, an isolated set of bipolar magnetic arches is situated in an isothermal atmosphere stratified by a uniform gravity. The magnetic field is separated from the unmagnetized plasma exterior by a boundary electric current surface. The interior has lower pressure and density as required by force balance. This region of low density is buoyant and is held down by the tension force of the rooted bipolar fields. Exact analytic solutions for this model show that the stability of the equilibrium states is controlled by a dimensionless parameter $C = F_0 \, (\sqrt{\Delta p} \, kT/mg)^{-1}$. This parameter is the ratio between the total magnetic flux F_0 of the bipolar field and an equivalent magnetic flux due to a field of strength $\sqrt{\Delta p}$ spread over a density scale height kT/mg, where Δp denotes the pressure difference at the base across the electric current surface. Consider a quasisteady evolution of the system due to a spreading of the bipolar magnetic footpoints, through successive equilibria with a fixed value of C. One may imagine that the evolution takes place with no magnetic flux entering or leaving the atmosphere, so that F_0 is constant and the boundary conditions hold the pressure difference Δp and temperature T constant (see Low, 1981b). If $C < 1$, the spreading of the bipolar magnetic footpoints takes the system through a sequence of stable equilibria that terminates abruptly in a marginally stable state having a critical maximum mean separation of the magnetic footpoints. The evolution may transit into a dynamical state at this terminal point. This result is akin to the nonequilibrium property demonstrated with evolving nonlinear force-free fields (Low, 1982a). If $C > 1$, the effect of gravity has a dramatic consequence. The spreading of magnetic footpoints again cannot proceed beyond a critical mean footpoint separation. As the critical separation is approached, the tops of the bipolar loops extend rapidly to unlimited great heights for successive equilibria. Any reasonable physical situation has finite MHD signal speeds. There would then be a point when the large plasma displacement at the top of the magnetic loops, required in order to pass from one equilibrium to the next, cannot be achieved fast enough, so that the quasisteady assumption breaks down. We then have a critical state where the magnetic tension force fails to balance the buoyancy force and the magnetic structure 'balloons' upward. Wolfson (1982) extended the simple model of Low to spherical geometry with a $1/r^2$ gravity. He found that the ballooning buoyancy instability depends not only on the mean footpoint separation, but also sensitively on the distribution of the magnetic flux at the base of the bipolar fields. It was suggested that magnetic buoyancy may be the origin of a class of transient phenomena in the corona (Low *et al.*, 1982). We shall discuss 'coronal transients' in Section 3.3.

Zweibel (1981) has formulated a sufficient criterion for local MHD instability based on the Bernstein *et al.* (1958) energy principle, including the presence of a uniform gravity. This criterion is applicable to two-dimensional equilibria, say, in the $y-z$ plane, with no x-component for the magnetic field. The most unstable modes are those with large wavenumbers in the x-direction (Asseo *et al.*, 1978; Gilman, 1970). Rigid boundary conditions are assumed. A simple rule of Zweibel's criterion is that any local maximum of the total pressure $p + B^2/8\pi$ located in the domain implies local instability. General classes of two-dimensional magnetostatic states can be shown to be locally unstable (Zweibel, 1981; Low *et al.*, 1984). An exception is the Kippenhahn–Schlüter prominence sheet, which has variation only perpendicular to the sheet. This prominence, taken in its idealized form in an infinite medium, is stable (Zweibel, 1982; Migliuolo, 1982). Note that if a two-dimensional equilibrium magnetic field has a nonvanishing x-component, the above perturbations need no longer be destabilizing, since excessive tangling of this field component with large wavenumber builds up the potential energy. The stabilizing effect of a magnetic field directed with a component along a prominence, identified with the direction of invariance in the above two-dimensional model, has been pointed out previously (e.g. Nakagawa, 1970). A problem of great interest is the stability of the ubiquitous X-ray coronal loops (Vaiana and Rosner, 1978). Work has just begun to generate these necessarily three-dimensional objects. A model for an inverted U loop found recently by Low (1982c) proves to be linearly stable in an isothermal atmosphere stratified by a uniform gravity.

Theorists have also been studying equilibrium states from a different fundamental consideration: what constitute the necessary conditions for equilibrium to exist? This is a question quite separate from whether a given equilibrium is stable or unstable. The following classical example of Parker (1972, 1979a) illustrates the point. Consider a magnetostatic equilibrium (neglecting gravity) which is symmetric, namely, it is invariant in the x-direction. Parker showed that no small perturbation can take the given system into a nearby equilibrium state that has a different symmetry, with, say, variation in the x-direction. An expansion in a small parameter ϵ can be shown rigorously to require the nearby equilibrium state, if it is to exist, to carry no dependence on x for all orders of ϵ. Parker's analysis assumes that the spectrum of nearby states is analytic about the $\epsilon = 0$ state. As pointed out by Rosner and Knobloch (1982) with examples, this does not preclude equilibrium states altogether because finite amplitude perturbations can take the system to states of different topologies. They also pointed out that magnetic surfaces are such that analyticity near $\epsilon = 0$ precludes the possibility of change in field symmetry. It follows that equilibrium states of different symmetries are disjoint sets in the sense that one state cannot evolve smoothly to another state of a different set. This remarkable property led Parker to suggest, in broad terms, why astrophysical magnetic fields are constantly in an active state. In nature, there is no reason to expect perturbations to have just the right symmetry relative to a given equilibrium state. This nonequilibrium property is unchanged if the equilibrium state is not static but carries a steady velocity flow, as established by Tsinganos (1982b).

To avoid the restriction posed by the use of expansion in small parameters, Parker's nonequilibrium theorem has been treated from a different point of view by Tsinganos *et al.* (1984). The formal exact analogy is first drawn between the physics of magnetostatics and Hamiltonian mechanics, relating the topology of a symmetric magnetic field in static equilibrium to the topology of integral surfaces of one and two-dimensional Hamiltonian systems in phase space. Translating the Kolmogorov–Arnold–Moser or KAM theorem (Arnold and Avez, 1968; Arnold, 1978) of Hamiltonian dynamics into the context of magnetostatic fields, it is then possible to prove that the magnetic surfaces of a symmetric equilibrium are necessarily destroyed by perturbations, not necessarily small amplitude and having arbitrary symmetry. Thus, Parker's result is extended and re-established without assuming the analyticity in ϵ.

Another way of looking at Parker's result is that a certain symmetry is required of a given magnetic field if it is to be in static equilibrium (Parker, 1972, 1977). A similar statement was conjectured by Grad (1967), namely, that three-dimensional magnetostatic states may not exist. Indeed, the only known magnetostatic equilibrium (without gravity) having no ignorable coordinate does have a symmetry, namely, the mirror symmetry (Lortz, 1970); this exceptional solution has the property that all its field lines are closed. Although what is meant by 'symmetry' is obvious in a system with an ignorable coordinate (say, invariance in the x-direction or axisymmetry), the symmetry becomes subtle when there is no obvious ignorable coordinate. To put it in specific physical terms, Low (1980b) considered Equation (3.8) for a general three-dimensional equilibrium. Since field lines are intersections of pairs of surfaces, a magnetic field can always be expressed in terms of two Euler potential surfaces U and V, at least in a localized region of space (Stern, 1966):

$$\mathbf{B} = F(U, V)\nabla U \times \nabla V, \qquad (3.25)$$

where F controls the field strength on the field lines defined by constant values of U and V. Automatically, $\nabla \cdot \mathbf{B} = 0$ as required. If we use, in this localized region, the curvilinear coordinates (U, V, ϕ) instead of (x, y, z), with $\phi = -gz$, the gravitational potential, it is easy to show that Equation (3.8) reduces to the components

$$\frac{1}{4\pi} \{\nabla \times [F \, \nabla U \times \nabla V]\} F \, \nabla V - \frac{\partial p}{\partial U} = 0, \qquad (3.26)$$

$$\frac{1}{4\pi} \{\nabla \times [F \, \nabla U \times \nabla V]\} F \, \nabla U + \frac{\partial p}{\partial V} = 0, \qquad (3.27)$$

$$\frac{\partial p}{\partial \phi} + \rho = 0. \qquad (3.28)$$

Equation (3.28) describes the purely hydrostatic equilibrium between pressure gradient and plasma weight along a field line. In contrast to systems with an ignorable coordinate, there are now two independent equations for force balance perpendicular to field lines. Eliminating the pressure between them, the complex relationship follows

$$\frac{\partial}{\partial U} [\{\nabla \times [F \, \nabla U \times \nabla V]\} F \, \nabla U]$$

$$+ \frac{\partial}{\partial V} [\{\nabla \times [F \, \nabla U \times \nabla V]\} F \, \nabla V] = 0. \qquad (3.29)$$

It is significant that this equation involves no pressure and density. It is the necessary condition for $\partial^2 p/\partial U \, \partial V = \partial^2 p/\partial V \, \partial U$ so that magnetostatic states can be constructed with p as a single-valued function of space. Translated into the geometric properties of the surfaces U and V, Equation (3.29) is the symmetry property we seek for a magnetic field to be consistent with equilibrium. It should be emphasized that this is a local property. Unfortunately, it is not possible to state in general physical terms what this symmetry is. For a system with one ignorable coordinate, the requirement of Equation (3.29) is trivially satisfied. In the recent construction of three-dimensional equilibria of Low (1982c) and Hu *et al.* (1984), Equation (3.29) is not trivial and leads to a hyperbolic partial differential equation that defines the admissible equilibrium magnetic surfaces. Finally, we note that the compatibility relation (3.29) has little to do with the presence of gravity. If gravity is absent, ϕ is just a dummy variable and Equation (3.28) reduces to requiring $p(x, y, z) = p(U, V)$ being constant along field lines. Equation (3.29) is a necessary but not sufficient condition for equilibrium and it originates in the geometric forms of the Lorentz force, pressure and gravity.

3.3. TIME-DEPENDENT PHENOMENA

A most remarkable property of coronal structures is that they can transit abruptly from quasisteady evolution into fully developed dynamical states. For example, a quiescent prominence can suddenly erupt in a spectacular manner. Sometimes the erupted prominence can form again, in the same location and regaining its pre-eruption configuration. This testifies to the complexity of the dynamical processes involved. The intense magnetic fields in the active region undergo restless slow evolution, building up stresses and breaking into various forms of eruption, the most impressive of which is, of course, the flare. The large-scale white-light corona itself erupts, on average once a day, ejecting masses of the order of 10^{15} g out of the corona. That these so-called coronal transients are an important component of the active corona has been realized during the past decade, through the use of space-borne coronagraphs (MacQueen, 1980; Sheeley *et al.*, 1980; House *et al.*, 1981).

The theoretical description of these time-dependent phenomena is much hampered by mathematical obstacles. Magnetohydrodynamics deals with a medium much more complex than that of classical hydrodynamics, in having not one but three characteristic signal speeds at each point in space. These are the speeds of traveling waves that travel to communicate between the different parts of a medium. These waves do not propagate isotropically. To study the rich, complex properties of this intrinsically anisotropic medium, most situations of interest require models having at least two spatial dimensions and the calculation carried out into the nonlinear time-dependent regime. Considerable simplification of the physics and mathematics is usually necessary to render the theoretical problem tractable. The flare and the coronal transient pose two distinct classes of MHD problems which we will survey below. The eruption of prominences are often associated with these two phenomena. The association will become clear as our discussion proceeds.

A primary question of flare studies concerns the dissipative mechanisms that can liberate some 10^{32} erg of energy with a rise time of less than 100 s. It is generally believed that this energy is first stored up in the electric currents associated with the active region magnetic fields (Gold, 1964). The high electrical conductivity of the atmosphere allows large currents to be built up quasisteadily by the inductive effect of photospheric motion. Conditions then develop under which current dissipation of various forms become important. At the heart of most MHD flare model is the idea, first proposed by Dungey (1958) and Sweet (1958), that current dissipation can take place rapidly at an X-type magnetic neutral point. This process is usually discussed in terms of the resistive reconnection of magnetic field lines. The simplest way to account for resistivity is to introduce an isotropic resistive coefficient η in the Ohm's law:

$$cE + v \times B = \eta \nabla \times B. \tag{3.30}$$

In the limit of $\eta = 0$, Equation (3.5) for an infinitely conducting medium is regained. All electric currents will dissipate in a resistive medium, of course. At a magnetic X-type neutral point, the dissipation is most dramatic because of the following effect. Plasma flows converging from opposite sides of the neutral point can transport magnetic fluxes of opposite signs into the diffusion zone to annihilate and reconnect by resistive diffusion in a narrow front. The reconnected field lines, by the action of the magnetic tension force, can evacuate the plasma away from the neutral point in the direction perpendicular to the converging flows. Thus, magnetic energy can be dissipated more rapidly than via simple resistive diffusion in a static medium. Sweet (1958) and Parker (1957) estimated the increased rate of dissipation, based on conservation laws applied to an assumed steady state. They found the reconnection speed to be a pessimistically small fraction of the characteristic Alfvén speed, implying that the process cannot account for the 100 s rise time of the flare. Optimism for reconnection as a flare mechanism was revived by Petschek (1964) with a modified dynamical model that reduces the size of the reconnection region considerably by the formation of standing Alfvén waves in an incompressible medium. An increased reconnection speed of a sizable fraction of the typical Alfvén speed is then possible. Sonnerup (1970) provided a formal self-consistent MHD solution for this type of model in the limit of a vanishingly small reconnection region. Work extending the theory to compressible medium have been reported by Yang and Sonnerup (1976), and more recently by Soward and Priest (1982).

The Petschek steady mechanism is a useful theoretical illustration of how magnetic energy can be converted rapidly into plasma energy through Ohmic heating, shock heating, and direct acceleration of bulk velocity. It is often suggested that the electric field induced at the X-type neutral point may provide a means of accelerating individual charged particles. The time development leading to rapid reconnection is a much more difficult problem to treat. Considerable work has been done for laboratory plasmas, beginning with the identification of the various linear resistive modes for simple systems (Furth et al., 1963). The evolution of these linear modes into the nonlinear regime has been treated numerically (Rutherford, 1973; Van Hoven and Cross, 1973; Spicer, 1981; Tajima et al., 1982). Unfortunately, in addition to the usual difficulty of carrying out a numerical solution, the present computer resources do not permit calculation of the nonlinear problem for magnetic Reynolds number anywhere large enough to be of direct relevance to the highly conducting solar atmosphere (e.g. Van Hoven and Cross, 1973; Sato and Hasegawa, 1982). Reviews that assess the current theoretical understanding of reconnection processes can be found in Vasyliunas (1975), Drake and Lee (1977a, b), Sonnerup (1979), and Parker (1979a).

In the real flare, the reconnection processes are many-fold more complex. At least during the impulsive phase, MHD breaks down as a valid approximation. The large-scale fluid behavior is strongly coupled to a multitude of microscopic plasma processes. The fluid properties generate conditions that excite the microscopic processes, which, in turn, influence the fluid properties by affecting the large-scale transport coefficients. This coupling between microscopic and macroscopic processes is exceedingly difficult to treat theoretically. An impressive accumulation of observational data for the flare now exists but theory is rather underdeveloped, with many fundamental processes not worked out or understood. The abundance of data has led to a considerable demand for phenomenological approaches.

Apart from the impulsive phase, a question of some interest concerns the slow large-scale evolution that sets the stage for a flare eruption. Formal calculations have been presented to demonstrate various MHD instabilities and resistive instabilities that can initiate a flare (see the reviews in Sturrock, 1980; Priest, 1982; Low, 1982a). These calculations are often taken to the point where an instability sets in, without pursuing them into the subsequent developments. This second step of the calculations is essential in order to demonstrate the flare as a nonlinear instability (Dungey, 1964; Sweet, 1964). The large amount of energy liberated by a flare implies that energy must first be stored. Then once the instability initiates, it liberates so much free energy that the system relaxes to an energy level well below the level where the instability becomes excitable (Sturrock, 1966; Parker, 1979a).

In his recent study of coronal heating, Parker (1983a, b, c) pointed out that magnetic flux tubes rooted to the same locality on the photosphere should be all twisted by convection in the same sense along the tubes. This field topology has the implication that a closed-packed set of such flux tubes cannot be in equilibrium unless properly chosen plasma pressures are imposed between the tubes to keep them apart. Parker suggested that nature is unlikely to provide these special pressures and narrow boundaries would naturally form for reconnection of locally opposite magnetic field lines. The reconnected field takes the form of pairs of oppositely twisted flux tubes, a result which follows from a simple topological consideration. This topological change eliminates all the intertube boundaries between locally opposite fields. The magnetic field relaxes to a lower energy state, and the dissipated energy is gained by the plasma. This process may perhaps operate, with an impulsive signature, in a flare. The attractive aspect of this model for the flare is that the energy build-up takes the simple form of flux tubes being twisted in the same sense, and the reconnection between the flux tubes can liberate the energy without implying any large-scale motion, such as is observed in the so-called compact flares.

Another question of interest concerns the MHD response of the system to the energy processes of the impulsive and postimpulsive phases. In the present preoccupation with the various radiative and heating processes in a flare, it has been popular to idealize the system with a one-dimensional gasdynamical model defined along a magnetic field line regarded as a rigid semicircular loop (e.g. Craig, 1982). The response of the magnetic field is completely suppressed. Some amount of physics can be learned with these one-dimensional models but it is uncertain whether the simplification adopted would have removed important physics. We saw in Section 3.1 that the interplay between demands of energy transport and force equilibrium is a formidable problem, even in the simpler setting of a static state. Because the time-dependent nonlinear flare dynamics does not appear likely to be tractable in the near future, physical intuition and phenomenological models are probably going to be the main source of insight into flare MHD. For example, the phenomenological model by Kopp and Pneuman (1976), which attributes the large two-ribbon flares to a magnetic reconnection process, has generated considerable interest.

It allows different aspects of the observed phenomena, in particular the associated eruptive prominence, to be fitted into a global scenario (see the review by Pneuman, 1982). However, the many fundamental physical processes making up this scenario have yet to be worked out on a firm basis.

A simpler MHD problem is posed by the coronal transient, at least insofar as we are asking how some 5×10^{15} g of magnetized plasma can be expelled from the gravitational well of the Sun. Coronal transients are usually observed with coronagraphs, instruments that record the photospheric white-light scattered by coronal electrons (MacQueen, 1980). The expelled plasmas often occur in the form of loop-like structures with length-scales of 10^5 km and larger as seen projected on the plane of the sky. These structures have leading edges moving radially outward with speeds in the range 100–1000 km s^{-1}. To read physical significance into these speeds, we note that the sound speed for a 2×10^6 K plasma is about 200 km s^{-1} and the gravitational escape speed at the base of the corona, set at $r = 10^{11}$ cm, is about 500 km s^{-1}. Over the above length and time-scales, a first approximation for a theoretical description would ignore energy dissipation so that the MHD Equations ((3.1)–(3.7)) form the starting point of analysis. Based on a mean speed of 500 km s^{-1}, the total kinetic energy is approximately 10^{31} erg. Observational estimates of the gravitational potential energy involved in expelling some 5×10^{15} g of matter suggest it is sometimes comparable to or larger than the estimated total kinetic energy (MacQueen, 1980). Gravity is therefore important for such events. Yet, there is a tendency for the expelled plasma to become inertial, with the leading edges of loops moving at constant speeds above $2\ R_\odot$ (Gosling *et al.*, 1976; MacQueen and Fisher, 1983). Moreover, this constant speed can be substantially lower than the local gravitational escape speed. In attempting to explain the dynamical properties of coronal transients, some progress has been made in our theoretical understanding of time-dependent MHD process in a $1/r^2$ external gravitational field.

There have been three main approaches to the theory of coronal transients. One approach interprets the coronal transient loops to be magnetic flux ropes subject to a Lorentz force that propels them against gravity, out of the corona. As such models are difficult to treat quantitatively, no attempt is made to solve the MHD equations. Instead, analytic approximations are made, with varying degrees of confidence, to explain the motion of the magnetic flux ropes in terms of the forces acting on them (e.g. Mouschovias and Poland, 1978; Anzer, 1978; Anzer and Pneuman, 1982). Notable among these models is that of Pneuman who extended the Kopp–Pneuman scenario for a two-ribbon flare to include not only the erupted prominence but also the coronal transient in a global process driven by magnetic reconnection. Phenomenological models have their value in providing quick insights into a complex process. However, the need to relate these models to basic processes should not be overlooked. Yeh and Dryer (1981) criticized the above flux rope models for neglecting the role of the plasma pressure. They pointed out that the Lorentz force at the axis of a flux rope vanishes for geometric reasons. Unless pressure forces are invoked, the plasma at the special location of the axis cannot be driven against the pull of gravity. The point can be taken further to note that the Lorentz force always acts perpendicular to the magnetic field. Thus, but for uninteresting exceptions, pressure forces are needed to drive plasmas against the inward component of gravity resolved along the direction of the local magnetic field. Physically, one expects pressure to play a crucial role even in a low beta plasma (MacQueen and Fisher, 1983).

The Lorentz force may be a dominant force in a localized region but not everywhere. This property is akin to the virial theorem that a static force-free magnetic field cannot confine itself but must interact somewhere with the plasma (Chandrasekhar, 1961). In a fully developed flow, reaction against the Lorentz force must build up pressure somewhere in the system.

Another theoretical approach employs numerical codes to treat the ideal MHD equations as an initial–boundary value problem. Numerical codes have the advantage that, within the special class of physical problems for which they are devised, one can freely prescribe the details of the initial and boundary conditions. On the other hand, a formal mathematical theory for most numerical algorithms used in numerical procedures often does not exist. Some amount of numerical experimentation is unavoidable, so that a numerically obtained solution is more properly to be regarded as a simulation rather than a true solution. Physical interpretation must procede with caution. In the laboratory, a numerical code is often devised to simulate a particular plasma machine. The two complement each other in the following way. At any one run of the machine, physical conditions can be controlled. To probe a process, only a selected set of physical parameters can be measured in turn at the sacrifice of other parameters, depending on the method of measurement. The numerical code generates the full set of parameters in a single simulation, of course. The comparison between numerical and experimental results provides both instructive insight into a physical process as well as a check on the numerical simulation. In astrophysics, we are handicapped by the impossibility of controlling the dynamics of a phenomenon and the difficulty of dictating the observational data we like to gather. The theoretical testing of numerical codes for their physical correctness, and the questioning of how well a particular code reflects the true environment of an observed phenomenon, are difficult but vital aspects of this type of work.

Many numerical simulations of the coronal transient have been presented for an axisymmetric magnetic atmosphere in spherical coordinates (e.g. Wu, 1980; Steinolfson et al., 1978; Dryer et al., 1979). Most of the calculations assume an initial state composed of a potential magnetic field and a hydrostatic atmosphere. (More recent work by Steinolfson (1982) considered a numerically generated initial state in the form of the Pneuman–Kopp (1971) type of steady solar wind around a helmet-streamer.) An energy pulse of the order of 10^{32} erg is introduced at the base of the corona, resulting in a wave-like nonlinear response of the ambient medium. The free parameters available in the numerical codes are adjusted to produce a wave structure whose density distribution simulates the coronal transient observed in white light. A principal result is that the density wave structure can move out readily only if the input pulse is initiated at the base of locally open magnetic field lines (Steinolfson et al., 1978). The key physical point seems to be that the initial state is a stable static equilibrium. In order to open up a potential magnetic field, a considerable amount of work has to be done by the plasma. It is simple to show that a potential dipole magnetic field rooted on a rigid sphere $r = r_0$ must increase its total energy in the region $r \geq r_0$ to twice its initial value if all its field lines are forced into a radial configuration. A modest field strength of 2 gauss at $r_0 = 10^{11}$ cm implies an increase of magnetic energy in $r \geq r_0$ by 10^{33} erg. Hence, no mass ejection is possible with the modest energy input of 10^{32} erg initiated under close field lines; instead, wave motions due to the oscillation of the system about its initial stable equilibrium would be a natural mode of response. It is therefore a crucial question whether coronal transients are in fact initiated in a stable atmosphere such as prescribed in the numerical model. A highly stressed static atmosphere contains free energy that can be substantial over large regions (Low, 1982d). In such a stressed atmosphere, an energy input of 10^{32} erg may elicit a response physically very different from that just described. The entire portion of an atmosphere, by virtue of its free energy may expand outward, opening up closed field lines as it does so. Even more realistically, the initial state may be taken to be a steady, magnetic solar wind, such as considered in the numerical study of Steinolfson (1982).

A well-known technique of treating time-dependent equations in gasdynamics is to assume self-similar evolution (Barenblatt and Zel'dovich 1972). This assumption isolates the subclass of solutions, if they exist, for which time-dependence appears not independently but only through some known combination of time and spatial variables. Considerable simplification of the mathematical problem is then achieved by transforming away the explicit time-dependence. Often, the problem is thus reduced to analytic tractability. In physical terms, self-similar solutions describe particular evolutions in which certain physical dimensions do not appear. The Sedov–Taylor point explosion, releasing energy E_0 in a cold medium of density ρ_0, is a classic example (Sedov, 1946; Taylor, 1950). Neglecting the radius of the initial fireball and the pressure of the ambient medium, no dimensions of length or time can be identified in the initial state. Thus, the spherically symmetric blast wave that forms carries spatial and time dependences through the well-known combination $\zeta = r(\rho_0/Ept^2)^5$, from which no characteristic length or time can be identified. From dimensional consideration, one would have been skeptical about the existence of self-similar solutions for the MHD equations with an $1/r^2$ gravity, except for cases of trivial magnetic geometries. As it turned out, nontrivial self-similar solutions for Equations (3.1)–(3.7) exist for a radial velocity field of the form $\mathbf{v} = (r/\Phi)$ $(d\Phi/dt)\hat{r}$, where $\Phi(t)$ is a known function with two free constant parameters (Low, 1982d). Defining the self-similar variable $\zeta = r/\Phi$, and setting the adiabatic index $\gamma = 4/3$, the entire set of MHD equations reduces to a single equation:

$$\frac{1}{4\pi} (\nabla \times \mathbf{B}) \times \mathbf{B} - \nabla p - \rho \left(\frac{GM}{r^2} + \alpha r/\Phi^3 \right) \hat{r} = 0, \qquad (3.31)$$

where α is some constant parameter. Transforming from real spacetime (r, θ, ϕ, t) into the similarity space (ζ, θ, ϕ), explicit dependence on time through $\Phi(t)$ can be removed and Equation (3.31) describes a 'magnetostatic' state where the usual inverse square gravity is modified with a force field that represents the rate of change of momentum in real space time. Although the velocity field is spherically symmetric, the magnetic field and plasma can depend on all three spatial coordinates. Various analytic axisymmetric solutions generated by Equation (3.31) have been presented to illustrate the expulsion of magnetized plasmas from a $1/r^2$ gravitational well (Low, 1982d, e; 1984a, b).

The $\gamma = 4/3$ self-similar solutions of Low are not the most general solutions of MHD, but they provide a rare opportunity to relate a complex process to first principles. They suggest that the white-light density structures of transients are the entrails of a pre-existing part of the atmosphere that has newly become gravitationally unbound and is expanding outward in a global flow. An infinite variety of plasma structures generated by the solutions of Equation (3.31) can be carried by the same radial flow $\mathbf{v} = \zeta(d\Phi/dt)\hat{r}$. In particular, an embedded prominence filament would erupt and be carried along in the outflow. In such a flow, the internal energy of the $\gamma = 4/3$ polytrope is a source of energy for the expansion against gravity. Depending on the nature of the magnetic field, the total magnetic energy may increase or decrease during the expansion. The former implies that the plasma has to do work against both gravity and the confining effect of its magnetic field. In the latter case, the magnetic field contributes free energy to work with the pressure against gravity. What is remarkable is that in all cases, the motion rapidly becomes inertial so that, although there is velocity dispersion, the speed of individual plasma parcels are constant in time. The asymptotic flow preserves force balance everywhere, and a 'heavy' prominence moves at constant speed only because it experiences a zero net force. The self-similar evolution preserves the large-scale coherence of the expanding structure. For loop-structures in such an outflow, the leading edge can thus have a constant speed of any value, even below the local gravitational escape speed. Moreover, quite different moving plasma structures from different events may exhibit the same velocity structure. Once set up, the same radial velocity can lift anything out of the gravitational well.

That a parcel of coronal plasma can become gravitationally unbound is not new. If not for the magnetic field, the corona would be expelling plasma everywhere in a state of spherically symmetric solar wind. Magnetic fields provide additional confinement with gravity. The generally downward tension force of bipolar magnetic fields holds down a pocket of coronal plasma first in locally bound state, which then transits into an outflow when equilibrium fails (Low, 1981b; Wolfson, 1982).

The corona is an active participant. It can store energy and can become unstable with the onset of a transient outflow. The flare, the prominence, and the coronal transient need not have a simple cause—effect relationship, but may be different manifestations of a global loss of equilibrium. Even where a flare is involved, a transient is likely to occur only if the corona is in a stressed state with stored free energy. Otherwise, stable closed bipolar field lines can confine the coronal plasma, heated by the flare, against an outward expansion.

There are many physical isues of coronal transient physics to be settled by future theoretical and observational work (see the discussion in Low, 1984b). We have obtained a glimpse of the basic physics. The analytic work developed so far puts us in a relatively secure position to venture into a new generation of numerical simulation aimed at more realistic MHD models. The analytic self-similar MHD solutions offer an opportunity to test numerical code. A known self-similar solution can be used to generate the initial—boundary value problem of which it is the proper solution. A reliable numerical code should reproduce this solution from the generated initial—boundary value problem. To achieve this reproduction, a high standard of accuracy is required because the possibility is there for a self-similar solution to be physically unstable. Instability here is taken in the sense that a slight physical disturbance of the time evolution can grow with time to overwhelm the self-similar character of the MHD flow.

3.4. DISCUSSION

The magnetic field plays an important role in most coronal processes but it cannot be readily observed. What we can know about the physical environments in the corona is limited to a partial set of spatially and temporally averaged plasma parameters that fall within the capability of observational techniques. Theory has the goal of providing models into which available observational data can be fitted to yield a physically consistent scenario. Such a goal can never be achieved in full, of course. In the course of testing and refining theoretical models with observation, progress is made in understanding both the phenomena and the underlying physical processes. Although concepts such as 'magnetic loops' and 'magnetic reconnection' have found their way into everyday data interpretation, the models we have at present remain rather crude. Our survey of theoretical problems in this section shows a slow but certain progress in identifying and studying MHD processes at a basic level. While these problems have been motivated by coronal phenomena, they tend to be formulated with the aim of illustrating the basic physics. This approach is useful for building up an intuition for the phenomenology and modeling of complex, realistic processes in the corona. The time seems ripe to attempt realistic modeling of phenomena. Numerical simulations, with large computer resources, are necessary, and need to be developed. Several long-standing problems come to mind as obvious candidates for immediate, renewed attention. Models of equilibrium magnetic fields with realistic geometry can suggest on theoretical grounds the magnetic field configurations one expects to find in a particular situation in the corona. In flare studies, force-free magnetic field computation should be carried out into the nonlinear regime. In prominence studies, the question of the three-dimensional structure of the prominence plasma in relation to the supporting magnetic field should be taken up. There is also the large class of problems on magnetostatic structures as determined by the joint effect of force balance and specific energy processes such as radiative cooling, heating, and thermal conduction. The study of linear stability should follow immediately as geometrically more realistic equilibrium states are made available. Much of previous work on stability treated essentially one-dimensional equilibrium states. It seems worthwhile to begin treating the more complex equilibrium states found by various workers in recent years. Finally, we have the problems of the fully developed, time-dependent, nonlinear flows. Some progress has been made in studying these problems in connection with the large-scale transient flows in the corona. We are thus only beginning to learn some basic physics. Similar studies should be initiated for the flare, which involves quite a different set of physical conditions and constraints. Numerical techniques are necessary for these new developments. As emphasized in Section 3.3, the testing and questioning of numerical codes for their physical reliability is an essential part of the work. In the broad field of plasma physics, MHD is an area where analytical tools have been used with an impressive degree of effectiveness. This is fortunate in that theoretical studies of fundamental effects can be placed on a firm basis. Physical understanding and intuition can, thus, be developed to a level at which we can judge and interpret the results of numerical computation undertaken to simulate realistic complex situations. Only when we are reasonably assured of the physical correctness of a numerical model, can there be a meaningful comparison between the model with data. Through such a comparison, we learn some physics and gain insight into the new generation of models to be built and observations to be made.

4. Energy and Momentum Balance of Open and Closed Coronal Structures

The existence of the corona and solar wind requires the transport of energy outward from the photosphere by some means in addition to thermal conduction or radiation. One can place observational and theoretical constraints on the form and magnitude of this energy transport, but these constraints are generally somewhat different for the various atmospheric regions − in particular, for magnetically closed and open regions. In this section we concentrate on an analysis of the energy supply to coronal holes and high-speed solar wind streams in order to illustrate some basic physical processes, and then we discuss the implications of this analysis for the energy supply to magnetically closed coronal regions.

4.1. CORONAL HOLES AND HIGH-SPEED STREAMS

The solar wind structures that require the largest energy supply to the corona are high-speed solar wind streams, which originate in solar coronal holes (see review by Hundhausen, 1977, and references therein). Because of the rapidly diverging magnetic fields in coronal holes, the solar wind accelerates quickly to supersonic speeds, leaving in its wake a very low-density solar corona (e.g. Kopp and Holzer, 1976). Yet the rapid divergence has little to do directly with the high asymptotic flow speed of the wind (see review by Leer *et al.*, 1982, and references therein). Instead, the high wind speed results from the distribution of extended energy deposition above the coronal base: viz. high-speed streams cannot arise from purely thermally driven winds, but require a substantial deposition of energy in the region of supersonic flow (e.g. Parker, 1958, 1963, 1964a, b, 1965; Holzer and Leer, 1980; Leer and Holzer, 1980).

This requirement makes hydromagnetic waves an attractive candidate for transporting energy outward from the coronal base. Alfvén (or intermediate mode) waves are the most difficult hydromagnetic waves to damp in the solar corona and thus deposit a large fraction of their energy in the supersonic flow (e.g. Hollweg, 1978a; Leer *et al.*, 1982). Fast-mode waves are not too strongly damped in coronal holes, so they may also transport a significant energy flux from the low corona to the region of supersonic flow; furthermore, these waves refract into regions of low Alfvén speed in coronal holes (unlike Alfvén waves, which transport energy along the magnetic field) and thus might produce high-speed streams from the central regions of coronal holes (e.g. Fla *et al.*, 1983); this would be consistent with inferences that can be drawn from comparisons of coronal hole and solar wind observations (Hundhausen, 1977). Slow-mode waves are much more readily damped in the corona and are less likely to play a significant role in transporting energy from the coronal base to the region of supersonic flow. Alfvénic fluctuations observed at 1 AU (Belcher and Davis, 1971) thus might well represent the remnants of a hydromagnetic waves flux propagating outward from the coronal base in all three modes, for the compressive (fast and slow) modes would not survive at 1 AU.

4.2. ALFVÉN WAVES IN THE LOWER SOLAR ATMOSPHERE

Hydromagnetic waves may be generated in the corona (e.g. Barnes, 1969), or they may be generated in the lower solar atmosphere and propagate into the corona. In the latter case, propagation through the transition region (between the chromosphere and corona) involves certain difficulties (e.g. Osterbrock, 1961): fast-mode waves are refracted strongly and can carry no significant energy flux through the transition region into the corona; slow-mode waves steepen rapidly in the transition region (if they are not already shocks in the upper chromosphere) and the resulting shock waves dissipate their energy below or very near the coronal base. In addition, observations of solar oscillations in the middle chromosphere (Athay and White, 1979a) place a lower limit on the upward energy flux density carried by slow-mode waves of about 2×10^4 erg cm^{-2} s^{-1} and an upper limit of about 10^5 erg cm^{-2} s^{-1} (Athay and White, 1979b), which is much less than the energy flux density required to heat the corona and drive the solar wind in coronal holes ($\approx 5 \times 10^5$ erg cm^{-2} s^{-1}), to say nothing of that required to heat the upper chromosphere. These observations (Athay and White, 1979a, b) can also be used to place limits on the energy flux carried by Alfvén waves, but a careful analysis of the behavior of Alfvénic disturbances in the lower solar atmosphere must first be carried out, and this is done below, following the analysis of Leer *et al.* (1982).

Because of the very small density scale height in the lower solar atmosphere, a moderate wave period (minutes to hours) of an Alfvénic disturbance is comparable to or longer than an Alfvénic travel time across a characteristic scale length of the medium (i.e. $\omega \lesssim dv_A/ds$, where ω is the disturbance frequency and v_A the Alfvén speed), so the WKB approximation, normally used to describe Alfvén-wave propagation, breaks down. Analysis of Alfvénic disturbances, therefore, requires a reconsideration of the conservation laws and Maxwell's equations. For slow oscillations of the plasma in the solar atmosphere, the mass and momentum conservation laws, Ohm's law, Ampere's law, and Faraday's law can be written

$$\frac{\partial \rho}{\partial t} + \nabla \cdot \rho \mathbf{u} = 0, \tag{4.1}$$

$$\rho \left(\frac{\partial}{\partial t} + \mathbf{u} \cdot \nabla \right) \mathbf{u} = -\nabla p - \rho \, \frac{GM}{r^2} \, \hat{r} + \mathbf{J} \times \mathbf{B}/c, \tag{4.2}$$

$$\nabla \times (\mathbf{E} + \mathbf{u} \times \mathbf{B}/c) = 0, \tag{4.3}$$

$$\nabla \times \mathbf{B} = \frac{4\pi}{c} \, \mathbf{J}, \tag{4.4}$$

$$\nabla \times \mathbf{E} = -\frac{1}{c} \, \frac{\partial \mathbf{B}}{\partial t} \,, \tag{4.5}$$

where $\rho = nm$. Let us consider toroidal Alfvénic disturbances in an axisymmetric background plasma. First, (4.1)–(4.5) yield two equations for the ϕ-components (in spherical coordinates) of $\mathbf{u}$ and $\mathbf{B}$:

$$\rho \left(\frac{\partial}{\partial t} + \mathbf{u} \cdot \nabla \right) u_\phi + \frac{\rho u_\phi}{r} \, (u_r + u_\theta \tan \theta) + \frac{1}{r \sin \theta} \, \frac{\partial p}{\partial \phi}$$

$$= -\frac{1}{4\pi} \left[\frac{1}{2r \sin \theta} \, \frac{\partial B^2}{\partial \phi} - (\mathbf{B} \cdot \nabla) B_\phi - \frac{B_\phi}{r} \, (B_r + B_\theta \cot \theta) \right], \tag{4.6}$$

$$\left(\frac{\partial}{\partial t} + \mathbf{u} \cdot \nabla\right) B_\phi + \frac{u_\phi}{r} (B_r + B_\theta \cot \theta)$$

$$= (\mathbf{B} \cdot \nabla) u_\phi + \frac{B_\phi}{r} (u_r + u_\theta \cot \theta) - B_\phi \left(\frac{\partial}{\partial t} + \mathbf{u} \cdot \nabla\right) \ln \rho. \tag{4.7}$$

Now, if we assume axial symmetry ($\partial/\partial\phi = 0$), take the background flow velocity and magnetic field to be radial, and assume that terms of second order in the disturbance quantities ($\delta v = u_\phi$, $\delta B = B_\phi$) can be neglected (allowing the neglect of $\partial p/\partial t$), we find, following Heinemann and Olbert (1980), that in the limit $\sqrt{u} \ll \sqrt{v_A}$

$$\left(\frac{\partial}{\partial t} - v_A \frac{\partial}{\partial r}\right) f = -\frac{1}{2} g \frac{dv_A}{dr} \,, \tag{4.8}$$

$$\left(\frac{\partial}{\partial t} + v_A \frac{\partial}{\partial r}\right) g = \frac{1}{2} f \frac{dv_A}{dr} \,, \tag{4.9}$$

where $\delta v = \delta v_0 (\rho_0/\rho)^{1/4} (g - f)/(g_0 - f_0)$, $\delta B = \delta B_0 (\rho/\rho_0)^{1/4} (g + f)/(g_0 + f_0)$, and $v_A^2 = B_r^2/4\pi\rho$.

If f, $g \propto \exp(-i\omega t)$, then (4.8) and (4.9) have simple solutions in the two limits $\omega \ll v_A'$ and $\omega \gg v_A'$ (where $v_A' = dv_A/dr$). In the low-frequency limit ($\omega \ll v_A'$) the first term on the left side in both (4.8) and (4.9) is negligible, and we have the two solutions

$$f_\pm = \pm g_\pm = f_0 \left(\frac{v_A}{v_{A0}}\right)^{\pm 1/2} e^{-i\omega t}, \tag{4.10}$$

$$\delta v_- = \delta v_0 \left(\frac{r}{r_0}\right) e^{-i\omega t}, \tag{4.11}$$

$$\delta B_- = \delta B_0 \left(\frac{r_0}{r}\right) \left(\frac{v_{A0} \, v_{A0}'}{v_A \, v_A'}\right) e^{-i\omega t} = -i \, \delta v_- \left(\frac{B_r}{2v_A}\right)\left(\frac{\omega}{v_A'}\right) = 0\left(\frac{\omega}{v_A'}\right), \tag{4.12}$$

$$\delta v_+ = \delta v_0 \left(\frac{r}{r_0}\right) \frac{\ln (\omega/v_A')}{\ln (\omega/v_{A0}')} - e^{-i\omega t} = i \, \delta B_+ \left(\frac{v_A}{B_r}\right)\left(\frac{\omega}{v_A'}\right) \ln\left(\frac{\omega}{v_A'}\right) = 0\left(\frac{\omega}{v_A'}\right) \tag{4.13}$$

$$\delta B_+ = \delta B_0 \left(\frac{r_0}{r}\right) \left(\frac{v_A' \, v_{A0}}{v_A \, v_{A0}'}\right) e^{-i\omega t}. \tag{4.14}$$

These solutions have a straightforward physical interpretation: if a toroidal oscillation is imposed on the atmosphere at $r = r_1$, the first ($-$) solution describes the rigid body oscillation of the lower density (higher v_A) region of the atmosphere (in $r > r_1$), and the second ($+$) solution describes the rigid-body oscillation of the higher density (lower v_A) region ($r < r_1$). In the outer region, only a negligible twist of the magnetic field is required to maintain rigid-body behavior (i.e. $\delta B/B_r \ll \delta v/v_A$), but in the inner region, rigid-body oscillation can only be maintained by a relatively large twist (i.e. $\delta B/B_r \gg \delta v/v_A$), because of the substantial inertia of that region. If the Alfvén speed continues to decrease with decreasing r in the inner region, eventually $\omega/v'_A > 1$, and rigid-body oscillation will give way to the propagation of a torsional Alfvén wave, as is described below. In the high-frequency limit ($\omega \gg v'_A$, the WKB limit), the coupling terms on the right sides of (4.8) and (4.9) are negligible, and these equations have solutions describing inward (f) and outward (g) propagating waves:

$$f = f_0 \, e^{-i(\omega t + \int k \; dr)} \tag{4.15}$$

$$g = g_0 \, e^{-i(\omega t - \int k \; dr)} \tag{4.16}$$

$$\delta v = \left(\frac{\rho_0}{\rho}\right)^{1/4} [a_1 \, e^{-i(\omega t + kr)} - a_2 \, e^{-i(\omega t - kr)}] \tag{4.17}$$

$$\delta B = \frac{B_{r0}}{v_{A0}} \left(\frac{\rho}{\rho_0}\right)^{1/4} [a_1 \, e^{-i(\omega t + kr)} - a_2 \, e^{-i(\omega t - kr)}] \tag{4.18}$$

where a_1 and a_2 are constants. In general, of course, all the terms in (4.8) and (4.9) must be included. For the special case of an atmosphere in which the Alfvén speed varies exponentially with height (i.e. $v_A = v_{A0} \exp[(r - r)/h_A]$, where h_A is a constant), (4.8) and (4.9) have the general solution (Ferraro and Plumpton, 1958)

$$\delta v = i \left(\frac{r}{r_0}\right) \left[a_3 J_0 \left(\frac{\omega}{v'_A}\right) + a_4 Y_0 \left(\frac{\omega}{v'_A}\right)\right] e^{-i\omega t}, \tag{4.19}$$

$$\delta B = \left(\frac{r}{r_0}\right) \frac{B_r}{v_A} \left[a_3 J_1 \left(\frac{\omega}{v'_A}\right) + a_4 Y_1 \left(\frac{\omega}{v'_A}\right)\right] e^{-i\omega t}, \tag{4.20}$$

where J and Y are Bessel functions of the first and second kinds, a_3 and a_4 are constants, and $\omega/v'_A = \omega h_A/v_A$.

Using an exponentially varying Alfvén speed, one can construct a very simple model of the solar atmosphere by taking $h_A = 200$ km in $r \leq r_0$, and $h_A = \infty$ in $r > r_0$, where r_0 represents the coronal base (i.e. the top of the transition region). In such a model, we have an analytic description of the behavior of toroidal Alfvénic disturbances: δv and δB are described by (4.19) and (4.20) in the lower solar atmosphere ($r \leq r_0$) and by (4.17) and (4.18) in the corona ($r > r_0$). We can thus gain a clear understanding of all relevant physical effects, while closely reproducing the results of Hollweg (1978b), who solved numerically for δv and δB, using essentially (4.9) and (4.20), in a solar atmosphere described by 16 exponential layers. If we require that there be no inward propagating wave in $r > r_0$ (i.e. $a_1 = 0$ and $a_2 = \delta v_0$), and that the Poynting flux and the velocity amplitude be continuous across r_0 (i.e. δv and δB are continuous across r_0), then (4.17)–(4.20) yield the real parts of δv and δB in $r < r_0$:

$$\text{Re } \delta v = \frac{\delta v_0}{J_0 Y_1 - J_1 Y_0} \left(\frac{r}{r_0}\right) \left\{ \left[Y_1 J_0 \left(\frac{\omega}{v'_A}\right) - J_1 Y_0 \left(\frac{\omega}{v'_A}\right)\right] \cos(-\omega t + k r_0) - \right.$$

$$\left. - \left[Y_0 J_0 \left(\frac{\omega}{v'_A}\right) - J_0 Y_0 \left(\frac{\omega}{v'_A}\right)\right] \sin(-\omega t + k r_0) \right\}, \tag{4.21}$$

$$\text{Re } \delta B = \left(\frac{B_r}{v_A}\right)\frac{\delta v_0}{J_0 Y_1 - J_1 Y_0}\left(\frac{r}{r_0}\right)\left\{\left[Y_0 J_1\left(\frac{\omega}{v_A'}\right) - J_0 Y_1\left(\frac{\omega}{v_A'}\right)\right]\cos(-\omega t + kr_0) + \right.$$

$$\left. + \left[Y_1 J_1\left(\frac{\omega}{v_A'}\right) - J_1 Y_1\left(\frac{\omega}{v_A'}\right)\right]\sin(-\omega t + kr_0)\right\}, \tag{4.22}$$

where Bessel functions without arguments explicitly stated are evaluated at $\omega/v_{A0}' = \omega h_A/v_{A0}$. As is clearly illustrated in Hollweg's (1978b) model (see his Table 1, which is based on Gingerich et al. (1971) and Vernazza et al. (1973)), the eight orders of magnitude density change from the photosphere to the corona leads to a nearly four order of magnitude change in the Alfvén speed; hence, there is a broad range of frequencies over which waves can propagate in the photosphere, but cannot propagate at higher levels and are thus reflected (i.e. $\omega/v_A' \gg 1$ at the photosphere, but $\omega/v_A' < 1$ higher up). For $n = 5 \times 10^{16}$ cm^{-3} and $B = 10$ G, $\omega/v_A' \gg 1$ if $\omega \gg 5 \times 10^{-4}$ s^{-1}, so waves with periods of several minutes or less propagate in the photosphere according to the short-wavelength (WKB) approximation. We can describe these waves by using the complex equivalents of (4.21) and (4.22) and taking the asymptotic form of the Bessel functions for large arguments ($\omega/v_A' \gg 1$):

$$\delta v = \frac{\delta v_0}{J_0 Y_1 - J_1 Y_0}\left(\frac{r}{r_0}\right)\left(\frac{v_A'}{2\pi\omega}\right)^{1/2} e^{-i(\omega t - kr_0)} \times$$

$$\times \left\{\left[(Y_1 + J_0) + i(Y_0 - J_1)\right] e^{i(\omega/v_A' - \pi/4)} + \right.$$

$$\left. + \left[(Y_1 - J_0) + i(Y_0 + J_1)\right] e^{i(\pi/4 - \omega/v_A')}\right\}, \tag{4.23}$$

$$\delta B = \left(\frac{B_r}{v_A}\right)\frac{\delta v_0}{J_0 Y_1 - J_1 Y_0}\left(\frac{r}{r_0}\right)\left(\frac{v_A'}{2\pi\omega}\right)^{1/2} e^{i(\omega t - kr_0)} \times$$

$$\times \left\{\left[(Y_1 + J_0) + i(Y_0 - J_1)\right] e^{i(\omega/v_A' - \pi/4)} - \right.$$

$$\left. - \left[(Y_1 - J_0) + i(Y_0 + J_1)\right] e^{i(\pi/4 - \omega/v_A')}\right\}. \tag{4.24}$$

(4.23) and (4.24) clearly show that at these relatively high frequencies the Alfvénic oscillations in the photosphere are composed of outward and inward propagating waves, the latter having arisen from reflection at higher levels (but below the coronal base). The reflection coefficient (i.e. the ratio of the energy flux density in the downward propagating waves to that in the upward propagating waves) characterizing the atmosphere overlying the photosphere is thus

$$R = \frac{(Y_1 + J_0)^2 + (Y_0 - J_1)^2}{(Y_1 - J_0)^2 + (Y_0 + J_1)^2}. \tag{4.25}$$

For waves that propagate in the short-wavelength limit from the photosphere to the top of the transition region, there is virtually no reflection:

$$R \approx 0 \qquad (\omega \gg v_{A0}'), \tag{4.26}$$

For a wave that must 'tunnel through' to the corona because its wavelength becomes greater than the local scale height well before the corona is reached, most of the energy is reflected:

$$R \approx 1 - 2\pi\omega h_A/v_{A0} \qquad (\omega \ll v'_{A0}), \qquad\qquad (4.27)$$

and a standing wave is produced in the lower solar atmosphere, as pointed out by Hollweg (1978b). It is clear that these standing waves are not caused by discontinuities in the density scale height, as has been suggested by Hollweg (1972, 1978b), but are simply due to the continuous variation of the refractive index that arises from the decrease of density with height. The features of such a standing wave pattern are readily determined using (4.21) and (4.22), and one example, for $\omega/v'_{A0} = 10^{-3}$, is described below.

The toroidal Alfvénic disturbances propagate outward as simple waves above the coronal base ($r > r_0$), give rise to rigid body oscillation of the atmosphere just below the coronal base, and produce a standing wave pattern in the lower atmosphere. The minima in the standing wave pattern are imperfect nodes (i.e. δv, $\delta B \neq 0$ at the nodes). The reflection caused by the Alfvén-speed gradient is very efficient at this frequency $\omega = 10^{-3}\, v'_{A0}$: reference to (4.27) indicates that only 0.6% of the energy carried by upward propagating waves in the photosphere reaches the corona; the rest is reflected. It has been noted (Hollweg, 1978b) that if the generation of the Alfvén waves takes place at some particular location near the top of the convection zone, then at certain 'resonant' frequencies a node will be located at the generation point, and a modest velocity perturbation (δv) will correspond to a very large magnetic perturbation (δB) and thus to a very large energy flux at that frequency. Of course, to produce such a large energy flux by driving the wave at a node there are three necessary requirements: (1) the wave driver must be phase-coherent over many wave periods (see Section 4.3 below); (2) the wave generation region must be narrow in height and remain at a fixed height over many wave periods; (3) the frequency for which the preceding two conditions are met must correspond to the presence of a node in the wave generation region. It seems unlikely that these three requirements, or their equivalent, will be met in the real solar atmosphere, and we conclude that the large 'resonant' energy fluxes discussed by Hollweg (1978b) are an artefact of the model considered and should not be invoked in discussing the transport of energy from the photosphere to the corona.

On the basis of the preceding analysis and of the observational results presented by Athay and White (1979b), we can place limits on the energy flux that can be transported from the lower solar atmosphere to the corona by Alfvén waves. At the atmospheric levels where observational information on δv is available, if $\omega \gg v'_A$ the energy flux density passing into the corona is given by

$$\phi_A = nm \langle \delta v^2 \rangle v_A (1 - R), \qquad\qquad (4.28)$$

where R is given by (4.27). Even if $\omega > v'_{A0}/2\pi$, (4.28) is approximately correct provided we take $R = 0$ (cf. (4.26)). If $\omega \ll v'_A$ there is rigid body oscillation everywhere above the level where δv is measured, so $\delta v_0 \approx \delta v$ and

$$\phi_A \approx n_0 m \langle \delta v^2 \rangle v_{A0}. \qquad\qquad (4.29)$$

When ω is comparable to v'_A, (4.28) and (4.29) can be used in conjunction to estimate ϕ_A. Applying the above procedure to the data provided in Table 1 of Athay and White (1979b) leads us to conclude that Alfvén waves generated in the upper chromosphere or below, with wave periods between about 30 and 3000 s, can provide an energy flux density of no more than 1×10^5 erg cm^{-2} s^{-1} to the corona. For wave periods of longer than 1 h, a large-scale Alfvén wave might give rise to a Doppler shift of spectral lines rather than simply to line broadening, which is the effect on which the preceding interpretation is based. For such long periods, however, the atmosphere will be in rigid body motion everywhere from the middle chromosphere to the corona, so that an energy flux entering the corona of more than 10^5 erg cm^{-2} s^{-1} at these periods would correspond to chromospheric velocities of more than 10 km s^{-1}. As such large chromospheric Doppler shifts are not observed, the upper limit given above would seem to apply to these longer periods as well. Finally, one might expect that very short period waves ($\tau \lesssim$ a few seconds) could carry larger energy fluxes to the corona, given the observational constraints on the velocity amplitude in the lower solar atmosphere, but these waves tend to be damped strongly by viscous and frictional effects in the photosphere and the chromosphere (Osterbrock, 1961). Of course, energy flux densities of the order 10^5 erg cm^{-2} s^{-1} in the corona generally imply nonlinear waves in the photosphere and lower chromosphere, and such waves should be strongly damped in these lower regions (Parker, 1960; Osterbrock, 1961). We conclude, therefore that an energy flux density of more than 10^5 erg cm^{-2} s^{-1} carried into the region of supersonic solar wind flow by Alfvén waves could not be transported upward from the photosphere or chromosphere by Alfvén waves alone and thus would have to arise, at least in part, from the generation of Alfvén waves in the transition region or the lower corona.

4.3. ENERGY SUPPLY TO MAGNETICALLY CLOSED CORONAL REGIONS

The observational constraints on the energy flux transported from the photosphere to the corona by Alfvén waves are the same for magnetically closed coronal regions as for open regions (see above), although this point seems not to be appreciated by several workers (e.g. Hollweg, 1978b, 1981; Leroy, 1980; Ionson, 1982). In an apparent attempt to circumvent this problem, it has been suggested that large Alfvén wave energy densities can be produced in the corona, because coronal loops can act, in principle, as resonant cavities (see Hollweg (1981) for a wave treatment and Ionson (1982) for an electrical circuit analog). A simple extension of the above analysis indicates that for a given loop length and a particular wave frequency, a coronal loop can indeed exhibit such characteristics. For a variety of reasons, however, it seems highly unlikely that such resonant effects can help to provide the required energy supply (e.g. Leer et al., 1983): (1) if the waves are driven at the resonant frequency, the driver must be phase coherent over many wave periods if substantial energy densities are to be built up in the standing waves (if the system is driven by a white noise spectrum, the time-scale for building up substantial energy densities increases dramatically over that appropriate to resonant driving); (2) the density, magnetic intensity, and length characterizing the loop must remain unchanged over many wave periods if the resonant frequency is not to change as the energy density is built up; (3) the predicted energy densities would produce observable effects (viz. in δv and δB) which are not observed; (4) the resonance phenomenon does not overcome the difficulties discussed above in meeting the coronal energy flux requirement.

Fortunately, in closed coronal regions the energy supply problem can be resolved by the slow twisting of flux tubes (e.g. Parker, 1979), which corresponds to energy transport by torsional Alfvén waves in the very low frequency (fully non-WKB) limit (and, in terms of the simple linear model just discussed, which corresponds to a coherence time for the source term $F(t)$ driving the system which is long when compared with natural period of the oscillator). In this case, δv and δB are not related simply, as in a normal WKB wave, and the observational constraints on δv for the subcoronal atmosphere cease to be a problem. This mechanism does not resolve the difficulty faced in coronal holes, because in open regions the magnetic field lines are not tied at both ends, and twists propagate rather than building up (see Leer *et al.*, 1983).

Acknowledgement

We thank Stefano Migliuolo for reading and commenting on the manuscript.

References

Altschuler, M. D. and Newkirk, G. Jr: 1969, *Solar Phys.* **9**, 131.
Anzer, U.: 1968, *Solar Phys.* **3**, 298.
Anzer, U.: 1978, *Solar Phys.* **57**, 111.
Anzer, U.: 1979, *IAU Coll.* **44**, 322.
Anzer, U. and Pneuman, G.: 1982, *Solar Phys.* **79**, 129.
Arnold, V. J.: 1978, *Mathematical Methods of Classical Mechanics*, Springer, New York.
Arnold, V. J. and Avez, A.: 1968, *Ergodic Problems of Classical Mechanics*, Benjamin, New York.
Asseo, E., Cesarsky, C. J., Lachieze-Rey, M., and Pellat, R.: 1978, *Astrophys. J.* **225**, 621.
Athay, R. G. and White, O. R.: 1979a, *Astrophys. J. Suppl.* **39**, 333.
Athay, R. G. and White, O. R.: 1979b, *Astrophys. J.* **229**, 1147.
Barenblatt, G. I. and Zel'dovich, Ya. B.: 1972, *Ann. Rev. Fluid Mech.* **4**, 285.
Barnes, A.: 1969, *Astrophys. J.* **155**, 311.
Barnes, C. W. and Sturrock, P. A.: 1972, *Astrophys. J.* **174**, 659.
Belcher, J. W. and Davis, L. Jr: 1971, *J. Geophys. Res.* **76**, 3534.
Bell, A. R., Evans, R. G., and Nicholas, D. J.: 1981, *Phys. Rev. Lett.* **46**, 243.
Bernstein, I. B., Frieman, E. A., Kruskal, M. D., and Kulsrud, R. M.: 1958, *Proc. Roy. Soc. London* **A244**, 17.
Birn, J., Goldstein, H., and Schindler, K.: 1978, *Solar Phys.* **57**, 81.
Bond, D. J.: 1982, *Phys. Lett.* **88A**, 144.
Braginskii, S. I.: 1965, *Rev. Plasma Phys.* **1**, 205.
Cary, J. R. and Littlejohn, R. G.: 1983, *Ann. Phys.* **151**, 1.
Chandrasekhar, S.: 1961, *Hydrodynamic and Hydromagnetic Stability*, Clarendon Press, Oxford.
Colgate, S.: 1978, *Astrophys. J.* **221**, 1068.
Coppi, B.: 1976, *Nucl. Fusion* **16**, 309.
Cowie, L. L. and McKee, C. F.: 1977, *Astrophys. J.* **211**, 135.
Cox, D. P. and Tucker, W. H.: 1969, *Astrophys. J.* **157**, 1157.
Craig, I. J. D. and Brown, J. C.: 1976, *Astron. Astrophys.* **49**, 239.
Craig, I. J. D.: 1982, in E. Priest (ed.), *Solar Flare Magnetohydrodynamics*, Gordon and Breach, New York, p. 277.
Drake, J. F. and Lee, Y. C.: 1977a, *Phys. Rev. Lett.* **39**, 453.

Drake, J. F. and Lee, Y. C.: 1977b, *Phys. Fluids* **20**, 1342.
Dryer, M., Wu, S. T., Steinolfson, R. S., and Wilson R. M.: 1979, *Astrophys. J.* **227**, 1059.
Dungey, J. W.: 1953, *Monthly Notices Roy. Astron. Soc.* **113**, 180.
Dungey, J. W.: 1958, *Cosmic Electrodynamics*, Cambridge Univ. Press, Cambridge.
Dungey, J. W.: 1964, in W. N. Hess (ed.), *Physics of Solar Flares*, NASA SP-50, Washington, D.C., p. 415.
Einaudi, G. and Van Hoven, G.: 1981, *Phys. Fluids* **24**, 1092.
Fla, T., Habbal, S., Holzer, T. E., and Leer, E.: 1983, *Astrophys. J.* **280**, 382.
Ferraro, V. C. A. and Plumpton, C.: 1958, *Astrophys. J.* **127**, 459.
Forslund, D. W.: 1970, *J. Geophys. Res.* **75**, 17.
Foukal, P.: 1978, *Astrophys. J.* **223**, 1046.
Freidberg, J. P.: 1982, *Rev. Mod. Phys.* **54**, 801.
Furth, H. P., Killeen, J., and Rosenbluth, M. N.: 1963, *Phys. Fluids* **6**, 459.
Galloway, D. and Weiss, N. O.: 1981, *Astrophys. J.* **243**, 945.
Gilman, P. A.: 1970, *Astrophys. J.* **162**, 1019.
Gingerich, O., Noyes, R. W., Kalkofen, W., and Cuny, Y.: 1971, *Solar Phys.* **18**, 347.
Gold, T.: 1964, in W. N. Hess (ed.), *Physics of Solar Flares*, NASA SP-50, Washington, D.C., p. 389.
Gosling, J. T., Hildner, E., MacQueen, R. M., Munro, R. H., Poland, A. I., and Ross, C. L.: 1976, *Solar Phys.* **48**, 389.
Grad, H.: 1967, *Phys. Fluids* **10**, 137.
Grad, H. and Rubin, H.: 1958, *Proc. Second UN Int. Conf. on the Peaceful Uses of Atomic Energy*, Vol. 31, p. 190.
Gray, D. R. and Kilkenny, J. D.: 1980, *Plasma Phys.* **22**, 81.
Hagyard, M. J., Cummings, N. P., and West, E. A.: 1982, *Solar Phys.* **80**, 33.
Harvey, J. W.: 1977, *Highlights of Astronomy* **4**, 223.
Heasley, J. N. and Mihalas, D.: 1976, *Astrophys. J.* **205**, 273.
Heinemann, M. and Olbert, S.: 1980, *J. Geophys. Res.* **85**, 1311.
Heyvaerts, J.: 1974, *Solar Phys.* **38**, 419.
Heyvaerts, J. and Priest, E. R.: 1983, *Astron. Astrophys.* **117**, 220.
Hinton, F. L. and Hazeltine, R. D.: 1976, *Rev. Mod. Phys.* **48**, 239.
Hollweg, J. V.: 1972, *Cosmic Electrodyn.* **2**, 423.
Hollweg, J. V.: 1978a, *Rev. Geophys. Space Phys.* **16**, 689.
Hollweg, J. V.: 1978b, *Solar Phys.* **56**, 305.
Hollweg, J. V.: 1981, *Solar Phys.* **70**, 25.
Holzer, T. E. and Leer, E.: 1980, *J. Geophys. Res.* **85**, 4665.
Hood, A. and Priest, E.: 1979, *Solar Phys.* **64**, 303.
House, L. L. and Smartt, R. N.: 1982, *Solar Phys.* **80**, 53.
House, L. L., Wagner, W. J., Hildner, E., Sawyer, C., and Schmidt, H. U.: 1981, *Astrophys. J.* **244**, L117.
Hu, W. R. Hu, Y. Q., and Low, B. C.: 1984, *Solar Phys.* **83**, 195.
Hundhausen, A. J.: 1977, in J. B. Zirker (ed.), *Coronal Holes and High Speed Wind Streams*, Colorado Assoc. Univ. Press, Boulder, p. 225.
Ionson, J. A.: 1978, *Astrophys. J.* **226**, 650.
Ionson, J. A.: 1982, *Astrophys. J.* **254**, 318.
Jockers, K.: 1978, *Astrophys. J.* **220**, 1133.
Khan, S. A. and Rognlien, T. D.: 1981, *Phys. Fluids* **24**, 1442.
Kippenhahn, R., and Schlüter, A.: 1957, *Z. Astrophys.* **43**, 36.
Knobloch, E.: 1980, *Vistas in Astronomy* **24**, 39.
Kopp, R. A. and Holzer, T. E.: 1976, *Solar Phys.* **49**, 43.
Kopp, R. A. and Pneuman, G.: 1976, *Solar Phys.* **50**, 85.
Kruskal, M. D. and Kulsrud, R. M.: 1958, *Phys. Fluids* **1**, 265.
Kulsrud, R. M.: 1967, in P. A. Sturrock (ed.), *Plasma Astrophysics*, Academic Press, New York, p. 46.
Kuperus, M., Ionson, J. A., and Spicer, D. S.: 1981, *Ann. Rev. Astron. Astrophys.* **19**, 7.
Lang, K. R.: 1980, *Astrophysical Formulae*, (2nd edn), Springer Verlag, pp. 226–8.
Leer, E. and Holzer, T. E.: 1980, *J. Geophys. Res.* **85**, 4681.

Leer, E., Holzer, T. E., and Fla, T.: 1982, *Space Sci. Rev.* **30**, 161.
Leer, E., Holzer, T. E., and Fla, T.: 1983, *Astrophys. J.* **273**, 808.
Lerche, I. and Low, B. C.: 1977, *Solar Phys.* **53**, 385.
Lerche, I. and Low, B. C.: 1980a, *Solar Phys.* **66**, 285.
Lerche, I. and Low, B. C.: 1980b, *Solar Phys.* **67**, 229.
Leroy, B.: 1980, *Astron. Astrophys.* **91**, 136.
Leroy, J. L., Bommier, V., and Sahal-Brechot, S.: 1984, *Solar Phys.* **83**, 135.
Levine, R. H.: 1975, *Solar Phys.* **44**, 365.
Lortz, D: 1970, *Z. angew. Math. Phys.* **21**, 196.
Low, B. C.: 1975a, *Astrophys. J.* **197**, 251.
Low, B. C.: 1975b, *Astrophys. J.* **198**, 211.
Low, B. C.: 1977a, *Astrophys. J.* **211**, 234.
Low, B. C.: 1977b, *Astrophys. J.* **217**, 988.
Low, B. C.: 1980a, *Astrophys. J.* **239**, 377.
Low, B. C.: 1980b, *Solar Phys.* **67**, 57.
Low, B. C.: 1981a, *Astrophys. J.* **246**, 538.
Low, B. C.: 1981b, *Astrophys. J.* **251**, 352.
Low, B. C.: 1982a, *Rev. Geophys. Space Sci.* **20**, 145.
Low, B. C.: 1982b, *Solar Phys.* **75**, 119.
Low, B. C.: 1982c, *Astrophys. J.* **263**, 952.
Low, B. C.: 1982d, *Astrophys. J.* **254**, 796.
Low, B. C.: 1982e, *Astrophys. J.* **261**, 351.
Low, B. C.: 1984a, *Astrophys. J.* **281**, 381.
Low, B. C.: 1984b, *Astrophys. J.* **281**, 392.
Low, B. C. and Nakagawa, Y.: 1975, *Astrophys. J.* **119**, 237.
Low, B. C. and Wu, S. T.: 1981, *Astrophys. J.* **248**, 335.
Low, B. C., Hundhausen, A., and Zweibel, E.: 1984, *Phys. Fluids* **26**, 2731.
Low, B. C., Munro, R. H., and Fisher, R. R.: 1982, *Astrophys. J.* **254**, 335.
MacQueen, R. M.: 1980, *Phil. Trans. Roy. Soc. London* **A297**, 605.
MacQueen, R. M. and Fisher, R. R.: 1983, *Solar Phys.* **89**, 89.
Manheimer, W.: 1977, *Phys. Fluids* **20**, 265.
Mason, R. J.: 1981, *Phys. Rev. Lett.* **47**, 652.
Matte, J. P. and Virmont, J.: 1982, *Phys. Rev. Lett.* **49**, 1936.
Max, C. E.: 1981, Lawrence Livermore Rept UCRL-53107.
Migliuolo, S.: 1982, *J. Geophys. Res.* **87**, 8057.
Milne, A. M., Priest, E. R., and Roberts, B.: 1979, *Astrophys. J.* **232**, 304.
Moffatt, H. K.: 1978, *Magnetic Field Generation in Electrically Conducting Fluids*, Cambridge Univ. Press, Cambridge.
Morozov, A. I. and Solov'ev, L. S.: 1980, *Rev. Plasma Phys.* **8**, 1.
Mouschovias, T. Ch.: 1974, *Astrophys. J.* **192**, 37.
Mouschovias, T. Ch. and Poland, A. I.: 1978, *Astrophys. J.* **220**, 675.
Nakagawa, Y.: 1970, *Solar Phys.* **12**, 419.
Nakagawa, Y. and Raadu, M. A.: 1972, *Solar Phys.* **25**, 127.
Newkirk, G. Jr: 1967, *Ann. Rev. Astron. Astrophys.* **5**, 213.
Nordlund, A.: 1982, in J. O. Stenflo (ed.) 'Magnetic Fields in The Sun and Stars', *IAU Proc.* **102**, 79.
Nowak, T. and Ulmschneider, P.: 1977, *Astron. Astrophys.* **60**, 413.
Orrall, F. Q. (ed.): 1981, *Skylab Workshop on Solar Active Regions*, Colorado Assoc. Univ. Press, Boulder.
Osterbrock, D. E.: 1961, *Astrophys. J.* **134**, 347.
Parker, E. N.: 1955, *Astrophys. J.* **121**, 491.
Parker, E. N.: 1957, *J. Geophys. Res.* **79**, 1558.
Parker, E. N.: 1958, *Astrophys. J.* **128**, 664.
Parker, E. N.: 1960, *Astrophys. J.* **132**, 821.
Parker, E. N.: 1963, *Interplanetary Dynamical Processes*, Interscience, New York.

Parker, E. N.: 1964a, *Astrophys. J.* **139**, 72.
Parker, E. N.: 1964b, *Astrophys. J.* **139**, 93.
Parker, E. N.: 1965, *Space Sci. Rev.* **4**, 666.
Parker, E. N.: 1966, *Astrophys. J.* **145**, 811.
Parker, E. N.: 1972, *Astrophys. J.* **174**, 499.
Parker, E. N.: 1975, *Astrophys. J.* **201**, 494.
Parker, E. N.: 1977, *Ann. Rev. Astron. Astrophys.* **15**, 45.
Parker, E. N.: 1979a, *Cosmical Magnetic Fields*, Clarendon Press, Oxford.
Parker, E. N.: 1979b, *Astrophys. J.* **244**, 631.
Parker, E. N.: 1979c, *Astrophys. J.* **244**, 644.
Parker, E. N.: 1933a, *Astrophys. J.* **264**, 635.
Parker, E. N.: 1983b, *Astrophys. J.* **264**, 642.
Parker, E. N.: 1983c *Geophys. Astrophys. Fluid Dynamics,* **24**, 79.
Petschek, H. E.: 1964, in W. N. Hess (ed.), *Physics of Solar Flares*, NASA SP-50, Washington, D.C., p. 425.
Pneuman, G.: 1982, in E. R. Priest (ed.), *Solar Flare Magnetohydrodynamics*, Gordon and Bleach, New York, p. 379.
Pneuman, G. and Kopp, R. A.: 1971, *Solar Phys.* **18**, 258.
Poland, A. and Anzer, U.: 1971, *Solar Phys.* **19**, 401.
Priest, E.: 1978, *Solar Phys.* **58**, 57.
Priest, E. R. (ed.): 1982, *Solar Flare Magnetohydrodynamics*, Gordon and Breach, New York.
Proctor, M. R. E. and Weiss, N. O.: 1982, *Rep. Prog. Phys.* **45**, 1317.
Raadu, M. A.: 1972, *Solar Phys.* **22**, 425.
Rae, I. C. and Roberts, B.: 1983, *Astron. Astrophys.* **119**, 28.
Ray, A. and Van Hoven, G.: 1982, *Solar Phys.* **79**, 353.
Rechester, A. B. and Rosenbluth, M. N.: 1978, *Phys. Rev. Lett.* **40**, 38.
Rosenbluth, M. R. *et al.*: 1966, *Nucl. Fusion* **6**, 297.
Rosner, R., Golub, L., and Vaiana, G. S.: 1984, *Astrophys. J.* (in press).
Rosner, R., Golub, L., Coppi, B., and Vaiana, G. S.: 1978, *Astrophys. J.* **22**, 317.
Rosner, R., and Knobloch, E.: 1982, *Astrophys. J.* **262**, 349.
Roussel-Dupre, R.: 1980, *Solar Phys.* **68**, 243.
Rutherford, P. H.: 1973, *Phys. Fluids* **16**, 1903.
Sakurai, T.: 1979, *Pub. Astron. Soc. Japan* **31**, 209.
Sakurai, T. and Levine, R. H.: 1981, *Astrophys. J.* **248**, 817.
Sato, T. and Hasegawa, A.: 1982, *Geophys. Res. Lett.* **9**, 52.
Schmidt, H. U.: 1964, in W. N. Hess (ed.), *Physics of Solar Flares*, NASA SP-50, Washington, D.C., p. 107.
Scudder, J. D. and Olbert, S.: 1979a, *J. Geophys. Res.* **84**, 2755.
Scudder, J. D. and Olbert, S.: 1979b, *J. Geophys. Res.* **84**, 6603.
Sedov, L. I.: 1946, *Prikl. Mat. Mekh.* **10**, 241.
Seehafer, N.: 1978, *Solar Phys.* **58**, 215.
Shafranov, V. D.: 1960, *Sov. Phys. JETP* **26**, 682.
Sheeley, N. R., Jr, Michels, D. J., Howard, R. A., and Koomen, M. J.: 1980, *Astrophys. J.* **237**, L99.
Shoub, E. C.: 1983a, *Astrophys. J.* **266**, 339.
Shoub, E. C.: 1983b, Stanford Univ. preprint, SUIPR 946.
Sonnerup, B. U. O.: 1970, *J. Plasma Phys.* **4**, 161.
Sonnerup, B. U. O.: 1979, in L. J. Lanzerotti, C. F. Kennel, and E. N. Parker (eds), *Solar System Plasma Processes* Vol. 3, North-Holland, Amsterdam, p. 45.
Soward, A. M. and Priest, E. R.: 1982, *J. Plasma Phys.* **28**, 335.
Spicer, D. W.: 1979, *Solar Phys.* **62**, 269.
Spicer, D. S.: 1981, *Solar Phys.* **71**, 115.
Spitzer, L., Jr: 1962, *Physics of Fully Ionized Gases*, Interscience, New York.
Spitzer, L. and Harm, R.: 1953, *Phys. Rev.* **89**, 977.
Spruit, H. C.: 1977, Ph.D. Thesis, Utrecht, Holland.

Steinolfson, R. S.: 1982, *Astron. Astrophys.* **115**, 39.
Steinolfson, R. S., Wu, S. T., Dryer, M., and Tandberg-Hanssen, E.: 1978, *Astrophys. J.* **225**, 259.
Stenflo, J. O.: 1978, *Rep. Prog. Phys.* **41**, 865.
Stern, D. P.: 1966, *Space Sci. Rev.* **6**, 147.
Sturrock, P. A.: 1966, *Phys. Rev. Lett.* **16**, 270.
Sturrock, P. A. (ed.): 1980, *Skylab Workshop on Solar Flares*, Colorado Assoc. Univ. Press, Boulder.
Sweet, P. A.: 1958, *IAU Symp.* **6**, 123.
Sweet, P. A.: 1964, in W. N. Hess (ed.), *Physics of Solar Flares*, NASA SP-50, Washington, D.C., p. 409.
Sweet, P. A.: 1971, *IAU Symp.* **43**, 457.
Tajima, T., Brunel, F., and Sakai, J.: 1982, *Astrophys. J.* **258**, L45.
Tandberg-Hanssen, E.: 1967, *Solar Activity* Blaisdell, Waltham.
Tandberg-Hanssen, E.: 1974, *Solar Prominences*, D. Reidel, Dordrecht.
Taylor, G. I.: 1950, *Proc. Roy. Soc. London* **A201**, 1975.
Taylor, J. B.: 1974, *Phys. Rev. Lett.* **33**, 1139.
Tsinganos, K.: 1981, *Astrophys. J.* **245**, 764.
Tsinganos, K.: 1982a *Astrophys. J.* **252**, 775.
Tsinganos, K.: 1982b, *Astrophys. J.* **259**, 832.
Tsinganos, K.: 1984, *Astrophys. J.* (to be submitted).
Tsinganos, K., Distler, J., and Rosner, R.: 1984, *Astrophys. J.* **278**, 409.
Tucker, W. H.: 1973, *Astrophys. J.* **186**, 285.
Unno, W., Ribes, E., and Appenzeller, J.: 1974, *Solar Phys.* **35**, 287.
Vaiana, G. S. and Rosner, R.: 1978, *Ann. Rev. Astron. Astrophys.* **16**, 393.
Vaiana, G. S., Krieger, A. S., Timothy, A. F., and Zombeck, M. V.: 1976, *Astrophys. Space Sci.* **39**, 75.
Vaiana, G. S., Reidy, W. P., Zehnpfenning, T., Van Speybroeck, L., and Giacconi, R.: 1968, *Science* **161**, 564.
Van Kampen, N. G.: 1976, *Phys. Reports* **24C**, 171.
Van Hoven, G. and Cross, M. A.: 1973, *Phys. Rev.* **A7**, 1247.
Van Hoven, G., Ma, S. S., and Einaudi, G.: 1981, *Astron. Astrophys.* **97**, 232.
Van Tend, W. and Kuperus, M.: 1978, *Solar Phys.* **59**, 115.
Vasyliunas, V. M.: 1975, *Rev. Geophys. Space Phys.* **13**, 303.
Vernazza, J. E., Avrett, E. H., and Loeser, R.: 1973, *Astrophys J.* **184**, 605.
Weiss, N. O.: 1966, *Proc. Roy. Soc. London* **A293**, 310.
Withbroe, G. L.: 1975, *Solar Phys.* **45**, 301.
Withbroe, G. L. and Noyes, R. W.: 1977, *Ann. Rev. Astron. Astrophys* **15**, 363.
Wolfson, R. L. T.: 1982, *Astrophys. J.* **255**, 774.
Wu, S. T.: 1980, *IAU Symp.* **91**, 443.
Yang, C. K. and Sonnerup, B. U. O.: 1976, *Astrophys. J.* **206**, 570.
Yang, C. K. and Sonnerup, B. U. O.: 1977, *J. Geophys. Res.* **82**, 699.
Yeh, H. and Dryer, M.: 1981, *Astrophys J.* **245**, 704.
Zweibel, E. G.: 1981, *Astrophys. J.* **249**, 731.
Zweibel, E. G.: 1982, *Astrophys. J.* **258**, L53.
Zweibel, E. G. and Hundhausen, A. J.: 1982, *Solar Phys.* **76**, 261.

R. Rosner,   *B. C. Low and T. E. Holzer,*
Harvard-Smithsonian Center   *High Altitude Observatory,*
   for Astrophysics,   *National Center for*
60 Garden Street,   *Atmospheric Research,*
Cambridge, MA 01236,   *Boulder, CO 80307,*
U.S.A.   *U.S.A.*

CHAPTER 12

MAGNETIC ENERGY STORAGE AND CONVERSION
IN THE SOLAR ATMOSPHERE

D. S. SPICER, J. T. MARISKA, AND J. P. BORIS

1. Introduction

Until man broke gravity's shackles, now more than 20 years ago, our knowledge of the multifaceted role played by magnetic fields on the Sun was meager. The past few decades of steadily improving X-ray and ultraviolet observations above the atmosphere have fostered a corresponding growth in our understanding of the importance of magnetic energy conversion and release and magnetohydrodynamic activity to virtually all components of solar variability which affect our environment here on Earth. Every experimental improvement has brought us a closer and better view of the dynamic, sometimes violent, MHD and plasma phenomena which are now generally agreed to control the solar transition region, corona, solar wind, flares, spots, bright points, and spicules.

The observational problems are difficult and the distance is great, but progress on the observational side has been rapid. Figures 1 and 2 show examples of recent observations of dynamic plasma–magnetic field interactions which have led the scientific community to recognize the importance of magnetic energy storage and release in the solar atmosphere. Figure 1 shows an eruptive prominence photographed in He II ($\lambda = 304$), whose scale is the radius of the Sun. The material is clearly being confined by the magnetic field associated with the prominence. The figure also shows much of the other large scale structures of the outer layers of the solar atmosphere. The bright regions on the disk are active regions, the overall bright–dark pattern is the chromospheric network–cell structure of the quiet Sun, and the dark regions at the poles are coronal holes. Figure 2 shows another example of an eruptive prominence, in this case observed with a coronagraph at visible wavelengths. Here the prominence material is following a coronal transient out into the outer corona. This may well be an example of the nonlinear evolution of the thermal instability which we discuss in Section 2.3. Our physical interpretation of phenomena such as these still lags appreciably, however, because we are observing phenomena for which we have little intuition and a very noisy background.

Today, in general, we do not have enough detailed, simultaneously obtained information about small-scale structures or dynamics to distinguish between many possibly similar interactions and configurations. We still lack good resolution of the 3D plasma, velocity, and magnetic field profiles on 100–1000 km spatial scales. While experimentalists, of course, will seek to improve this situation, our theoretical understanding of what has been observed does not even encompass many of the gross properties that have become well established in the last two or three decades.

Peter A. Sturrock (ed.), Physics of the Sun, Vol. II, pp. 181–248.
© 1986 *by D. Reidel Publishing Company.*

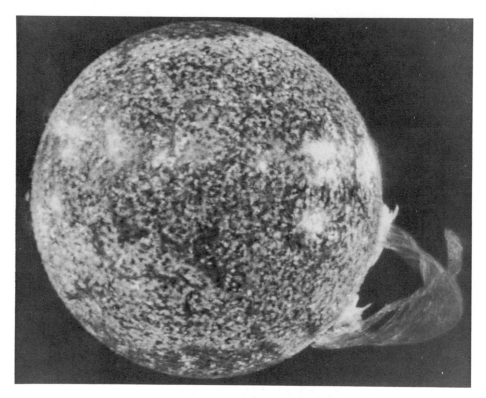

Fig. 1. Eruptive prominence photographed in He II 304 Å line by the NRL spectroheliograph on Skylab.

In the past, the problem of storage and conversion of magnetic energy on the Sun has been investigated primarily in the context of solar flares. Since we now know that magnetic energy conversion is relevant to other forms of 'counter-entropic' solar activity such as coronal heating (Kuperus *et al.*, 1981), the present review, while emphasizing many aspects of the physics of flares, will also point out and develop some applications to other solar phenomena. Solar flares are highly energetic transient events, so we can isolate structures better and hence can be more sure of what we are observing than in other solar phenomena such as spicules, coronal heating, or solar wind acceleration. Even though flares form a rather natural basis for discussion, our concern will be the general processes of magnetic energy storage and conversion. We could not hope to review all the specific flare models in which these processes occur. Further, a comprehensive objective review of this controversial topic already exists (Sturrock, 1980).

Chapters by Gilman and Noyes elsewhere in this work treat the generation of magnetic fields by convective dynamo action and the general aspects of solar and stellar magnetic activity and thus complement our discussion of the specific configurations, processes, and interactions involved in storing and converting this energy to the radiation and particle fluxes which so profoundly influence our Earth.

Astrophysicists have only recently come to understand that all low-density, high-temperature solar phenomena seem to stem from or be related to the generation and

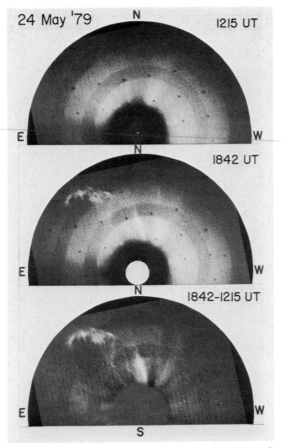

Fig. 2. Eurptive prominence following a coronal transient into the outer solar corona. Photograph taken with the NRL white-light coroagraph aboard the P78–1 spacecraft.

dissipation of magnetic fields (Book, 1981; Noyes, Chapter 19 in this work). In the absence of such 'counter-entropic' phenomena we expect stars to be hottest in the center and progressively cooler at larger and larger radii. The hot, tenuous corona, now known to be associated with many stars of several spectral types in addition to the Sun, actually heats the surface by thermal conduction. The original views on coronal heating favored upward propagating acoustic waves from the photosphere. Recent estimates of acoustic wave heating (Bruner, 1978; Athay and White, 1978, 1979), however, have shown that this mechanism does not provide sufficient energy to heat the corona, and have led to the conclusion that magnetic activity is responsible for the elevated coronal temperatures. The mechanisms and configurations are still very much in doubt (as we shall see), but the crucial role of magnetic fields is generally accepted.

There are a number of ways that a discussion of our current view of magnetic energy storage and release on the Sun could be structured. Observationally one is led to divide and subdivide along the lines of describing and explaining certain classes of phenomena such as flares, prominences, spicules, coronal heating, etc. This approach is attractive because it has a historical tradition and corresponds to large-scale features which have been seen. Separating phenomena along discipline lines according to whether they are principally magnetohydrodynamic (MHD) or microplasma-dynamic in origin is also attractive. Since the underlying MHD and plasma interactions seem to be similar in many different solar phenomena, however, it is also valuable to think in terms of a taxonomy which emphasizes simpler configurations and interactions and then extends and combines these simpler processes to explain particular complicated observations. We adopt this third perspective. We identify particularly important simple configurations of magnetic field and plasma and then concentrate on understanding the large-scale dynamic processes and transformations which these systems can undergo.

There are four good reasons for taking this approach. First, the use of simple examples realistically coupled to the solar environment is the best way to express positively what we do know about the complex magnetic energy storage and conversion mechanisms in the solar atmosphere. Second, this approach minimizes redundancy by concentrating on distinct fundamental physical processes which appear to be acting within many different solar contexts. These processes include magnetic flux emergence, tearing, anomalous field-aligned current dissipation, etc. Third, it is a good vehicle for discussing the storage, conversion, and release of magnetic energy because it permits the rather distinct active and passive roles for the magnetic field to be identified with correspondingly distinct configurations and processes. Finally, this approach generally concentrates on simple, macroscopically observable properties of the systems under study and thus it is well matched to the level and quality of observations available. Let us now identify several different idealized configurations and processes which span the observational and physical spectrum and yet are sufficiently simple that theory and computation can hope to bridge the information and understanding gaps which exist after the observations are analyzed. These gaps must be closed before we can honestly claim that the observations have been properly interpreted. Table I lists ten dynamic situations with their corresponding configurations, whether plasma or MHD properties are of primary importance, and several observed situations where these processes are thought to play a determining role in the evolution of the system. Five of these situations involve passive magnetic influences on the process in question. The magnetic field defines the geometry, makes the local plasma transport anisotropic, or channels energy flow in the atmosphere. The second five situations involve the active conversion and release of energy stored predominantly in local magnetic fields. Because magnetic energy is volumetric and the volumes are necessarily large, the macroscopic rate of energy transport to the conversion site is extremely important — as we shall see below.

It is not surprising that magnetic flux tubes with trapped plasma figure predominantly as the likely configuration. In fact, the active and passive processes, from the magnetic field point of view, clearly correspond to phenomena taking place perpendicular and parallel to the flux tubes respectively.

Section 2 develops and discusses some of the fundamental issues of magnetic energy conservation, transport, and electrodynamic coupling which are common to the various processes and simple scenarios we are considering. Rather than concentrate first on the processes of Table I, we begin by considering the special properties of plasma magnetically confined to flux tubes. The electrodynamic coupling viewpoint is then presented. This approach originates in the ionospheric–magnetospheric literature (e.g. Roederer, 1979) and is a lumped-parameter equivalent circuit approach capable of reproducing many of

TABLE I

Magnetodynamic processes in the solar atmosphere

Passive, magnetically controlled energy conversion (Section 3)

Process	Predominantly	Configuration	Could help explain
Plasma transport modification	Plasma	Transition region flux tubes	Transition region, flare evolution, solar wind
Transition region structure and dynamics	Plasma	Flux tubes	Transition region, condensations, spicules, jets
Plasma channeling and compression	MHD	Flux tubes	Spicules, coronal heating, network downflows
Wave channeling and dissipation	MHD	Flux Tubes	Spicules, coronal heating, solar wind, flares
Field-aligned current dissipation	Plasma	Flux tubes	Flares, coronal heating, bright points

Active, magnetohydrodynamic energy conversion (Section 4)

Process	Predominantly	Configuration	Could help explain
Magnetic flux emergence	MHD	Flux tubes	Bright points, coronal heating, flares
Dynamic geometric readjustments	MHD	Kinking or crossing flux tubes	Flares, bright points, coronal heating
Reconnection and magnetic tearing	Plasma	Crossing tubes, sheared fields	Flares, solar wind, spicules, bright points, heating
Particle and plasma acceleration	Plasma	Flux converging towards a neutral sheet	Flares, solar wind, cosmic rays, magnetotail
Flare trigger mechanisms	MHD and Plasma	Flux tubes, sheared fields	Flares, coronal heating, prominences

the transient variations and modeling most current comprehensive views of magnetic energy storage (cf. Van Hoven *et al.*, 1980). By accounting for the global electrodyamic coupling of the solar atmosphere, we are able to examine the fundamental problems associated with the concept of *in situ* versus remote magnetic energy storage. For modeling the detailed physical local phenomena such as the ten processes of Table I, the approach is limited by its lumped-parameter nature. Used in conjunction with detailed theories and models, however, electrodynamic coupling provides a very clean and satisfying way of connecting relatively idealized local scenarios, which we are forced to consider because of theoretical, computational and observational limitations, into a globally consistent 'real' solar environment.

Section 3 focuses on the five passive magnetic field roles identified in Table I and the plasma dynamics parallel to B on magnetic flux tubes. We emphasize physical understanding of these mechanisms, as opposed to mathematical rigor. Section 4 presents the five active magnetic energy release processes which we cavalierly have called magnetohydrodynamic, even though some aspects of particle acceleration are clearly microplasmakinetic in origin rather than fluid dynamic.

In addition, there are several underlying themes in this chapter we would like the reader to be aware of. We hope, first, that the reader will take away an appreciation of the creativity and hard work that has gone into bringing us the close-up view we now have of this rich panoply of dynamic phenomena, virtually undreamed of before the space age. However, it will also be obvious that this view is still too blurry to identify unambiguously all the operative mechanisms or to form a quantitative data base for calibrating predictive models. Our second theme is that the complex dynamics which follows from the interaction of many different processes occurring simultaneously on the macroscopic, MHD, and plasma time-scales means that the entire coupled system, including energy sources and boundary conditions at large distances from the region of interest, must be considered as a unit before quantitative predictions will be possible. Our final theme is that classical fluid concepts require substantial modification when applied to the dynamics of plasmas on solar flux tubes. The reader will become aware that many intuitive notions we hold from our earthbound fluid dynamics experience about how plasma trapped on flux tubes should behave are simply wrong. Not only are the magnetic field interactions extremely complex, with strong dependences on coupling to remote regions, the basic local properties of the medium are strange and nonintuitive.

Section 5 completes the chapter with a discussion of several outstanding questions still to be answered, including the new questions which have arisen as a result of our steadily evolving views on these complex solar processes and interactions.

2. Fundamental Concepts

In this section we introduce some of the fundamental MHD and plasma concepts needed to understand magnetic energy storage and transport in the solar atmosphere, while a more complete review can be found in Parker (1979) and Priest (1982). These basic theoretical notions are then applied to our selected processes and configurations in Sections 3 and 4. We first review how currents are generated and transported and the conditions that need to be satisfied for the storage of magnetic energy. Next, in Section 2.2, some of the basic concepts of ideal MHD theory are discussed including the energy principle. Other important properties of the medium are discussed in Section 2.3, followed by an examination of the origins of 'anomalous' resistivity in a collisionless plasma (Section 2.4). Finally, in Section 2.5 we introduce the concept of global electrodynamic coupling in the solar atmosphere, and examine how electrodynamic coupling affects both the magnetic energy storage process and the macroscopic stability of those plasma–magnetic field configurations in which this storage might occur.

2.1. MAGNETIC ENERGY GENERATION

Magnetic fields arise as a result of electric charges in motion; that is, currents. Magnetic energy storage arises when a current system, driven by external sources of electromotive force − i.e. voltage − does work against the voltages induced by the build-up of the current system itself. This work is stored in the magnetic field associated with the currents, and can be regained by allowing the currents to decay. To maintain the stored magnetic energy, the particular configuration must have the ability to generate new magnetic energy or to permit its transport into the system at a rate which is faster than dissipation can cause its conversion to other forms of energy. To understand how magnetic energy is stored in the solar atmosphere, therefore, we must address the question of how currents are generated and subsequently transported to the storage volume. In one way of viewing a flare, dynamic motions in the photosphere generate currents which are subsequently transported into the higher atmosphere. However, these currents cannot simply stop somewhere in the atmosphere, but must flow globally in such a way as to satisfy the requirement of current continuity $\nabla \cdot \mathbf{J} = 0$[1], where $\mathbf{J}$ is the current density. If $\mathbf{J}$ is resolved into a field-aligned current, $\mathbf{J}_{\|}$, and a current perpendicular to the magnetic field, $\mathbf{J}_{\perp}$, we can relate $\mathbf{J}_{\|}$ to $\mathbf{J}_{\perp}$ by $\nabla \cdot \mathbf{J}_{\|} = -\nabla \cdot \mathbf{J}_{\perp}$. That is, $-\nabla \cdot \mathbf{J}_{\perp}$ acts as a source for the field-aligned currents. Hence, $\mathbf{J}_{\|}$ communicates between adjacent regions of the solar atmosphere: the hot fully ionized coronal plasma is coupled electrodynamically with the colder partially ionized chromospheric and photospheric plasmas.

The physical processes that lead to the generation of $\mathbf{J}$ have their origins in the different motions electrons and ions experience in the direction perpendicular to the magnetic field. These differences result from either the small electron to ion mass ratio or from the difference in the sign of the charge. Using the single-fluid momentum equation (and neglecting gravity),

$$\rho \frac{d\mathbf{v}}{dt} = -\nabla p + \frac{\mathbf{J} \times \mathbf{B}}{c}, \tag{2.1}$$

and taking the vector cross product with $\mathbf{B}$ yields

$$\mathbf{J}_{\perp} = \frac{c\mathbf{B} \times \nabla p}{B^2} - \frac{c\rho}{B^2} \frac{d\mathbf{v}}{dt} \times \mathbf{B}, \tag{2.2}$$

where p is the total gas pressure, ρ the mass density, and $\mathbf{v}$ the bulk velocity of the plasma. Note that (2.1) reveals nothing about $\mathbf{J}_{\|}$. Equation (2.2) demonstrates that $\mathbf{J}_{\perp}$ arises in regions with pressure gradients and convective flow fields. Using $\nabla \cdot \mathbf{J}_{\|} = -\nabla \cdot \mathbf{J}_{\perp}$ and (2.2), we find (Sato and Iijima, 1979)

$$\mathbf{J}_{\|} = B_0 \int_0^s \left\{ \frac{c\rho}{B} \frac{d}{dt} \left(\frac{\mathbf{B} \cdot \nabla \times \mathbf{v}}{B^2} \right) + \frac{\mathbf{J}_{\perp} \cdot \nabla B}{B^2} - \frac{1}{\rho B} \left[\mathbf{B} \times \left(\frac{\rho}{B^2} \frac{d\mathbf{v}}{dt} \right) \cdot \nabla \rho \right] \right\} ds, \tag{2.3}$$

---

[1] We are justified in assuming $\nabla \cdot \mathbf{J} = 0$ if we consider phenomena for which the displacement current can be neglected, as is the case here.

where the integration is performed over the field-aligned coordinate s and incompressible flow is assumed. Equation (2.3) illustrates the various means by which $-\nabla \cdot \mathbf{J}_\perp$ gives rise to a $\mathbf{J}_\parallel$ in a fluid theory. The vorticity term (the leftmost term in the curly brackets) is the source usually thought to generate the currents believed to cause flares (e.g. Stenflo, 1969; Heyvaerts, 1974b). Note that if the integrand in (2.3) were zero, $\mathbf{J}_\parallel = \alpha_0 B$ would result, where α is the constant of integration for a specific field line. Thus, force-free currents must communicate between regions producing $-\nabla \cdot \mathbf{J}_\perp$.

Force-free fields are equilibrium solutions to a reduced set of magnetostatic equations based on the assumption that the pressure gradient either vanishes or is negligibly small; thus corresponding to a low β plasma—magnetic field configuration, where $\beta \equiv 8\pi P / B^2$ (e.g. Longmire, 1963).[2] The relevant force-free equations are

$$\nabla \times \mathbf{B} = \alpha_0(r)\mathbf{B} \tag{2.4}$$

and

$$\mathbf{B} \cdot \nabla \alpha_0 = 0. \tag{2.5}$$

The scalar function α_0 is constant on magnetic surfaces. The use of magnetostatic equilibria solutions in a clearly time-varying problem (energy storage requires time) is justified only if the magnetic field evolves in a time much longer than an MHD transit time, the time it takes an MHD wave to propagate over the longest magnetically determined spatial scale of the plasma—magnetic field configuration. The system thus appears to evolve through successive stages of quasistatic force-free equilibria each being in a higher energy state if storage is to occur. Since magnetic energy storage takes $\sim 10^6$ s in the case of a solar flare, this assumption is reasonable.

From the virial theorem it can be shown that no plasma—magnetic field configuration can be globally force free, only locally so (e.g. Longmire, 1963; Shafranov, 1966; Schmidt, 1979). Indeed, a fundamental result of the virial theorem is that the total pressure outside the region occupied by the force-free magnetic field must be greater than the mean pressure inside the force-free region. In laboratory magnetic confinement experiments where force-free fields are involved, such as reversed field pinches, this external pressure is provided by a highly conducting immoveable wall. Such a wall does not exist in the solar atmosphere, however, implying that a force-free field must be surrounded either by a high-pressure gas blanket, which could only occur in the low atmosphere, or by a potential field whose normal component matches that of the force-free field across their common boundary. This implies that the magnitude of the potential field is comparable to that of the force-free field. If the release of 10^{32} erg by a large flare is to be explained with force-free fields, we require $\gtrsim 500$ gauss potential fields distributed over volumes much greater than 10^{29} cm^3. An alternative is to assume that the principal field is potential with parallel currents (Syrovatskii, 1966; Barnes and Sturrock, 1972; Sakurai and Uchida, 1977; Uchida and Sakurai, 1977).

If α_0 is constant then the curl of (2.4) yields

$$\nabla^2 \mathbf{B} + \alpha_0^2 \mathbf{B} = 0, \tag{2.6}$$

---

[2] We emphasize that beta is a local quantity and varies by orders of magnitude within a given solar plasma—magnetic field configuration.

the vector Helmholtz equation. This equation is linear and has a general solution in terms of poloidal and toroidal components (Ferraro and Plumpton, 1966). Various authors have utilized (2.6) to compute the structure of force-free fields, and thus the magnetic energy stored in the atmosphere, using magnetograph line-of-sight field measurements and adjusting α_0 independently to achieve at least gross agreement with observed structures (Raadu and Nakagawa, 1971; Nakagawa and Raadu, 1972; Nakagawa, 1973, 1974; Levine, 1975, 1976; Barbosa, 1978; Nakagawa, 1978; Nakagawa et al., 1978).

Force-free field calculations represent an intermediate approach to understanding magnetic field structure on the sun. While line-of-sight field measurements can be used together with simple potential field theory to predict the general appearance of coronal structures, the inclusion of the α_0 function generally improves agreement with the observed coronal structure. The immensely more complex fully dynamic case would require a far better understanding of the solar plasma and magnetic fields than is likely to be available for some time. In fact force-free field calculations are just now achieving a fuller potential with the advent of vector magnetograph measurements to determine $\alpha_0 = (\nabla \times \mathbf{B}) \cdot \mathbf{B}/B^2$ with higher precision.

2.2. IDEAL MHD THEORY

The fundamental distinction between ideal MHD and dissipative MHD theory is that magnetic flux is a conserved quantity in the ideal MHD case. The conservation of flux in ideal MHD is a consequence of Faraday's law,

$$\frac{1}{c} \frac{\partial \mathbf{B}}{\partial t} = - \nabla \times \mathbf{E}, \tag{2.7}$$

which derives from the fact that the electric field, $\mathbf{E}$, integrated around any closed contour, is the negative of the rate of change of the magnetic flux ϕ through that contour. ϕ is defined by

$$\phi = \frac{1}{c} \int \mathbf{B} \cdot d\mathbf{S}. \tag{2.8}$$

In the nonrelativistic limit, Faraday's law is consistent with the following transformation of the electric fields $\mathbf{E}_f$ and $\mathbf{E}_m$ (Jackson, 1962):

$$\mathbf{E}_f = - \frac{\mathbf{v} \times \mathbf{B}}{c} + \mathbf{E}_m, \tag{2.9}$$

where $\mathbf{E}_f$ and $\mathbf{E}_m$ are measured in fixed and moving frames of reference, respectively, and $\mathbf{v}$ is the velocity of the moving frame relative to the fixed frame. Using (2.7), (2.8), and (2.9), the rate of change of flux through any moving contour is

$$\frac{d\phi}{dt} = - \oint dl \cdot \left(\mathbf{E}_f + \frac{\mathbf{v} \times \mathbf{B}}{c} \right), \tag{2.10}$$

where $d/dt = \partial/\partial t + \mathbf{v} \cdot \nabla$. For a perfectly conducting plasma, Ohm's law requires that

$$\mathbf{E}_f = - \frac{\mathbf{v} \times \mathbf{B}}{c}; \tag{2.11}$$

that is, the electric field in the frame moving with the plasma, $\mathbf{E}_m$, is zero. Hence, (2.10) reduces to $d\phi/dt = 0$, which means that the magnetic flow, ϕ, is convected along with the perfectly conducting plasma and is a constant of the motion.

In Sections 3.4 and 4 we consider in greater detail the hydrodynamic aspects of MHD and the new types of waves (Alfvén and magnetosonic) which the combined plasma–magnetic field system can support.

Next, consider how the motion of a perfectly conducting plasma induces currents, which in turn can produce heating. Assume that a sudden pressure increase occurs at the center of a cylindrical plasma in which only an axial magnetic field, B_z, is imbedded. As the central plasma pressure is increased, the resultant radial force inbalance drives an outward radial flow with velocity v_r. This flow velocity produces an azimuthal electric field, $-v_r B_z/c$, which reduces the axial field according to Faraday's law. This reduction in the B_z profile induces azimuthal currents, thus providing an inward $J_\theta B_z$ force which restores force balance at the higher plasma pressure. In the absence of dissipation, the plasma overexpands and then oscillates spatially about the new equilibrium. If dissipation is present, these oscillations are damped by resistive (collisional or collisionless) and/or viscous dissipation so that the oscillation energy reappears as heat. In this manner an MHD process can transport energy from a small volume into a larger volume without invoking diffusive transport mechanisms and can deposit the energy in a large volume if dissipation is present.

As a general rule, ideal MHD instabilities occur when a perturbation does not bend or stretch the magnetic field lines. Such perturbations do not induce corresponding magnetic restoring forces. Since these are necessary for restoration of equilibrium, the perturbation continues to grow. To identify, in a more definitive manner, those effects which are capable of causing instability and those which are capable of stabilizing, we utilize the energy principle (Bernstein et $al.$, 1958). By dividing the potential energy δW into positive and negative parts, we can identify the stabilizing and destabilizing terms. We write δW in the form (Furth et $al.$, 1966)

$$\delta W = \delta W_F + \delta W_v + \delta W_s, \tag{2.12}$$

where δW_F, the change in potential energy resulting from the perturbation of the plasma, is given by

$$\delta W_F = \frac{1}{2} \int d^3x \left\{ \underbrace{\frac{\delta B_\perp^2}{4\pi}}_{\text{(Alfvén)}} + 4\pi \underbrace{\left| \frac{\delta \mathbf{B}_\parallel}{4\pi} - \frac{B_0 \boldsymbol{\xi} \cdot \nabla P_0}{B_0^2} \right|^2}_{\text{(Magnetosonic)}} + \underbrace{\gamma_0 P_0 (\nabla \cdot \boldsymbol{\xi})^2}_{\text{(Sound)}} \right.$$

$$\left. - \underbrace{\frac{\mathbf{J}_0 \cdot \mathbf{B}_0}{B_0^2} (\mathbf{B}_0 \times \boldsymbol{\xi}) \cdot \delta \mathbf{B}}_{\text{(kink)}} - \underbrace{2\boldsymbol{\xi} \cdot \nabla p_0 \boldsymbol{\xi} \cdot \boldsymbol{\kappa}}_{\substack{\text{(interchange)} \\ \text{(ballooning)}}} \right\}; \tag{2.13}$$

δW_v, the change in potential energy resulting from the perturbation of any vacuum magnetic field, is given by

$$\delta W_v = \frac{1}{2} \int d^2x \, \frac{|\delta B|^2}{4\pi}; \tag{2.14}$$

and δW_s, the change in potential energy associated with any surface currents present, is given by

$$\delta W_s = \frac{1}{2} \int dS \cdot \left[\left[\boldsymbol{\nabla} \ P_0 + \frac{B_0^2}{8\pi} \right] \right] (\mathbf{n} \cdot \boldsymbol{\xi})^2 , \qquad (2.15)$$

where $\boldsymbol{\xi}$ is the fluid displacement, γ_0 is the ratio of the specific heats, n is a unit vector normal to the equilibrium magnetic surface, δB is the perturbed magnetic field, and

$$G = \frac{1}{2B_0^4} \ [B_0 \times \boldsymbol{\nabla}(8\pi P_0 + B_0^2)] \times \mathbf{B}_0 \qquad (2.16)$$

is the curvature of the magnetic field. δW_v always provides stabilization, and δW_s vanishes if no surface currents exist.

The first three terms in δW_F are stabilizing while the last two are destabilizing. The first of the two potentially destabilizing terms results from currents flowing parallel to B_0, and the second destabilizing term is due to the interaction of the pressure gradient with the field curvature. Notice that these terms arise from $\mathbf{J}_\parallel$ and $\mathbf{J}_\perp$, respectively. In solar situations β, the ratio of plasma to magnetic field energy $(\beta = 8\pi NkT/B^2)$, can vary by orders of magnitudes, so that $\mathbf{J}_\perp$ may play an important role in one region while $\mathbf{J}_\parallel$ predominates in others. However, just the opposite is true inside neutral sheets, where $\beta \to \infty$ since $B \to 0$ and $\mathbf{J}_\perp$ becomes very important. Interchanges are expected in neutral sheets where finite curvature effects exist according to Uchida and Sakurai (1977), and might play an important role in the energetics of neutral sheet reconnection during a flare.

The first and second stabilizing terms in δW_F arise because energy is required to stretch and shift lines of force where the direction of the magnetic field is changed by the perturbation. Contained within these terms are the global magnetic shear, the average shear over the entire magnetic surface, and the local shear, the amount a field line must be stretched if it is to exactly replace a neighboring field line in the course of the perturbation (Ware, 1965). Magnetic shear provides stabilization because the direction of the magnetic field changes with position in a sheared magnetic field. It is very difficult to interchange two neighboring field lines which are oriented at an angle to one another unless a line is bent and/or stretched. Since bending and stretching require that work be done on the field, δW_F increases rather than decreases; hence shear is stabilizing. We emphasize that shear is stabilizing in the ideal MHD limit because flux is conserved, i.e. field lines cannot be broken. However, dissipative effects such as resistivity allow the lines of force to break and reconnect. Thus, shear in the presence of dissipation is less effective as a stabilizing influence than in the ideal MHD approximation (see Sections 4.3 and 4.4).

Two compression terms exist in (2.13), both of which are stabilizing: $\gamma_0 p_0 (\boldsymbol{\nabla} \cdot \boldsymbol{\xi})^2$ and $(B_0 \boldsymbol{\xi} \cdot \boldsymbol{\nabla} p_0)^2 / B_0^2$. This follows because they are a measure of the net energy absorbed by the configuration in compressing the plasma and the magnetic field. As with stretching, a finite amount of compression is necessary if one field line is to replace another exactly. Notice that both terms become ineffective in the limit of $\beta \to 0$.

The destablizing term $\mathbf{J}_0 \cdot \mathbf{B}_0 (\mathbf{B}_0 \times \boldsymbol{\xi}) \cdot \delta \mathbf{B}$ is responsible for driving kink instabilities in force-free fields, by means of forces resulting from the interaction of the currents parallel to $\mathbf{B}_0$ and $\delta \mathbf{B}$ (Voslamber and Callebaut, 1962; Green and Johnson, 1962; Anzer,

1968; Raadu, 1972; Spicer, 1976; Van Hoven et al., 1977; Van Hoven, 1981). Energy is released by lowering the net current along the magnetic field. The constraint that the magnetic flux within a given flux surface be conserved is satisfied by bending and stretching the field lines into a helical or screw shape (Kruskal and Kulsrud, 1958).

The term $2\boldsymbol{\xi} \cdot \nabla P_0 \boldsymbol{\xi} \cdot \boldsymbol{\kappa}$ is related to the curvature and, thus, the tension of the lines of force, and is responsible for driving the interchange instability. This tension results in a force which is proportional to B^2, so that work must be done to move lines of force against this tension. One additional stabilizing effect, line-tying, is not obvious from the discussion here. In line-tying the fact that magnetic lines of force may penetrate a dense relatively immoveable conducting medium means that they are correspondingly more difficult to displace everywhere because of the globally coupled nature of magnetic field structures (see Sections 4.3 and 4.4).

2.3. NON-MHD PROPERTIES

In ideal MHD, dissipation and diffusion effects are ignored and the mean free paths of particles comprising the fluid are assumed to be short. Real plasmas do not satisfy these conditions so we must expect rather major departures in the behavior of plasma in magnetic flux tubes from the classical fluid or MHD pictures which replicate our earthbound intuition and experience. Radiation losses and strong electron thermal conduction are crucial phenomena omitted from usual MHD models. As we will show in sections 3 and 4, the behavior of plasma in the direction along a flux tube is usually dominated by these mechanisms, a situation often leading to behavior which is counter-intuitive. Not only must radiation losses be somehow replaced for a time averaged steady state to exist, but the composite profiles must be modified to transport the replacement energy continually to the regions of maximum radiation.

Thermal conduction parallel to the magnetic field in solar transition region and coronal plasmas is well represented by the nonlinear energy transport equation

$$\frac{\partial E}{\partial t} + \nabla \cdot (E + P)v = \nabla \cdot [1.1 \times 10^{-6} T_e^{5/2} \nabla T_e] + \text{Sources} - \text{Losses}, \qquad (2.17)$$

where E is the energy density in units of erg cm^{-3} and the electron temperature is measured in kelvins. The coefficient of thermal conduction increases as the temperature to the 5/2 power and is uaully nonnegligible for temperatures above 10^6 K, where it is so strong that constant temperature is often a good approximation.

Because the plasma has this very strong diffusion term threaded up its spine, transients can propagate and transport energy faster than the sound speed (for Alfvén or magneto-acoustic speeds). Since the sound speed

$$C_s = \left(\frac{\gamma P}{\rho}\right)^{1/2} \qquad (2.18)$$

is the determining characteristic speed in gasdynamics, it is not surprising that our medium, even on the macroscopic time and space scales, behaves very differently. There is, for example, a nonlinear thermal conduction 'shock' which Equation (2.17) admits when the density is constant. This is a profile which propagates at constant velocity and fixed shape as shown in Figure 3. The temperature at $Z = -\infty$ is $T_=$. T_0 and X_0 are characteristic temperature and length scales for the profile. In the frame of the shock the conducting plasma is in motion to the right at a speed

$$V_\kappa = 4kT_0^{7/2}/15P_0X_0 = 2.93 \times 10^{-7}T_0^{7/2}/P_0X_0 \qquad (2.19)$$

which is just the ratio of a characteristic thermal conduction energy flux to a characteristic thermal energy density.

THERMAL CONDUCTION "SHOCK"

Seek propagating temperature profile of fixed shape

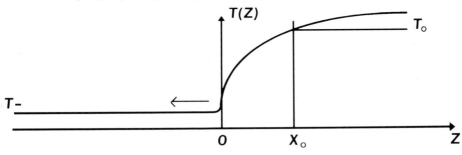

Fig. 3. The temperature profile through a propagating thermal conduction shock. Heating on the right drives a conduction wave into the cool or cold material (temperature T_-). Since the thermal conduction is a nonlinear function of the temperature, the profile becomes very steep when the temperature is low.

This characteristic velocity increases with the temperature as the 7/2 power and linearly with gradient strength $1/X_0$. Taking the ratio of this thermal conduction velocity to the sound speed gives a thermal conduction shock 'Mach' number

$$V_\kappa/C_s = 2 \times 10^{-11}T_0^3/P_0X_0. \qquad (2.20)$$

In the typical solar transition region $X_0 \sim 1$ km, $T_0 \sim 10^5$ K, and the pressure $P_0 \sim 0.1$ erg cm^{-3}. Using Equation (2.20), $V_\kappa/C_s \sim 2$; that is, the thermal conduction shock travels twice as fast as the sound speed. At higher temperatures, or in steeper gradients, the sound speed C_s may be even smaller relative to V_κ and thermal conduction will then dominate.

It should not be surprising that these additional effects, strong thermal conduction and radiation, introduce an important mode of 'condensational' instability to the system. Figures 1 and 2 show condensations of plasma in the corona, for instance. The linear properties of condensational instabilities in solar plasmas are simple to understand. A plasma medium such as the solar corona is optically thin in the principal components of its radiation field. The volumetric energy loss rate may be written as $L \sim \rho^2 \phi(T)$, where ρ is the local density and $\phi(T)$ is a positive function of temperature. If the pressure is constant and there is a steady heat source, S, some regions of the plasma cool slowly and others are heated according to the balance of L and S. As an element of the plasma cools, the density increases so that pressure balance is maintained with the hotter surroundings. When L increases faster with ρ than it decreases with the lower temperature, there is a strong tendency for material to condense radiationally; this is called the condensational instability.

Now consider a volume of plasma which is hotter than its surroundings. As it expands, the energy addition rate increases its dominance over the radiation cooling process. This runaway temperature effect, called the thermal instability, is equivalent to the condensational instability with a temperature perturbation of the opposite sign. These types of instabilities are exponential for small perturbations and are generally damped by thermal conduction for perturbations with a short enough scale length.

A substantial amount of research has been done to understand the linear condensational (thermal) instability in fluids at astrophysical temperatures and pressures. Perhaps the most complete analysis is that of Field (1965) who added substantially to the prior work of Parker (1953), Zanstra (1955a, b), and Weymann (1960). Field's work clarified the stability criteria and extended the linear analyses to include effects other than radiation loss such as magnetic fields, thermal conduction, rotation, and density stratification. Recently, Antiochos (1979) has extended this work to consider the stability of a stratified fluid for which the heating source term has a temperature dependence.

The nonlinear behavior of condensational instability, the way the system evolves and eventually establishes a new dynamic equilibrium where radiation and thermal conduction are balanced, has recently been explored computationally by Oran et al. (1982) for the solar atmosphere. They observe that the nonlinear evolution of the condensational instability results in a bifurcated system. Most of the mass resides in a small but dense 'condensation' surrounded by a hot, tenuous plasma atmosphere which conducts energy into the condensation at the rate at which it can be radiated away. It is clear from these calculations that it is possible for the condensation to sit stably in a hot plasma without evaporating. The bifurcated system which results in the simulations looks very much like the chromosphere–corona system with a sharp transition region between that is stable. In the next section we shall consider the solar transition region in greater detail since it is strongly influenced by magnetic structures.

Equation (2.17), in addition to showing the strong influences of thermal conduction and radiation, hints at some of the effects of the long collisional mean free path on our fluid equations. The temperature which enters is the electron temperature, not the total temperature, because electron conduction is some 25 times more effective than ion conduction in transporting energy along the magnetic field. In fact, many solar problems require electrons and ions to be treated as separate but coupled interpenetrating fluids with their own energy equations and separate temperatures. Collisional coupling delays between electrons and ions in the corona can be longer than either thermal or sonic characteristic times. Further, since radiation and ionization are atomic processes driven by collisions, it comes as no surprise that nonequilibrium radiation and ionization effects are often important. In some cases departures from a Maxwellian distribution function for the electrons can develop, leading to a high-velocity tail. These more energetic electrons can then substantially alter electron–ion collision rates and produce departures from equilibrium in the ionization balance of elements in the transition region (Shoub, 1982).

2.4. THE CONCEPT OF 'ANOMALOUS' RESISTIVITY

In hot plasmas such as the solar corona the collisional electrical resistivity is quite small for the same reason that the thermal conduction is so large. The plasma has long mean free paths and generally prefers to support collective modes as a means of coupling electrons and ions in the relative absence of collisions. These new plasma modes exist entirely outside MHD and fluid frameworks. Three effects we will consider are anomalous Joule heating, reconnection (neutral sheets, current sheets, tearing modes, resistive kinks), and double layers (current interruption). Here we deal with the general subject of resistivity in a hot plasma.

As magnetic field lines move through a plasma, resistive damping of the induced currents results in Joule heating, thus, energy is removed from the field and appears in the plasma as heat. The energy per unit volume lost from the field in a time τ_L is $\eta J^2 \tau_L$, so that the magnetic energy dissipation time is expressed approximately by

$$\tau_L = \frac{B^2}{8\pi\eta J^2}, \tag{2.21}$$

where η is the electrical resistivity. In much the same way that radiation is a local mechanism for removing thermal energy from the system, the resistivity provides a local mechanism for transforming magnetic field energy into heat.

To understand the physical origins of resistivity, consider the equation of motion of a test particle of charge $-e$ and mass m_e, drifting with a velocity v_D with respect to a stationary ion background:

$$m_e \frac{dv_D}{dt} = -eE - m_e v_D \nu(v), \tag{2.22}$$

where E is the macroscopic electric field, taken to be constant; and $\nu(v)$, the collision frequency, is given by (Spitzer, 1962)

$$\nu(v) = \frac{4\pi n_e e^4 \ln \Lambda}{m_e^2 v^3}. \tag{2.23}$$

Here n_e is the background electron number density; v, the particle's net velocity; and $\ln \Lambda = \ln(4\pi n_e \lambda_{D_e}^3)$, with λ_{D_e} representing the electron Debye radius. In the absence of collisions, electrons are accelerated freely, relative to the ions, such that

$$v_D = \frac{-eEt}{m_e}. \tag{2.24}$$

Conversely when the electrons undergo collisions with the more massive ions or with slow-moving field electrons, a steady state can be achieved, characterized by

$$v_D = \frac{-eE}{m_e \nu(v)}. \tag{2.25}$$

The electron velocities are randomized during these collisions so that the energy associated with the drift velocity is converted to Joule heat. Using the following relationship between the current density and the electron drift velocity $J = -n_e e v_D$, and (2.25), we find a local relationship between J and E:

$$J = \frac{n_e e^2 E}{m_e \nu(v)} = \frac{E}{\eta}; \tag{2.26}$$

where η, the electrical resistivity, is given by

$$\eta = \frac{4\pi\nu(v)}{\omega_{pe}^2}, \tag{2.27}$$

and the plasma frequency $\omega_{pe} = (4\pi n_e e^2 / m_e)^{1/2}$.

If the drift velocity of the electrons is less than the electron thermal velocity $v_{T_e} = (k_b T_e / m_e)^{1/2}$, we have $v = \sqrt{v_D^2 + v_{T_e}^2} \approx v_{T_e}$, so that (2.23) reduces to

$$\nu = \frac{4\pi n_e e^4 \ln \Lambda}{m_e^{1/2} (k_b T_e)^{3/2}} \tag{2.28}$$

the classical result (Spitzer, 1962). Using the definition $\lambda_{D_e} = v_{T_e}/\omega_{pe}$, (2.28) can be expressed as

$$\nu = \frac{\ln \Lambda}{\Lambda} \omega_{pe}. \tag{2.29}$$

Notice that the drag force on the electrons for which $v_D < v_{T_e}$, is $m_e v_D \nu \sim v_D/v_{T_e}^3$ and increases with v_D so the steady-state condition given by (2.25) can be achieved. For electrons with drift velocities in the regime $v_D > v_{T_e}$, the drag force decreases as v_D^{-2}, so a steady state cannot be attained. The following physical picture thus emerges. In the low-velocity regime ($v_D < v_{T_e}$), where the drag force dominates the electric force, the electron motion is essentially random and a steady state with drift can be achieved. However, as the dynamical friction becomes weaker at higher velocities, there is a critical velocity beyond which electrons are accelerated faster than collisions can decelerate them. As a result, these electrons continually gain more energy, because the friction they feel is reduced still further with increasing velocity. Eventually, the friction becomes sufficiently negligible that they are freely accelerated by the dominant electric force, until some other energy and momentum loss mechanism, such as radiation or an instability, becomes dominant. The effect of these additional loss mechanisms invariably appears as a cutoff in the distribution function at higher energies. These freely accelerated electrons are called 'runaway' electrons. The following expression for the critical velocity, v_c, at which electrons start to run away is obtained by balancing the two opposing forces on the RHS of (2.22) using (2.23):

$$v_c^2 = \frac{4\pi n_e e^3 \ln \Lambda}{m_e |E|}.$$ (2.30)

The Dreicer electric field, E_D (Dreicer, 1959), is defined as the electric field at which a thermal electron will run away (i.e. for which $v_c = v_{T_e}$). We have

$$E_D = \frac{e \ln \Lambda}{\lambda_{De}^2}.$$ (2.31)

Physically, this is the electric field at which the electron energy gained in one collision time is equal to the thermal energy. Using (2.30) and (2.31), we find that runaway occurs for any electron with a velocity

$$v \geqslant v_c = \left[\frac{E_D}{E}\right]^{1/2} v_{T_e}.$$ (2.32)

So far, we have treated single particle motion only. However, the behavior of a bulk electron distribution changes remarkably in the presence of an external electric field, due to the reduction of $\nu(v)$ with higher velocities. In the collision-dominated portion of the velocity distribution, we expect the electron distribution to be a slightly skewed Maxwellian drifting relative to the ions with a velocity given by (2.25). The entire electron distribution thus consists of a skewed and drifting Maxwellian, containing most of the population, with a very long and highly anisotropic tail antiparallel to E. This anisotropic, drifting distribution is a source of excess free energy, capable of exciting various collective plasma microinstabilities which may, in turn, inhibit the free acceleration of the tail of the distribution.

If the drift velocity of the skewed Maxwellian lies in the range $(v_{T_i}, c_s) \lesssim v_D \lesssim v_{T_e}$, where $c_s = (kT_e/m_i)^{1/2}$ is the ion sound velocity, and v_{T_i} the ion thermal velocity, various current-driven collective microinstabilities can be excited. These instabilities are driven by the bulk of the current carrying electrons. Other microinstabilities can be excited by the runaway electron population, if their distribution possesses a bump. The plasma can then be unstable to the generation of waves which are normal modes of the plasma. Electrostatic waves, in particular, grow at the expense of the free energy associated with the drift energy of the electrons. Scattering of the drifting electrons by these turbulent wave electric fields causes an enhanced momentum and energy loss, and so the term 'anomalous resistivity'.

There are many possible plasma instabilities, particularly in the presence of time varying magnetic fields, which can contribute to anomalous resistivity. Which instability and its strength and importance depend very much on the specific configuration and circumstances. Nevertheless, the general techniques used to estimate the macroscopic transport properties of the plasma are important, irrespective of which configuration or instability is being considered.

Various approaches to computing the level of plasma turbulence, and thus resistivity, from a given plasma instability have been applied to solar physics problems. For example, the mode-independent approach of Galeev and Saydeev (1979) is based on the conservation of energy and momentum between waves and particles at marginal stability. The assumption of marginal stability in solar problems is a particularly powerful technique for computing turbulence levels because the growth times and saturation times of instabilities which produce anomalous resistivity are $\sim 10^6$ times shorter than any macroscopic time, such as a tearing mode growth time or the lifetime of a flare. Thus, if anomalous resistivity is to be important in solar problems, the instability causing the anomalous resistivity must be at or near marginal stability and driven continually during the course of a flare by a source of free energy external to the instability. The marginal stability approach (Boris et al., 1970; Manheimer and Boris, 1972, 1977; Manheimer and Flynn, 1974; Manheimer et al., 1976; Manheimer, 1979; Manheimer and Antonsen, 1979), has yielded theoretical predictions which are consistent with laboratory experiments and astrophysical observations.

In general a micro-unstable plasma is assumed to exist in a linearly unstable state. The anomalous transport coefficients are determined by the r.m.s. fluctuation level, which is limited by local nonlinear effects such as reasonance broadening (Dupree, 1967) or mode coupling (Manheimer et al., 1976; Cohen et al., 1976). To obtain correct values of transport coefficients, it is essential to utilize a nonlinear theoretical treatment of all relevant mechanisms. In the Manheimer approach, it is assumed that the relaxation of the macroscopic plasma parameters to the linearly stable state is the most effective stabilization mechanism, so that the plasma presumably is at or near marginal stability even if some mechanism continually drives it towards instability. In this way the nonlinear theory of the relevant plasma mechanism plays a far less important role in the analysis of a marginally stable plasma than for a plasma that is linearly unstable. This is not to say that nonlinear effects play no role at all, because some instability saturation mechanism must be in effect to limit the turbulence level; nevertheless, the marginal stability approach does not depend on the detailed nonlinear evolution of this mechanism. In principle, the marginal stability approach is similar to the quasilinear theory of turbulent transport (cf. Galeev and Saydeev, 1979) but with the additional assumption that the characteristic time over which the instability develops and fluctuation amplitudes vary is short compared to the time-scale of the macroscopic fluid quantities.

We wish to note that thermal conduction, mass diffusion, and viscosity can also have anomalous components driven by plasma microinstabilities. Because the magnetic field-current system plays a major role in energy storage and transport, however, it is natural that resistivity is usually considered most important. In Sections 3 and 4 below we consider specific examples of anomalous resistivity in the solar context, with emphasis on the configurations and processes of Table I.

2.5. GLOBAL ELECTRODYNAMIC COUPLING

In magnetically permeated plasmas, such as occur on the Sun, the globally interconnected magnetic field lines satisfy $\nabla \cdot \mathbf{B} = 0$. This means that it is generally incorrect to look only at the local region in studying any dynamic phenomenon. For any change which affects the magnetic field locally, a corresponding external change must occur "somewhere else' to keep $\mathbf{B}$ divergence free. On macroscopic time-scales the electric current density J is also divergence free, further complicating the coupling between the

local region of interest and somewhere else. In trying to understand a localized phenomenon, therefore, we are interested in those global properties of 'somewhere else'
which are easy to specify and relatively independent of special assumptions.

Because the magnetized plasma is an electrodynamically active medium, it is natural
to use a lumped-parameter circuit-element approach to describe these global interactions
between the magnetic field loops and plasma structures of the chromosphere and corona
and the underlying flows and current sources in the photosphere. Global considerations
not only determine how much magnetic energy is available in any given configuration,
they also influence the rate at which energy can be transferred from a storage site to a
site of dissipation and heat release.

Although magnetic energy can be transported at any velocity up to that of light, the
Alfvén velocity is one important characteristic velocity associated with the dielectric
properties of a magnetized plasma, and is important for transmitting low-frequency
signals of voltage, current, and inductance change. For example, when a voltage source
is switched on suddenly in a circuit made up of a magnetized plasma and a load, the
current surge propagates through the circuit with a dispersion of speeds determined by
the frequency spectrum excited and the wave modes which can be supported by the
dielectric properties of the magnetized plasma. The specific mode which will transport
the bulk of the power associated with the current surge depends on the power spectrum
associated with the switching voltage. One of these modes may be the Alfvén-wave mode
or a mode that propagates at the speed of light. In fact, it is well known that switching
operations or short circuits in transmission lines used in utilities, which cause a change
from steady state, result in current surges accompanied by a self-consistent voltage
surge that propagate at light speeds. However, such utility circuits cannot support Alfvén
waves because they are not normal modes of such circuits, whereas a magnetized plasma
can support Alfvén waves as well as higher velocity waves.

From a physical point of view, electrodynamic coupling arises when the resistivity
tensor, η, varies from some finite value to effectively zero along a given field line in the
presence of an external force. This can be understood in the solar context as follows.
Consider two domains connected by a magnetic field (Figure 4). The first domain,
denoted by PC (perfectly conducting) schematically corresponds to the corona and
satisfies Ohm's law in the ideal MHD approximation,

$$\mathbf{E} + \frac{\mathbf{v} \times \mathbf{B}}{c} = 0, \qquad (2.33)$$

which implies that the corona is essentially flux preserving.[3] The second domain, denoted
by R (resistive), corresponds to the photosphere and convection zone of the Sun and is
characterized by an Ohm's law

$$\mathbf{E} + \frac{\mathbf{v} \times \mathbf{B}}{c} = \eta \cdot \mathbf{J}. \qquad (2.34)$$

Contained in η are components that connect neighboring field lines, and thus allow the
discharge of excess charges or the flow of currents perpendicular to $\mathbf{B}$. Note that some

---

[3] Here we ignore for now regions where the ideal MHD approximation can break down, such as in
neutral sheets or in tearing layers (cf. Vasyliunas, 1976).

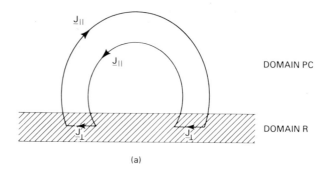

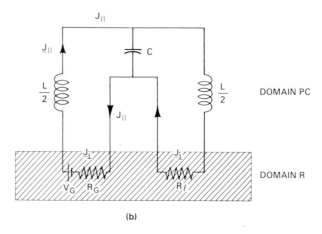

Fig. 4. (a) A schematic for a closed current loop threading the coronal (*PC*) and photosphere (*R*). (b) An equivalent lumped parameter circuit as described in the text.

magnetic loops may exist entirely in the resistive domain. Assuming that some external force exists in *R* such that $\nabla \cdot \mathbf{J}_\perp \neq 0$, then the field-aligned currents $\mathbf{J}_\parallel$ will link domains *R* and *PC*, the photosphere and corona. If $\nabla \cdot \mathbf{J} = 0$ is to be satisfied, then a $\mathbf{J}_\perp$ must exist. Since $\mathbf{J}_\perp$ is nonzero, a Lorentz force ($\mathbf{J} \times \mathbf{B}$) in the corona attempts to force the plasma there into co-motion. Thus, a plasma in the corona will become comoving with the resistive plasma in the photosphere, if the resistive plasma is under the influence of an external force.

Because energy storage involves both local and global quantities — e.g. current density *J* and total current *I* — and because we can use global equivalent circuits to illustrate several concepts about energy storage, we next investigate the relationship between such global quantities such as total inductance *L*, total resistance R_0, and total current *I*, and such local quantities as resistivity and current density. Consider first the equation of motion for the electron component of the plasma

$$n_e m_e \frac{d\mathbf{v}_e}{dt} = -n_e e \left(\mathbf{E} + \frac{\mathbf{v}_e \times \mathbf{B}}{c} \right) - \nabla P_e + n_e e \boldsymbol{\eta} \cdot \mathbf{J}, \qquad (2.35)$$

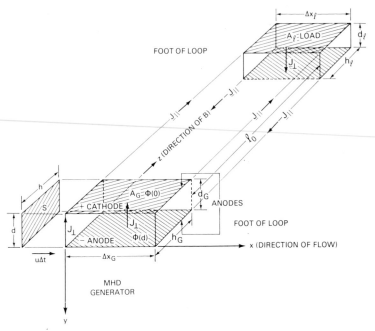

Fig. 5. This figure relates the schematic diagrams of Girue 4 to real physical quantities in a coronal loop: lengths, areas, and velocities.

where P_e is the electron pressure, n_e the electron density, and v_e the electron drift velocity. Neglecting the electron inertia (l.h.s. of (2.35)), setting $\mathbf{E} = -\nabla\phi - (1/(\partial A/c)\partial t)$, and integrating (2.35) yields

$$\int_1^2 \mathbf{E}_0 \cdot d\mathbf{l} = \int_1^2 (\eta \cdot \mathbf{J}) \cdot d\mathbf{l} + \int_1^2 \nabla\phi \cdot d\mathbf{l} + \frac{1}{c} \int_1^2 \frac{\partial \mathbf{A}}{\partial t} \cdot d\mathbf{l}, \qquad (2.36)$$

where

$$\mathbf{E}_0 = \frac{1}{n_e e} \nabla P_e + \frac{\mathbf{v}_e \times \mathbf{B}}{c}; \qquad (2.37)$$

the integrations in (2.36) are performed over the current path between two points 1 and 2 in the plasma. The l.h.s. of (2.36) is the applied voltage V, and the terms on the r.h.s. of (2.36) are, respectively, the resistive, capacitive, and inductive voltage drops. $\mathbf{E}_0$ contains both a solenoidal and an electrostatic terms. A relatively small displacement of electrons (with negligible currents) will cause the capacitive term to cancel the electrostatic part of $\mathbf{E}_0$, proportional to ∇P_e. However, a much larger current is driven continuously through the plasma by the solenoidal part of $\mathbf{E}_0$, $\mathbf{v}_e \times \mathbf{B}$. If the capacitive term in (2.36) is neglected for the moment, and the integration path is assumed to be that of a simple loop circuit, Equation (2.36) yields:

$$V = IR_0 + \frac{d}{dt}(IL), \qquad (2.38)$$

where

$$V = \int_1^2 \mathbf{E}_0 \cdot d\mathbf{l},$$

(2.39)

$$IR_0 = \int_1^2 (\eta \cdot \mathbf{J}) \cdot d\mathbf{l},$$

(2.40)

and

$$\frac{d}{dt}(IL) = \frac{1}{c}\int_1^2 \frac{\partial \mathbf{A}}{\partial t} \cdot d\mathbf{l}.$$

(2.41)

Equations (2.39), (2.40), and (2.41) demonstrate that the contribution of local quantities along the current path determines R_0, L, and I. A more subtle conclusion is that the evolution of the relevant circuit depends not only on the local parameters but also on the global aspects of the circuit.

Resistive dissipation in the photosphere is expected to dominate resistive dissipation in the corona (ignoring for now the possibility of instabilities occurring in the corona that both limit and dissipate currents), so that magnetic energy storage is strongly dependent on what happens in the photosphere. This latter point can be clarified further by considering a simple lumped circuit with a battery at potential V, an inductor with self-inductance L, and a resistor with total resistance R_0. At $t = 0$ a switch is closed. At this instant, there is no current in the circuit and the voltage drop across the resistor is zero. However, there is voltage drop across the inductor given by $L\, dI/dt$ and equal to V, where I is the net current. As soon as the current begins to flow, a voltage begins to appear across the resistor, resulting in a decreased voltage drop across the inductor and a decrease in dI/dt. The final steady-state value of the current, I_0, is determined by R_0 and the effective battery e.m.f. In a steady state the complete e.m.f. V is across the resistor, since dI/dt is zero. Hence, the maximum magnetic energy stored, $\frac{1}{2}LI_0^2$, is strongly dependent on R_0, since $I_0 = V/R_0$. In addition, because the net current grows and decays in time as L/R_0, we see that not only the total energy stored, but also the rate at which the magnetic energy is stored, are dependent on the resistivity of the photosphere.

The electrodynamic coupling of the solar atmosphere also plays a fundamental role in stabilizing potential instabilities that can occur in the corona and thereby altering the ability of the current system to store magnetic energy. Consider a plasma which is disconnected from the dissipation of the photosphere. If an external force F is applied, the plasma will undergo a force drift $V_F = cF \times B/B^2 e$, which in turn causes a force-generated current density $J_F = cnF \times B/B^2$. The plasma then will become polarized in such a way as to sustain an electric field which results in an acceleration exactly equal to F/ρ_0. However, if this plasma is now connected to the more resistive photosphere, the polarization charges will drain into or draw neutralizing charges from below in the form of parallel currents, $J_\parallel$, so that the effect of F on the plasma will be altered. For example, if an ideal MHD interchange instability were to occur in the corona driven by gravity,

the resultant perpendicular polarization electric fields can be short circuited by neutralizing parallel currents drawn from the photosphere. This is the electrodynamic view of
line-tying.

The current patn obviously plays an important role in magnetic energy storage. Referring again to Figure 4, only those paths taken by currents in closed coronal loops can
result in systematic magnetic energy storage, because the stresses resulting from the
current closure condition can be balanced by reactive inertial stresses in the photosphere.
If the currents followed paths leaving the lower corona, electrodynamic energy generated
by a photospheric dynamo might simply propagate out into the solar wind, with no
storage. Current paths which close back into the photosphere in simple magnetic loop
structures are least likely to allow energy to propagate out to the exosphere of the Sun.

To estimate the magnitude of the flow field velocity as well as the time required to
build up and store an amount of magnetic energy $E_F = \frac{1}{2}LI_0^2$, we refer to Figure 5 and
note that the maximum available electrical power is the kinetic energy density of the
flow field delivered through the surface area S between the anodes; that is

$$P_{max} = \frac{1}{2}nm_iu^3S. \tag{2.42}$$

To store E_F in a time Δt_s requires

$$\frac{\epsilon}{2}nm_iu^3S\,\Delta t_s \approx E_F, \tag{2.43}$$

where the factor ϵ represents the efficiency of flow to electric energy conversion. Since
$\Delta t_s \approx L/R_T$, we find u must satisfy the inequality

$$u \gtrsim \left(\frac{I_0^2\eta_\perp^g}{10\epsilon(1-\alpha_e)m_in\,\Delta x_gh_g^2}\right)^{1/3}, \tag{2.44}$$

where $R_T = R_1 + R_G = R_G/(1-\alpha_e)$, $\alpha_e = (1+R_G/R_1)^{-1}$, and $R_G = \eta_\perp^g d_g/\Delta x_gh_g$.
Equation (2.44) clearly requires $1 - \alpha_e \lesssim 1$ since otherwise u will become intolerably
large although the energy storage time will shorten since $\Delta t_s = L(1-\alpha_e)/R_G$. To estimate
the required u and Δt_s for a large flare, e.g. with $E_F \approx 10^{32}$ erg output, we note that
$L \approx I_0/c^2$ and that $\Delta x_g \approx 10^{10}$ cm and $I_0 \approx 10^9$ cm are typical for such a flare. Using
$\eta_\perp^g \approx 10^{-10}$ s, $h_g \approx 5 \times 10^8$ cm, $d_g \approx 10^9$ cm, and $n \approx 10^{17}$ cm^{-3} as typical temperature
minimum values, we find $\Delta t_s \approx 5 \times 10^6(1-\alpha_e)$ s and $u > 3.4 \times 10^5[\epsilon(1-\alpha_e)]^{-1/3}$ cm
s^{-1}, and a total current $I_0 \approx 10^{22}$ stat-amps (from $E_F = LI_0^2/2$). If $\epsilon \approx 0.1$ and $\alpha_e = \frac{1}{2}$,
then $\Delta t_s \approx 2.5 \times 10^6$ and $u \gtrsim 10^5$ cm s^{-1}. Both of these values, while rough estimates,
appear reasonable for the preflare state.

One puzzling aspect of magnetically driven flare models is the large currents, and thus
large magnetic fields, required to drive the flare. These fields typically range from a few
hundred gauss to 1000 gauss or more. In fact, the observed energy densities of flare
plasmas require fields of 500 gauss or more to contain them (Widing and Spicer, 1981).
Part of the solution to this problem is obtained by noting that $\nabla \cdot \mathbf{J} = 0$ implies that the
net current must vanish globally. From Ampere's equation in integral form,

$$\oint \mathbf{B} \cdot d\mathbf{l} = \frac{4\pi}{c}\int \mathbf{J} \cdot d\mathbf{S},$$

we note that a surface integral taken outside the volume containing antiparallel currents must vanish and $\nabla \cdot \mathbf{J} = 0$ is required for MHD equilibrium. Hence, large currents and fields can, in principle, exist in such solar magnetic configurations. In addition, if these currents are locally force free, the mutual repulsion that arises from antiparallel currents will be locally suppressed.

3. Magnetically Controlled Energy Conversion

This section treats those processes and configurations where the magnetic field plays a predominantly passive role in shaping and modifying the plasma. We are not concerned here with the release of magnetic energy *per se*; the more active role for the magnetic field are considered in Section 4.

Section 3.1 considers the anisotropic modifications of the local plasma transport properties which result from immersing a hot plasma, which is essentially collision free, in a magnetic field. In the solar context this leads naturally to a discussion of the structure and variability of the transition region (Section 3.2), including systematic fluid flows which can arise along magnetic loops as a result of simple asymmetries. More violent channeling, compression, and acceleration of dense plasmas to supersonic speeds along the magnetic field are considered in Section 3.3 and the corresponding channeling and dissipation of MHD waves is treated in Section 3.4. We conclude, in Section 3.5, with a discussion of anomalous dissipation of field-aligned currents by plasma microinstabilities.

3.1. MAGNETIC MODIFICATIONS OF PLASMA TRANSPORT

Although it has been convenient and profitable to use fluid and MHD models to describe the plasma in the solar atmosphere, the collisionless nature of the hot plasma is a complicating factor which we cannot afford to neglect. Even when the temporal and spatial variations in the system are sufficiently weak that the 'anomalous' processes of the last section do not occur, the fact that ions and electrons flow essentially freely through each other in a plasma leads to modification of almost all aspects of the medium's behavior. When a magnetic field is included, the charged particles can still move freely through each other with only small angle deflections along the field. The electrons and ions are both constrained by the magnetic field to execute gyromotion in the directions transverse to B. This magnetic rigidity is not dissipative, however, but rather reactive. Characteristic Larmor frequencies enter the problem and entirely new waves and modes are possible.

In the previous section we focused on one of the most important of these transport effects, electron thermal conduction, to illustrate how different the fluid properties can be from what we expect when new modes of propagation and degrees of freedom are available. When a magnetic field is present, the picture is even more complicated. The NRL Plasma Formulary (Book, 1980) summarizes the transport properties of plasma, compiled and reviewed first by Braginskii (1965). In a strong magnetic field the ratio of parallel to perpendicular electron thermal conduction is

$$\frac{\kappa_\parallel}{\kappa_\perp} \approx 0.6 \, \omega_{ce}^2 \tau_e^2, \tag{3.1}$$

where ω_{ce} is the electron gyrofrequency ($1.76 \times 10^7 \, B$) and τ_e is a characteristic electron collision time,

$$\tau_e \sim 3.44 \times 10^5 \, \frac{T_e^{3/2}(\text{eV})}{n_e \lambda_c}. \tag{3.2}$$

For conditions characteristic of the transition zone $B \sim 100$ gauss, $T \sim 10$ eV, the Coulomb logarithm $\lambda_c \sim 10$, and the electron density $N_e \sim 10^{10}$ cm^{-3}. The corresponding value of $\kappa_\parallel / \kappa_\perp$ is 2.5×10^{10}, an enormous disparity. Because the particle motion across the magnetic field is effectively curtailed, the cross-field transport terms are greatly reduced making the magnetized plasma a highly anisotropic medium.

In magnetic flux tubes in the solar atmosphere, therefore, three-dimensional problems are often reduced to one dimension of variability – along the magnetic field. Transport of plasma and heat across the field becomes so slow that we may consider different field lines as being isolated from one another. Further, since the field lines generally drop into the photosphere somewhere, considerations of magnetic islands, mode rational surfaces, etc., which arise in periodic configurations such as laboratory toroidal confinement devices, are of little relevance on the Sun, where generally none of the particles can transit the system along a field line. Instead there are 'collisionless' thermoelectric effects where heat fluxes beget electric fields and electric fields draw heat fluxes. The currents which result from the electric fields (or perhaps cause them – depending on your perspective) give rise to additional magnetic fields which introduce another complicating factor and may also be important.

At the plasma and magnetic field conditions which prevail in flux tubes on the Sun, the geometry, flows, and electric current paths are essentially determined by the magnetic field. Because the plasma pressure above the solar surface is generally smaller than the magnetic pressure, the fluid is carried around by the motions of the magnetic field but the field is affected by the plasma only globally as a result of integrated currents. We noted earlier the usefulness of force-free magnetic field configurations as a model for describing the solar magnetic field. This interest stems from the fact that the tenuous plasma which resides in the corona is incapable of providing a sufficient pressure gradient force to dominate the strong magnetic forces.

If a magnetic field is introduced into the problem of anomalous resistivity, we encounter new physical effects which alter some of the arguments given in Sections 2.4 and 2.5. In particular, an electric field perpendicular to the magnetic field will not drive runaway electrons. It gives rise only to an $E \times B$ drift velocity of the plasma as a whole; hence, there is no difficulty in maintaining a stationary electron distribution. However, a current flowing parallel to the magnetic field presents theoretical problems, particularly when $\Omega_{ce}/\omega_{pe} \gtrsim 1$, which is true throughout a large fraction of the solar atmosphere. Here Ω_{ce} is the electron gyrofrequency and ω_{pe} is the plasma frequency. The difficulty arises because the electron motion is essentially one-dimensional unless the effective collision frequency of the turbulence, ν_{eff}, satisfies $\Omega_{ce}/\nu_{\text{eff}} \ll 1$. Otherwise, the electrons will behave adiabatically. If a magnetized electron moves through a turbulent region where $\Omega_{ce}/\nu_{\text{eff}} \gtrsim 1$, it enters and leaves the region of turbulence with the same magnetic moment so that its perpendicular energy remains unchanged. Consequently, the electron velocities parallel to the magnetic field simply increase in the presence of the electric field driving the current, instead of being randomized in all directions as is required by the standard definition of a resistivity mechanism. In these circumstances, we should expect the entire electron distribution to run away and not be restrained by any resistive process. If this occurs, the concept of a local Ohm's law is no longer appropriate, because the drift velocity of the current is determined by the global electric field at each point along the current path rather than by local properties of the plasma. Thus, to regain a local Ohm's law, the bulk of the electron population that carries the current must be trapped while at the same time the parallel electron velocities must be thermalized; that is, the electrons' velocities perpendicular to the magnetic field are increased at the expense of their parallel velocities.

3.2. TRANSITION REGION STRUCTURE AND FLOWS

The chromosphere–corona transition region plays a major role in determining the structure and energy balance of the outer solar atmosphere. The conventional picture of this region in the quiet Sun (e.g. Withbroe and Noyes, 1977) is one of a thin layer of plasma whose structure is determined by a balance between local radiative losses and the energy provided by thermal conduction from the overlying corona, possibly with some additional energy deposition directly in the layer. Since this heat flux is the major energy loss from the quiet corona, the ability of the chromosphere and transition region to radiate that energy determines how much material can be supported in the corona. Changes in the coronal heat source can therefore produce significant readjustments in density and temperature in both the transition region and the overlying corona.

A wealth of observations of EUV emission lines formed in the transition region and inner corona exhibit brightness changes, nonthermal line broadening, and apparent systematic mass flows (e.g. Withbroe and Noyes, 1977). These dynamic phenomena may, in part, be due to the adjustment of the transition region to variations in coronal heating. It is clear, however, that these phenomena make equilibrium ionization balance models of limited value for interpreting EUV emission lines. It has been shown by Mariska *et al.* (1982) that the equilibrium ionization balance assumption leads to significant departures from the actual nonequilibrium ionization balance in only mildly disturbed flux tubes.

High-resolution EUV observations also show that the transition region in the quiet Sun is highly structured. Since the ultimate transport mechanism for much of the coronal magnetic energy is radiation and most of the radiation in UV wavelengths comes from the transition region, it is crucial to understand why measurements indicate the existence of much more transition region material than a simple laminar theory will allow. The observed concentration of emission in the network regions (e.g. Reeves, 1976), where the magnetic field is also concentrated, has led to simple magnetic field models for the quiet transition region (e.g. Gabriel, 1976). These models have magnetic field lines that diverge rapidly from the chromospheric network, become nearly radial in the low corona, and then close on a larger scale, resulting in the observed large-scale coronal structure.

On the other hand, the large amount of fine-scale structure observed in the network at transition region temperatures could be due to much smaller closed magnetic loop structures. Lyman-α filtergrams indicate that the chromospheric network at the base of the transition region may be composed of many small loops (Bonnet *et al.*, 1980). These loops would reach to heights of only a few thousand kilometres, the characteristic width of the network in transition region lines. Simple static models, such as those of Rosner *et al.* (1978), would then indicate that for typical quiet Sun pressures, the maximum temperatures in these structures would be only a few times 10^5 K.

Previous calculations of the formation of the transition region and its strong dynamical stability by Oran *et al.* (1980, 1982) indicate that a single laminar transition region is considerably thinner than some integrated observations would imply (e.g. Mariska *et al.*, 1978). A conglomeration of ragged, patchy transition regions extending over many times the ideal laminar thickness is one possible way to reconcile the observations and calculations. Antiochos (1979) and An *et al.* (1980) explored the possibility that condensational instability in the transition region caused this raggedness. The current understanding is that the combined corona–transition region–chromosphere system is stable. This means that the observed fluctuations in the transition region plasma have to be attributed to external driving effects like changes in the heating rate or magnetic flux tube geometry. The fluctuations appear not to result from any inherent turbulent tendency in the transition region structure itself. The uniformity and repeatability of laser-plasma experiments, where the thin ablation layer is closely analogous to the transition region, supports this theoretical conclusion indirectly.

In the transition region the temperature and, hence, the electron thermal conductivity decrease rapidly downward requiring very short temperature gradient scale lengths to maintain the downward directed heat flux. At the typical quiet Sun coronal pressures of about 0.2 dyne cm^{-2}, temperature scale heights $h < 1$ km are required in the $1-5 \times 10^4$ K region. An important part of the downward directed heat flux high in the corona is conducted into the dense but thin transition region plasma at temperatures between $\sim 10^4$ and $\sim 3 \times 10^4$ K and is radiated away as Lyman-α.

The leakage also means the lower transition region is constantly 'stressed' or connected by a significant heat flux well down into material normally considered chromospheric. The profiles of temperature, density, and hence radiation output in the hotter $3-15 \times 10^4$ K plasma just above are correspondingly altered to acquire steeper gradients, which then transport the additional downward directed heat flux that is required to feed the Lyman-α emission. This moving heat flux connects separated regions in the transition region and lower corona by a communication mechanism, electron thermal conduction, which moves faster than sound and therefore tends to stabilize purely fluid motions.

In section 2 we saw how electron thermal conduction could transmit energy much faster than the sound speed. The Alfvén speed,

$$V_A \approx \frac{B}{\sqrt{4\pi\rho}}, \tag{3.3}$$

is generally an order of magnitude or so faster than the sound speed in the corona because magnetic pressure typically dominates the plasma pressure (i.e. $\beta = 8\pi P/B^2 \ll 1$) but Equation (2.20) shows that V_κ, the thermal conduction speed, could still be much larger than V_A on magnetic loops where the temperature approaches or exceeds 10^6 K. This means that the parallel structure along the field can change and adjust itself faster than the magnetic field structure changes. Even though the magnetic structure may be evolving as magnetic energy is released, the structure is essentially stationary insofar as parallel thermal conduction is concerned. The quiescent prominence model of Steinolfson (1981) is an example where such an ordering of characteristic velocities is appropriate, and radiation, thermal conduction, and coronal heating are assumed to be in balance.

When a downward heat flux beyond that needed for the Lyman-α raditation readjustment is supplied, the temperature rises in an energy conserving way until the heat flux to lower layers, plus the increased radiation from the higher temperatures and densities, can handle the increased input. Chromospheric material evaporates off into the corona until a new, higher pressure equilibrium is established.

Persistent flows of plasma at chromospheric and transition region temperatures have been observed by a number of experiments (e.g. Gebbie *et al.*, 1980; Brueckner *et al.*, 1980). If we assume that the moving plasma is confined to magnetic loops and that mass exchange with the regions surrounding each loop is not important, there are basically two general mechanisms to produce the observed flows. One is impulsive mass ejection events, such as spicules or jets in quiet solar regions, providing upflow followed by the return of the material as a steady downflow. The other is a steady flow of material up one leg of the flux tube and back down the other. In either case, a steady flow of plasma is a potential source of information about the heating mechanism which must be its ultimate cause. A steady flow of plasma along a closed magnetic flux tube is easily produced by an effect as simple as asymmetric heating.

If the heating rate change in a full loop is accomplished by localizing some or all of the energy deposition asymmetrically, then each side of the loop will respond differently to the change. Figure 6 shows a symmetric loop being heated off-center. The geometry, as shown, is semicircular and the flux tube is assumed to have constant cross-section, but these assumptions are not crucial to the argument. Mariska and Boris (1981, 1982) have performed a number of detailed numerical simulations and do not see supersonic flows over the top of the loop at any time during these calculations; nor was it possible to generate standing shocks as postulated in the literature. The usual models for flow in closed loops generally also neglect the smoothing, stabilizing effects of thermal conduction. It has been found, however, that systematic flows result quite naturally from the asymmetric coronal heating illustrated. In general, the flow is away from regions of excess heat deposition and directed in such a way as to equalize radiative losses from the two transition regions.

These flows are usually saturated in the sense that the flow is as fast as it can be without overbalancing the energy distribution in the opposite direction to the nonphysical case where more energy is being dissipated on the side of the loop farthest from the heat source. Once the asymmetry becomes appreciable, the velocity saturates at about 5 km s^{-1}. This is the velocity necessary to redistribute the energy evenly via the enthalpy flux.

It would be tempting to try to relate these flows to effective pressure differences across the footpoints of the loop (e.g. Cargill and Priest, 1980; Glencross, 1981, and references therein) but a pressure difference along the flux tube from the transition region at one end to the transition region at the other cannot be sustained by the complete chromosphere–corona–chromosphere system (Mariska and Boris, 1983). The hot corona in a low-lying loop effectively equalizes the pressure when a continual source of coronal heat is present. Thus, the major flow is an expansion of the chromosphere at the high-pressure end of the loop and a corresponding compression at the low-pressure

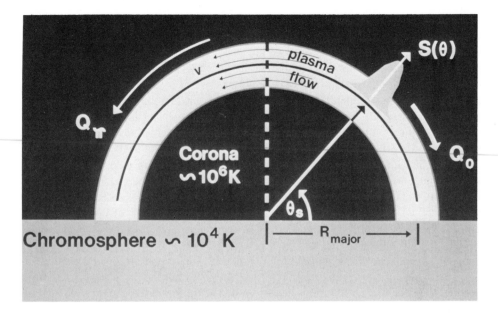

Fig. 6. Section of a symmetric semicircular magnetic loop in which asymmetric heating drives a steady flow of plasma from the hotter towards the cooler end.

end until the two pressures have essentially equalized. The transition regions and corona readjust quickly and smoothly to follow the damped acoustic oscillations of the two chromosphere.

In the steady state which eventually ensues (with or without systematic flow) there can be no net pressure difference without a corresponding acceleration of fluid from one end of the loop to the other. Since there is no net momentum increase in the overall system, the pressure difference at a given altitude can only arise from masses of overlying material. Since the coronal density is very low, the transition region pressures are essentially equal.

3.3. CHANNELING AND ACCELERATION OF PLASMA

In section 2.3 the condensational instability and the nonlinearly stable condensations which evolve from it were discussed. In a strong magnetic field the condensed material is channeled into one-dimensional motion along the field lines. When the field strength increases in one direction along the field, the lines of force are converging. The plasma will have to compress, moving towards stronger fields, but otherwise simple one-dimensional approximations work well. When two sides of a condensation interpret different thermal conduction energy fluxes, the condensation accelerates slowly as a whole away from the strongest heat flux. When two condensations are close together, they seem to attract each other and coalesce because less energy is being deposited in the gap than in the larger volume outside. These dynamic coalescence results probably have relevance to interstellar and intergalactic condensations as well.

The coalescence of condensations is a manifestation on stellar scales of the ablative acceleration phenomenon which occurs in laser fusion. The laser ablation layer is a very similar structure to the idealized transition region on the Sun. In the regime of solar parameters, the longer time-scale permits radiation to balance conduction; thus ablation of the colder chromospheric material is better described as evaporation on the Sun. When condensations accelerate and coalesce, only a small fraction of the incident energy goes into acceleration. Most of the deposited energy is still being radiated away. This makes it very difficult to accelerate condensations to velocities approaching the sound speed in the hot tenuous component by the remote deposition of heat.

On the Sun there are many phenoman that may be associated with the formation of stable condensations and the channeling and acceleration of plasma by the magnetic field. Coronagraph observations, such as those shown in Figure 2, have revealed the presence of relatively cool, Hα-emitting material, moving through the corona at heights up to 10 solar radii (Sheeley *et al.*, 1981). Solar prominences are another vivid example of both the channeling of material by the magnetic field and the presence of cool dense material surrounded by a much hotter medium. On a smaller spatial scale, visible wavelength observations reveal relatively cool material in spicules being ejected to coronal heights. In ultraviolet emission lines similar events are observed to occur at both high and low velocities. All these events show unmistakable evidence for channeling by magnetic fields; the role of the field in accelerating the plasma is less clear.

Recent exciting observations of small-scale (~ 1000 km), high-velocity (~ 400 km s^{-1}) bullets or jets accelerating rather uniformly upward out of the chromosphere, reported by Brueckner (1980) and Brueckner and Bartoe (1982), show that supersonic as well as subsonic condensations can exist. As pointed out by Karpen *et al.* (1981, 1982), this has two consequences. First, the acceleration mechanism for the jets must be magnetic; and second, the rapidly accelerating condensed jet must heat up. Computations showed that jets accelerating at the observed rate would heat to over 2×10^5 K in a minute and become invisible in the C IV resonance line used in the HRTS observations. This one-minute visibility limit also characterized the observations. Pneuman (1981) has also considered the relationship of the jets to magnetic field phenomena in the corona.

A simple lumped parameter model for what is happening as these jets heat up and disappear can be derived by integrating the momentum and energy density conservation equations over the region of the rapidly accelerating plasma. One obtains several simple ordinary differential equations whose solution shows that the acceleration produced by an applied force is reduced from that expected using dynamical arguments alone. Some of the applied force is required to snowplow coronal material and provide for radiation losses in the bullet. Typical reductions are a factor of 2 or 3 and give solutions which are consistent with the observed heating rates.

3.4. CHANNELING AND DISSIPATION OF MHD WAVES

While MHD waves are not likely to be the source of energy for a solar flare, they have figured prominently in theories for coronal heating. The whole topic of waves and oscillations is dealt with in more detail by Rhodes and Brown elsewhere in this work. Orrall and Pneuman also examine the role of waves in coronal heating. Here we briefly discuss the role that the magnetic field plays in the channeling and dissipation of MHD waves.

Since loop-like magnetic field structures can channel plasma motions, it should come as no surprise that MHD waves and the energy that they carry can also be channelled over long distances. Further, since nonlinear MHD waves exert strong body forces on the plasma, rapid acceleration is possible in principle. Hollweg (1979, 1981) considered the possibility that upward propagating MHD waves might produce mass ejections in the solar atmosphere. These studies are focused on a single impulsive ejection of plasma, rather than continuous acceleration lasting for a minute or longer. Furthermore, the final velocities derived from his calculations are much smaller than those observed for the coronal jets but very similar to those of spicules. Hence, this worm may not be directly applicable to highly accelerated phenomena. Nevertheless the MHD wave mechanism is attractive because magnetic restructuring is not an inherent feature. Consequently, complex (and often ill-understood) magnetic field changes need not be considered as intrinsic to the acceleration itself, although they might appear as by-products of the energizing process.

Since magnetic field lines have tension and can store energy when they are bent or twisted (Section 2.2), MHD systems support more normal modes of oscillation than the three that the fluid component alone could sustain. In general there are seven roots, the usual two sonic modes and one entropy mode for each wavevector k, plus two Alfvén modes and two magnetoacoustic modes. The Alfvén modes involve transverse displacement of the magnetic field and no compression of the fluid. The corresponding wave phase velocity is $V_A = B^2/4\pi\rho$, where the gas pressure does not enter because the plasma is not compressed. The fast magnetoacoustic modes are also principally transverse displacement modes where compression of both plasma and magnetic field combine to give a high wave velocity, $V_{MA} = (C_S^2 + V_A^2)^{1/2}$.

Our brief consideration of MHD waves is limited to fluctuations of modest amplitude, so that it is meaningful to consider an essentially stationary magnetic field configuration which structures and guides the waves. The principal results for the dissipation of hydromagnetic waves on a radially varying flux tube with sheared magnetic field can best be expressed in terms of the equilibrium function

$$\omega_A^2(r) = \frac{k^2 B_z^2(r)}{4\pi\rho(r)}, \qquad (3.4)$$

the square of the fluid "Alfvén frequency". A given profile of $\omega_A^2(r)$ determines the characteristic normal mode structure in the flux tube. Figure 7 summarizes the results pictorially. Consider an initially constant profile of $\omega_A^2(r)$ deformed as shown on the right in the successive views (a), (b), (c), and (d). The lowest eigenmode for each of the two eigenmode subsequences is sketched on the left for each of the four profiles (Boris, 1968),

The least damped, most highly resonant MHD waves have eigenfrequencies which cluster at the extrema of the Alfvén frequency in the presence of dissipation. Typically there will be separate low and high Alfvén frequency subsequences of modes which form about the global maximum and the global minimum of $\omega_A^2(r)$. When resistive dissipation is small, the damping of eigenmodes near maxima and minima is much smaller than elsewhere: Thus, it is these modes, localized near the surface and in the center of flux tubes, in general, which can be driven to largest amplitude resonantly and, hence, couple

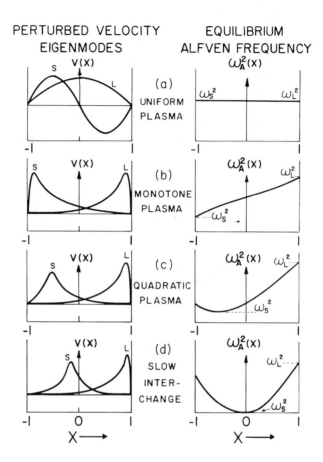

Fig. 7. Qualitative pictorial summary of the principal results for the two hydromagnetic mode subsequences. The lowest order eigenfrequencies are indicated on the $\omega_A^2(x)$ profiles shown on the right. The relative widths of the resistive layers scale roughly as $\eta^{1/3}$ and occur near the global minimum and maximum of $\omega_A^2(x)$.

the most energy into heating the plasma near ideal MHD singular surfaces where $\omega = \omega_A(r)$.

Recently, there has been considerable work on wave modes that depend on the observed coronal structuring for their existence. These 'surface waves' appear where there is a discontinuous change in the Alfvén speed, such as at the boundary between one magnetic flux tube and a second tube with a much smaller or larger gas pressure. Investigation of these waves (Ionson, 1978) suggests that they may be important for heating coronal structures. The heating produced by these waves would take place in a thin sheath surrounding a flux tube and, hence, high-resolution observations of the radial thermal structure of coronal loops could provide a crucial test of the theory.

3.5. ANOMALOUS DISSIPATION OF FIELD-ALIGNED CURRENTS

Anomalous Joule heating assumes a pivotal role in numerous flare and coronal heating theories due to the rapid rate of heating and magnetic energy dissipation associated with this process (cf. Kuperus, 1976; Norman and Smith, 1978). Anomalous Joule heating occurs in plasmas in which collisionless transport, rather than collisional transport, dominates. In a collisionless plasma, the plasma evolution is determined by microturbulent electric and magnetic fields excited by various plasma instabilities with scale lengths much smaller than MHD scale lengths. Energy and momentum are transferred from plasma currents to electric and magnetic field oscillations, and back again to the plasma particles as 'thermal' or randomized energy. The reader is referred to the numerous reviews which have appeared within the last few years (Mozer, 1976; Davidson and Krall, 1977; Papadopoulos, 1977, 1979), a detailed treatment being beyond the scope of this review. In the past, two classes of flare models have been developed which utilize anomalous Joule heating: those driven by $\mathbf{J}_\perp$, and those driven by $\mathbf{J}_\parallel$. For quasistatic dissipation of the currents to occur, the magnetic field structure must change continually in order to maintain the currents. We shall still treat the field configuration as fixed in this section, however, whether parallel or perpendicular currents are being dissipated. Our interest here is the local energy release processes and not the global readjustment of the magnetic field and plasma required to provide this energy and transport it to the dissipation site.

The two primary advantages of an anomalous Joule heating flare mechanism are the large heating rate associated with this process and the fact that anomalous Joule heating gives rise to a very hot thermal plasma (≈ 10 keV), which provides a thermal explanation for a specific class of impulsive X-ray bursts (Spicer, 1981b). A distinct drawback of such models is the large current densities required; this problem is more severe for $J_\parallel$-driven anomalous Joule heating and double layer models than for $J_\perp$ neutral sheet models.

A purist's definition of double layers and anomalous resistivity would state that a double layer is a collisionless laminar structure, much like a collisionless laminar electrostatic shock (cf. Tidman and Krall, 1971), and that anomalous resistivity results from its highly turbulent substructure. An examination of the theoretical studies of these two phenomena, however, suggests a more confusing picture in which double layers consist of a turbulent structure superimposed on a large-scale laminar structure, while anomalous resistivity may require approximately laminar, large-amplitude ion density waves with similar turbulent structure. In addition, both mechanisms lead to trapping of large numbers of current-carrying electrons, while requiring similar magnitudes of the current drift speed for excitation. This latter point implies that double layers, which are highly localized along the current path, may be embedded in a larger region characterized by anomalous resistivity, as discussed by Smith and Goertz (1978).

We note that it is not clear how a double layer dissipates the stored magnetic energy, since it appears to lead to primarily monoenergetic beams without any local irreversible dissipation of the current, except possibly through beam-driven instabilities or where the beam is stopped, for example, in the footpoints of a loop.

In this regard Rowland et al. (1981) have described the evolution of a double layer in a magnetized plasma, in which the beam formed by the double layer becomes two-stream unstable, leading to the formation of a quasilinear plateau in the particle distribution. At the same time this beam excites a cyclotron-resonance instability. A transfer of parallel momentum into perpendicular momentum results isotropizing the velocities of particles trapped in the double layer and thereby dissipates the current. Future research should clarify the interconnections between double layers and anomalous current dissipation — perhaps these phenomena are just two manifestations of a more general $J_{\parallel}$-driven phenomenon in magnetized plasmas.

In recent years, numerous experimental and theoretical studies have investigated the properties of double layers, with particular emphasis on their application to the Birkeland currents ($J_{\parallel}$ currents) which couple electrodynamically the ionosphere and the magnetosphere. Because several excellent reviews on double layers already exist (Block, 1975, 1978; Goertz, 1979; Carlquvist, 1979a; Torvén, 1979), we present here only a synopsis of the important aspects of double layers, with particular focus on applications to solar flares.

A double layer consists of two oppositely charged, essentially parallel but not necessarily plane, laminar space-charge layers, which trap a large fraction of the current-carrying electron population and accelerate the remainder (Block, 1978). The potential, electric field, and space-charge density vary qualitatively within the layer, as shown in Figure 8. A double layer is believed to occur when a potential difference in a finite length plasma is

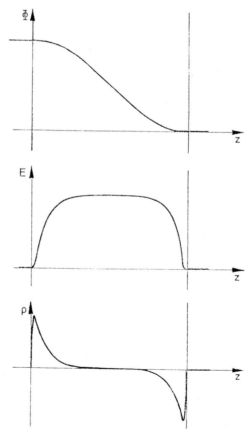

Fig. 8. Schematic of the potential (Φ), electric field (E), and charge density (ρ) of a single double layer.

concentrated in a shock-like localized region, rather than distributed over the entire length of the system. Such double layers are expected to produce monoenergetic particle distributions. Four conditions must be fulfilled for double layers to occur (Block, 1978):

(1) the potential difference ϕ_0 across the layer must satisfy $|\phi_0| > k_b T_e / e$;
(2) the electric field must be much stronger inside than outside the double layer, so that the integrated positive and negative charges nearly cancel;
(3) quasineutrality is violated locally in both charge layers; and
(4) the collisional mean free path must be much greater than the double layer thickness, because the formation of a double layer is a collisionless phenomenon.

The first application of double layer anomalous dissipation of parallel currents to solar flares was carried out by Alfvén and Carlqvist (Alfvén and Carlqvist, 1967; Carlqvist, 1969, 1979b). In their model, many double layers, driven by a high-inductance current system, are formed in numerous current filaments, each with very large current densities, traversing the solar atmosphere. Alfvén and Carlqvist argued that inductively stored magnetic energy is released rapidly as accelerated particles, plasma heating, and bulk plasma motion, as a consequence of the formation of double layers. The Alfvén–Carlqvist model was dismissed by Smith and Priest (1972), who maintained that the Buneman instability (Buneman, 1959) or the ion-acoustic instability would lead to plasma turbulence, and, thus, to the occurrence of anomalous resistivity rather than double layers. Goertz and Joyce (1975) showed that a strong double layer requires a current density given by $|J| = neV_{T_e}$, where v_{T_e} is the electron thermal speed. This current density is sufficient to excite the current-driven Buneman instability. There is considerable evidence now, however, that local plasma evacuation, required for the formation of a double layer, can be caused by the Buneman instability or the ion-acoustic instability, in certain circumstances, as enumerated by Carlqvist (1973), Smith and Goertz (1978), and Raadu and Carlqvist (1979). As a result, the double layer or current interruption model of Alfvén and Carqvist has recently invoked renewed interest.

To utilize a current-driven double layer for solar flare modeling we require that the system consisting of a current driver and a double layer the at marginal stability (i.e. in steady state) during some fraction of the flare duration. For this reason the entire current system must be analyzed to understand the nature of the current generator, the inductive magnetic storage reservoir, and the load. Hence, double layers, as well as $J_\parallel$-driven anomalous resistivity systems, require large global time constants in order to keep the current roughly constant.

An important constraint on a double layer model is obtained by noting that double layers lead primarily to particle acceleration. Intense Joule heating, if present, occurs only in a very small volume, $\approx 10^{15}$ cm^3. A double layer must be able to accelerate $\approx 10^{36} - 10^{38}$ 25 keV electrons during the course of a flare, to be consistent with a nonthermal hypothesis for typical hard X-ray bursts (Hoyng et al., 1976; Brown et al., 1979). Since particle acceleration is collisionless through a double layer of thickness L_{DL}, the rate γ of electrons accelerated freely through L_{DL} by a potential $e\phi_0$ is

$$\gamma^{-1} = \left[\frac{L_{DL}^2 m_e}{e \phi_0} \right]^{1/2} , \tag{3.5}$$

where m_e is the electron mass.

The production rate of accelerated electrons per unit volume is given approximately by

$$\frac{dn_b}{dt} = \gamma n_e - \frac{n_b}{\tau},\tag{3.6}$$

where n_b represents the nonthermal number density and τ^{-1} the rate at which fast electrons are thermalized due to collisions in the acceleration volume. If we assume that γ, n_e, and τ are roughly constant and $t \gg \tau$, then (3.6) yields $n_b \approx n_e \gamma \tau$.

Excitation of a double layer by the current drift speed requires $V_D \gtrsim V_{T_e}$. Furthermore, it is necessary to maintain $V_D \gtrsim V_{T_e}$ to keep the double layer at nearly steady state throughout an X-ray burst. Hence, using Ampere's equation, we find that the current density must exist in a channel with a thickness of roughly $\delta r \approx (c/\omega_{pe})\beta^{*-1/2}$. Taking a cylindrical shell of area $2\pi r \delta r$ for the current channel cross-section and noting that the double layer thickness is (Hubbard and Joyce, 1979)

$$L_{DL} \approx \left(\frac{e\phi_0}{k_b T_e}\right)^{1/2} 6\lambda_{De},\tag{3.7}$$

the total number of electrons accelerated by one double layer is

$$N_T \approx n_e \tau \left(\frac{e\phi_0}{m_e}\right)^{1/2} 2\pi r \frac{c}{\omega_{pe}}\beta^{*-1/2},\tag{3.8}$$

where r is the radius of the current shell. Assuming $n_e \approx 10^9$ cm^{-3}, $e\phi_0 \approx 25$ keV, $r \approx 5 \times 10^8$ cm, $\beta^* \approx 10^{-2}$, and $T_e \approx 1$ keV, we find that $N_T \approx 10^{32}$ electrons are accelerated in a volume of $V \approx 2\pi r\, \delta r L_{DL} \approx 10^{15}$ cm^3. Hence, approximately $10^4 - 10^6$ double layers, dispersed throughout a typical flare volume of $\approx 10^{29}$ cm^3, are required by the nonthermal hypothesis to explain a flare. Such a large number of double layers seems unlikely. However, one must remember that these estimates are based on double layer theory that is very much in its infancy and on the assumption of a nonthermal electron distribution. Note also that $10^4 - 10^6$ double layers constitute only $10^{-10} - 10^{-8}$ of the entire flare volume!

Anomalous resistivity may also result from perpendicular currents. $J_\perp$-driven instabilities which cause anomalous resistivity can be divided into two categories: those that are important when $\Omega_{ce}/\omega_{pe} \ll 1$, i.e. the Buneman and ion acoustic instabilities; and those which predominate when $\Omega_{ce}/\omega_{pe} \gg 1$, i.e. the beam–cyclotron, lower hydbrid-drift, modified two-stream and ion–cyclotron drift instabilities (cf. Davidson and Krall, 1977; Papadopoulos, 1979). In the unmagnetized $J_\perp$-driven instabilities, the turbulent waves excited can easily randomize the perpendicular electron and ion velocities, thus leading to heating in all directions. For the magnetized cases, heating occurs in directions normal to the magnetic field. The resultant particle distribution may be fully isotropized, however, by the concurrent excitation of secondary instabilities, such as whistler and electromagnetic ion–cyclotron modes (Spicer, 1976).

To illustrate the marginal stability approach, we consider a magnetic loop in which a parallel current flows such that the net current is roughly constant for a time $\tau \approx L/R$ and that $J_\parallel$ exceeds the threshold of the ion-acoustic instability (e.g. Krall and Trivelpiece, 1973) along some fraction, Δs, of the loop length, L_0; hence growing ion-acoustic waves are present. We expect the level of turbulence to lie in the range between the equipartition value and the saturated value as determined by the marginal stability condition. Phys-ically, three time-scales are of interest: the anomalous Joule heating time, t_J; the cooling time, t_c, due to radiative losses or diffusive transport, whichever process yields the smallest t_c; and the hydrodynamic expansion time, t_h, due to expansion either parallel or perpendicular to the magnetic field, whichever is faster. For solar flare plasmas, $t_J < t_c \ll t_h$. With these time-scales in mind, we expect the following sequence of events:

(1) Rapid anomalous Joule heating, due to the presence of unstable ion-acoustic waves, induces modifications of the threshold condition for the ion-acoustic instability; the instability ceases to operate and the transport coefficients decay to their classical values. This occurs within a few Joule heating times.

(2) Conductive cooling, both parallel and perpendicular to the magnetic field, and radiation eventually allow the threshold conditions to be satisfied so that the instability is switched on again.

(3) Sequences (1) and (2) occur many times (subcycle) before hydrodynamic expan-sion occurs.

(4) Finally, hydrodynamic expansion causes cooling and reduces the density in the heated volume until hydrodynamic equilibrium is restored. The final plasma temperature is higher than the initial temperature and, if expansion along the magnetic field takes place, the drift velocity V_D increases to maintain a constant net current.

Thus a dynamic balance is struck between anomalous heating and cooling, with the current–plasma system residing at, or perhaps oscillating about, the marginal stability point. Since these oscillations about the marginal stability point occur many times on a hydrodynamic time-scales, t_h, the combination of heating and cooling can be treated as a steady-state process on this longer time-scale. The fundamental quantity to determine, then, is not the level of turbulence but the temperature and density profiles. Once $J_\parallel$ and these profiles are known, we can calculate the amount of energy transported. Given the amount of energy transported, we can compute the turbulent fluctuation level using simple quasilinear theory.

We must examine two physical quantities, before proceeding, in order to clarify the importance of these flare mechanisms in dissipating stored magnetic free energy: the characteristic local and global magnetic energy dissipation times, τ_L and τ_G, respectively. The energy per unit volume lost from the field in a time τ_L is $\eta J^2 \tau_L$, so that the magnetic energy dissipation time is expressed approximately by

$$\tau_L = \frac{B^2}{8\pi\eta J^2} . \tag{3.9}$$

Using Ampere's equation, (3.9) is transformed to

$$\tau_L = \frac{2\pi(\delta l)^2}{\eta c^2}, \tag{3.10}$$

where δl is the characteristic scale length of the magnetic field, B, associated with the induced plasma current, J. Since τ_L is a local quantity, it does not reveal anything about the characteristic global magnetic energy dissipation time, $\tau_G = L/R$, unless we integrate τ_L along the entire current path, where L is the total inductance and R the total resis-tance. For example, consider a current-carrying loop of length L_0 with a load resistance

R_T in its feet. If a sharp increase in resistivity, in series with the current, were introduced along some fraction of the loop length, Δs, the global time constant would be expressed approximately by

$$\tau_G = \frac{L}{[R_T + (\Delta s/\tau_L c^2)]} . \tag{3.11}$$

Two categories of models for flare energy storage are thus suggested. If $R_T \ll \Delta s/\tau_L c^2$, then $\tau_G = L_0 \tau_L c^2/\Delta s$, and all the globally stored energy, $LI^2/2$, is dissipated across the enhanced impedance within Δs. To explain a flare that releases $\approx 10^{32}$ erg in $\approx 10^3$ s requires $LI^2/2 \approx 10^{32}$ erg and $\tau_G \approx 10^3$ s. Thus the entire reservoir of stored magnetic energy is released at a rate determined by the global time constant. Conversely, if $R_T \gg \Delta s/\tau_L c^2$, only a small fraction of the globally stored energy is released. This fraction is expressed approximately by

$$\frac{\Delta s L I^2}{\tau_L c^2 R_T} . \tag{3.12}$$

To explain a flare energy release of $\approx 10^{32}$ erg in $\approx 10^3$ s requires $LI^2/2 \gg 10^{32}$ erg. Hence, only a small fraction of the stored energy is dissipated, and the global time constant plays no role because it remains relatively unchanged, as does the total magnetic energy reservoir. This distinction between types of storage models is important for evaluating the feasibility of many of the flare energy release mechanisms proposed. To illustrate: two of the flare mechanisms reviewed here, double layers and anomalous resistivity, require essentially constant currents to operate throughout a flare, because they will switch off very quickly if the net current decays sufficiently. The distinction between classes of storage models also may help explain homologous flares, particularly if future observations show that homologous flares repeat in times much shorter than the energy storage times.

Returning to Equation (3.10), we see that basically two mechanisms can decrease the local dissipation time: either the scale length of the spatial variation of B is decreased (i.e. the field gradients are steepened), or the resistivity is increased, or both. As shown in Section 4.3, reconnection mechanisms decrease τ_L by driving the scale length δl to smaller values, while anomalous Joule heating mechanisms further decrease τ_L by also increasing the effective resistivity. The reduction of τ_L is the key to understanding reconnection and anomalous Joule heating flare mechanisms. However, the role of τ_L in the double layer mechanism is unclear, as was discussed above.

We introduced the local quantity β as a measure of the ratio of the internal energy density to the magnetic energy density; we also pointed out that, for low β equilibrium configurations, the magnetic field is essentially force-free ($J_\parallel$-dominated) or potential. Note that β also plays an important role in determining the Joule heating rate of a plasma. If β^* represents the β associated with the free magnetic energy, then the Joule heating time can be expressed as $t_J \approx nkT/\eta J^2 \approx 2\beta^* \tau_L$. For a fixed τ_L, the Joule heating time is shorter for smaller values of β^*. Hence, plasma heating is faster in a low-β^* system. In addition, if β^* is small, the plasma can be Joule heated through dissipation only a small fraction of the available magnetic free energy. These arguments suggest that the most useful plasma–field configuration for energy dissipation and storage is characterized by a low β^*, as is generally expected in force-free configuration.

4. Magnetohydrodynamic Energy Conversion

When sufficient energy is rapidly made available to a plasma-laden magnetic flux tube, the magnetic field and plasma move freely perpendicular to B as well as parallel. The characteristic velocities for the parallel motion are the same as before, V_κ and C_S, but other characteristic velocities now appear: the Alfvén speed, $V_A = B^2/4\pi\rho$, and the magnetoacoustic speed, $V_{MA} = (C_s^2 + V_A^2)^{1/2}$, as considered in Section 3.4. In the low solar corona where flares occur, these new MHD velocities are generally much higher than the local sound speed though they are still somewhat smaller than the thermal conduction speed V_κ. Therefore, measurements of large amplitude MHD phenomena from the Earth and satellites are correspondingly poorly resolved in time compared to the slower phenomena discussed in the last section. Our database on these dynamic, high-energy content processes needs substantial improvement, particularly with respect to simultaneous, spatially well-resolved measurements of temperature, pressure, velocity components, and magnetic field.

Figures 1 and 2, described earlier, show eruptive prominences and coronal transients in which these dynamic MHD phenomena are clearly taking place. Here the B fields are sufficiently small and the distance scales sufficiently large that resolution of the phenomena in time and space is possible. For small, strong loops in the lower corona the Alfvénic and magnetoacoustic time-scales cannot be resolved easily, so much of our knowledge about possible phenomena has to come from theory and numerical simulation. The interested reader is referred to the body of work by Dryer, Nakagawa, Steinolfson, and Wu (e.g. Steinolfson and Dryer, 1978; Wu *et al.*, 1978) in this regard.

In this section we discuss several of the MHD mechanisms and configurations which seem to play a major role in energy conversion and release. The emergence of magnetic flux tubes through the photosphere and chromosphere into the corona is treated in Section 4.1. Once flux tubes have adopted at least a quasistatic behavior, the possibility of exponential instability leading to dynamic geometric rearrangements of the flux and plasma above the solar surface must be considered. In Section 4.2 we treat this topic from the ideal MHD point of view and in Section 4.3 consider reconnection and magnetic tearing, the most important of the dissipative MHD processes. Section 4.4 continues the topic of reconnection and tearing in presenting the collisionless acceleration of plasma particles to very high energies in a neutral sheet during the process of fast magnetic reconnection. Our discussion of MHD energy conversion concludes with a discussion of flare trigger mechanisms in Section 4.5.

4.1. MAGNETIC FLUX TUBE EMERGENCE

Beneath the transition region the scale height of the solar atmosphere is relatively small i.e. ~500 km. Magnetic flux imbedded in the convectively rising fluid moves with the rather slow fluid velocities when $\beta \gg 1$ and becomes structured on the scales of the fluid turbulence. Expansion of the plasma–magnetic flux medium as it approaches the solar surface follows from the rapid drop in pressure with height in the photosphere and chromosphere. Therefore, as the magnetic flux 'breaks the surface', a wide range of conditions and subsequent dynamics is possible. Characteristic velocities are at least as large as the convective flows ($1-2$ km s^{-1}) and can be much larger, perhaps approaching the magnetosonic speed, when the magnetic fields and bouyancy are strong and the emergence becomes explosive. If rapid emergence occurs on a relatively local scale (~1000 km), Book (1981) has estimated that the magnetic energy released by the expansion $P\, dV$ work is more than ample to supply the energy requirements of the solar wind and the corona.

Figure 9 shows a greatly simplified schematic of one particular geometric configuration in which magnetic flux, in the form of a loop, will emerge explosively. The central

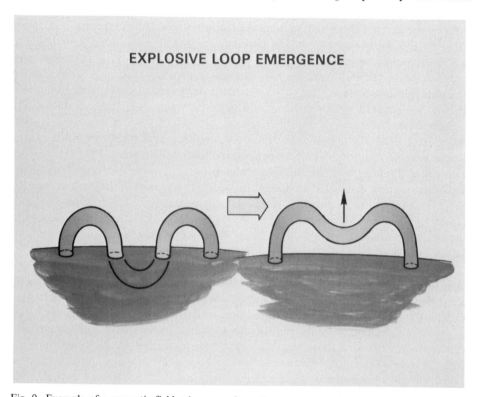

Fig. 9. Example of a magnetic field–plasma configuration where a portion of a flux tube can emerge explosively through the solar surface.

portion of the loop on the left half of the figure is rising somewhat more slowly than the two segments of the loop already shown above the surface. As the central portion rises through the surface, the substantial drag of dense photospheric and chromospheric material falls behind and the trapped central portion of the flux tube is freed. The very strong dynamic tension in the curved magnetic field takes over, rapidly accelerating the flux and the remaining trapped plasma upward. Of particular interest in this configuration is the fact that neither reconnection nor MHD instability is involved. As the expanding flux tube rises from the lower chromosphere to the corona, its energy balance gradually changes from domination by internal gas pressure, to equipartition between magnetic and thermal energy densities, then to domination by magnetic pressure. This magnetic bubble expands into the corona at about the Alfvén speed, while carrying a decreasing amount of material but, perhaps, driving a shock ahead of it. Furthermore, simple energy-balance arguments demonstrate that nearly all the magnetic energy in a segment of a flux tube is released when the tube emerges from the phosphere (Book, 1981). This substantial amount of energy is available, potentially, for accelerating or heating the overlying plasma. Only 5–10% of the magnetic energy stored in the flux below the surface is available for release through reconnection at a later time after the tube was risen well above the surface. Of this 5%, moreover, 90% is transformed into internal energy (i.e. heat) and only 10% goes into kinetic energy (Spicer, 1981a). Magnetic reconnection seems to be an extremely inefficient mechanism for bulk acceleration of discrete masses. In situations where magnetic reconnection and restructuring occur simultaneously, most of the resultant energy flow will probably be contributed by the large-scale readjustment process. Further, the intrinsic complexity of magnetic fields in turbulent fluids suggests that many other potentially explosive configurations also exist.

Unfortunately, our observational database is not yet detailed enough to allow us to say with certainty exactly how the magnetic flux does emerge. In addition, the availability of energy from global conservation considerations does not ensure, or even suggest, how the magnetic energy gets converted to heat. MHD shocks are one possible class of mechanisms but geometric changes may also induce secondary MHD instabilities (Section 4.2) or magnetic reconnection and tearing (Section 4.3).

The configuration of Figure 9 also suggests a simple model of a flare using the magnetic flux emerging from the photosphere as the flare driver. Assume that the emerging magnetic flux, ϕ_0, is associated with a field pressure that greatly exceeds the ambient atmospheric pressure. If no previously existing magnetic barrier exists above the emerging flux, the configuration lowers its energy by expanding at roughly the Alfvén velocity, V_A. The acceleration is V_A^2/R, where R is the radius of curvature of the field lines (Spicer et al., 1981; Book, 1981). The work thus performed on the surroundings is of order $\Delta W \approx \Delta V B^2/8\pi = L_i \phi_0^2/A_i \, 8\pi$, were ΔV is the initial volume of the emerging flux; B, the initial magnetic field; L_i, the initial length of the emerging flux; and A_i, the initial area. Taking $A_i \approx 10^{16}$ cm^2, $L_i \approx 10^9$ cm, and $B \approx 10^3$ G, we find $\Delta W \approx 4 \times 10^{29}$ erg, enough to power a small flare. In addition, the emerging flux ejects a mass $\Delta m \approx \rho \, \Delta V$ (Book, 1981). For $\Delta V \approx 10^{25}$ cm^{-3} and $\rho \approx 10^{-10}$ g cm^{-3}, $\Delta m \approx 10^{15}$ g. The total power involved in expansion initially is $P_e \approx \rho V_A^3 \Delta V/R$, which yields $P_e \approx 10^{26}$ erg s^{-1} for $R \approx \pi L_i$. This power output increases rapidly at first, then drops back to zero when the flux system reaches, or perhaps oscillates about, pressure equilibrium with the ambient atmosphere. Since the emerging flux moves at velocities comparable to the local Alfvén speed, the flux may be expected to drive collisionless shocks in front of it. These shocks, in turn, may be the irreversible mechanism needed to explain the observed flare heating and particle acceleration. The temperature changes due to the collisionless shock are estimated to be $\Delta T \approx 9T_0/8 \, (V_A/C_s)^2$ for large Alfvénic Mach numbers, where C_s is the local sound speed and T_0 is the initial temperature. Since $V_A \gtrsim 10 C_s$ was assumed, changes of $\Delta T \approx 100 \, T_0$ can be expected to result.

The simple results obtained above illustrate three important points:

(1) no macroscopic instability is necessary to explain a flare with this model;

(2) remote, photospherically stored magnetic energy can drive a flare; and

(3) ideal MHD motions can indirectly drive perpendicular currents which, in turn, can excite cross-field anomalous resistivity effects to bring dissipation into the collisionless shocks which result.

In our earlier discussions of electrodynamic coupling we considered circuits that are coupled conductively by flowing currents. However, two or more loops also can couple strongly with one another inductively if they are sufficiently close together. The mutual inductance which exists between two loops can be easily extended to multiloop systems in which some loops may be newly emerging and, hence, very energy rich. This concept of inductively coupling distinct magnetic configurations in the solar atmosphere has new implications for flare and coronal heating theory and needs further study. Consider the following possible scenarios:

(1) The magnetic flux variation resulting from a one-loop flare induces large e.m.f.'s and, thus, currents in a neighboring loop. This could trigger a flare in the neighboring loop if sufficient magnetic energy is stored there (Section 4.5).

(2) Emerging magnetic flux induces temporal variations of the global field configuration in which magnetic energy is already stored. These inductive field changes produce large e.m.f.'s and currents, thus causing the global field configuration to release its energy in the form of a flare. No mechanical coupling of the two configurations is required, as is generally employed in emerging flux flare models (e.g. Heyvaerts et al., 1977).

(3) A time-varying magnetic field due to emerging flux or a flare loop can also couple inductively to an inverted-Y field configuration. The induced e.m.f. produces accelerated streams of electrons and protons that move into the photosphere or out into the solar wind along that portion of the inverted-Y configuration with open field lines. Such a scenario could produce radio sources such as Types III, IV, and V bursts, with no transport of particles across field lines required.

Because of the high electrical conductivity in the solar atmosphere, the skin depth, $\Delta x_S \approx c/\omega_{pe}\sqrt{\nu_{ei}\,\Delta t}$, where ν_{ei} is the electron–ion collision time and ω_{pe} the plasma frequency, can be very small if the time-scale of flux variation, Δt, is small. Hence, very large current densities can be produced as well as high-energy particle streams (Spicer, 1981b). These large current densities can then relax, due either to MHD restructuring (instability) which reduces the free magnetic energy or by nonideal MHD and plasma effects which break the flux conservation constraint.

4.2. GEOMETRIC REARRANGEMENTS

Here we consider MHD energy release which comes from sources other than buoy ˄t/ explosive flux tube emergence. Potential instability mechanisms abound in the contorted solar atmosphere even in the presence of a number of stabilizing effects. Figure 10 shows a configuration considered by Barnes and Sturrock (1972) where photospheric flow twists one end of a magnetic flux tube relative to the other (left half figure). When a suitable threshold, is reached, the magnetic energy twisted into the system reappears, at least partially, as the macroscopic MHD kink instability shown. The magnetic field knots much as a twisted rubber band would do, converting magnetic energy to flow and pressure. Where the 'knot' crosses itself a local neutral sheet is formed, so further dissipation and heating in the form of tearing and reconnection can occur. Magnetic energy storage and release can also be driven in the system directly by photospheric flows, effects which we shall consider later in this section.

There are several MHD instabilities relevant to magnetic energy release and solar

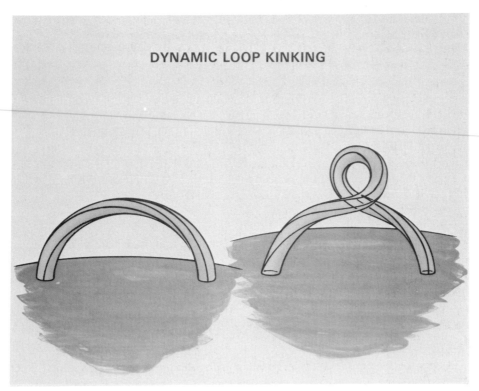

DYNAMIC LOOP KINKING

Fig. 10. Example of a coronal flux tube where photosphere flows at the loop footpoints, as shown, take the loop from a stable configuration to one which is unstable to the MHD kink instability.

flares. There are basically two equilibrium sources of driving energy for both ideal and resistive MHD instabilities: currents perpendicular and parallel to the magnetic field. Currents perpendicular to B, due to pressure gradients, are responsible for driving the so-called interchange instabilities. These instabilities cause one portion of a plasma to exchange places with another portion and depend only on the local conditions near the line of force. Therefore, they constitute local instabilities or modes, as opposed to global modes. In general, the presence of interchange instabilities does not necessarily imply global instability; some level of local instability is tolerable and usually appears as convective turbulence. This turbulence may be manifested in various solar features such as prominences.

The second source of driving energy, currents parallel to B such as shown in Figure 10, drive macroscopic instabilities which spread themselves out over the plasma volume. These 'kinks', or helical instabilities, can be subdivided further into free-boundary kinks (external kinks) and internal kinks. The external kink involves motions of the entire plasma–magnetic field configuration. In the solar context, a classical example of this instability appears to be the erupting prominences (Sakurai, 1976; Spicer, 1976). Internal kinks, on the other hand, involve distortions and motions within the plasma–magnetic field configuration that are not necessarily visible to an external observer.

The investigation of the MHD stability of coronal loops and prominences is a problem first examined by Anzer (1968) and Raadu (1972). More recent studies (Van Hoven *et al.*, 1977; Chiuderi *et al.*, 1977; Hood and Priest, 1979; Silleen and Kattenberg, 1980; Hasan, 1980) were precipitated by the Skylab observations which suggest that the so-called compact flares originate in loops (cf., e.g., Sturrock, 1980, for reviews) and by the loop flare model developed by Spicer (1976, 1977a, b, 1981a, b) and Colgate (1978). Attempted theoretical analyses of potentially unstable magnetic loop and flux tube configurations must make many simplifications to proceed which are questionable in the solar context. Such results as can be obtained are important, but they can only tell a limited part of the story. Van Hoven *et al.* (1977), Chiuderi *et al.* (1977), and Giachetti *et al.* (1977) present an MHD marginal stability analysis of a loop using as a model a cylindrically symmetric loop. Their loop model is essentially a weakly non-force-free field embedded in a high-β force-free environment. Silleen and Kattenberg (1980) have determined growth rates for kink instabilities driven by a Bessel function force-free field which was assumed to be embedded in a low-β potential field. Their results are in agreement with requirements imposed by the virial theorem. Hasan (1980) investigated the stability of a cylindrically symmetric force-free field with constant pitch and found it always unstable, consistent with the results of Anzer (1968).

Of major concern in these analyses are stabilization mechanisms such as line-tying and shear stabilization, discussed briefly in Section 2.2. Most present-day models, in one way or the other, attempt to include line-typing of the loop's ends in their stability analyses. Hood and Priest (1979), in particular, make the strong claim that line-tying, as opposed to shear stabilization, is the dominant MHD stabilizing feature of coronal loops. In fact line-tying and magnetic shear stabilization are closely related to each other. Line-tying imposes the condition that the field lines have a finite length and are embedded in a plasma with high conductivity at each end and has an effect similar to shearing the field lines. Line-tying limits the maximum wavelength of a perturbation to the length of the system and inhibits the long-wavelength interchange of field lines. Tying the field lines tends to increase the stability of the system by forcing stabilizing shear to accompany any disturbance from the equilibrium. What is perhaps most surprising about the problem of line-tying is that a general and sufficient condition for stability due to line-tying in three dimension already exist (Solov'ev, 1975).

To store *in situ* magnetic energy sufficient to explain a flare in an exponential atmosphere requires that the energy be stored within a rather large volume ($> 10^{28}$ cm^3) in the solar atmosphere. Storage in a small volume would cause the energy density of the magnetic field to be so large that the field would simply expand at roughly the Alfvén speed to achieve global pressure balance. To build up and store magnetic energy requires that the magnetic configuration should not develop any internal instabilities lowering its stored energy before a sufficient amount of energy is accumulated to explain a flare. MHD expansion of a magnetic field is the most likely way a magnetic configuration will take to lower its energy density. Lowering the magnetic energy by expansion requires the field line to shorten (the field line can untwist if it is twisted) and the flux tube area to increase. Thus storage of *in-situ* magnetic energy requires very special conditions to be satisfied to prevent expasion — conditions which must be understood if *in-situ* storage, and the flare and coronal heating mechanisms which depend on it, are to be valid.

MHD stability concepts must be applied very carefully to solar magnetic configurations since it is not clear what constitutes an unstable plasma–magnetic field configuration in the context of an exponential atmosphere and of our still-limited ability to observe these configurations. We are not really interested in whether a configuration is stable at all times, but only whether the configuration holds together long enough for sufficient energy to be stored in it.

Two other unresolved questions arising in stability studies of loops are: 'What are the appropriate boundary conditions?' and 'What are the appropriate geometries?' The simplest approach is to straighten a magnetic loop into a cylinder and assume the field lines are tied to perfectly conducting plates. This approach has two drawbacks which must be kept in mind. First, ideal perfectly conducting line-tying isolates the configuration from disturbances propagating into the volume between the plates, while helping to short circuit disturbances generated within the volume. While it is understandable that the dense, highly conducting photosphere can damp disturbances propagating into it from the corona, we must also consider disturbances generated in or below the photosphere which propagate into the corona. Waves and shear flows are known to exist in the photosphere, so we must, at some point, consider their effect on the configurations in the corona. In other words, from the 'lower' side we need transmitting boundary conditions.

A second drawback to assuming cylindrical geometry for a loop is the neglect of curvilinear terms in the metric coefficients and inhomogeneities in more than one spatial direction. As these effects can alter the stability of a system, observing and analyzing the correct geometry is important. If a cylindrically symmetric loop were always assumed, modes which occur only when a cylinder is bent — such as ballooning modes or tilting modes — would never occur. The ballooning mode is an interchange mode that would be localized to a region at the top of a loop where there exists magnetic fields curved unfavorably towards the plasma (Schmidt, 1979), and is driven by this magnetic curvature effect. It is difficult for a mode to localize itself along a field line, since it will set up shear Alfvén waves which are stabilizing. As the plasma density and, hence, β are increased, the Alfvén waves slow down and their stabilization effect is reduced. Analysis (Spicer, 1976) yields an approximate β necessary for stability of the loop against ballooning modes,

$$\beta \approx \frac{V_{Ti}^2}{V_A^2} \lesssim \frac{2RB_\theta^2}{rB_z^2}.$$

(4.1)

Equation (4.1) is identical to the result one obtains from line-tying interchange modes (Kulsrud, 1967), because line-tying localizes the mode to a maximum wavelength corresponding to the length of a loop, while ballooning modes are localized because of the variation of the toroidal component of the field with position.

We conclude that the actual complex magnetic field configurations in the low solar atmosphere are liable to be much more stable to gross macroscopic MHD instabilities than would be supposed from considerations of ideal MHD alone in geometries simple enough to be dealt with convincingly. On the other hand, macroscopic flows, granule-scale turbulence, differential rotation, and large-scale MHD readjustments will all give rise to configurations and situations where magnetic shear layers containing relatively localized currents are compressed. These situations are certain to produce magnetic tearing and reconnection, the subject of Section 4.3. Furthermore, the strong localized electric fields and current filamentation structures which result are sure to accelerate electrons and protons in the collisionless plasma to very high energies. It is almost certainly this structuring of the coronal and chromospheric plasma on successively finer and finer scales that leads to the apparently counter-entropic hot corona lying over the cooler chromosphere and photosphere.

4.3. RECONNECTION AND MAGNETIC TEARING

Reconnection, the most popular flare mechanism currently being considered, is attractive for two fundamental reasons: it allows magnetic energy which has been stored globally to be converted locally into thermal energy and convective flows, while simultaneously giving rise to large, induced electric fields which may lead to particle acceleration parallel to the magnetic field or in the plane of the neutral sheet. Because most of the recent developments in reconnection theory that are pertinent to solar flares and other forms of magnetic energy conversion are associated with the tearing instability, the bulk of our review is focused on that instability. The following review assumes that reconnection is collisionally dominated, at least in the linear growth phase, but this simplification is probably not very serious because we expect the external, macroscopic transport of magnetic flux and plasma to control the rate of reconnection. If a collisionless dissipation mechanism is operative, we expect the local conditions to vary in the region where reconnection is taking place, but we expect the effects of this local region on the overall configuration and energy balance to be relatively insensitive to the specific mechanism or chain of mechanisms which are actually responsible for the dissipation.

A bare bones explanation of reconnection begins with a resistive boundary layer mechanism that violates the conservation of magnetic flux, the fundamental constraint of ideal MHD theory. This new degree of freedom for the system leads to topological changes in the magnetic flux surfaces that change the path for current closure. These topological changes are manifested by what are called neutral points and magnetic islands. A neutral point is denoted as either an X point or an O point (Figure 11); a magnetic island consists of at least one X point and one O point. Neutral points represent sites where large current densities can be produced without being opposed by $J \times B$ forces, as first demonstrated by Dungey (1953). Neutral points therefore provide ideal sites for producing large $J_\parallel$ currents which may be manifested as accelerated particle streams. However, topological changes in magnetic flux surfaces can occur only under rather special conditions. In particular, the flow field which drives reconnection must be an odd function of position with respect to the 'neutral' mode rational surface, $k \cdot B_0 = 0$, so that the two fluid elements on opposite sides of the surface are flowing towards each other. In this respect, two basic types of treatments of reconnection are found in the literature: those in which steady-state flow fields on opposite sides of the $k \cdot B = 0$ surface always face each other in the presence of one X point; and those which assume a dynamic flow field that includes flow elements on opposite sides of the neutral surface flowing either towards or away from each other in the presence of one or more X points. The latter type of flow characterizes reconnection by the tearing mode instability (Furth *et al.*, 1963; Drake and Lee, 1977; Mahajan *et al.*, 1979; Hassam, 1980). The former type of flow characterizes steady-state reconnection theories (see Vasyliunas, 1975, for

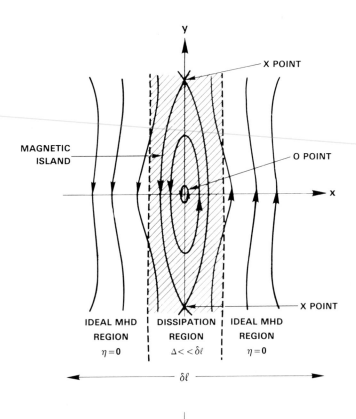

Fig. 11. Schematic of a single magnetic island with one O point and two X points; ideal and resistive MHD regions are delineated.

a review review.) Presently a clear relationship does not exist between the dynamic and steady-state theories of reconnection, although presumably a steady-state reconnection model describes the saturated state of one magnetic island consisting of one X point and one O point.

The role of the neutral surface in reconnection is illustrated by the following example. After Manheimer and Lahmore-Davies (1984), consider a two-dimensional incompressible flow field of a perfectly conducting fluid in the $x-y$ plane as depicted schematically in Figure 12. Here $V_x(-x) = -V_x(x)$ and $V_y(-x) = V_y(x)$ such that $\nabla \cdot \mathbf{V} = 0$. Δ represents the thickness of the resistive boundary layer and is assumed here to be arbitrarily small, and k is a wavenumber in the y-direction. This type of flow field is called a tearing flow. An examination of the figure shows that a very rapid flow exists in the y-direction, where $|x| < \Delta$. The flow streamlines shown in Figure 12 illustrate that, as the fluid elements with $y > 0$ approach the $x = 0$ plane from the positive and negative direction, they continuously turn away from the $x = 0$ plane into the y-direction and then continuously back into the negative and positive x-directions, respectively. Those fluid elements with $y < 0$ flow in a similar manner, but move away from the $x = 0$ plane in the $-y$-direction

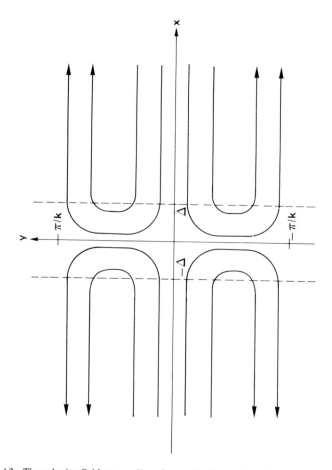

Fig. 12. The velocity field stream lines for tearing flow (after Manheimer, 1979b).

and then back into the negative and positive x-directions. The fluid element initially located at $y = 0$ does not develop a velocity component in the y-direction, but piles up at the stagnation point at $x = y = 0$.

If the perfectly conducting flow field is now imbedded in an equilibrium magnetic field of the form

$$\mathbf{B}_0 = B_0 \frac{x}{\delta l} \mathbf{e}_y, \tag{4.2}$$

which results from a current sheet of the form

$$J_x = \frac{c}{4\pi} \frac{B_0}{\delta l}, \tag{4.3}$$

where δl is the magnetic field shear length such that $\Delta \ll \delta l$ (Figure 12), the magnetic field will begin to be contorted by the flow field because the magnetic field is frozen into the flow. The time history of the magnetic field is shown in Figure 13. The magnetic

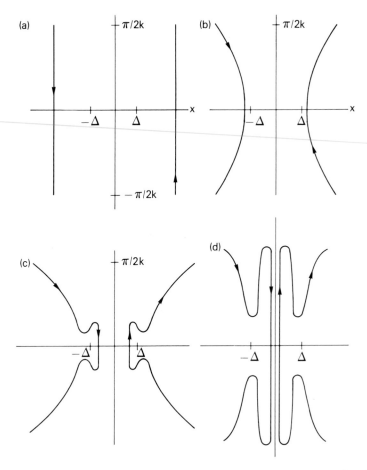

Fig. 13. The evolution of the magnetic field lines at different times in the presence of tearing flow.

field becomes more and more contorted on either side of the $x = 0$ plane as time increases, and the bending of the magnetic field lines causes larger and larger forces which attempt to restore the magnetic field lines to their original form. Notice that no magnetic restoring forces are generated where $\mathbf{k} \cdot \mathbf{B_0} = 0$ (i.e. at $x = 0$), and that the magnetic field is straigth and antiparallel within $|x| < \Delta$ on opposite sides of the $x = o$ plane. Eventually, the magnetic restoring forces equal the forces exerted by the flow field and a steady-state is achieved. These arguments qualitatively demonstrate that tearing flow is stable in the ideal MHD approximation at least in slab geometry. An analytical proof can be found in Furth et al. (1963), who first treated the theoretical aspects of the tearing mode for slab geometry.

If the perfectly conducting constraint is relaxed and finite dissipation (nonzero resistivity) is permitted, the foregoing picture changes. In both cases, the gradients in the magnetic field, and thus the $\mathbf{J_\perp}$ currents, increase as the magnetic field is distorted. In the ideal MHD case, the resultant large currents cannot be dissipated, but simply build up in magnitude. With finite dissipation, however, these currents can be dissipated, or 'annihilated' in the form of Joule heating and particle acceleration. The magnetic restoring forces are reduced and lead to the collapse of the steep current profile developing across the $x = 0$ plane that was built up by the flow field.[4] In addition, long antiparallel magnetic field lines exist on opposite sides of the $x = 0$ plane within the resistive boundary layer Δ due to the tearing flow (see Figure 13). In the presence of finite dissipation, these antiparallel magnetic field lines can diffuse together and merge to form neutral points of X and O type (Figure 11). Since the addition of dissipation permitted the lowering of magnetic energy, tearing flow leads to an instability, called the tearing mode, in the presnece of finite dissipation.

Let $\tau_A = (kV_A)^{-1}$ and $\tau_L = 4\pi(\delta l)^2/\eta c^2$ be the characteristic hydromagnetic and resistive diffusion times for the problem. The magnetic Reynolds number, $S = \tau_L/\tau_A$ is $> 10^{10}$ for solar plasmas. Inside the neutral sheet, gradients are steep enough that dissipation from the resistive term is dominant within the layer of thickness Δ about the neutral sheet in Figure 12. Outside this neutral layer, the resistive term is no longer important and ideal MHD analysis is valid. Figure 12 illustrates the splitting between resistive-dominated and ideal MHD-dominated regions.

Division into resistive and ideal MHD regions suggests that the growth rates and thickness Δ can be found using a boundary layer analysis (Furth et al., 1963) where the resistivity is assumed to be negligible everywhere except within the resistive region $|x| < \Delta$. For slow tearing modes (cf. below), Δ is given (Furth et al., 1963; Drake and Lee, 1977) by

$$\Delta = 2 (\gamma\tau_A^2/\tau_L)^{1/4}\delta l. \tag{4.4}$$

The rather complicated solution for the growth rate of slow tearing modes in slab goemetry is

$$\gamma = \left(\frac{\Gamma(\frac{1}{4})\Delta'\delta l}{\pi\Gamma(\frac{4}{5})}\right)^{4/5}\frac{S^{2/5}}{\tau_L}, \tag{4.5}$$

where $\Delta' = d\Delta/dx$ is a measure of the available magnetic free energy which can be dissipated (Adler et al., 1980). From (4.5), we see that instability occurs when $\Delta' > 0$, marginal stability when $\Delta' = 0$, and stability when $\Delta' < 0$. Hence, the tearing mode grows in such a way as to cause Δ' to vanish. This instability causes the current density profile to be flattened as the width of magnetic islands, W, increases, so that saturation occurs when W equals the thickness of the current channel, δl, as illustrated in Figure 14.

The occurrence of either steady-state or dynamic reconnection is determined by the nature of the flow field in the presence of a dissipative boundary layer located at a neutral surface. To evaluate the importance of reconnection in solar phenomena, the source of the flow field, or equivalently the free energy supply that produces the requisite flow field, must be identified. There are two possible sources of magnetic free energy:

---

4 Reconnection can be viewed as a mechanism by which flow energy is converted to magnetic energy and then into particle kinetic energy. It should be obvious that these flow structures will be ubiquitous in any system where turbulent fluid flows because structures such as shown in Figure 13 will occur whenever two clumps or jets of fluid impact and deflect each other. Thus, reconnection in and below the photosphere is at least as likely as reconnection in the corona.

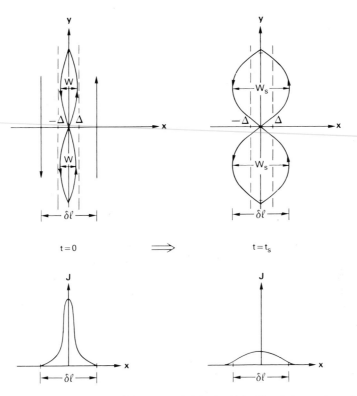

Fig. 14. Schematic of the evolution of the magnetic island width, W, relative to the resistance layer 2Δ thick. The evolution of the current profile is also schematically shown.

perpendicular and parallel currents (or correspondingly the flows which generate and maintain them). As indicated above, reconnection can be driven by a perpendicular current. It is impossible to explain a flare with energy stored in the form of an equilibrium pressure gradient, as can be proven by a simple estimate of the pressure required and how it would manifest itself observationally. Thus, for a flare to be initiated by $J_\perp$-driven reconnection in the corona, a tearing flow must exist in the corona or photosphere just prior to and during the flare process. Various ways of achieving these flow fields have been proposed. The two sources of flow fields that are most common in flare theory are solar wind-driven inverted-Y neutral sheet theories (Carmichael, 1964; Sturrock, 1966b, 1967, 1972, 1974, 1980; Kopp and Pneuman, 1976) and the emerging-flux-driven flow field (Priest and Heyvaerts, 1974; Canfield $et\ al.$, 1974; Heyvaerts $et\ al.$, 1977; Tur and Priest, 1978; Krivsky, 1968).

Figure 15 shows a simplified schematic of this latter type of configuration where one flux tube has emerged beneath another. As the latter loop pushes upward and expands, the two loops come into contact and compress the magnetic flux and plasma in the vicinity of the contact point. Both the magnetic field lines and the currents can reconnect from one loop to the other at the contact point in the presence of any dissipative mechanism which locally disrupts ideal MHD flux conservation. A third configuration where tearing and reconnection may be very important is shown in Figure 16. The convective

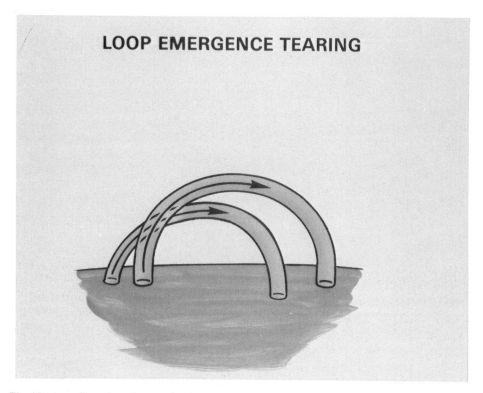

Fig. 15. A configuration where magnetic tearing and the reconnection of field lines and current paths occurs as two magnetic loops come into contact.

downflow into the chromospheric network will sweep out the supergranule cell centers of magnetic flux and concentrate it in the loop footpoints, two of which are shown in the upper half of the figure, in the network. The sheet current being compressed between the two initially distinct flux tubes will filament and reconnect when the tearing instability turns on, as shown in the lower half of the figure. Such configurations might, for example, account for spicules.

A tearing flow field can be driven indirectly by a parallel current in several ways, and is the basis of the loop flare model of Spicer and Colgate (Spicer, 1976, 1977a; Colgate, 1978; Sturrock, 1980). To illustrate $J_{\parallel}$-driven reconnection, note that kink modes are driven by a parallel current. In particular, consider the $m = 1$ kink, which is uniquely a curvilinear effect that cannot occur in slab goemetry. Figure 10 shows a magnetic loop arching into the corona with a representative magnetic field line shown spiraling around the tube of magnetic field and plasma from one footpoint to the other. A twisting at one footpoint, as noted, will drive a parallel current along the loop which eventually gets large enough to cause the loop to kink.

As shown in the second half of this figure, the nonlinear limit of the kinking brings two regions of magnetic field against each other at an oblique angle which gives rise to a $k \cdot B_0 = 0$ neutral surface ideal for tearing modes. Further, the dynamic energy release from the kinking itself (Section 4.2) drives the local tearing flow of Figure 12 directly. Thus, the magnetic energy stored in the parallel current can drive tearing flow in the

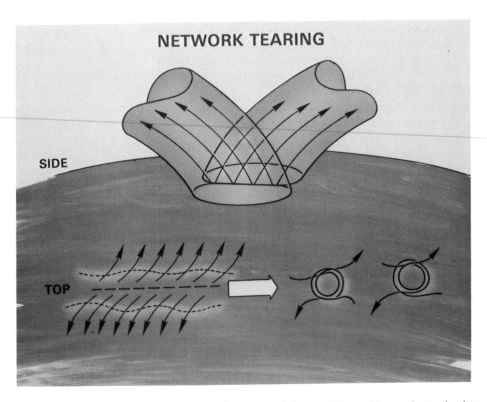

Fig. 16. Magnetic tearing at the footpoints of two magnetic loops which are driven against each other by the convective flows in the chromospheric network. Such tearing, as indicated in the lower portion of the figure, could account for a row of spicule-like structures above the network.

form of a $m = 1$ kink perpendicular to the magnetic flux surfaces, which appears as large perpendicular currents. The large perpendicular currents can be dissipated in turn as reconnection occurs via an $m = 1$ tearing mode.

From the above discussions it is clear that reconnection only occurs when a plasma–magnetic field configuration is stable to tearing flow in the ideal MHD approximation. If it were unstable to tearing flow, ideal MHD motions would lower the magnetic energy by expansion before appreciable tearing could take place. A more subtle conclusion is that tearing flow is a means by which a parallel current can drive a perpendicular current across field lines. In this way, tearing flow can cause neighboring antiparallel currents to link via a perpendicular current which short circuits the current system and lowers the total inductance. Reconnection causes a time rate of change of the inductance which enters the equivalent circuit equations as an effective impedance term that can be much larger than the initial impedance of the circuit. Hence, the rate at which reconnection occurs can greatly affect the response of the global current system to reconnection.

As shown earlier, the growth rate of a tearing mode is characterized by the magnetic Reynolds number, S. The linear growth rate for any tearing mode can be written as

$$\gamma = \alpha_1 S^p / \tau_L, \tag{4.6}$$

where α_1 is proportional to the driving energy of the tearing mode evaluated at the $k \cdot B = 0$ neutral surface, and p, which is always less than unity, is mode dependent. In cylindrical geometry growth rates scale as $S^{2/3}$ (Coppi et al., 1976; Drake, 1978) compared to the $S^{2/5}$ dependence in (4.5) above. The cylindrical tearing mode has a rate of growth $\sim S^{4/15}$ greater than a slab tearing mode; that is, ≈ 400 times greater if $S \approx 10^{10}$.

The difference in linear growth rates is also manifested in radically different nonlinear evolution. The growth time of the tearing mode is large compared to the time required to achieve pressure balance by shocks or magnetosonic waves. Hence, approximate pressure balance must be assured at all times, in the region outside the $k \cdot B = 0$ surface. Conversely, inclusion of the inertial term implies that pressure balance is not achieved in the region outside the $k \cdot B = 0$ surface. Tearing modes for which the inertial term must be included are called 'fast' tearing modes. Fast tearing modes include the $m = 1$ and double tearing modes (e.g. Coppi et al., 1976; Drake, 1978; Pritchett et al., 1980; Schnack and Killeen, 1978, 1979) and were applied to loop models of flares by Spicer (1976, 1977, 1981a). Tearing modes for which the inertial term can be neglected are called 'slow' tearing modes, and were first applied to flares by Jaggi (1964) and Sturrock (1966a, b, 1967, 1972, 1974). Pritchett et al. (1980) show that the growth time of the fast tearing mode, γ_F^{-1}, is equal to the diffusion time (sometimes called the skin time) associated with the boundary layer; that is,

$$\gamma_F^{-1} = \frac{4\pi\Delta^2}{\eta c^2}. \tag{4.7}$$

The growth time of the slow tearing mode, γ_S^{-1}, on the other hand, is much greater than the diffusion time of the layer. Consequently, the flux perturbations generated by slow tearing modes have time to communicate across the boundary layer during their growth, while the flux perturbations generated by the fast tearing modes occur too rapidly to communicate across the layer. This result has important implications for flares because the electric fields, $\delta E_{\parallel}$, which are parallel to the magnetic field at the X points induced by the fast tearing process do not have time to penetrate into the magnetic islands produced by the tearing process (see below). This means that the plasma volume accelerated by $\delta E_{\parallel}$ remains the same as that produced by a slow tearing mode in the linear regime. On the other hand, the $\delta E_{\parallel}$ produced by a slow mode has sufficient time to penetrate into the islands, causing $\delta E_{\parallel}$ to weaken as the magnetic island widens.

Returning to the growth rate defined by Equation (4.5), note that γ is defined entirely in terms of local quantities. This is not too surprising since the linear growth rate of a boundary layer instability is not affected greatly by global conditions. The global configuration, however, is expected to dominate the nonlinear evolution of the instability even though local effects may dictate local dissipation rates that influence the global evolution. Further, because reconnection is a boundary layer phenomenon, the standard conservation jump relationships must apply and mass flux, magnetic flux, and energy flux are conserved through the layer (Parker, 1963). Although the dissipation process is localized to a thin layer, the global conservation relations are preserved. The nonlinear evolution of the reconnection process must be consistent with global conditions. Therefore, the entire global current system ultimately determines the evolution of the reconnection process. We further emphasize that because the inductive properties of a laboratory plasma–magnetic field configuration with a magnetic Reynolds number of 10^4 differ radically from those of a typical solar Reynolds number $\sim 10^{10}-10^{12}$, comparisons between solar and laboratory plasma experiments or numerical simulations with low S only represent useful guides. The Sun is, in fact, our best laboratory for studying these phenomena.

In the above discussions, we have outlined the evolution of a plasma–magnetic field configuration that is initially tearing unstable and then evolves until saturation, at which time a chain of magnetic islands is fully developed. Because each island represents a current filament, with current flowing in the same direction, they attract one another. Furthermore, an equilibrium is established because each island feels equal and opposite attractive forces from its counterpart on each side. However, if one of the islands is displaced towards the right, for example, the islands to the right of the displaced island feel a greater attractive force to those on the left of the displaced one. As the displaced island begins to move towards its neighbor, however, the magnetic flux between the two islands is compressed; the local magnetic pressure is thus increased, tending to force the islands apart. If the compression force is dominant, the plasma is stable, whereas if the attraction force is dominant, the islands coalesce. Finn and Kaw (1977) first examined this phenomenon and found that the attractive force is dominant and the islands coalesce. Pritchett and Wu (1979) simulated this effect in the ideal MHD approximation, and found that compression ultimately stops the island coalescence when the magnetic pressure becomes sufficiently large. However, when finite resistivity is permitted, coalescence proceeds to completion. Spicer (1976, 1977a) used island coalescence to explain a class of impulsive electromagnetic bursts associated with flares.

The above synopsis of nonlinear effects during reconnection clearly illustrates that the reconnection process in any solar configuration is likely to be quite complex with rapid dynamic phenomena occurring on a number of scales at once. The complexity and rapidity of these interactions make it correspondingly difficult to untangle our current observations of magnetic energy release phenomena in the solar atmosphere with any real assurance of the uniqueness of a given model.

4.4. PARTICLE ACCELERATION AT A NEUTRAL SHEET

In the preceding section magnetic reconnection and tearing at a 'neutral' sheet $(\mathbf{k} \cdot \mathbf{B}_0 = 0)$ was considered both locally and globally. It is not necessary that the magnetic field go to zero, simply that the phase fronts lie locally parallel to B_0. Because the magnetic fields in the vicinity of the 'neutral' sheet are changing with time, there can be electromagnetic as well as electrostatic components of the electric field along the zero-order magnetic field.

To illustrate how tearing modes lead to parallel electric induction fields, we express the perturbed electric field parallel to the magnetic field as

$$\delta E_{\parallel} = -\frac{\gamma \delta A_{\parallel}}{c} - ik_{\parallel}\,\delta\phi, \tag{4.8}$$

where $\delta A_{\parallel}$ is the perturbed component of the vector potential parallel to the equilibrium magnetic field and $\delta\phi$ is the perturbed electrostatic potential. Since $\delta A_\| = \delta\psi e_z$, (4.8) becomes

$$\delta E_\| = -\gamma \frac{\delta\psi}{c} - ik_\| \, \delta\phi. \tag{4.9}$$

The parallel induction field $-\gamma\,\delta\psi/c$ causes the electrons to flow with a velocity $\delta v_{e\|}$. However, because the induction field leads to charge separation between ions and electrons, a parallel electrostatic field $-ik_\|\delta\phi$ is also produced, which shorts out the induction field for sufficiently large $k_\|$, thus $\delta E_\| = 0$ except where $k_\| \to 0$, where $\delta E_\| = -\gamma\delta\psi/c$. Within the resistive layer, therefore, large parallel electric fields are produced and particle acceleration can occur (Drake and Lee, 1977) only in the vicinity of $k_\| = 0$.

For both fast and slow tearing modes, electric induction fields estimated using solar parameters always exceed the Dreicer electric field defined by (2.31). Estimates of parallel electric fields generated by slow tearing modes (Van Hoven, 1979) indicate that the predicted electric field is ~3 times the Dreicer field, while estimates of the parallel electric field from the fast tearing mode (Spicer, 1981a) always exceeded the Dreicer field by orders of magnitude if the saturated value of the electric field is assumed. The physical basis for these results appears to be as follows: since solar current systems are inductively dominated due to the large solar diffusion times, the current will attempt to remain constant. At the same time, the small resistive skin depth, characteristic of a solar plasma, constrains the electric induction field to small channels about the X points. These two effects together require the electric fields to be large. Collision-dominated tearing modes thus may develop into semicollisionless or collisionless tearing modes (Drake and Lee, 1977) and $J_\|$-driven anomalous resistivity may be produced, thereby altering the local nonlinear evolution of the tearing mode even further (Spicer, 1981a).

Although it is not generally felt that a substantial fraction of the low-grade convective energy in these reconnection scenarios goes into particles at very high energy (> 30 keV), the existence of these particles as a natural consequence of relatively ubiquitous MHD interactions provides a very important observable diagnostic of what is going on. These particles can propagate for long distances because of their long mean free path in even relatively dense plasma and their radiation signature is unmistakable. The general subject of superthermal particle acceleration was felt to be so important that an entire chapter in this work (by Forman, Ramaty, and Zweibel) is devoted to the subject.

4.5. FLARE TRIGGER MECHANISMS

Flare models have assumed, historically, that a flare instability must be excited by some external perturbation or 'trigger' (cf. Sturrock, 1966; Sweet 1969, Van Hoven *et al.*, 1980; Spicer and Brown, 1981). as opposed to directly driving the flare, for example, by externally applied flow fields (Heyvaerts *et al.*, 1977). Sturrock (1966) has noted that there are two basic types of instability onsets: explosive or nonexplosive. The explosive onset occurs when a system is near marginal stability, and is linearly stable to infinitesimal perturbations but nonlinearly unstable to finite perturbations. Conversely, nonexplosive onset occurs when a system is linearly unstable to infinitesimal perturbations. Sturrock also contended that only instabilities with explosive onsets are viable as flare trigger mechanisms. This point of view is justified as follows: for preflare storage of energy to occur, energy must accumulate in the plasma–magnetic field configuration without being dissipated continuously through instabilities excited by infinitesimal perturbations. Rather, the energy must accumulate up to some level beyond which a finite perturbation can drive the configuration to instability, thus releasing the accumulated energy.

Recent research in collisionless plasmas by Dupree (1981) has shown that nonlinear excitation of the ion-acoustic instability – for example, by relatively slow drifts *below* the linear instability threshold – occurs and can lead to large fluctuations and, hence, anomalous dissipation. There is every reason to believe that many complex dynamic systems display this property as a system parameter, say the total current in the flux tube, is reduced from linearly unstable values into the stable regime. In our example, there is still free magnetic energy in a current-carrying flux tube even though linear analysis may indicate exponential MHD stability. Provided the system has sufficient degrees of freedom and finite amplitude perturbations, the nominally stable system is actually nonlinearly unstable. The growth is initially very slow when the perturbations are small but the intrinsic nonlinearity makes it explosive at later times. Such nonlinear instabilities, which go to large amplitude in finite time, have just the properties required of a trigger mechanism. Long times are required and available during the perturbation growth phase for excess energy to be stored. The conversion or release of the magnetic energy will be faster than exponential when it does occur.

We have considered three mechanisms which can be driven directly by both $J_\perp$ or $J_\parallel$: reconnection, double layers, and anomalous Joule heating. We also argue that $J_\perp$ mechanisms require externally imposed flow fields. Hence, $J_\perp$-driven mechanisms do not require triggers and are excited directly once the flow field exceeds the necessary speeds. However, $J_\parallel$-driven mechanisms can be either driven directly or triggered. A directly-driven $J_\parallel$ mechanism can result from the natural evolution of the current profile that accompanies the time-dependent, preflare magnetic energy storage process. Conversely, a $J_\parallel$-driven mechanism may be initiated by a trigger that leads to an alteration of the current profile even after the maximum amount of magnetic energy has been stored. Nevertheless, because the preflare magnetic energy storage processes apparently occur in times much greater than an MHD transit time, the preflare state can be examined under the assumption of a quasi-equilibrium. One can also assume that the preflare magnetic configuration slowly stores energy during its evolution, until the configuration is one the verge of losing its quasi-equilibrium (Jockers, 1976; Low, 1977a, b; Anzer, 1978; Birn *et al.*, 1978: Heyvaerts *et al.*, 1979; Priest and Milne, 1979; Hood and Priest, 1979). This approach is related to bifurcation theory. As the relevant parameters are varied, there is a point of bifurcation at which the equilibrium either can go unstable or can enter a more stable regime. This approach is attractive primarily because the treatments usually involve more than one dimension and allow for global electrodynamic coupling through more reasonable boundary conditions. However, the approach is limited, at present, because analyses have had to utilize some symmetry condition to reduce the extreme mathematical complexity of the problem. This has the unfortunate consequence of removing the degrees of freedom available to the magnetic configuration for lowering its energy state. In addition, these analyses do not account for transport phenomena which, in real magnetized plasmas, help control the evolution of the existing current profiles. For example, preflare magnetic energy storage involves induction currents and these induction currents may flow in very narrow channels. Thus, the current profile is dictated, in part, by the resistivity profile in the magnetized plasma, as well as the resistivity profile at the boundaries.

Accounting for transport effects naturally leads to examination of the role of thermal instabilities in altering the current profile, and thus in triggering a flare (see Section 2.3). Thermal instabilities were studied first by Parker (1953) and Field (1965), while Sweet (1969) and Kahler and Kreplin (1970) first pointed out the importance of thermal instabilities in flares. The thermal instabilities ensue when classical thermal conduction is reduced in the presence of a strongly temperature-dependent radiation function such as occurs in plasmas with solar abundances. In a magnetized plasma, this process corresponds to preventing thermal conduction parallel to the magnetic field and can be accomplished by a long-wavelength perturbation parallel to the magnetic field or short-wavelength magnetohydrodynamic turbulence.

The electromagnetic superheating (overheating) instability (Kadomtsev, 1966), which is a thermal instability applicable to both neutral sheets and sheared magnetic fields, may be particularly important for triggering flares. It has been utilized in several flare models (Coppi and Friedland, 1971; Heyvaerts, 1974a; Coppi, 1975; Spicer, 1976, 1977a). The superheating instability requires a current flowing parallel to a magnetic field driven by a parallel, roughly equipotential, electric field that may differ in strength on different field lines. Because $J_\parallel = E_\parallel/\eta$ and $\mathbf{E} \cdot \mathbf{J} = E_\parallel^2/\eta$, any perturbations that yields an increase in temperature will reduce η, increase $J_\parallel$, increasing the temperature again, and so on. This process results in an increase of both the current density and the temperature. The superheating instability will occur if

$$\frac{J_{0\parallel}^2 \eta_0}{n_0 k_b T_0} > -\frac{3}{2n_0 k_b} \frac{dQ_R}{dT_0}, \qquad (4.10)$$

where the subscript 0 specifies the equilibrium quantities, and Q_R, the radiation loss function. Hence, if $dQ_R/dT_0 < 0$ and (4.10) is satisfied, this instability occurs. The superheating instability physically manifests itself in the form of plasma striations, composed of long-wavelength filaments parallel to the magnetic field with high current density and temperature, alternating with filaments of low current density and temperature.

The superheating instability is important because it promotes build-up of the current density in locations where $\mathbf{k} \cdot \mathbf{B}_0 = 0$, while a tearing mode attempts to flatten the localized current density profile. These mechanisms work in opposition to one another and possibly a nonlinear balance may be achieved between the two competing instabilities. Thus steady-state reconnection may occur in sheared fields, but only if cross-field transport does not limit the further growth of the superheating instability before a tearing instability occurs. This is yet another example of the complexity we face in treating a given situation completely.

At present, the superheating instability has only been treated in the local approximation, analogous to the manner in which the stability analysis of loops has been treated locally (cf. Section 4.2), and no accounting for the role of global coupling has been given.

Various triggers for $J_\parallel$-driven double layers, and also for anomalous Joule heating mechanisms, are discussed by Carlqvist (1979b). Since a double layer is driven by a high-inductance current system and requires a current drift speed of $\gtrsim v_{T_e}$, local reduction in the density provides the basic trigger of a double layer. The triggering results from the requirement of a temporally constant net current. In the presence of a density decrease, the current drift speed must increase sufficiently to keep the current constant. According to Carlqvist (1979b), this density reduction may occur in several ways. The requisite heating could be generated by the superheating instability, the tearing mode, or anomalous Joule heating resulting from a $J_\parallel$-driven anomalous resistivity mechanism with a much lower threshold than a double layer. An example of this could be the electrostatic ion–cyclotron instability (cf. Kindel and Kennel, 1971). Since hydrodynamic expansion parallel to the magnetic field results in a constant pressure along the field lines (neglecting gravity), an increase in temperature of 10–100 causes the density to be reduced by a corresponding amount provided chromospheric evaporation is choked.

An alternate trigger for a double layer can occur in a high-inductance parallel current system, as in an emerging flux loop. As a flux loop expands, the density drops until the drift speed of the current exceeds threshold for either an anomalous Joule heating mechanism or a double layer. For this triggering mechanism to apply, the net current cannot change appreciably during the expansion process.

Other effects which may be important as trigger mechanisms, and which have not been examined in the context of solar flares, as yet, are:

(1) Investigations by Dobrott *et al.* (1977) and Pollard and Taylor (1979) have shown that equilibrium diffusion flows have an important effect on the stability of a magnetic configuration to tearing modes.

(2) Lau and Liu (1980) have examined the stability of shear flow in a magnetized plasma slab. They found that velocity shear alone cannot produce an MHD instability that qualitatively alters the initial laminar state of the plasma. However, they suggest that a small amount of magnetic field curvature, such as occurs in solar configurations, may reverse this conclusion.

Flares must also be considered as flare triggers. Homologous sequences of flares, regularly spaced in time, appear to come from the same volume in the solar atmosphere and are quite similar in form (Ellison, 1963; Svestka, 1976). The interval between such flares is observed to increase with importance, up to more than 10^6 s for large events. According to the single loop circuit model presented earlier in this section, the largest flare requires $\Delta t_s > 6 \times 10^6$ s for energy storage to occur. For small flares this time certainly is shorter. For a sequence of flares to occur, spaced roughly by Δt_s, the basic current system, and thus the original magnetic field configuration, must remain intact. If the fields vent out the top of the atmosphere, however, the current system is changed and the value of the inductance will depend on how the current closes in the new field configuration. If the current closes at infinity, the new current field configuration cannot repeat the flare in a time Δt_s, because the energy storage time constant tends to infinity as the inductance becomes infinite. The rate at which retrievable magnetic energy is stored goes to zero because an infinite time is required to transport energy to infinity. From these arguments, we conclude that the phenomenon of 'homologous' flares demands that the current system close at a distance above the photosphere not greater than 10^{10} cm, to ensure that the time between sequences of flares is sufficiently short.

5. Outstanding Questions

In the previous sections we reviewed current understanding of the fundamentals of magnetic energy storage and conversion in the outer layers of the solar atmosphere. In this final section we raise a number of questions focused on clearing up the major remaining unknowns.

Perhaps the most important question that needs to be addressed is: 'What are the dynamics of flux emergence?' We argued in Section 4.1 that a significant amount of energy is made available when flux emerges into the transition region and corona. Whether that magnetic energy is converted to heat, how that conversion takes place, and at what rate are unknown. Higher spatial and temporal resolution observations of flux emergence at the base of the chromosphere will help answer this question as will better observations of the evolution of the coronal tracers of magnetic field lines. Numerical simulations will also play an important role. We already have an excellent understanding of the basic temperature, pressure, and density structure of the background atmosphere into which the magnetic flux emerges. Thus simulations of the various possible flux emergence scenarios outlined in Section 4.1 may provide important clues about the real process on the Sun.

A second important question that must be addressed is: 'What is the rate of plasma–magnetic field interpenetration?' Convective motions in and below the photosphere are constantly stressing the flux tubes which protrude into the corona. Sheared magnetic fields, where tearing can occur, are continually being compressed in some regions, and adverse curvature regions where fluting can occur are automatically generated elsewhere. Since these MHD unstable regions are continuously being regenerated, we have to expect the instabilities in these regions to have progressed into a nonlinear turbulent state. Small scale fluting allows magnetic field lines to pull straight through the plasma. The rate of interpenetration is essentially unknown, however, so the dynamic friction and hence plasma heating are unknown.

By the same arguments magnetic tearing can be expected to occur in nonlinear turbulent states spanning a range of size scales. The macroscopic result is again the conversion of magnetic field energy to kinetic energy and heat accompanied by the interpenetration of magnetic field and plasma. These two dissipative mechanisms, along with a host of others identified above, all break the captured flux concept of ideal MHD and thus are crucial to an understanding of the evolution and dynamics of macroscopic MHD systems. Further, because the space and time scales on which this interpenetration occurs are smaller than the general flux tube scales, direct observation is correspondingly more difficult, so reliance on theory and computation will have to be all the greater.

A third outstanding question is: 'What is the location where the magnetic energy believed to power a solar flare is stored?' The prevalent view of the preflare state maintains that the flare-free energy must be stored in coronal magnetic fields because it appears to be released there (Svestka, 1976; Van Hoven *et al.*, 1980) as opposed to storage in a remote photospheric site; hence the term '*in situ* storage'. There are two principal arguments for this view. The first is based on the observational fact that no large changes in the photospheric structure have been detected to date throughout the duration of most flares (cf. Svestka, 1976, for a review). The second argument is essentially theoretical: the free energy required to drive a large flare is $\approx 10^{32}$ erg and the only source of this energy that can be identified observationally is the atmospheric magnetic field (Parker, 1957). To release $\approx 10^{32}$ erg of atmospheric magnetic field requires relaxing ≈ 400–500 gauss within a volume of $\approx 6 \times 10^{28}$ cm^3. Note that the magnetic pressure associated with such fields is two to three orders of magnitude greater than the ambient gas pressure. While both arguments suffer from observational and theoretical difficulties, the proposition of *in situ* storage of magnetic energy has motivated a majority of the theoretical studies related to flare physics. These studies have generally utilized the concept of force-free fields, that is, fields for which the Lorentz force vanishes. The utilization of force-free fields is motivated by the need to build up, slowly, magnetic energy *in situ* by stressing the magnetic fields ($J_\perp \times B \neq 0$) while maintaining a relatively stable large-scale magnetic configuration in a low-pressure atmosphere. Under these conditions, large-scale equilibrium magnetic configurations must be either potential with no free energy (current sources are external to the atmosphere, that is, below the photosphere) or force-free, because the plasma pressure gradients in the atmosphere are smaller than the magnetic pressure by up to four orders of magnitude and, therefore, cannot support $J \times B$ forces (e.g. Van Hoven *et al.*, 1980).

While the concept of *in situ* energy storage is very appealing physically, there remains a great deal of work to be done before it becomes an unambiguously demonstrated concept. In summary, the weaknesses associated with *in situ* coronal storage are:

(1) Large coronal fields, distributed over large volumes, are required.
(2) Expansion of the magnetic field must be prevented: the suggestion that potential fields will prevent expansion remains to be verified.
(3) We must attempt to understand the effect on energy storage of photospheric disturbances propagating into the energy storage volume.

The principle strength of *in situ* storage is that the flare energy is stored exactly where it is needed to explain the flare. However, this is only an apparent strength because it rests on a possibly dubious interpretation of observations: that no large changes in the photospheric structure are observed throughout the duration of most flares (cf. Svestka, 1976). We have used the term 'dubious' because sheared flow fields tangential to the solar surface, which are required for both *in situ* and remote storage as well as directly driving the flare, are difficult to detect observationally. In addition, because it is clearly possible to transport electromagnetic energy at speeds up to that of light in high-inductance systems, there is no *a priori* justification for ignoring the possibility of remote inductive storage. Nevertheless, a storage site still suffers the same theoretical difficulties as an *in situ* storage site because an inductive current source requires the existence of large, essentially constant (as a function of time) net currents coupling various segments of the solar atmosphere. The primary advantage of a remote storage site is that the bulk of the inductive storage volume will be photospheric, as opposed to coronal, thus reducing the requirement that large fields must be distributed throughout a large coronal volume.

The fourth question concerns the most controversial aspect of flare research, choosing one of the mechanisms reviewed as the explanation for the sudden release of magnetic energy. This controversy is unwarranted in one sense, however. Studies of both laboratory and near-Earth space plasmas have shown that the normal situation in an unstable magnetized plasma is for many instabilities to be operating simultaneously. For example, an MHD instability, such as a kinking prominence, may drive a shock in front of it. If this shock is associated with $J_\perp$-driven anomalous transport mechanisms (Davidson and

Krall, 1977), a highly anisotropic perpendicular and parallel temperature ratio can be produced leading to excitation of various loss cone mechanisms (Spicer, 1976). As a second example, a constant net current flowing in a loop may excite the superheating instability, which steepens the current density gradient, thus exciting an $m = 1$ ideal or resistive kink. The resultant kink may drive large perpendicular and parallel currents, both of which can induce anomalous resistivity mechanisms, and so on (Spicer 1976). These examples illustrate that the flare process is probably quite complex, simultaneously involving many instabilities and local configurations and requiring a careful examination of how each process can either excite or interact with the others to produce a flare.

Despite the complexity of the flare problem, we can attempt to identify which of the three instabilities reviewed is most likely to occur first. In the case of $J_\perp$-driven reconnection and anomalous transport, this question can be answered in part. If the flow field initially convects new magnetic flux towards a $\mathbf{k} \cdot \mathbf{B}_0 = 0$ surface faster than it can be dissipated by a tearing mode, the current steepens until $J_\perp$-driven anomalous resistivity mechanisms are excited and dissipate the current build-up. However, if the new flux initially is convected at a rate less than the tearing mode dissipation rate, the current is dissipated by the reconnection process itself and anomalous resistivity is unlikely to occur, except possibly through the induced electric field associated with the reconnection.

In the case of $J_\parallel$-driven mechanisms, the current density or the current density profile is the important threshold criterion for excitation of double layers, anomalous resistivity, or MHD processes (ideal or resistive). To determine which MHD process occurs first, a full understanding of the global three-dimensional current system is required — not an easy task. It is generally believed that MHD processes precede double layers or anomalous resistivity mechanisms (Van Hoven, 1981). However, the universal applicability of this may be questioned because MHD processes are very geometry dependent. The Birkeland current, made up of parallel currents flowing along the potential magnetic field of the Earth, is capable of driving anomalous resistivity mechanisms or double layers (Wolf, 1975) even though the overall system is MHD stable. Hence, it is not at all obvious which of the $J_\parallel$-driven mechanisms occurs first. The immediate task of the experimental efforts should be, as we have argued, to resolve accurately in time and space the detailed dynamic evolution of some particular realizations of the simple processes and configurations we focused on earlier; that is, flux tube emergence, MHD geometry changes, global and local aspects of magnetic tearing, current, plasma, and thermal instabilities, and the propagation and dissipation of jets and MHD waves. Without actual close-up measurements of the flailing, unwinding, and searing reconnections which explosively emerging flux tubes on the Sun most likely undergo, we must rely on unbenchmarked theories and computational models. No matter how sophisticated or detailed, these are only projections and are unlikely to be calibrated convincingly in terrestrical experiments because of the extreme scale changes, if for no other reason. At least a substantial fraction of the observational effort should go into trying to map completely many aspects of selected events by a battery of simultaneous high temporal and spatial resolution measurements using different instruments. Flow fields and magnetic field strengths and directions are particularly important to extract from the data, so a substantial part of the theoretical and computational effort should go into devising algorithms for this new kind of multi-instrument profile reconstruction.

For a focus of the theoretical efforts on magnetic energy storage and conversion, we can visualize a sequence of dynamic processes and phenomena beginning with mag-netodynamic action in the solar convective zone at one end and ending with solar wind interacting with the Earth's magnetosphere at the other. The specific phenomena can be thought of as individual links in a chain of predictability which potentially could be linked up to extend all the way from the Sun to the Earth. At present we know a lot about many of the links, and the experimental program is geared towards providing even better descriptions in the future.

However, for a complete predictive capability starting with dynamo action in the convective zone and ending with particle and radiation fluxes at the Earth, we have clearly a very long way to go — if the goal is in fact attainable. We *are* in a good position to begin dealing quantitatively and accurately with the interconnections between some of the individual processes. Further, it is extremely important that we begin doing this now. Because our ability to collect data at any stage of the causal sequence of magnetically active processes is limited, it is crucial to know how dependent the results of a model prediction are to the variations in initial conditions which can be expected from experi-mental inaccuracies or coarseness of the data set. These considerations have been applied to terrestrial weather modeling efforts for decades, but only recently has our under-standing and database on the MHD and plasma phenomena in the solar atmosphere become sufficient for such limit-of-predictability analyses to have any real validity. In the case of our weather, it has been estimated that local forecasts for more than five days are made impossible by the physics of weather and its quantitative monitoring, not by the computational power at hand. Similar calculations for the Sun might tell us how far in advance it is possible to predict solar flares given plausible monitoring nets; how deep into the convective zone must we reach to 'predict' flux tube emergence and the dynamics of a single bipolar magnetic region; and whether the rear side of the Sun must be continually monitored to predict coronal holes and sector boundary locations accurately.

We can also expect a big payback for the experimental program. Considerations of the predictability from one link in the chain to the next will allow the experimental community to work the problem backward. They can estimate the measurement of accuracy that is required to distinguish between two possible models of a particular process under investigation.

Acknowledgements

The authors appreciate useful comments, criticisms, and discussions, concerning various topics reviewed in the manuscript, with Drs D. Book, J. F. Drake, J. Feddars, J. Guillory, W. Manheimer, P. Palmadesso, and R. A. Smith. The authors also would like to make special thanks to Drs J. Karpen and C. R. DeVore for comments, criticisms, and excellent editing of the manuscript. Special thanks also goes to Joyce Brown, Loris Sizemore, Darlene Miller, and Fran Rosenberg who typed the many drafts of the manuscript.

The authors are supported, in part, by grants from the National Aeronautics and Space Administration and the Office of Naval Research.

References

Adler, E. A., Kulsrud, R. M., and White, R. B.: 1980, *Phys. Fluids* **23**, 1375.
Alfvén, H. and Carlqvist, P.: 1967, *Solar Phys.* **1**, 220.
An, C.-H., McClymont, A. N., and Canfield, R. C.: 1980, *Bull. Amer. Astron. Soc.* **12**, 913.
Antiochos, S. K.: 1979, *Astrophys. J.* **232**, L125.
Anzer, U.: 1968, *Solar Phys.* **3**, 298.
Anzer, U.: 1978, *Solar Phys.* **57**, 111.
Athay, R. G. and White, O. R.: 1978, *Astrophys. J.* **226**, 1135.
Athay, R. G. and White, O. R.: 1979, *Astrophys. J.* **229**, 1147.
Barbosa, D. D.: 1978, *Solar Phys.* **56**, 55.
Barnes, C. W. and Sturrock, P. A.: 1972, *Astrophys. J.* **174**, 659.
Bernstein, I. B., Freeman, E. K., Kruskal, M. D., and Kulsrud, R. M.: 1958, *Proc. Roy. Soc. London* **A244**, 17.
Birn, J. Goldstein, H., and Schindler, K.: 1978, *Solar Phys.* **57**, 81.
Block, L. P.: 1975, in B. Hultqvist and L. Stenflo (eds.). *Physics of Hot Plasmas in the Magnetosphere,* Plenum, New York, p. 229.
Block, L. P.: 1978, *Astrophys. Space Sci.* **55**, 59.
Bonnet, R. M. Bruner, E. C. Jr, Acton, L. W., Brown, W. A., and Decandin, M.: 1980, *Astrophys. J.* **237**, 47.
Book, D. L.: 1980, *The NRL Plasma Formulary* (revised).
Book, D. L.: 1981, *Comments on Plasma Physics and Controlled Fusion, Comments on Modern Physics: Part E* **6**, 193.
Boris, J. P.: 1968, 'Resistively Modified MHD Modes', Thesis, Princeton Univ. University Microfilms, Ann Arbor.
Boris, J. P. and Mariska, J. T.: 1982, *Astrophys. J.* **258**, L49.
Boris, J. P., Orens, J. H., Dawson, J. M., and Roberts, K. V.: 1970, *Phys. Rev. Lett.* **25** (II), 706.
Braginskii, S. I.: 1965, *Reviews of Plasma Physics*, Vol. 1, Consultants Bureau, New York, p. 205.
Brown, J. C., Melrose, D. and Spicer, D. S.: 1979, *Astrophysics. J.* **228**, 592.
Brown, J. C. and Smith, D. F.: 1980, *Rep. Prog. Phys.* **43**, 125.
Brueckner, G. E.: 1980, *Highlights of Astronomy* **5**, 557.
Brueckner, G. E., Bartoe, J.-D. F., and Dykton, M.: 1980, *Bull. Amer. Astron. Soc.* **12**, 907.
Brueckner, G. E. and Bartoe, J.-D. F.: 1982, in preparation.
Bruner, E. C., Jr.: 1978, *Astrophys. J.* **226**, 1140.
Buneman, O.: 1959, *Phys. Rev.* **115**, 503.
Canfield, R. C., Priest, E. R., and Rust, D. M.: 1974, *Flare Related Magnetic Field Dynamics*, HAO, Boulder, p. 361.
Cargill, P. J. and Priest, E. R.: 1980, *Solar Phys.* **63**, 251.
Carlqvist, P.: 1969, *Solar Phys.* **7**, 503.
Carlqvist, P.: 1973, Tech. Rept. TRITA-EPP-73-05, Dept of Plasma Phys. Royal Inst. of Tech., Stockholm.
Carlqvist, P.: 1979a, in P. J. Palmadesso and K. Papadopoulos (eds.), *Wave Instabilities in Space Plasma*, D. Reidel, Dordrecht, p. 83.
Carlqvist, P.: 1979b, *Solar Phys.* **63**, 353.
Carmichael, H.: 1964, in W. M. Hess, (ed.) *AAS-NASA Symp. on Physics of Solar Flares*, NASA SP-50, Washington, D.C., p. 451.
Chiuderi, C., Giachetti, R., and Van Hoven, G.: 1977, *Solar Phys.* **54**, 107.
Cohen, B. I., Krommes, J., Tang, W. M., and Rosenbluth, M. N.: 1976, *Nucl. Fusion* **16**, 971.
Colgate, S. A.: 1978, *Astrophys. J.* **221**, 1068.
Coppi, B. and Friedland, A. B.: 1971, *Astrophys. J.* **169**, 379.
Coppi, B.: 1975, *Astrophys. J.* **195**, 545.
Coppi, B., Galvao, R., Pellat, R., Rosenbluth, M. N., and Rutherfold, P. H.: 1976, *Sov. J. Plasma Phys.* **2**, 533.

Davidson, R. C. and Krall, N. A.: 1977, *Nucl. Fusion* 17, 1313.

Dobrott, D. R., Prager, S. C., and Taylor, J. B.: 1977, *Phys. Fluids* 20, 1850.

Drake, J. F. and Lee, Y. C.: 1977, *Phys. Fluids* 20, 1341.

Drake, J. F.: 1978, *Phys. Fluids* 21, 1777.

Dreicer, H.: 1959, *Phys. Rev.* 115, 238.

Dungey, J. W.: 1953, *Phil. Mag.* 44, 725.

Dupree, T. H.: 1967, *Phys. Fluids* 10, 1049.

Dupree, T. H.: 1981, *Bull. A. P. S.* 26, 1060.

Einaudi, G. and Van Hoven, G.: 1981, *Phys. Fluids* 24, 1092.

Ellision, M. A.: 1963, *Quart. J. Roy. Astron. Soc.* 4, 62.

Ferraro, V. C. A. and Plumpton, C.: 1966, *An Introduction to Magneto-Fluid Mechanisms*, Clarendon Press, Oxford.

Field, G. B.: 1965, *Astrophys. J.* 142, 531.

Finn, J. M. and Kaw, P. K.: 1977, *Phys. Fluids* 20, 72.

Furth, H. P., Killeen, J., and Rosenbluth, N. M.: 1963, *Phys. Fluids* 6, 459.

Furth, H. P., Killeen, J., Rosenbluth, M. N., and Coppi, B.: 1966, *Plasma Physics and Controlled Nuclear Fusion Research*, Vol. 1, p. 617, IAWA, Vienna.

Gabriel, A. H.: 1976, *Phil. Trans Roy. Soc. London* A281, 339.

Galeev, A. A. and Saydeev, R. Z.: 1979, in M. A. Leontovich (ed.), *Reviews of Plasma Physics*, Vol. 7, Consultants Bureau, New York, p. 1.

Gebbie, K. B., *et al.*: 1980, *Bull. Amer. Astron. Soc.* 12, 907.

Giachetti, R., Van Hoven, G., and Chiuderi, C.: 1977, *Solar Phys.* 55, 371.

Glencross, W. M.: 1981, *Solar Phys.* 73, 67.

Goertz, C. K. and Joyce, G.: 1975, *Astrophys. Space Sci.* 32, 165.

Goertz, C. K.: 1979, *Rev. Geophys. Space Phys.* 17, 418.

Green, J. M. and Johnson, J. L.: 1962, *Phys. Fluids* 5, 510.

Hasan, S. S.: 1980, *Solar Phys.* 67, 267.

Hassam, A. B.: 1980, *Phys. Fluids* 23, 2493.

Heyvaerts, J.: 1974a, *Astron. Astrophys.* 37, 65.

Heyvaerts, J.: 1974b, *Solar Phys.* 38, 419.

Heyvaerts, J., Priest, E. R., and Rust, D. M.: 1977, *Astrophys. J.* 216, 123.

Heyvaerts, J., Lasry, J. M., Schatzman, J., and Witomsky, G.: 1979, *IAU Coll.* 44, 174.

Hollweg, J. V.: 1979, *Solar Phys.* 62, 227.

Hollweg, J. V.: 1981, *Solar Phys.* 70, 25.

Hood, A. and Priest, E. R.: 1979, *Solar Phys.* 64, 303.

Hoyng, P., Brown, J. C., and Van Beek, H. F.: 1976, *Solar Phys.* 48, 197.

Hubbard, R. and Joyce, G.: 1979, *J. Geophys. Res.* 84 (A8), 4297.

Ionson, J. A.: 1978, *Astrophys. J.* 226, 650.

Jackson, J. D.: 1962, *Classical Electrodynamics*, Wiley, New York.

Jaggi, R. K.: 1964, in W. Hess (ed.), *AAS-NASA Symp. on Physics of Solar Flares*, NASA SP-50, Washington, D.C., p. 419.

Jockers, V.: 1976, *Solar Phys.* 50, 405.

Kadomtsev, B. B.: 1966, in M. A. Leontovich (ed.) *Reviews of Plasma Physics*, Vol. 2, Consultants Bureau, New York, p. 153.

Kahler, S. W. and Kreplin, R. W.: 1970, *Solar Phys.* 14, 372.

Karpen, J. T., Oran, E. S., Boris, J. P., Mariska, J. T., and Brueckner, G. E.: 1981, *Bull. Amer. Astron. Soc.* 13, 913.

Karpen, J. T., Oran, E. S., Boris, J. P., Mariska, J. T., and Brueckner, G. E.: 1982, *Astrophys. J.* 261, 375.

Kindel, J. M. and Kennel, C. G.: 1971, *J. Geophys. Res.* 76, 3055.

Kopp, R. A. and Pneuman, G. W.: 1976, *Solar Phys.* 50, 85.

Krall, N. A. and Trivelpiece, A. W.: 1973, *Principles of Plasma Physics*, MacGraw-Hill, New York.

Krivsky, L.: 1968, *IAU Symp.* 35, 465.

Kruskal, M. D. and Kulsrud, R. M.: 1958, *Phys. Fluids* 1, 265.

Kulsrud, R. M.: 1967, in P. A. Sturrock (ed.), *Plasma Astrophysics*, Academic Press, New York, p. 46.

Kuperus, M.: 1976, *Solar Phys.* 47, 79.

Kuperus, M., Ionson, J., and Spicer, D. S.: 1981, *Ann. Rev. Astron. Astrophys.* 19, 7.

Lau, Y. Y. and Liu, C. S.: 1980, *Phys. Fluids* 19, 1644.

Levine, R. H.: 1975, *Solar Phys.* 44, 365.

Levine, R. H.: 1976, *Solar Phys.* 46, 159.

Longmire, C. L.: 1963, *Elementary Plasma Physics*, Interscience, New York.

Low, B. C.: 1977a, *Astrophys. J.* 212, 234.

Low, B. C.: 1977b, *Astrophys. J.* 217, 288.

Mahajan, S. M., Hazeltine, R. D., Strauss, H. R., and Ross, D. W.: 1979, *Phys. Fluids* 22, 2147.

Manheimer, W. M. and Boris, J. P.: 1972, *Phys. Rev. Lett.* 28 (II), 659.

Manheimer, W. and Boris, J. P.: 1977, *Comm. Plasma Physics Cont. Fusion* 3, 15.

Manheimer, W. and Flynn, R. W.: 1974, *Phys. Fluids* 17, 409.

Manheimer, W. M. Colombant, D., and Flynn, R. W.: 1976, *Phys. Fluids* 19, 1354.

Manheimer, W. M.: 1979, *J. Physique* 40, C7-269.

Manheimer, W. M. and Lahmore-Davies, C.: 1984, *MHD Instabilities in Simple Plasma Configurations.*

Manheimer, W. and Antonsen, T. M.: 1979, *Phys. Fluids* 22, 957.

Mariska, J. T. and Boris, J. P.:1981, *Bull. Amer. Astron. Soc.* 13, 836.

Mariska, J. T. and Boris, J. P.: 1982, in M. S. Giampapa and L. Golub (eds.) *Dynamic Phenomena in Coronal Flux Tubes,* Proceedings of the Second Cambridge Workshop on Cool Stars, Stellar Systems, and the Sun,p. 53.

Mariska, J. T. and Boris, J. P.: 1983, *Astrophys. J.* 267, 409.

Mariska, J. T., Boris, J. P., Oran, E. S., Young, T. R. Jr, and Doschek, G. A.: 1982, *Astrophys. J.* 255, 783.

Mariska, J. T., Feldman, U., and Doschek, G. A.: 1978, *Astrophys. J.* 226, 698.

Mozer, F. S.: 1976, in B. M. McCormac (ed.), *Magnetospheric Particles and Fields*, D. Reidel, Dordrecht, p. 125.

Nakagawa, Y. and Raadu, M. A.: 1972, *Solar Phys.* 25, 127.

Nakagawa, Y.: 1973, *Astron. Astrophys.* 27, 95.

Nakagawa, Y.: 1974, in Y. Nakagawa and D. M. Rust (eds), *Flare Related Magnetic Field Dynamics*, HAO, Boulder, p. 53.

Nakagawa, Y.: 1978, *Astrophys. J.* 219, 314.

Nakagawa, Y., Steinolfson, R. S., and Wu, S. T.: 1978, *Solar Phys.* 47, 193.

Norman, C. A. and Smith, R. A.: 1978, *Astron. Astrophys.* 68, 145.

Oran, E. S., Boris, J. P., Mariska, J. T., and Doschek, G. A.: 1980, *Bull. Amer. Astron. Soc.* 12, 910.

Oran, E. S., Mariska, J. T., and Boris, J. P.: 1982, *Astrophys. J.* 254, 349.

Papadopoulos, K.: 1977, *Rev. Geophys. Space Phys.* 15, 113.

Papadopoulos, K.: 1979, in S.-I. Akasofu (ed.), *Dynamics of the Magnetosphere*, D. Reidel Dordrecht, p. 289.

Parker, E. N.: 1953, *Astrophys. J.* 117, 431.

Parker, E. N.: 1957, *Phys. Rev.* 107, 830.

Parker, E. N.: 1963, *Astrophys. J. Suppl.* 8, 177.

Packer, E. N.: 1979, *Cosmical Magnetic Fields*, Clarendon Press, Oxford.

Pneuman, G.: 1981, submitted *Solar Phys.*

Pollard, R. K. and Taylor, J. B.: 1979, *Phys. Fluids* 22, 126.

Priest, E. R. and Heyvaerts, J.: 1974, *Solar Phys.* 36, 433.

Priest, E. R. and Milne, A. M.: 1979, *Solar Phys.* 65, 315.

Priest, E. R.: 1982, *Solar Magnetohydrodynamics*, D. Reidel Dordrecht.

Pritchett, P. C. and Wu, C. C.: 1979, *Phys. Fluids* 22, 2140.

Pritchett, P. C., Lee, Y. C., and Drake, D. F.: 1980, *Phys. Fluids* 23, 1368.

Raadu, M. A. and Nakagawa, Y.: 1971, *Solar Phys.* 20, 64.

Raadu, M. A.: 1972, *Solar Phys.* 22, 425.

Raadu, M. A. and Carlqvist, P.: 1979, TRITA-EPP-79—21, Dept of Plasma Phys. Royal Inst. of Tech., Stockholm.

Reeves, E. M.: 1976, *Solar Phys.* **46**, 53.

Roederer, J. G.: 1979, in C. F. Kennedy, L. Lanizerotti, and E. N. Parker (eds), *Solar System Plasma Physics*, Vol. 2 p. 1.

Rosner, R., Tucker, W. H. and Vaiana, G. S.: 1978, *Astrophys. J.* **220**, 643.

Rowlands, H., Palmadesso, P., and Papadopoulos, K.: 1981, *Geophysical Research Letts* **8**, 1257.

Sakarai, K.: 1976, *Publ. Astron. Soc. Japan* **28**, 177.

Sakarai, K. and Uchida, Y.: 1977, *Solar Phys.* **52**, 397.

Sato, T., and Iijima, T.: 1979, *Space Sci. Rev.* **24**, 347.

Schmidt, G.: 1979, *The Physics of High Temperature Plasmas* (2nd edn), Academic Press, New York.

Schnack, D. D. and Killeen, J.: 1978, in *Theoretical and Computational Plasma Physics*, IAEA, Vienna.

Schnack, D. D. and Killeen, J.: 1979, *Nucl. Fusion* **19**, 877.

Shafranov, V. D.: 1966, *Rev. Plasma Physics* **2**, 103.

Sheeley, N. R. Jr, Michels D. J., Howard, R. A., and Koomen, M. J.: 1981, *Eos, Trans. AGU* **62**, 153.

Shoub, E.: 1983, *Astrophys. J.* **266**, 339.

Silleen, R. K. J. and Kattenberg, A.: 1980, *Solar Phys.* **67**, 47.

Smith, D. F. and Priest, E. R.: 1972, *Astrophys. J.* **176**, 487.

Smith, R. A. and Goertz, C. K.: 1978, *J. Geophys. Res.* **83**, 2617.

Solov'ev, L. S.: 1975, *Rev. Plasma Physics* **6**, 239.

Spicer, D. S.: 1976, *An Unstable Arch Model of a Solar Flare*, RNL Formal Report 8036.

Spicer, D. S.: 1977a, *Solar Phys.* **53**, 305.

Spicer, D. S.: 1977b, *Solar Phys.* **51**, 431.

Spicer, D. S.: 1977c, *Solar Phys.* **54**, 379.

Spicer, D. S.: 1981a, *Solar Phys.* **70**, 149.

Spicer, D. S.: 1981b, *Solar Phys.* **71**, 115.

Spicer, D. S., Benz, A. O., and Huba, J. D.: 1981, *Astron. Astrophys.* **105**, 221.

Spicer, D. S. and Brown, J. C.: 1981, in S. Jordan (ed.), *The Sun as a Star,* NASA/CNRS, p. 413.

Spitzer, L.: 1962, *Physics of Fully Ionized Gases*, Interscience, New York.

Steinolfson, R. S. and Dryer, M.: 1978, *J. Geophys. Res.* **83**, 1576.

Steinolfson, R. S.: 1985, *Solar Physics* (In press).

Stenflo, J. O.: 1969, *Solar Phys.* **8**, 115.

Sturrock, P. A.: 1966a, *Phys. Rev. Lett.* **16**, 270.

Sturrock, P. A.: 1966b, *Nature* **211**, 695.

Sturrock, P. A.: 1967, *IAU Symp.* **35**, 471.

Sturrock, P. A.: 1972, in P. A. McIntosh and M. Dryer (eds), *Solar Activity Observations and Predictions*, MIT Press, Cambridge, p. 173.

Sturrock, P. A.: 1974, *IAU Symp.* **57**, 437.

Sturrock, P. A.: 1980, P. A. Sturrock (ed.), *Solar Flares*, Colorado Assoc. Univ. Press, Boulder, p. 411.

Svestka, Z.: 1976, *Solar Flares*, Reidel, Dordrecht.

Sweet, P. A.: 1969, *Ann. Rev. Astron. Astrophys.* **7**, 149.

Syrovatskii, S. I.: 1966, *Astron. Zh.* **43**, 340.

Tidman, D. A. and Krall, N. A.: 1971, *Collisionless Shock Waves*, Wiley, New York.

Torvén, S.: 1979, in P. J. Palmadesso and K. Papadopoulos (eds), *Wave Instabilities in Space Plasmas*, D. Reidel, Dordrecht, p. 109.

Tur, T. J. and Priest, E. R.: 1978, *Solar Phys.* **58**, 181.

Uchida, Y. and Sakurai, T.: 1977, *Solar Phys.* **51**, 413.

Van Hoven, G., Chiuderi, C., and Giachetti, R.: 1977, *Astrophys. J.* **213**, 869.

Van Hoven, G.: 1979, *Astrophys. J.* **232**, 572.

Van Hoven, G. *et al.*: 1980, in P. A. Sturrock (ed.), *Solar Flares*, Colorado Assoc. Univ. Press, Boulder, p. 177.

Van Hoven, G.: 1981, in E. R. Priest (ed.), *Solar Flare Magnetohydrodynamics*, Gordon and Breach, New York and London, p. 217.

Van Hoven, G., Ma, S. S., and Einaudi, G.: 1981, *Astron. Astrophys.* 97, 232.
Vasyliunas, V. M.: 1975, *Rev. Geophys. Space Phys.* 13, 303.
Vasyliunas, V. M.: 1976, in B. M. McCormac (ed.), *Magnetospheric Particles and Fields*, D. Reidel, Dordrecht, p. 99.
Voslamber, D. and Callebaut, D. K.: 1962, *Phys. Rev.* 128, 2016.
Ware, A. A.: 1965, Culham Lab. Rept. CLM-M53.
Weymann, R.: 1960, *Astrophys. J.* 132, 425.
Widing, K. G. and Spicer, J. D. S.: 1981, *Astrophys. J.* 142, 1243.
Withbroe, G. L. and Noyes, R. W.: 1977, *Ann. Rev. Astron. Astrophys.* 15, 363.
Wolf, R. A.: 1975, *Space Sci. Rev.* 17, 537.
Wu, S. T., Nakagawa, Y., and Dryer, M.: 1977, in M. A. Shea, D. F. Smart, and S. T. Wu (eds.), *Study of Travelling Interplanetary Disturbances*, D. Reidel, Dordrecht, p. 43.
Zanastra, H.: 1955a, J. M. Burgers and H. C. Van de Hulst (eds), *Gas Dynamics of Cosmic Clouds*, North-Holland, Amsterdam, Chap. XIII.
Zanstra, H.: 1955b, A. Beer (ed.), *Vistas in Astronomy*, Pergamon Press, New York, p. 256.

E. O. Hulburt Center for Space Research
and
Laboratory for Computational Physics,
Naval Research Laboratory,
Washington, DC 20375
U.S.A.

CHAPTER 13

THE ACCELERATION AND PROPAGATION OF SOLAR FLARE ENERGETIC PARTICLES

M. A. FORMAN, R. RAMATY, AND E. G. ZWEIBEL

1. Introduction

Acceleration of energetic particles is a widespread phenomenon in nature, one that occurs at a variety of sites ranging from the Earth's magnetosphere to distant objects such as supernovae, active galaxies, and quasars. There are in fact many explosive phenomena in astrophysics, solar flares among them, in which energetic particles are routinely produced and often contain a large fraction of all the available energy.

It is widely believed (e.g. Syrovatskii, 1981) that solar flares draw their energy from the annihilation of magnetic fields. The strong electric fields which accompany such annihilation should accelerate particles, and in addition the rapid energy deposition following the annihilation should be an important source of shocks and turbulence. As proposed by Fermi (1949), charged particles can be accelerated to high energies in repeated reflections from magnetized clouds, or, in the more recent view, from hydromagnetic turbulence and shocks. This mechanism must be important in solar flares where shocks are known to exist and turbulence is expected to be produced by both shocks and other mechanisms.

Because the Sun is close to the Earth, particle acceleration in flares can be observed in considerable detail. The relevant observables are electromagnetic radiations produced by the accelerated particles in the solar atmosphere and energetic particles that escape from the Sun and are directly detected in interplanetary space. From the analysis of electromagnetic radiations and energetic particles observed from flares it follows that not all solar flare particle species are accelerated to their final energies at the same time. There appears to be a first-phase particle acceleration which produces mostly nonrelativistic electrons and a second phase which accelerates ions and relativistic electrons. It is not clear, however, whether these two phases are manifestations of different acceleration mechanisms, or whether they are due to the evolution in space and time of the same mechanism modified by different energy loss and particle transport processes. Also, recent gamma-ray observations have shown that energetic ions can be produced very promptly, within a few seconds of the nonrelativistic electrons.

The energy spectrum resulting from solar flare particle acceleration can be determined down to a few Mev per nucleon from the maximum flux at each energy, as observed in interplanetary space. This spectrum is clearly not a power law; simple shock and stochastic acceleration models appear to produce energy spectra that can fit the data. The chemical composition of solar energetic ions observed in interplanetary space, while grossly similar to the composition of the solar atmosphere, varies from one flare to

Peter A. Sturrock (ed.), Physics of the Sun, Vol. II, pp. 249–289.

another, and occasionally departs dramatically from the photospheric composition. These variations probably result from the existence of acceleration thresholds and injection mechanisms which depend on the charge and mass of the particles. The ionic states of energetic particles give some indication that the particles observed in interplanetary space are accelerated in the corona; they also seem to imply that selective heating mechanisms are occasionally important.

We review the observations and their implications in Section 2; we treat acceleration by turbulence, by shocks, and in direct electric fields in Section 3; we discuss the problem of the determination of solar flare particle energy spectra from interplanetary observations in Section 4; and we summarize the chapter and provide an outlook for future research in Section 5.

2. Energetic Particles in Solar Flares

The solar flare is capable of producing a rich and complex particle population characterized by a broad spectrum of particle energies and containing many different chemical, isotopic, and ionic species. Of these, the most commonly accelerated particles are electrons of energies in the range from about 10 to 100 keV. Information on these particles is obtained from hard X-ray observations (Hudson, 1979; Kane, et al., 1980), from microwave and Type III radio observations (Lin, 1974), and from direct electron detections in interplanetary space (Lin, 1974; Ramaty et al., 1980).

Solar flares also accelerate protons, nuclei, and relativistic electrons. The nucleonic component is observed in interplanetary space (McDonald et al., 1974); it is occasionally detected on the ground by neutron monitors (Shea and Smart, 1973; Duggal, 1979), and its interactions at the Sun are manifest in gamma-ray lines (Chupp et al., 1973; Ramaty et al., 1975; Chupp, 1982). Relativistic electrons are observed directly in interplanetary space (Simnett, 1971; Datlowe, 1971; Evenson et al., 1981) and their interactions at the Sun produce gamma-ray continuum (Suri et al., 1975) and Type IV radio emission (Ramaty et al., 1980).

In the present section we review the most pertinent data on solar energetic particles and discuss their implications. Electromagnetic emissions are discussed in Section 2.1 and the direct particle observations in 2.2. Since much of the information on solar radio emission is treated by Smith and Goldman in Chapter 15 of this volume, we discuss this topic only briefly here. Solar hard X-ray observations have also been reviewed in considerable detail recently (Hudson, 1979; Kane et al., 1980), and the gamma-ray observations have been reviewed by Ramaty in Chapter 14.

2.1. ELECTROMAGNETIC RADIATIONS

The electromagnetic emissions that contain the least ambiguous information on solar energetic particles are radio emissions, X-rays, and gamma-rays. Other emission, such as EUV and white-light radiations, are not discussed here, but the reader can refer to other recent reviews (e.g. Kane et al., 1980; Ramaty et al., 1980).

It has been known for some time (Wild et al., 1963; Frost and Dennis, 1971; Lin, 1974; Ramaty et al., 1980) that at least two phases of particle acceleration can occur in solar flares. The first phase is characterized by impulsive bursts of hard X-rays and microwaves and by Type III radio emission, while the second phase is manifest in Type II bursts, microwave and metric Type IV emission, flare continuum radio emission, and gradual hard X-ray. Gamma-ray emission is probably due to particles accelerated in both the first and second phases. The first phase has typical rise times of seconds or less and durations as short as a few seconds, while the second phase, in general, has longer rise times and durations. We proceed now to discuss these emissions in more detail.

2.1.1. *Radio Emissions*

As just mentioned, first-phase acceleration in the radio band is characterized by impulsive microwave bursts and Type III radio bursts. Impulsive microwave bursts are generally believed to be due to gyrosynchrotron radiation of electrons of energies in the range from several tens to several hundreds of kilo-electronvolts (Takakura, 1960; Ramaty, 1969). Type III bursts result from the conversion into electromagnetic emission of Langmuir waves excited by beams of electrons of energies in the 10–100 keV range (Wild *et al.*, 1963). The same energy electrons also produce hard X-rays (see below). These electrons have access to both closed field lines of low loops, where they radiate microwaves and hard X-rays, and open field lines where they produce Type III bursts. The electrons on open field lines can also be directly observed in interplanetary space (e.g. Lin, 1974).

Second-phase acceleration is characterized by Type II radio bursts, microwave Type IV bursts, flare continuum and moving Type IV emission. Type II bursts, presumably also due to the conversion of Langmuir waves, are indicators of shock fronts (Wild *et al.*, 1963) in the corona. Particle acceleration by these shocks or by the turbulence behind them, is consistent with many second-phase phenomena including the protons and nuclei observed in interplanetary space. This conclusion is based on the good correlation between proton events and Type II bursts (Svestka and Fritzova, 1974). The Type II emission itself is probably produced by 10–100 keV electrons that are also accelerated by the shock.

Type IV emission, in general, is due to gyrosynchrotron radiation of relativistic electrons. This emission includes microwave Type IV bursts, produced in low coronal loops where the magnetic field is relatively high, and moving Type IV bursts which are generally observed in the corona. The microwave Type IV emission can be distinguished from impulsive microwave bursts by their delayed rise to maximum intensity. The moving Type IV bursts are due to synchrotron radiation of ~1 MeV electrons in coronal magnetic fields of a few gauss. Flare continuum emission is also observed in the corona at the time of the passage of a Type II burst. As opposed to moving Type IV bursts, this emission is stationary and is believed to be due to plasma radiation (Dulk *et al.*, 1978).

2.1.2. *Hard X-Rays*

Hard X-rays in solar flares result from the bremsstrahlung of electrons in the energy range from about 10 keV to several hundred. In the X-ray band, first-phase acceleration is characterized by impulsive hard X-ray bursts. The correlation between these bursts and impulsive microwave bursts (e.g. Lin, 1974) indicates that both emissions should be produced by essentially the same electron population. As already mentioned, these electrons also produce Type III radio bursts. The first-phase mechanism is thought to be responsible for the impulsive acceleration of 10 to several hundred keV electrons in flares (e.g. Ramaty *et al.*, 1980).

The most detailed information on first-phase acceleration of electrons comes from hard X-ray observations. This information, however, is model dependent. Two limiting cases exist regarding the nature of the electrons. They could belong to a nonthermal population whose number density is lower than that of the relatively cool ambient medium with which it interacts, or they could form a quasithermal hot plasma. In the former case, the radiation yield is the ratio of the bremsstrahlung production rate to the nonradiative collisional loss rate. Since in the X-ray band this yield is very small ($\sim 10^{-5}$ at 25 keV), hard X-ray production in flares should be accompanied by the deposition of large amounts of energy into the solar atmosphere. On the other hand, in thermal models (Chubb, 1970; Crannell *et al.*, 1978), a larger fraction of the electron energy content could in principle, be emitted as hard X-rays. It is difficult, however, to confine the hot electrons to dense regions for times comparable to their radiation loss time (Smith and Auer, 1980) and therefore the radiation yield of the thermal model is probably not much higher than that of the nonthermal model.

There are also a variety of particle interaction models which can be crudely classified as thin or thick target (e.g. Kane *et al.*, 1980). In the thin-target model (Datlowe and Lin, 1973) the X-rays are produced by electrons which escape from the interaction region at the Sun, while in the thick-target model (Brown, 1971; Hudson, 1972), X-ray production takes place as the electrons slow down in the solar atmosphere. Clearly, the radiation yield is lower for a thin target than for a thick target, because for the latter, in addition to the collisional losses, the electrons also carry away kinetic energy. However, from the comparison of number of electrons observed in interplanetary space with that needed to produce the impulsive hard X-rays at the Sun, it follows (Lin, 1974) that the majority of the 10–100 keV electrons remain trapped at the Sun and produce X-rays in thick-target interactions.

Various estimates exist of the energies deposited in the solar atmosphere by electrons accelerated in the first phase. From the summary of Ramaty *et al.* (1980) we have that in the thick-target and nonthermal model, the energy deposited by electrons above 25 keV is $\sim 2 \times 10^{32}$ erg for the August 4, 1972 flare, which was one of the strongest events observed. The energy deposited in other cases ranges from about 5 to 50×10^{29} erg in typical strong events, and from about 2 to 20×10^{28} erg in typical small events. The fact that these energies are comparable to the total observed flare energies has led to the suggestion (Lin and Hudson, 1976) that most flare phenomena could be accounted for, at least energetically, by the interaction of the first-phase electrons with the solar atmosphere.

First-phase acceleration must be very efficient, both in the amount of energy that it converts into accelerated particle energy, as we have just seen, and the number of ambient particles that it accelerates. The mechanism, however, accelerates fewer protons to 10–100 keV than electrons, as indicated by the absence of nonthermal wings of hydrogen Ly-α (Canfield and Cook, 1978). We discuss below the recent gamma-ray evidence for the impulsive acceleration of MeV protons. As we shall see, the energy content in these particles is also lower than that in the 10–100 keV electrons. Mechanisms for first-phase acceleration have been reviewed by Smith (1979).

In addition to the impulsive component, solar hard X-rays occasionally exhibit a gradual component (duration ~10 min) which is believed to be a manifestation of second-phase acceleration. Hudson (1979) has reviewed the evidence for this component. It can be found in hard X-ray observations from flares located behind the limb of the Sun (Hudson *et al.*, 1981) and from flares on the visible solar hemisphere which produce extended X-ray bursts (Frost and Dennis, 1971; Hoyng *et al.*, 1976). These X-rays are produced in the corona and their spectrum is generally harder than that of the impulsive X-rays. This indicates additional acceleration. But the energy deposited in the solar atmosphere by the 10–100 keV electrons responsible for the gradual X-rays is lower by at least an order of magnitude than that deposited by the electrons of the impulsive phase (Hudson, 1979).

2.1.3. *Gamma-Rays*

Gamma-ray lines, produced in energetic particle reactions in the solar atmosphere (Ramaty *et al.*, 1975), are tracers of the nucleonic component accelerated in flares. Many narrow and broad lines are produced (e.g. Ramaty *et al.*, 1979). The strongest narrow line is at 2.22 MeV from neutron capture on hydrogen. This line has been seen from a number of flares so far, as have some of the other strong narrow lines, e.g. at 4.44 MeV from ^{12}C, at 6.13 MeV from ^{16}O, and at 0.51 MeV from positron annihilation (Chupp, 1982; Ramaty, Chapter 14 in this volume). In addition to being observed individually, nuclear lines make a significant contribution to the total gamma-ray emission above 1 MeV (Ramaty *et al.*, 1980). In particular, in the 4 to 7 MeV band, nuclear radiation appears to be the dominant emission mechanism (Ramaty *et al.*, 1977; Ibragimov and Kocharov, 1977). On the other hand, gamma-ray emission up to 1 MeV is almost entirely due to electron bremsstrahlung and, hence, is a good tracer of the relativistic electrons in this energy range. Nuclear reactions in the solar atmosphere occasionally produce high-energy neutrons which can travel as far as the Earth resulting in detectable neutron fluxes (Lingenfelter and Ramaty, 1967). High-energy neutrons were observed from a large flare in 1980 (Chupp *et al.*, 1982).

Gamma-ray observations can provide information on the timing of the acceleration of protons, nuclei, and relativistic electrons; they can determine the energy deposited by the nucleonic component in the solar atmosphere; and they can place constraints on the energy spectra of the protons and nuclei. Through Doppler shifts, gamma-ray lines could provide unique information on the beaming of the energetic particles (Ramaty and Crannell, 1976; Kozlovsky and Ramaty, 1977; Zweibel and Haber, 1983).

We first address the question of the timing of the nucleonic acceleration and its relationship to the timing of the acceleration of the 10–100 keV electrons in the first phase. This question was studied (Bai and Ramaty, 1976; Lin and Hudson, 1976) for the August 4, 1972 flare which was the first flare from which nuclear gamma-rays were seen (Chupp *et al.*, 1973).

The time dependences of hard X-rays, gamma-ray continuum, high-frequency microwave emission, and the 2.22 MeV line for the August 4, 1972 flare are shown in Figure 1 (from Bai and Ramaty, 1976). As can be seen, the emissions produced by relativistic electrons (> 0.35 MeV gamma-ray continuum and high-frequency microwave emission) has a time history different from that of the X-ray emission produced by < 100 keV

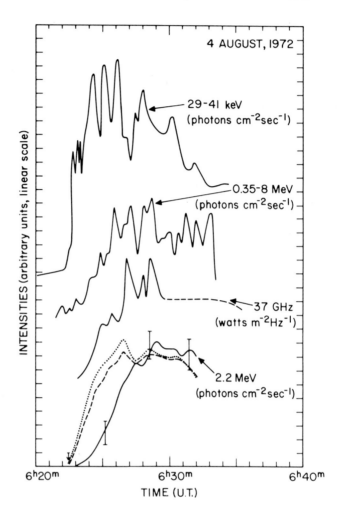

Fig. 1. Time dependences of X-rays, gamma-rays, and microwave radiation from the August 4, 1972 flare (from Bai and Ramaty, 1976).

electrons. This difference is manifest in that the gamma-ray continuum and microwave emission reached peak strength a few minutes later than the X-rays. In addition, as can be seen from Figure 1, the time profile of the 2.22 MeV line could be better explained when the time profile of neutron production was assumed to be similar to the time profile of the gamma-ray continuum rather than that of the X-rays. (See Chapter 14 by Ramaty, in this volume, for a detailed discussion on the delay of the emission of 2.22 MeV photons owing to the finite capture time of the neutrons.) This led Bai and Ramaty (1976) to conclude that the >0.35 MeV electrons and >10 MeV protons are accelerated in the second phase. However, as can be seen in Figure 1, even though the build-up of the higher energy emissions in the August 4, 1972 flare was slower than that of the hard X-rays, the two emissions were not clearly separated in time.

As opposed to the gradual build-up of the nucleonic component on a time-scale of a few minutes seen in the August 4 event, the recent SMM (Chupp, 1982) and HEAO-3 (Prince et al., 1982) observations have provided examples of very prompt acceleration of protons and nuclei. The most useful information in this regard comes from the comparison of the time history of the prompt gamma-rays in the energy band from 4 to 7 MeV (which is dominated by nuclear gamma-rays) with that of the hard X-rays. For the June 7, 1980 flare (Chupp, 1982) this comparison (Bai, 1982) implies that the <100 keV electrons and >10 MeV nuclei were accelerated in very close time proximity (less than a few seconds). There are, nevertheless, differences between the two time profiles: there is a delay of approximately 2 s between individual hard X-ray and gamma-ray peaks, and the gamma-rays reach maximum strength at least 10 s after the X-rays. The small but finite lag of gamma-rays behind the hard X-rays in the June 7 flare would be yet another manifestation of a second acceleration phase. This type of second-phase acceleration – i.e. one which is very closely correlated with the first-phase acceleration – has been referred to as "second-step acceleration" (Bai and Ramaty, 1979; Bai, 1982).

We next discuss the energy content in the nucleonic component and its relationship to the energy deposited by electrons accelerated in the first phase. The nucleonic energy content is the sum of the energy deposited during gamma-ray line production and the energy carried away by particles escaping from the Sun.

We consider numerical values for two flares (August 4, 1972 and June 7, 1980) for which the gamma-ray data have been analyzed in considerable detail. These are given in Table I. Here W_d (>1 MeV nucleon^{-1}) and W_{esc} (>1 MeV nucleon^{-1}) are, respectively,

TABLE I

Flare	W_d (>1 Mev nucleon^{-1})	W_{esc} (>1 MeV nucleon^{-1})	W_e (>20 keV)	$\bar{N}_p$ (>10 MeV)	$N_{esc,p}$ (>10 MeV)
August 4, 1980	2.5×10^{30} erg	3×10^{31} erg	2×10^{32} erg	1.3×10^{34}	3×10^{35}
June 7, 1972	2×10^{29} erg	2×10^{27} erg	4×10^{30} erg	8.5×10^{32}	10^{31}

the energies deposited and carried away by the nucleonic component, W_e (>20 keV) is the energy deposited by electrons above 20 keV, and $\bar{N}_p$ (>10 MeV) and $N_{esc,p}$ (>10 MeV) are the numbers of protons above 10 MeV that interact and escape from the Sun, respectively. For the August 4 flare, W_d and $\bar{N}_p$ are from Ramaty (see Chapter 14), W_{esc} and $N_{esc,p}$ are from Lin and Hudson (1976), and W_e has been given earlier in this section. For the June 7 flare, W_d and $\bar{N}_p$ are again from Ramaty (see Chapter 14), W_{esc} and $N_{esc,p}$ are from Von Rosenvinge et al. (1981), and W_e is from A. Kiplinger (private communication, 1981).

Considering the numerical values of Table I, we first note that the escape conditions for the nucleonic component can vary considerably from one flare to another. While for the June 7 flare only a small fraction of the particles escaped from the Sun, for the August 4 flare the number of protons observed in space exceeded the number that interacted at the Sun by more than an order of magnitude. Evidently, in certain cases particles have access to open field lines while in others they do not. In both cases, however, the total energy in the nucleonic component is less than the energy deposited by the electrons accelerated in the first phase. Shocks, turbulence, or both, produced during energy deposition in the first phase, could thus be responsible for second-phase acceleration.

The gamma-ray observations also set constraints on the energy spectra of the protons and nuclei. The derived energy spectra (see Chapter 14) turn out to be consistent with those obtained from direct particle observations in interplanetary space (McGuire *et al.*, 1981, and Section 2.2 below). This agreement implies that the same acceleration mechanism could be responsible for both the escaping and trapped nucleonic component, even though the ratio between the numbers of particles in these two populations can vary dramatically from flare to flare. On the other hand, we cannot rule out the possibility that these two populations are produced by different mechanisms.

For a discussion of additional implications of the solar gamma-ray observations (e.g. beaming, photospheric ^{3}He abundance, limb-darkening effects) the reader is referred to Chapter 14 of this volume.

2.2. ENERGETIC PARTICLES

Complementary to the electromagnetic radiations discussed in the previous subsection are the direct particle observations in interplanetary space and in the Earth's atmosphere. Spacecraft observations (e.g. Van Hollebeke, 1979; Gloeckler, 1979) can determine the energy spectra of the various particle species including electrons, the abundances of essentially all elements from H through Ni, and isotopic and ionic abundances for a few abundant elements. Ground-based neutron monitor observations (e.g. Duggal, 1979) extend the spectrum of the nucleonic component of solar flare particles to about 20 GeV. Particle observations, however, can only provide limited information on the timing of the acceleration. We proceed now to discuss the most pertinent of these observations.

2.2.1. *Energy Spectra and Electron–Proton Correlations*

Detailed studies of solar flare proton spectra that attempted to eliminate the effects of coronal and interplanetary transport were made by Bryant *et al.* (1965), Reinhard and Wibberenz (1974), and Van Hollebeke *et al.* (1975). In the study of Van Hollebeke *et al.* the effects of propagation on the observed spectra were minimized by considering only particle events from flares that were well connected magnetically to the observing spacecraft and by constructing the proton energy spectra at times of maximum intensity at each energy (see Section 4). Because of instrumental limitations, however, the resultant energy spectra extended only over the narrow energy range from 20 to 80 MeV and therefore could not differentiate between various possible spectral fits such as power laws or exponentials.

A more recent study (McGuire *et al.*, 1981) has extended considerably the particle energy range under investigation (1–400 MeV for protons) and has also provided data on α-particle spectra. By using the same techniques for minimizing the propagation effects as Van Hollebeke *et al.* (1975), McGuire *et al.* (1981) find two spectral forms that provide good fits to the data. These are the Bessel function spectrum (Ramaty, 1979, §IIIa)

$$\frac{dJ}{dE}(E) \propto \beta K_2 \left[2(3\beta/\alpha T)^{1/2}\right], \tag{1}$$

and an exponential spectrum in rigidity

$$\frac{dJ}{dE}(E) \propto \exp\left(-\frac{R}{R_0}\right)\frac{dR}{dE} \quad . \tag{2}$$

Here dJ/dE is differential intensity measured in particles $(cm^2\,s\,sr\,MeV\,nucleon)^{-1}$, $v = c\beta$, E and R are particle velocity, energy per nucleon and rigidity, respectively, and K_2 is the modified Bessel function of order 2 (Abramowitz and Stegun, 1965). The parameters αT for the Bessel function and R_0 for the exponential characterize the shape of the particle energy spectra. Equation (1) was shown to be the solution of a transport equation for stochastic acceleration with acceleration efficiency α and loss time T (M. A. Lee, private communication, 1978; Ramaty, 1979, §IIIa).

An example of the results of McGuire *et al.* (1981) is shown in Figure 2. Here the

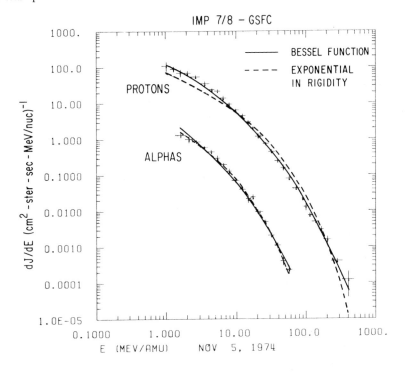

Fig. 2. Bessel function (see Section 3.1) in velocity, and exponential in rigidity fits to proton and α-particle spectra for the November 5, 1974 flare particle event. Fit parameters are αT (proton) = 0.024, αT (alpha) = 0.015, R_0 (proton) = 73 MV, R_0 (alpha) = 80MV.

data points show the observed particle intensities from the November 5, 1974 event, and the solid and dashed curves are given by Equations (1) and (2), respectively. The spectral parameters for this event are $\alpha T = 0.024$ and $R_0 = 73$ MV, for protons, and $\alpha T = 0.015$ and $R_0 = 80$ MV for α-particles. A very important result of McGuire *et al.* (1981) is that power laws in energy $(dJ(E)/dE \propto E^{-\gamma}$ with γ a constant) do not fit the data over the entire observed energy range, as can be clearly seen from Figure 2.

The variability of the spectral parameters αT and R_0 from one flare to another was also studied by McGuire *et al.* (1981). They find, as did Van Hollebeke *et al.* (1975) before, that for the well-connected flares the spectral parameters are confined to rather narrow ranges. For protons, $\alpha T = 0.025 \pm 0.011$ and $R_0 = 70 \pm 27$ MV. The α-particle-spectra are generally steeper than the proton spectra. The ratios of the proton to α-particle αT's and R_0's are 1.6 ± 0.2 and 0.7 ± 0.1, respectively (R. McGuire, private communication, 1981).

We pointed out in Section 2.1 that constraints on the energy spectra of the protons and nuclei that interact at the Sun can be set by gamma-ray line ratio observations. Unlike the interplanetary observations, these data are not influenced by propagation effects, but they depend on the assumed interaction model. For thick-target interactions, Ramaty (see Chapter 14) finds that for seven flares for which line ratios are available, αT (assumed to be the same for all particle species) is in the range 0.014–0.020. This relatively narrow range is consistent with that found by McGuire *et al.* (1981) from interplanetary observations. As already mentioned, the implication of this result is that a single mechanism could be responsible for both the escaping and trapped nucleonic component.

We turn now to a discussion of electron energy spectra and correlations between electrons and protons of various energies. Energy spectra of electrons accelerated in solar flares were analysed by Lin (1971, 1974) and recently by Lin *et al.* (1982) who have minimized the propagation effects in the same fashion as was done for the protons and α-particles discussed above. An energy spectrum of a large electron event is shown in Figure 3 (from Lin *et al.*, 1982). The break at ~100 keV, characteristic of all solar flare electron spectra, is probably due to the acceleration process itself (Lin *et al.*, 1982), since the time of maximum treatment is expected to minimize the propagation effects, and Coulomb losses in the solar atmosphere would produce a continuous flattening rather than a single break. For small electron events the energy spectra are steeper than the spectrum of Figure 3, both below and above the break. Whereas in large events, relativistic electrons are occasionally seen above 10 MeV (Datlowe, 1971), in small events, electrons cannot be seen above a few hundred keV.

Correlations between electrons and protons were studied by Ramaty *et al.* (1980) and more recently by Evenson *et al.* (1981). Figure 4 (from Ramaty *et al.*, 1980) shows the correlation between 0.5–1.1 MeV electrons and 10 MeV protons. As can be seen, for large events the two populations are well correlated, but for smaller events there is an overabundance of electrons. This effect is possibly a manifestation of the two acceleration phases, with the first-phase producing more electrons than protons in small events. The good correlation seen for larger events indicates that relativistic electrons just below 1 MeV and ~10 MeV protons could be accelerated by the same mechanism. This could be second-phase acceleration. The correlation between these two particle populations is supported by the gamma-ray data. As discussed in Section 2.1 and seen in Figure 1, the >0.35 MeV continuum produced predominantly by <1 MeV electrons, and nuclear gamma-rays from $\geq$ 10 MeV protons, have similar time histories.

Evenson *et al.* (1981) have recently examined the relationship between protons and relativistic electrons at nearly the same energy (~10 MeV). They find that these two particle populations are very poorly correlated. In particular, the majority of the proton events have very low ($\lesssim 10^{-3}$) electron-to-proton ratios at ~10 MeV. This is in contrast

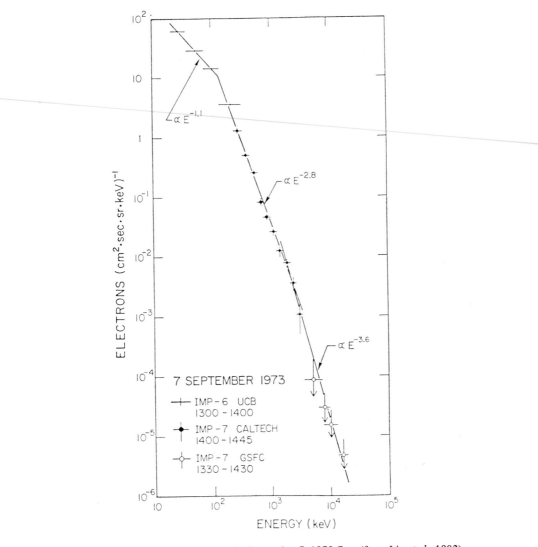

Fig. 3. Electron energy spectrum from the September 7, 1973 flare (from Lin *et al.*, 1982).

to the correlation seen in Figure 4 where all proton events are accompanied by 0.5–1.1 MeV electrons. A few of the events, however, show larger electron-to-proton ratios, and some of them are as high as 0.2 at 10 MeV. Gamma-rays (lines or continuum) were seen from all of these events. But it is not clear at the present time whether these electron enrichments are caused by the acceleration mechanism or whether they reflect different escape condition for protons and electrons. Oppositely directed beams of ions and electrons produced in a direct electric field could yield gamma-ray lines at the Sun and relativistic electrons in space. Such electric fields are discussed in Section 3.3.

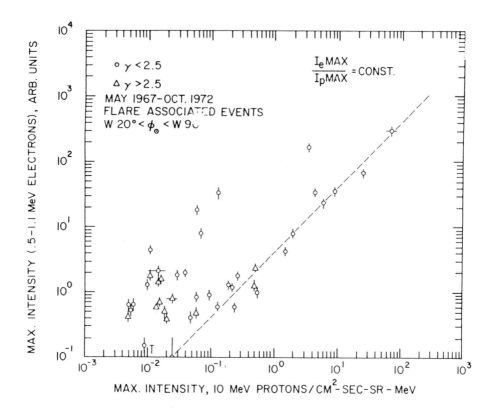

Fig. 4. Correlation between solar flare electron and proton intensities observed in interplanetary space (from Ramaty *et al.*, 1980).

2.2.2. *Chemical Compositions*

Nuclei heavier than He in solar energetic particles were first detected by Fichtel and Guss (1961) and since then many measurements of such particles have been made (see reviews by McDonald *et al.*, 1974; Fan *et al.*, 1975; Ramaty *et al.*, 1980). While the earlier results indicated rough agreement of the energetic particle composition with photospheric composition (e.g. Bertsch *et al.*, 1969), with more recent results, it became obvious that these two sets of abundances can differ drastically from each other.

The first indications for large abundance anomalies in flare-accelerated particles came from the observations of Price *et al.* (1971) which showed large enhancements at low energies of iron-group nuclei over photospheric abundances. Further studies (Mogro-Campero and Simpson, 1972; Teegarden *et al.*, 1973) have also revealed such enhancements for Mg and Si.

The most dramatic departure of a solar energetic particle abundance from its photospheric value is that of ^{3}He (Garrard *et al.*, 1973). Here very large enhancements are occasionally observed in the ^{3}He/^{4}He ratio above its likely photospheric value (see Ramaty *et al.*, 1980, for a review of the data).

Enrichments of ^{3}He in energetic particle populations (for example, the galactic cosmic-rays) have been generally attributed to nuclear reactions between the energetic particles and the ambient medium. But, as first pointed out by Garrard *et al.* (1973), this interpretation of the solar ^{3}He enrichments, in its simplest form, is inconsistent with much of the ^{3}He data. If the ^{3}He enrichments are due to nuclear reactions of the energetic particles, then they should be accompanied by similar enrichments in ^{2}H and, to a lesser degree, in ^{3}H. Such enrichments, however, are not observed.

Several schemes based on nuclear reactions have been proposed to overcome this difficulty. These rely on the kinematics and angular distributions of the reaction products which favor the preferential escape of ^{3}He (Ramaty and Kozlovsky, 1974; Rothwell, 1976) and the thermonuclear destruction of ^{2}H and ^{3}H in a model in which the energetic products of the nuclear reactions are confined to thin filaments and interact with each other (Colgate *et al.*, 1977). But, as proposed by Fisk (1978) and Kocharov and Kocharov (1978), the enhanced ^{3}He abundance in solar energetic particles could be due to preferential heating of ambient ^{3}He. Provided that the acceleration mechanism has an injection threshold (Section 3), such heating would greatly enhance the number of accelerated particles.

Several systematic studies of solar energetic particle composition have been carried out recently (McGuire *et al.*, 1979, Cook *et al.*, 1980; Mason *et al.*, 1980; Reames and Von Rosenvinge, 1981). The results of these studies are summarized in Figure 5. Here ratios of energetic particle abundances to photospheric abundances are shown for a variety of elements as well as for ^{3}He and ^{4}He.

For each element or isotope plotted in Figure 5 we consider flares for which the energetic particles are rich in ^{3}He and flares from which no ^{3}He is observed. The data for these two groups are separated by the dashed vertical lines, with ^{3}He-rich flares to the right of the lines and flares with no ^{3}He to their left.

The closed circles are the data of McGuire *et al.* (1979) who measured the energetic particle composition of eight large non-^{3}He-rich solar flares in the energy range from 6.7 to 15 MeV nucleon^{-1}. Such compositions were also measured by Cook *et al.* (1980) from a few to 15 Mev nucleon^{-1} and by Mason *et al.* (1980) near 1 MeV nucleon^{-1}. The values given by crosses and stars are the average particle abundances measured by Cook *et al.* (1980) and Mason *et al.* (1980), respectively, for their sample of large solar flares. As can be seen, these averages are consistent with the data of McGuire *et al.* (1979).

To the right of the dashed lines are data for ^{3}He-rich flares. The open circles represent the recent data of Reames and Von Rosenvinge (1981) for six flares, while the values given by the diamonds represent abundances labelled by Mason *et al.* (1980) as anomalous. In particular, diamond-1 is for their carbon-poor flares and diamond-2 is for the October 12–13, 1977 flares. Both these sets of energetic particles are rich in ^{3}He.

The distinction between ^{3}He-rich and normal events follows mainly from the ^{3}He abundance itself. The ^{3}He/O ratio of the former exceeds the upper limits on this ratio for the latter by at least an order of magnitude. The other abundances, however, are not too different for the two classes of events. As can be seen in Figure 5, ^{3}He-rich events can be both rich or poor in H (see values of diamonds 1 and 2 for H), although it has been noticed (e.g. Ramaty *et al.*, 1980) that ^{3}He-rich flares have, on average, low ^{1}H/^{4}He ratios. It also follows from Figure 5 that ^{4}He and C tend to be somewhat suppressed in ^{3}He-rich events although it is not necessarily true that all ^{3}He-rich events are C poor. The

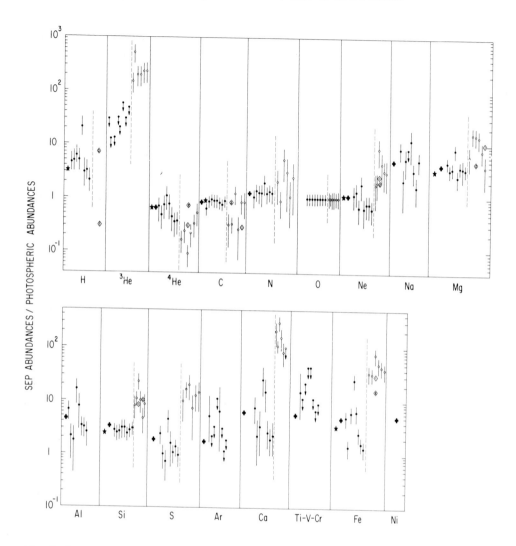

Fig. 5. Composition of solar energetic particles relative to the photosphere (see Section 2.2 for details).

variability from event to event in ^{4}He, C, and other elements is larger in ^{3}He-rich events than in normal events.

Both normal and ^{3}He-rich events show enrichments in heavy elements, although these are significantly larger for the ^{3}He-rich events. The abundance enhancements in Ca and Fe are, in fact, almost as large as those of ^{3}He. The correlation between ^{3}He and Fe enhancements in ^{3}He-rich events was first noticed by Anglin *et al.* (1977).

Various models have been proposed to account for these abundance variations. Meyer (1981) has pointed out that some of the less dramatic abundance variations could simply reflect coronal abundances which are different from those of the photosphere. Mullan and Levine (1981) suggested that differences in Coulomb energy losses (Section 3.1) of partially stripped heavy ions could lower the injection thresholds of these ions and, hence, increase the number of particles that are accelerated. Eichler (1979) has pointed out that in acceleration by shocks of finite widths (Section 3.2), partially stripped ions with large gyroradii could be preferentially accelerated because they are capable of sampling more of the compression at the shock. This assumes a rigidity dependent mean free path (Section 3.2) and that all ions have the same initial velocities. Axford (1982) points out that, aside from this effect, the shock may selectively heat ions of different mass-to-charge ratios by plasma processes, so that the ions will be injected in a manner which depends on this ratio. In Fisk's (1978) model (see also Mason $et\ al.$, 1980), the electrostatic ion–cyclotron waves which heat the ^{3}He also heat partially stripped heavy ions which have A/Z near 3 (e.g. $^{16}O^{+5}$, $^{56}Fe^{+17}$) by resonance with the second harmonic of the particle cyclotron frequency. Only mechanisms which can selectively preheat certain ions, such as the mechanisms of Fisk (1978) and Kocharov and Kocharov (1978), combined with injection thresholds which are large compared to the thermal speeds of the normal ions in the source, can explain the observed ^{3}He enhancements.

2.2.3. *Isotopic and Ionic Compositions*

In addition to the ^{3}He observations discussed above, the only elements whose isotopic abundances have been observed in solar energetic particles are Ne and Mg (Dietrich and Simpson, 1979, 1981; Mewaldt $et\ al.$, 1979, 1981). For Ne, it has been found that $^{22}Ne/^{20}Ne = 0.13 \pm 0.003$ and $^{21}Ne/^{22}Ne < 0.01$. These abundances are consistent with those of Neon-A in meteorites (Podosek, 1978) which is believed to represent the primordial Ne isotopic abundance at the time of the formation of the solar system. The isotopic abundances of Ne in the photosphere are not known. In the solar wind, the $^{22}Ne/^{20}Ne$ ratio of 0.07 ± 0.002 (Geiss, 1973) is significantly lower than in the solar energetic particles. The origin of this discrepancy is not yet understood. But because the Mg isotopic abundances in solar energetic particles, $^{25}Mg/^{24}Mg = 0.14$ (+0.05, −0.02), $^{26}Mg/^{24}Mg = 0.15$ (+0.04, −0.03) (Mewaldt $et\ al.$, 1981) seem to be consistent with primordial solar system material, it is possible that Ne isotopic fractionation takes place in the solar wind. Mg isotopic abundances have not yet been measured in the solar wind and are not known in the photosphere.

The first measurements (Gloeckler $et\ al.$, 1976) of the ionic states of solar energetic particles revealed a charge distribution consistent with that of a gas in ionization equilibrium at about 1 to 2×10^6 K. Subsequent measurements (Sciambi $et\ al.$, 1977) found that the charge states of C and O in several solar particle events essentially do not vary with energy (from 0.037 to 1 MeV per charge) in time and from event to event. These results are consistent with particles being accelerated in coronal material and traversing little material during and after their acceleration. The recent observation (Hovestadt $et\ al.$, 1981) of singly charged He ions in solar energetic particles indicates that cooler material is also accelerated. This could be due to temperature inhomogeneities in the corona or the injection of chromospheric material into the acceleration region.

Measurements of charge states have important implications on the heating mechanisms discussed above for producing the ^{3}He and heavy element abundance anomalies. Ma Sung $et\ al.$ (1981) find that $^{16}O^{+5}$ and $^{56}Fe^{+17}$ are indeed present in a ^{3}He-rich flare, as would be expected in Fisk's (1978) model. However, the existence of appreciable concentrations of such ions in the ambient medium, requires that particles be accelerated from regions spanning a broad range of temperatures (4×10^5 to 5×10^6 K).

Recently, Kleckler *et al.* (1982) have reported that the mean charge of Fe during events with large ^{3}He and Fe enrichments is significantly larger than the mean charge of Fe during flares of normal composition. Since the charge of Fe during normal events is not expected to be as high as +17, this result is consistent with the heating model of Fisk (1978) which preferentially enhances Fe^{+17}.

3. Mechanisms of Solar Flare Particle Acceleration

Based on the observations described in the previous section, a solar flare acceleration mechanism or combination of mechanisms must fulfill the following requirements: the acceleration mechanism must be capable of imparting a large fraction of the available flare energy to energetic particles and it must be possible to accelerate a significant number of ambient particles; the dominant mechanism or the combination of mechanisms should be capable of producing the spatial and temporal evolution of the different energetic particle populations, in particular the rapid rise of the various electromagnetic emissions; the mechanisms must have acceleration thresholds, which, with appropriate injection or preacceleration mechanisms, should be capable of accounting for the observed compositions and their variations; and the mechanisms must be able to produce particles with the observed energy spectra.

The observations also indicate that the occurrence of solar flare ions in space is associated with coronal shocks and by implication with turbulence produced by the shock. Therefore, it is reasonable to invoke stochastic acceleration and direct shock acceleration for solar flare ion production. On an even more basic level, solar flares are associated with the release of magnetic energy by the collapse and reconnection of magnetic fields, and these processes may generate large transient electric fields which could accelerate ions and electrons directly.

In Section 3.1 we discuss stochastic acceleration and in Section 3.2 we discuss shock acceleration with emphasis on recent ideas of diffusive shock acceleration. For both mechanisms we consider energy spectra, the problem of injection, and the acceleration times of the particles. In Section 3.3 we briefly discuss magnetic reconnection and electric fields and their possible role in particle acceleration.

The results of Sections 3.1 and 3.2 are most relevant to ion acceleration in solar flares. The reader is referred to several recent reviews of first-phase electron acceleration (e.g. Smith, 1979; Ramaty *et al.*, 1980; Brown and Smith, 1980). Our treatment of particle acceleration, as well as the treatments that we review, do not discuss the generation and decay of the shocks and turbulence, and do not, for the most part, take into account the effect of the accelerated particles on the turbulence or on the shocks which accelerate them. This important aspect of acceleration theory remains a subject for future research.

3.1. STOCHASTIC ACCELERATION

Processes in turbulent plasmas which cause particles to change their energy in a random way with many increases and decreases in energy lead to stochastic acceleration. In the original stochastic Fermi mechanism (Fermi, 1949), the process was reflection from randomly moving magnetized clouds. Stochastic acceleration can also result from resonant pitch-angle scattering from Alfvén waves with wavelength of the order of the particle gyroradius. To accelerate paticles these waves must propagate both parallel and antiparallel to the average magnetic field (Skilling, 1975). Other modes of stochastic acceleration, called magnetic pumping and transit-time damping, occur through interaction with magnetosonic waves whose wavelengths are much longer than the particle gyroradius (Kulsrud and Ferrari, 1971; Melrose, 1980; Achterberg, 1981). These modes require additional pitch-angle scattering to keep the particles isotropic. Langmuir (plasma) waves with phase velocities of the order of the particle speed will also accelerate particles stochastically (Melrose, 1980).

When the random energy increments are small compared to the particle energy, stochastic acceleration results in a diffusive current in momentum space, $S_p = -D_{pp}(\partial f/\partial p)$, where p is the magnitude of the momentum, $f(p)$ is the number of particles per unit volume in phase space, and S_p is measured in cm^{-3} momentum^{-2} s^{-1}. Particles injected at some momentum p_0 will diffuse in momentum to larger and smaller p. In terms of f, the differential particle intensity per unit energy per nucleon is given by $dJ/dE = A\,p^2\,f$, where A is the nuclear mass number. Additional nondiffusive energy changes can be added to S_p,

$$S_p = -D_{pp}\,\frac{\partial f}{\partial p} + \frac{dp}{dt}\,f, \tag{3}$$

where dp/dt represents convection in momentum space due to processes which change the energy of all particles (e.g. ionization or Coulomb losses).

The momentum diffusion coefficient D_{pp} depends on the nature of the stochastic process. If the process is simple scattering with mean free path λ, then D_{pp} can be derived (Parker and Tidman, 1958; M. A. Lee, private communication, 1978) from the Boltzmann equation. This yields

$$D_{pp} = p^2\,(\delta V)^2/3v\lambda, \tag{4}$$

where $(\delta V)^2$ is the mean square velocity of the scatterers and v is particle speed.

If the stochastic acceleration is due to resonant pitch-angle scattering from Alfvén waves, the momentum diffusion coefficient obtained from quasilinear theory is (Skilling, 1975)

$$D_{pp} = \frac{2p^2\,V_A^2}{v^2}\int \frac{D_+\,D_-}{D_+ + D_-}\,d\mu \tag{5}$$

where V_A is the Alfvén speed, $D_+(D_-)$ is the pitch-angle scattering coefficient due to forward (backward) propagating Alfvén waves, and μ is the cosine of the particle angle in the mean field. It is clear from this equation that there is no stochastic acceleration due to Alfvén waves unless the waves propagate in both directions. This requirement occurs because the electric fields of Alfvén waves in one direction can be Lorentz-transformed away and so cannot accelerate particles. For example, Alfvén waves generated by the streaming of energetic particles (Wentzel, 1969) propagate only in the direction of the streaming and hence do not accelerate the particles.

By comparing Equations (4) and (5) we can define an effective mean free path for stochastic acceleration by Alfvén waves, $\lambda^A = p^2 V_A^2/3vD_{pp}^A$. This λ^A is, in general, a function of particle rigidity which is determined by the power spectrum of the Alfvén waves. Let $W_\pm(k)$ be the energy density per wavenumber k in waves propagating in the + or − direction. Then, from quasilinear theory (Hasselmann and Wibberenz, 1968; Jokipii, 1971; Luhmann, 1976),

$$D_\pm(\mu) = v(2\pi Ze/pc)^2 \; W_\pm(1/\mu r_c)(1 - \mu^2)/|\mu|, \tag{6}$$

where Ze is the particle charge and r_c its gyroradius in the average magnetic field, B. For example, if $D_+ = D_-$ and $W(k) \propto k^{n-2}$,

$$\lambda^A(R) = \frac{B^2 r_c^2}{96\pi^2 W(1/r_c)} (2 - n)(4 - n) \propto R^n. \tag{7}$$

For stochastic acceleration due to long-wavelength magnetosonic waves (assuming adequate particle pitch-angle scattering) the expression for D_{pp} from quasilinear theory is

$$D_{pp}^M = \xi \frac{p^2 V_A^2}{3v} \frac{\langle \delta B^2 \rangle}{B^2} \langle k \rangle \tag{8}$$

(adapted from Achterberg, 1981), where $\langle \delta B^2 \rangle$ is the mean square of the fluctuations in the field magnitude and $\langle k \rangle$ is the mean wavenumber of the magnetosonic waves. The number ξ depends on the angular distribution of the waves and it is usually assumed that $\xi \sim 1$. Note that the λ^M corresponding to Equation (7) is $B^2 \langle k \rangle^{-1}/\langle \delta B^2 \rangle$ and is independent of particle momentum or charge. It is not necessary for the magnetosonic waves to propagate in both directions to accelerate particles, but as already mentioned a certain level of pitch-angle scattering is required to isotropize the particles. This condition is $D_+ + D_- > D_{pp}^M/p^2$ (Achterberg, 1981). Unlike the acceleration by Alfvén waves, the waves which do this scattering are not required to propagate in both the + and − directions.

The relative importance of acceleration by magnetosonic and by Alfvén waves is given by D_{pp}^M/D_{pp}^A. In order of magnitude, this ratio is

$$\frac{D_{pp}^M}{D_{pp}^A} \sim \langle k \rangle r_c W^M \left(\frac{1}{k_r W_+(k_r)} + \frac{1}{k_r W_-(k_r)} \right), \tag{9}$$

where $k_r = r_c^{-1}$ is the resonant wavenumber and W^M is the total energy density in the long-wavelength magnetosonic waves. Since by assumption $\langle k \rangle r_c \ll 1$, acceleration by Alfvén waves dominates when their energy density is comparable to that in magnetosonic waves, but only when there is appreciable power in Alfvén waves propagating in both directions.

The momentum diffusion coefficient for isotropic Langmuir turbulence, D_{pp}^L, is given by Melrose (1980; Equation 8.13). This results in a $\lambda^L \propto (A/Z)^2$ times a function of particle velocity.

The acceleration models which we now discuss in detail consider the physical mode of stochastic acceleration only through the momentum diffusion coefficient D_{pp}. Neglecting spatial convection, averaging over some volume of space and introducting an escape time T from this volume, particle conservation results in the transport equation

$$\frac{\partial f}{\partial t} + \frac{1}{p^2} \frac{\partial}{\partial p} (p^2 S_p) + \frac{f}{T} = Q(p, t), \tag{10}$$

where $Q(p, t)$ is the particle source in momentum space. Equation (10) has been applied to the acceleration of solar energetic particles by Barbosa (1979), Ramaty and Lee (Ramaty, 1979), and Mullan (1980).

In Barbosa's (1979) model (hereafter referred to as model B) the acceleration is by Alfvén waves with a power law spectrum for $W(k)$, T is assumed to be a power law in velocity, $T \propto v^m$, and dp/dt is neglected. In Mullan's model (1980, referred to as model M) the scattering is from neutral sheets in turbulent motion leading to a constant λ; numerical integration is used in order to incorporate realistic losses due to Coulomb scattering and adiabatic deceleration in an expanding turbulent region, followed by an energy-independent escape. The Ramaty–Lee model (Ramaty, 1979, model RL) is a special case of the other two, since it assumes that both the mean free path and escape time are energy-independent and that dp/dt is negligible when acceleration is effective. These are rather special assumptions which need to be physically justified. Nevertheless, the predicted energy spectrum of this model fits the interplanetary observations of proton and α-particle spectra at 1 AU very well (see Section 2.2). Because of this and because of its inherent simplicity, we proceed first to discuss the RL model.

With a steady source of q particles cm^{-3} s^{-1} at momentum p_0, Equation (10) becomes

$$\frac{\partial f}{\partial t} - \frac{1}{p^2} \frac{\partial}{\partial p} \left(\alpha \frac{p^4}{3\beta} \frac{\partial f}{\partial p} \right) + \frac{f}{T} = \frac{q \, \delta(p - p_0)}{4\pi p_0^2}, \tag{11}$$

where the acceleration parameter $\alpha = (\delta V)^2/\lambda c$. The steady-state solutions of this equation are characterized by the dimensionless, energy-independent constant αT. For ultrarelativistic particles

$$f = \frac{q(p/p_0)^{-3/2 \pm (9/4 + 3/\alpha T)^{1/2}}}{4\pi p_0^3 \alpha (1 + 4/3\alpha T)^{1/2}}, \tag{12}$$

where the plus sign applies to $p < p_0$ and the minus sign to $p > p_0$. For nonrelativistic particles and $p > p_0$

$$f = \frac{6q}{4\pi p_0^2 mc\alpha} \frac{p_0}{p} I_2 (2(3p_0/mc\alpha T)^{1/2}) K_2 (2(3p/mc\alpha T)^{1/2}), \tag{13}$$

where m is the ion mass, and I_2 and K_2 are the modified Bessel functions of order 2 (Abramowitz and Stegun, 1965). The arguments of I_2 and K_2 are interchanged for $p < p_0$. The corresponding intensity per energy per nucleon is $dJ/dE = Ap^2 f$.

The spectrum of dJ/dE is not a power law. It has an energy-dependent slope $\gamma(E) \equiv -d \ln (dJ/dE)/d \ln E$ which approaches zero at low energies. From the asymtotic expression for $K_2(x) \propto x^{-1/2} \exp(-x)$, we find that at energies $E \gg 3.26 \, \alpha^2 \, T^2$ Mev nucleon^{-1}

$$\frac{dJ}{dE} \propto E^{3/8} \exp(-(E/(3.26(\alpha T)^2))^{1/4}). \tag{14}$$

This spectrum steepens with increasing energy up to the fully relativistic domain where Equation (12) applies and $\gamma(E) = \frac{1}{2} (9 + 12/\alpha T)^{1/2} - \frac{1}{2}$.

As discussed in Section 2.2, McGuire $et\ al.$ (1981) have found that Equations (13) or (14) fit the spectra of protons and α-particles in interplanetary space quite well with αT around 0.025 for protons. Different values of αT, however, are required for these two particle species; in many events the values of αT which fit the helium spectra are less than those for protons. The solar energetic particle spectra, therefore, are not strictly velocity dependent as implied by a constant λ. The possibility of a rigidity dependent diffusion mean free path is considered in the model of Barbosa (1979) which we discuss next.

As already mentioned, the B model uses the quasilinear theory of particle resonant scattering to evaluate D_{pp} from a given spectrum of Alfvén waves. The steady-state spectrum of nonelativistic particles in this model, with $\lambda \propto R^n$ (cf. Equation (7)) and $T \propto v^m$, is

$$\frac{dJ}{dE} \propto v^{(1+n/2)} K_v \left(\frac{1}{s} \left(\frac{3v\lambda}{V_A^2 T} \right)^{1/2} \right), \tag{15}$$

where $s = (n + 1 - m)/2 > 0$ and $v = (2 - n)/2s$. By using the asymptotic expression for $K_v(x)$, Equation (15) can be written as

$$\frac{dJ}{dE} \propto E^{(3+n+m)/8} \exp\left(-\left(\frac{A}{Z}\right)^{n/2} \left(\frac{E}{3.26K^2}\right)^{(1+n-m)/4} \right), \tag{16}$$

which is valid if the argument of the exponential is much larger than unity. The constant K in Equation (16) for the B model is equivalent to αT in Equation (14) for the RL model. For $n = m = 0$, Equations (15) and (16) reduce to Equations (13) and (14), respectively, i.e. the B model reduces to the RL model.

For $n > 0$, Equation (16) predicts α-particle spectra, which, when expressed as functions of energy per nucleon, are steeper than the proton spectra for the same K. This feature of the B model, not present in the RL model, is qualitatively consistent with the observations, as discussed above. But the comparison of Equation (16) with the solar flare proton spectra for a variety of m's and n's shows that the best fit is obtained for $m = n = 0$. Other values make the slope of the spectrum vary more rapidly with energy than observed. For example, if $m = 0$, we find that Equation (16) fits the observed proton spectrum shown in Figure 2 only if $n \lesssim 0.1$. For such small values of n, the α-particle spectrum is only slightly steeper than the proton spectrum with the same K, while the observed proton and α-particle spectra shown in Figure 2 are considerably different. Thus, while the B model allows rigidity-dependent spectra, the fit of the calculations to the data implies that this rigidity dependence is not very pronounced.

In the M model, the scattering elements are assumed to move with velocities δV of the order of shock velocities in the corona. Since λ is assumed to be energy independent, D_{pp} is given by the same expression as in the RL model, Equation (4). The $\mathrm{d}p/\mathrm{d}t$ term in Equation (3), however, is taken into account. In particular, a time-dependent combination of Coulomb and adiabatic deceleration losses followed by energy-independent escape is used. Because of this complexity, the resultant energy spectra can only be given graphically. They are approximately constant below 1 MeV nucleon^{-1}, very steep above 100 MeV nucleon^{-1}, and have $\gamma \sim 3$ at ~ 50 MeV nucleon^{-1}. This is in general agreement with observed spectra, although detailed comparisons over an extended energy range have not yet been made.

A very important question in all particle acceleration theories, including stochastic acceleration, is that of injection. We first note that the basic concept of stochastic acceleration assumes that the energy changes are small compared with the particle energy and therefore the particle velocity must be much greater than δV. Furthermore, for resonant scattering, ions must have $v > V_A$ to scatter from Alfvén waves and electrons must have $v > 43 V_A$ to scatter from whistlers (Melrose, 1974).

An additional injection condition is set by the requirement that the systematic acceleration rate due to diffusion in momentum space be larger than the ionization and Coulomb energy loss rates of the particles. The systematic acceleration rate is (e.g. Ramaty, 1979)

$$\left(\frac{\mathrm{d}p}{\mathrm{d}t}\right)_{\mathrm{acc}} = (vp^2)^{-1} \frac{\partial}{\partial p} (vp^2 D_{pp}), \tag{17}$$

and for D_{pp} from Equation (4)

$$\left(\frac{\mathrm{d}E}{\mathrm{d}t}\right)_{\mathrm{acc}} = \frac{4\alpha}{3} (E(E + 2 Mc^2))^{1/2}. \tag{18}$$

Here M and E are the proton mass and kinetic energy per nucleon for nuclei, and the electron mass and kinetic energy for electrons.

Energy loss rates due to ionization in a neutral medium and Coulomb losses in a fully ionized medium were summarized by Ramaty (1979). These loss rates, together with the systematic acceleration rates for nuclei and electrons (Equation (18)), are plotted in Figures 6 and 7 for neutral and ionized media, respectively. Particles can be accelerated only if the rate of systematic energy gain exceeds the rate of energy loss. Depending on the ratio of the acceleration efficiency, α, to the ambient density, n, the energy gain curve may or may not intersect the energy loss curve. In the former case an injection mechanism is required which pre-accelerates the particles to $E \sim E_0$, where E_0 is the energy where the two curves intersect.

The values of α/n indicated in Figures 6 and 7 were chosen such the E_0 for electrons is around 0.1 MeV, an energy at which there seems to be a transition from first-phase to second-phase acceleration (Section 2). For protons, however, these α/n's are such that the systematic gain is larger than the loss at all energies, and in this case it is possible to accelerate ambient particles directly. However, stochastic acceleration still requires that the particle velocities be larger than δV. Assuming that δV is of the order of the Alfvén

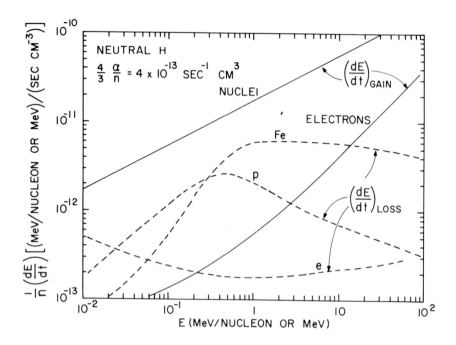

Fig. 6. Energy loss rates of protons, Fe nuclei, and electrons in neutral H (dashed curves) and energy gain rates from Equation (18) (from Ramaty, 1979).

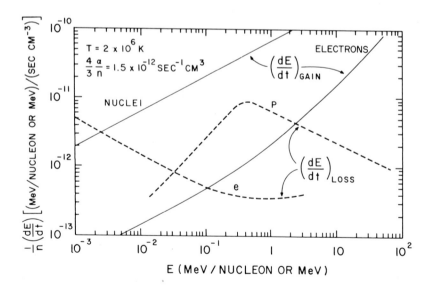

Fig. 7. Energy loss rates of protons and electrons in an H plasma of 2×10^6 K (dashed curves). The loss rates for other ions scale as Z^2/A. The energy gain rates are from Equation (18) (from Ramaty, 1979).

velocity, the particles must be accelerated either from a high β plasma (β = thermal pressure/magnetic pressure) or possibly from the nonthermal tail of a low-β plasma. Such nonthermal tails are observed in the solar wind (Ogilvie *et al.*, 1978; Scudder and Ogilvie, 1979). If the Alfvén speed is the threshold, then [3]He-rich events cannot originate in high-β plasmas.

The solar energetic particle abundances and their variability must be strongly dependent on the injection process as we have discussed in Section 2.2. Injection questions have not yet been fully investigated and they remain outstanding problems for future research.

The acceleration time of stochastic acceleration can be studied from the time-dependent solutions of Equation (10). Such solutions were obtained in the RL model for impulsive injection of particles at $t = 0$ and $p = p_0$, with $T \to \infty$ and λ = const. In Figure 8

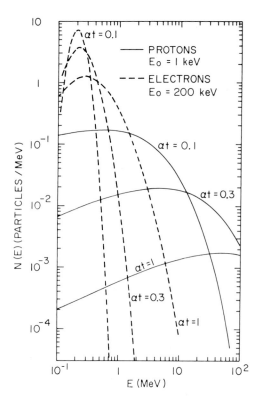

Fig. 8. Time-dependent proton and electron spectra from stochastic acceleration with impulsive injection and no losses, E_0 is the injection energy (from Ramaty, 1979).

(from Ramaty, 1979) we show the differential proton and electron number densities ($N(E) = 4\pi p^2 f/v$) at various times t after injection. As can be seen, in a given time, protons are accelerated to much higher energies than electrons. This result is the direct consequence of the acceleration rates shown in Figures 6 and 7 which assume the same α for protons and electrons.

From the gamma-ray observations (Section 2.1) it follows that, in at least some flares, protons and nuclei are accelerated to energies greater than 10 Mev nucleon^{-1} in about 1 s. As can be seen from Figure 8, acceleration of a significant fraction of the protons to such energies requires that $\alpha t \sim 0.1$. Then if $t \simeq 1$ s., $\alpha \simeq 0.1$ s^{-1}. Because δV can probably not be greater than 3×10^8 cm s^{-1}, λ must be less than 3×10^7 cm. Since this value is 50 times the gyroradius of a 20 MeV proton in a 1 gauss field, we could expect such short mean free paths in the strong turbulence of solar flares. But there is no direct observational evidence for such turbulence.

The characteristic acceleration time in the B model is of the same form as in the RL model. Barbosa (1979) estimates that the acceleration time to $\sim$10 MeV can be as short as $\sim$10 s. As above, this requires strong turbulence on scales $\sim$10^6 cm. Mullan (1980), on the other hand, estimates from numerical solutions that the time to accelerate keV protons to $\sim$5 MeV is about 1 h. This time is too long for the acceleration of the gamma-ray producing protons and nuclei, but it could be adequate for the acceleration of the solar flare particles observed in interplanetary space.

3.2. SHOCK ACCELERATION

Solar flare shocks propagate upward through the solar corona at speeds of about 500–2000 km s^{-1}, as indicated by Type II radio bursts (Smith and Goldman in Chapter 15 of this volume, and Section 2.1), and laterally through the chromosphere where they are seen as Moreton waves (Uchida, 1974). The occurrence of solar energetic particles in space is strongly correlated with flares having Type II bursts (Svestka and Fritzova, 1974). Flare shocks are observed to accelerate particles in interplanetary space (Richter and Keppler, 1977), as are corotating shocks (Barnes and Simpson 1976; McDonald *et al.*, 1976) and planetary bow shocks (Asbridge *et al.*, 1968; Lin *et al.*, 1974; Scholer *et al.*, 1980; Zwickl *et al.*, 1981). A flare shock can transport particles in an energy-independent manner through the corona until they escape onto open field lines. Shock acceleration has been recently reviewed by Toptygin (1980) and Axford (1982) and applied to solar flares by Achterberg and Norman (1980), Decker *et al.* (1981), and Lee and Fisk (1982). See also McDonald (1981) for review.

There are basically two types of shock acceleration: scatter-free, in which particles gain energy by reflection in a single shock encounter (Sonnerup, 1973; Chen and Armstrong, 1975; Pesses *et al.*, 1982) and diffusive, in which particles gain energy by repeated scattering between the converging plasmas on each side of the shock (e.g. Axford *et al.*, 1977; Axford, 1982). The scatter-free mechanism can enhance the particle energy by about an order of magnitude if the shock is nearly perpendicular (i.e. the magnetic field is nearly perpendicular to the shock normal), but in that case, only particles with speeds which are already much greater than the shock speed can be reflected. Further acceleration, however, requires multiple reflections. These are possible if there is particle scattering in the fluid flow or if the particles are trapped between converging shocks in a flare loop (Wentzel, 1965).

Acceleration by diffusive scattering across the shock is a first-order Fermi process, in the sense that every shock crossing results in an energy gain. It is, in principle, more efficient than stochastic acceleration because it derives energy directly from the compression of the flow at the shock. For this mechanism to be effective, however, there must be adequate particle scattering both upstream and downstream of the shock. The passage of the shock is expected to generate turbulence downstream, which will scatter the particles. Scattering upstream, however, is more problematic (Holman et al., 1979). Observations (Tsurutani and Rodriguez, 1981) of interplantetary shocks and planetary bow shocks show that when they are nearly parallel there is a very turbulent foreshock region capable of scattering particles. Such a region could be produced by the accelerated particles themselves (Achterberg and Norman, 1980; Lee, 1982).

In the simplest example of a plane shock where the only losses are due to convection of the particles away from the shock downstream, the energetic particle density in phase space is given by a power law, $f \propto p^{-3V/\Delta V}$, where V is the shock speed and ΔV the discontinuity in the plasma speed at the shock. In terms of the compression ratio, r, at the shock, $V/\Delta V = r/(r - 1)$. For a strong shock in a nonrelativistic fluid, $r = 4$ and hence $f(p) \propto p^{-4}$. For weaker shocks, $4 > r > 1$ and therefore the power law is steeper. Blandford and Ostriker (1980) have modelled cosmic-ray acceleration and propagation in the Galaxy on the basis of this result, including acceleration by shocks with a spectrum of compression ratios. They have been able to reproduce the observed cosmic-ray spectrum which approximates a power law over a wide range of energies.

Unlike galactic cosmic-rays, solar flare particles do not have power law spectra (e.g. Figure 2). In fact, none of the energetic particle populations which are observed to be accelerated by shocks in interplanetary space have power law spectra. Diffusive shock acceleration produces such curved spectra rather than power laws when particle energy losses or escape losses are significant, or when the shock has a finite size or lifetime, compared to the natural scales of the shock acceleration. For example, diffusive escape losses from the finite bow shock of the Earth (Eichler, 1981; Forman, 1981b; Ellison, 1981a; Lee, 1982) and adiabatic losses near interplanetary shocks (Fisk and Lee, 1980; Forman, 1981a) have been shown to reproduce the observed spectra quite well.

In the present subsection we describe the general methods used to obtain energetic particle spectra from shock acceleration, and provide a simple solution of the appropriate transport equation which provides a spectrum that fits the solar flare data quite well.

The transport equation which describes diffusive shock acceleration is similar to Equation (10) without the stochastic acceleration term proportional to D_{pp}, but with additional terms due to convective transport, spatial diffusion and adiabatic compression of particles in the plasma flow. This equation is given (Axford, 1982) by

$$\frac{\partial f}{\partial t} + \mathbf{V} \cdot \nabla f - \nabla \cdot (k \cdot \nabla f) - \frac{\nabla \cdot \mathbf{V}}{3} p \frac{\partial f}{\partial p} + \frac{f}{T} + \frac{1}{p^2} \frac{\partial}{\partial p} \left(p^2 \left(\frac{dp}{dt} \right) f \right) = Q(p, r, t), \quad (19)$$

where f, T, and dp/dt have been defined in connection with Equation (10), $\mathbf{V}$ is the plasma velocity, and k the spatial diffusion tensor which couples the energetic particles to the plasma converging at the shock. The terms containing $\mathbf{V}$ and k are essential for the description of acceleration across the shock front. The injected particles are explicitly introduced in Equation (19) by the source term Q; the injection may also be treated as a boundary condition such that f approaches a given value f_0 far upstream. The losses due to particle escape can be treated via the escape time T, or as diffusive escape. In the latter case the scattering becomes negligible at a finite distance from the shock.

The usual method for deriving a steady-state $(\partial f/\partial t = 0)$ particle spectrum is to first solve Equation (19) separately on each side of the shock and then to match the two solutions at the shock by imposing boundary conditions. These conditions are that both the energetic particle density and the normal component of the spatial streaming $(S = -4\pi p^2 (Vp(\partial f/\partial p) + k \cdot \nabla f))$ of these particles be continuous at the shock. Toptygin (1980) has shown that this is an appropriate approach even though Equation (19) is not valid very close to the shock.

We present here a steady-state solution of Equation (19) by assuming that dp/dt is negligible and T is constant (as in the RL model for stochastic acceleration), that only diffusion along the shock normal is important (hence k can be treated as a scalar), that the shock is an infinite plane at $x = 0$ and infinitely thin, and that the particles are steadily injected at the shock, i.e. $Q = q\delta(x)\delta(p - p_0)/(4\pi p_0^2)$, where q is measured in particles cm^{-2} s^{-1}. For these conditions, the solution on each side of the shock is given by

$$f(x, p) = f(0, p) \exp(-\beta_i |x|), \tag{20}$$

where upstream

$$\beta_i = \beta_1 = (V + (V^2 + 4k_1/T)^{1/2})/2k_1 \tag{21}$$

and downstream

$$\beta_i = \beta_2 = (-(V - \Delta V) + ((V - \Delta V)^2 + 4k_2/T)^{1/2})/2k_2. \tag{22}$$

The streaming continuity at the shock gives the equation

$$\Delta V \frac{p}{3} \frac{\partial f}{\partial p} (0, p) + (k_1 \beta_1 + k_2 \beta_2) f(0, p) = \frac{q\delta(p - p_0)}{4\pi p_0^2}. \tag{23}$$

The exact nonrelativistic solution of this equation is

$$f(0, p) = \frac{3q}{4\pi p_0^3 \Delta V} \left(\frac{p}{p_0}\right)^{-3V/\Delta V} \exp\left(-\frac{3(VI_1 + (V - \Delta V)I_2)}{\Delta V}\right), \tag{24}$$

where

$$I_i = (1 + \eta_i)^{1/2} - 1 - \operatorname{ctn} h^{-1} (1 + \eta_i)^{1/2} - \ln (\eta_i/4)^{1/2}, \tag{25}$$

with $\eta_i = 4\lambda_i p/(3m V_i^2 T)$. Since $I_i \to 0$ for $\eta_i \to 0$, the distribution function is a power law for $v \ll 3V_i^2 T/4\lambda_i$ with the same spectral index $(3V/\Delta V)$ as for the case of a plane shock with no losses. However, when v becomes comparable to the lower $3V_i^2 T/4\lambda_i$, the shape of the spectrum is determined by the exponential term in Equation (24). In other words, the spectrum is determined by the losses on the side of the shock where V_i^2/λ_i is lower. This is probably the upstream side.

Assuming for simplicity that V^2/λ is the same on both sides of the shock, and that

$3V/\Delta V = 4$, as for a strong shock, the differential intensity per energy per nucleon, for $v \gg 3V^2 T/4\lambda$, is

$$\frac{dJ}{dE} \propto E^{-1} \exp\left(-\left(\frac{E}{0.422(\alpha T)^2}\right)^{1/4}\right), \qquad (26)$$

where $\alpha T = V^2 T/(\lambda c)$. Equation (26) for shock acceleration is quite similar to Equation (10) for stochastic acceleration in the RL model. These two spectra are compared in Figure 9 using values of αT which fit typical data in both cases. As can be seen, the difference between them is most pronounced at the lowest energies where the solar flare spectrum is unknown because of adiabatic energy losses in the interplanetary medium (Section 4).

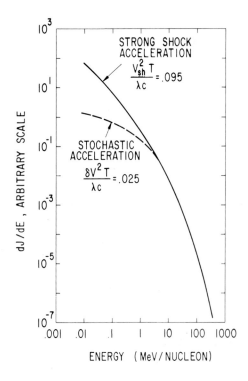

Fig. 9. Comparison of the particle spectra produced by steady-state stochastic acceleration (Equation (13)), with shock acceleration (Equation (24)), for models in which the scattering mean free path, λ, and the escape time, T, are energy independent. V_{sh} is the shock speed and $(\delta V)^2$ the mean square turbulent velocity.

As for stochastic acceleration, and for shock acceleration, the question of injection is very important. Ionization and Coulomb energy losses to the ambient medium have the same role in determining injection conditions in shock acceleration as they do in stochastic acceleration. In addition, for diffusive shock acceleration, particles downstream must have sufficient velocity to overtake the shock. This is at least $(V - \Delta V)$ directed towards the shock, and increases as $\cos^{-2} \psi$, where ψ is the angle between the downstream field and the shock normal. The velocity $V - \Delta V$ is at least as great as V_A or δV, and with the additional $\cos^{-2} \psi$ factor, the threshold for shock acceleration is expected to be higher than for stochastic acceleration.

Another injection condition is set by the finite width of the shock which could depend on many parameters, including the pressure of the accelerated particles. When this pressure is taken into account (Axford et al., 1977; Eichler, 1979; Drury and Volk, 1981, Drury et al., 1982) all or part of the velocity change ΔV is smoothed out over a length scale $\sim \bar{k}/V$, where $\bar{k}$ is the diffusion coefficient of the energetic particles averaged over their energy spectrum. As pointed out in Section 2.2, this smoothing is expected to affect the composition of the accelerated particles (Eichler, 1979). Drury et al. (1982) show analytically that when k is independent of energy, the smoothing causes the dominant accelerated species (i.e. protons) to have a steeper spectrum than in the case of an infinitely thin shock. Minor species which are only partially stripped could have larger diffusion coefficients than the protons if the diffusion mean free path is rigidity dependent, and therefore for them $k > \bar{k}$. Drury et al. (1982) show that the spectrum of such minor species is flatter than for protons and approaches that of an infinitely thin shock.

Ellison (1981b) has studied these effects with a Monte Carlo calculation appropriate for a parallel shock. In this treatment, the plasma flow and the energetic protons are required to conserve mass, momentum, and energy flux, the mean free path is assumed to be proportional to gyroradius, and particles are injected from the shocked plasma downstream in which the temperatures of different ions are proportional to their mass. Ellison (1981b) finds that the shock is indeed broadened and that there are modest enhancements of energetic particles which increase with the mass-to-charge ratio.

The acceleration time in shock acceleration can be obtained from the time-dependent solution of Equation (19). Such solutions have been obtained for various initial conditions and geometries by Fisk (1971), Forman and Morfill (1979), and Topygin (1980) and have been reviewed by Axford (1982). The general result is that

$$\frac{dE}{dt} = \frac{\Delta V}{c}\left(\frac{\lambda_1}{V_1} + \frac{\lambda_2}{V_2}\right)^{-1} (E(E + 2Mc^2))^{1/2}. \tag{27}$$

This expression shows explicitly that if the mean free path on either side of the shock is very large, the rate of energy gain is very small. Since, as already mentioned, shocked gas downstream is expected to be turbulent, the efficiency of the acceleration depends critically on the presence of turbulence upstream.

In the absence of upstream turbulence, particles can still be accelerated stochastically by the shock-generated turbulence downstream. By comparing Equations (18) and (27) we find that the ratio of the shock acceleration rates and the stochastic acceleration rate downstream is of the order $V \Delta V/(\delta V)^2 (\lambda_2/(\lambda_1 + \lambda_2))$, where δV is the velocity of the turbulent elements and λ_2 and λ_1 are the downstream and upstream mean free paths, respectively. Thus, if the mean free paths are comparable, shock acceleration will be more rapid since V and ΔV are expected to be larger than δV.

3.3. ACCELERATION IN DIRECT ELECTRIC FIELDS

Sections 3.1 and 3.2 have dealt with mechanisms that can be best applied to the acceleration of ions in solar flares. In addition to stochastic and shock acceleration, there is also the possibility of accelerating particles in direct electric fields. Such fields are associated with magnetic reconnection in the vicinity of magnetic neutral points and current sheets (Syrovatskii, 1981), where they appear perpendicular to the magnetic field. Particle acceleration is also possible in electric fields parallel to the magnetic field (Colgate, 1978). Parallel electric fields arise from the interruption (due to plasma instabilities) of the parallel currents associated with twisted magnetic flux tubes and from the formation of double layers of electric charge (e.g. Spicer, 1982).

The application of direct electric fields to the acceleration of nonrelativistic electrons in solar flares has been discussed most recently by Spicer (1982) – see also Smith (1979), and Heyvaerts (1981). Here we wish to emphasize the role that direct electric fields could play in the very rapid production of ions and the acceleration of relativistic electrons.

As discussed in Sections 3.1 and 3.2, both stochastic and shock acceleration could accelerate ions quickly enough to account for the gamma-ray observations, but it is not clear that the mean free paths are, in fact, sufficiently short and the turbulent and shock velocities are large enough to account for the rapid acceleration. On the other hand, in an electric field model (such as that of Colgate, 1978), the acceleration time could be as short as 0.1 s for a loop of length $\sim 10^9$ cm.

Relativistic electrons around 10 MeV, however, cannot be produced fast enough and in sufficient quantities in stochastic and shock acceleration because the particle energy gain per collision is proportional to particle mass. But there seems to be a good correlation between interplanetary electrons of such energies and gamma-ray flares (Evenson et al., 1981; Section 2.2). Since the gamma-rays are produced by ions of MeV nucleon^{-1}, this correlation could indicate electron and ion acceleration to the same energy by an electric field.

In this section we review particle acceleration processes in current sheets and in the vicinity of magnetic neutral points. For discussions of acceleration by parallel electric fields, the reader should consult Colgate (1978), Smith (1979), Heyvaerts (1981), and the review of Spicer (1982).

Current sheets and magnetic neutral points are involved in a wide variety of contexts in solar physics. The primary energy release in solar flares may occur by magnetic reconnection (e.g. Kahler et al., 1980; Sturrock, 1980). Parker (1979, and references therein) has emphasized the near-singular conditions required for magnetostatic equilibrium, and has proposed that the reconnection of tangled magnetic fields occurs routinely and inevitably in nature. Syrovatskii (1981) reviewed the formation of current sheets. In view of the important role suggested for current sheets and reconnection in the solar atmosphere it is natural to consider particle acceleration mechanisms associated with these phenomena. Reconnection and associated particle acceleration have been extensively studied in the Earth's magnetosphere (Stern and Ness, 1982).

In models of steady-state reconnection, magnetic fields of opposite polarity are brought together by a flow of fluid. A detailed discussion of the fluid theory of steady-state reconnection has been given by Vasyliunas (1975). Early descriptions of the reconnection process and its application to solar flares were given by Parker (1957, 1963) and

Sweet (1958). Two of the best-known examples of reconnection are the models of Petschek (1964) and Sonnerup (1970) which involve X-type neutral points. The magnetic field lines are sketched in Figures 10(a) and 10(b) for these two models. These field

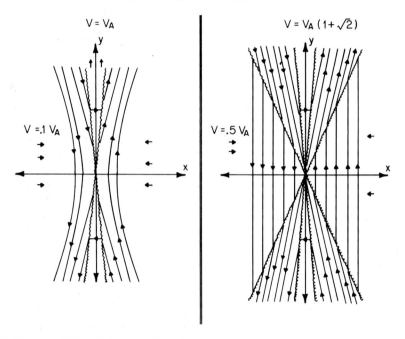

Fig. 10. Magnetic field and flow configuration in (a) and Petschek (1964) and (b) the Sonnerup (1970) reconnection models.

lines lie in a plane (say, the $x-y$ plane) and there is a constant electric field, **E**, perpendicular to that plane. The electric field is related to the magnetic field **B**, flow velocity **V**, and current J in the plasma by Ohm's law

$$\mathbf{E} + \mathbf{V} \times \mathbf{B}/c = \eta \mathbf{J}, \tag{28}$$

where η is the resistivity. The inclusion of other terms in Equation (28) is discussed by Vasyliunas (1975). The resistive term is unimportant everywhere except in a small region near the X-type neutral point where B becomes small and **J** is large (because the gradients of **B** are large). Here the frozen in condition of the magnetic field is violated. Rather than being transported by the fluid, the field diffuses through the semistagnant plasma. The region, therefore, is referred to as the "diffusion region".

As shown in Figure 10, cold fluid flows inward along the x-axis. Most of the fluid never enters the diffusion region, but passes through a shock front (or pair of fronts, in Sonnerup's model) and flows out parallel to the y-axis. In Petschek's model, the inflow occurs at a small fraction ($\sim$0.1) of the Alfvén speed, while the outflow is exactly Alfvénic. This acceleration takes place at the expense of magnetic energy. In Sonnerup's model, the inflow can be faster than in Petschek's model, in which case, less acceleration of the bulk flow occurs. In both models, however, the thermal speed of the ions increases across the shock front to about the Alfvén speed. The increase of electron thermal energy depends on the structure of the shock and the nature of the coupling between electrons and ions.

Similar results have been obtained by Hayashi and Sato (1978) who have simulated time-dependent, compressible reconnection by solving the fluid equations numerically. In their model, an initially planar neutral sheet is compressed into an X-type neutral line by imposed fluid flow directed inward, perpendicular to the initial field lines. They assume that when the current density J exceeds some critical value J_c, anomalous resistivity develops and increases as a power of $J - J_c$. Thus, reconnection of the magnetic field lines and Joule heating of the plasma near the neutral line begin when the neutral sheet has been sufficiently compressed by the incoming flow. They find that, as in the analytic models of reconnection, the regions of inflow and outflow are separated by shocks and that the outflow speed is of the order of the Alfvén speed.

If ions gain flow velocities of the order of the Alfvén velocity, their energy per nucleon is $2.5 \times 10^4 \, B^2/n$, where B is in gauss and n in cm^{-3}. While for most combinations of B and n in the solar atmosphere this energy is not as high as the observed particle energies, this bulk flow acceleration could nonetheless be important as an injection mechanism. In certain exceptional cases, however, B could be high enough and n low enough for the particles to achieve energies as high as 10 MeV nucleon^{-1} (e.g. Sonnerup, 1973) which would suffice, for example, for the production of nuclear gamma-rays.

In addition to the acceleration and heating of the plasma as it passes through the reconnection region, the possibility exists of direct particle acceleration in the electric field at the magnetic null line.

An example of such acceleration was discussed by Speiser (1965). He considered a magnetic field with field lines parallel to the y-axis with field reversal at $x = 0$,

$$\mathbf{B} = \mathbf{e}_y B_1 \frac{x}{d}, \ |x| < d$$

$$\mathbf{B} = \mathbf{e}_y B_1 \frac{x}{|x|}, \ |x| > d, \tag{29}$$

where $\mathbf{e}_y$ is the unit vector in the y-direction and B_1 is a constant. Here the field equals B_1 for $|x| > d$ and goes linearly to zero for $|x| < d$. A constant electric field $\mathbf{E} = V B_1 \mathbf{e}_z/c$ is also present and assumed to be continuous at $x = 0$. Everywhere in space except near the field reversal region at $x = 0$, the particle orbits consist of gyration in the x–z plane (the usual Larmor motion) combined with the $\mathbf{E} \times \mathbf{B}$ drift in the x-direction, towards the plane $x = 0$. Since for $|x| > d$ the motion in z is oscillatory, the average energy gain is zero in this region. However, once a particle has drifted to near the $x = 0$ plane where the field is very weak, it does not gyrate — it is simply accelerated in the z-direction. The final energy, E, of the particle depends on the extent, L, of the current sheet in the z-direction. Numerically, E in MeV equals $10^{-6} \, BLV$, where B is in gauss, L in km, and V km s^{-1}. Thus particles can be accelerated to tens of Mev if B, V, and L are large enough.

Whether neutral sheets have sufficient lengths, and, whether or not particles can remain in the sheet for such lengths, are problems which have not yet received clear-cut answers for solar flares.

In order to estimate the distance the particles stay in the neutral sheet, particle orbits in various magnetic geometries have been calculated by Speiser (1965), Friedman (1969), Bulanov and Sasorov (1976), Bulanov (1980), and Syrovatskii (1981). The results depend strongly on assumed geometries, and analytic treatments can only be made for the simplest field configurations. Time-dependent electric and magnetic fields have also been considered (Burke and Layzer, 1969; Levine, 1974). Levine extended Burke and Layzer's work by trying to estimate the effect of Coulomb collisions between the test particles and field particles in the ambient medium. The existence of an energy loss mechanism which competes with acceleration sets an upper limit on the time constant for collapse of the neutral sheet such that particle acceleration can still occur. Recently, Mullan and Levine (1981) have studied the implications of the collapsing magnetic neutral sheet model for the composition of solar flare accelerated ions (see Section 2.2).

Attempts have also been made to calculate the energy spectrum of particles accelerated in neutral sheets. In the absence of stochastic processes such as Coulomb collisions or wave–particle interactions, the phase space distribution function f of the test particles satisfies Liouville's theorem; the density of particles in phase space is constant on phase space trajectories. Therefore, given an initial spatial distribution of injected particles and solutions for the particle orbits, the momentum distribution function of particles leaving the acceleration region can be calculated (Bulanov and Sasorov, 1976; Bulanov, 1980). They find that when the initial distribution of particles is uniform in space, the spectrum is an exponential in energy or in a fractional power of the energy. Energy spectra have also been calculated numerically by Bulanov (1980) who finds good agreement with the analytic results. Friedman (1969) has used numerical techniques to calculate final energies for test particles in Petschek's (1964) reconnection model. A generally recognized problem with particle acceleration in neutral sheets is that only very few particles are accelerated.

4. Solar Flare Particle Spectra in Interplanetary Space

The spectra of particles observed at any one time in interplanetary space are not exactly the same as the spectra released at the Sun because of velocity dispersion caused by energy-dependent diffusion, and because the solar wind convects and decelerates energetic particles as it expands. If these solar wind effects are small, the effect of velocity dispersion can be eliminated in prompt, magnetically well-connected events by using the time-of-maximum (TOM) energy spectrum formed by the maximum intensity at each energy (Lin, 1970; Van Hollebeke et al., 1975). We justify this method below.

At energies less than about 1 MeV nucleon^{-1} for ions and 1 MeV for electrons, convection and adiabatic deceleration will make the TOM spectrum flatter than the spectrum at the Sun, but the precise effect on the spectrum is uncertain. In addition, the TOM spectrum becomes less reliable for such low-energy particles because their propagation becomes more sensitive to inhomogeneities in the solar wind and magnetic fields (e.g. shocks).

The important questions which need to be resolved before interplanetary spectra can be used to test models of the acceleration and release process at the Sun are: 'What is the most convenient way to estimate the spectrum at the Sun from spacecraft observations of energetic flare particles?; above what energy is that spectrum reliable?; and what is the approximate nature of the change in the spectrum below that energy?' First, we show why the TOM spectrum in simple well-connected events is a good approximation to the spectrum at the Sun.

The observed $f(p, r, t)$ is related to the spectrum injected into interplanetary space by the transport equation.

$$\frac{\partial f}{\partial t} - \nabla \cdot (k \cdot \nabla f) + V \cdot \nabla f - \frac{\nabla \cdot V}{3} p \frac{\partial f}{\partial p} = F_0(p, \theta, t) \frac{\delta(r - r_0)}{4\pi r^2}, \tag{30}$$

which is analogous to Equation (19) for shock acceleration. Here $4\pi p^2 F_0(p, \theta, t)/v$ is the number of particles released per (energy-sec) at momentum p, longitude θ, time t, and radial distance r_0 from the Sun, and V is the solar wind velocity.

It is clear that F_0 can be most reliably recovered from $f(p, r, t)$ when F_0 and the interplanetary parameters k and V are as uncomplicated as possible. Therefore, we consider only events in which the effects of delayed and longitude-dependent release at the Sun, longitudinal transport in interplanetary space and all interplanetary propagation complexities (e.g. shocks) are minimal. These are the magnetically well-connected and prompt events at solar longitudes in the range $60° \pm 40°$ west of central meridian. In addition, the spectrum can be derived with confidence only for those events in which the particle fluxes have weak anisotropies during and after the time of maximum (indicating sudden release) and which have smooth time profiles (indicating that the injection is uniform over the connection longitudes). In such simple events, the particle propagation can be treated as spherical diffusion along each field line.

When, furthermore, $k \gg Vr$ and for times $t \ll r/V$, the convection and adiabatic deceleration can be neglected. Then Equation (30) reduces to simple time-dependent spherical diffusion from a point source. If the outer boundary is far away, the solution of Equation (30) for k independent of r is (Parker, 1963)

$$f(p, r, t) = \frac{N_0(p)}{4\pi p^2} \frac{\exp(-r^2/4tk(p))}{2\sqrt{\pi}(tk(p))^{3/2}}, \tag{31}$$

where $N_0(p) \, dp$ is the number of particles between p and $p + dp$ released at the Sun. Clearly, if the assumptions so far are valid, $f(t)$ will have the time dependence in Equation (31) and a fit of particle data to that form will confirm the assumptions and determine $N_0(p)$.

A more direct approach is to note that Equation (31) has a maximum at $t_m = r^2/6k(p)$, and that for particles at their $t_m(r, p)$,

$$f(p, r, t_m(r, p)) = \frac{(6/e)^{3/2}}{2\sqrt{\pi}r^3} \frac{N_0(p)}{4\pi p^2} = \frac{0.93}{r^3} \frac{N_0(p)}{4\pi p^2}. \tag{32}$$

This is the basis for the assertion that the TOM spectrum is the same as that released at the Sun. If $k \propto r^b$ with $b < 2$, a similar relation holds and $N_0(p) \sim 4\pi p^2 f(p, r, t_m) r^3$.

Many of the assumptions leading to the validity of the TOM spectrum break down at low ion energies because of the long time to maximum (~ 1 day at 1 MeV nucleon^{-1}) and because of the stronger coupling to the solar wind, which is highly variable on time scales of 1 day. In addition, the condition $Vr \ll k$, which allows convection and adiabatic deceleration to be neglected, is equivalent to $v \gg 3Vr/\lambda$. With $\lambda \sim 0.1$ AU as in Figure 11 (Palmer, 1982), this velocity is $\sim 10^9$ cm s^{-1}, or energies $\gg 1$ MeV nucleon^{-1}.

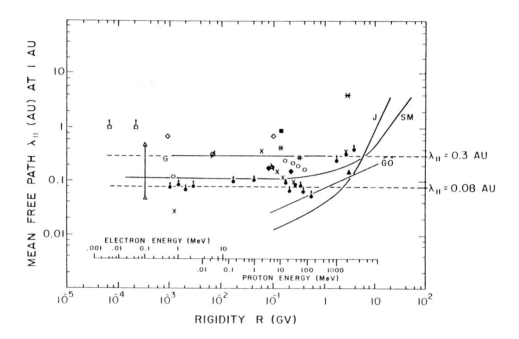

Fig. 11. Mean free paths of solar electrons and protons in interplanetary space deduced by fitting time profiles in many events. (From Palmer, 1982 – *Rev. Geophys. Space Sci.* **20**, 336, © by the American Geophysical Union.)

The effect of adiabatic deceleration on the TOM spectrum can be estimated several ways, which give similar results. Scholer (1976) has shown by numerical simulations of particle trajectories that when $k \lesssim Vr$ and k depends on energy, convection and adiabatic deceleration not only make the time of maximum flux of particles observed at energy E earlier than $r^2/6k(E)$, but make the discrepancy larger at lower energies. This is because particles observed at energy E had higher energy near the Sun and so diffused there faster; the effect increases at lower energies because the relative energy change is larger at lower energies. The energy change must be less than the energy shift in the $t_m(E)$ curve compared to $r^2/6k(E)$. This behavior of t_m is discussed by Kurt *et al.* (1981) who show that while the t_m versus energy curves for well-connected flares vary as $1/v$ (as expected for $\lambda \sim$ const) above an energy E_1 (~ 1 MeV), the variation is weaker below E_1 in the manner of Scholer's (1976) simulations. The break energy E_1 varies slightly from flare to flare, but is always near where $Vr = k$, on the basis to the t_m at high energies. Kurt *et al.* (1981) also show that the TOM spectra of these flares are flatter below E_1 than above E_1, and they discuss the apparent energy change, and its effect on the TOM spectrum.

Scholer (1976) found the energy change for each particle by integrating the adiabatic deceleration law $dp/dt = -2Vp/3r$ along the computed spatial trajectories of the particles. Kurt *et al.* (1981) approximate the trajectories with diffusion speed $U(p, r)$ such that $dr = U\,dt$. Then with $U = 3ak/r$, where $k \propto E^a$, the deceleration law can be integrated. It can be shown that $U = 2k/r$ is a better choice, consistent with certain analytic results on the effect of convection and adiabatic deceleration on the time of maximum and the particle flux at that time (Fisk and Axford, 1968; Forman, 1971). The energy change law then becomes $dp/dr = -Vp/3k$. When λ is constant, this results in a change in the speed of a nonrelativistic particle, given by $\Delta v = -vr/\lambda$ between the Sun and r. For the typical value of $\lambda = 0.1$ AU, this is $\Delta v \sim (-4 \times 10^8\ r(\text{AU}))$ cm s^{-1}.

Figure 12 shows the effect of this adiabatic energy change on the spectrum.

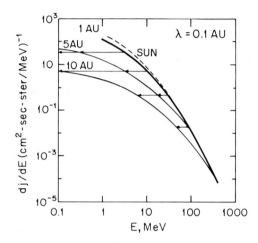

Fig. 12. Illustration of the probable effect of adiabatic deceleration on the spectra of solar flare ions at the time of maximum. The spectrum at the Sun (shown as a dashed line) is shifted to lower energies.

We conclude that the TOM spectra of solar flare particle fluxes in interplanetary space is a good first-order estimate of the spectrum released at the Sun, for reasonably well-connected flares showing fairly rapid particle release, down to energies where convection and adiabatic deceleration are significant. The TOM spectrum should be used because it eliminates the velocity dispersion caused by the energy-dependence of the diffusion coefficient; the spectrum varies slowly at that time; it is relatively insensitive to finite injection times; and it is early enough that convection and adiabatic deceleration effects are minimal. The TOM spectrum begins to deviate from the spectrum injected at the Sun at energies where $k(E) \lesssim Vr$, that is $E < E_1 \sim 3(r/10\lambda)^2$ MeV nucleon^{-1} for ions and $\lesssim 1.5\ (r/10\lambda)^2$ keV for electrons, or when the time of maximum is longer than the solar wind transit time divided by the Compton–Getting factor. Typical values of λ make $E_1 \sim 1$ MeV nucleon^{-1} for ions and ~ 1 keV for electrons.

The effect of adiabatic deceleration and convection on ion spectra with realistic propagation parameters below a few MeV per nucleon is not known exactly. We have presented some rules which describe approximately how the TOM spectrum is related to the spectrum released at the Sun. Since, as we have shown in Section 3, the spectrum below 1 MeV nucleon^{-1} differs greatly between acceleration theories which explain the spectrum above 1 MeV nucleon^{-1} equally well, a more exact method of deducing the accelerated spectrum below 1 MeV nucleon^{-1} from that observed in interplanetary space is needed to decide among such theories. However, because the propagation of such low-energy ions is extremely sensitive to interplanetary conditions, it is problematic whether this can be practically done with observations at 1 AU.

5. Summary and Outlook

We have reviewed observations and theories of particle acceleration in solar flares. Flare accelerated particles that remain trapped at the Sun produce a variety of electromagnetic emissions which provide important information on the timing of the acceleration and on the number and energy of the accelerated particles. While it is clear that not all particle species are accelerated to their final energies at the same time, the distinction between first-phase and second-phase acceleration is no longer as clear as it had been prior to recent gamma-ray observations. There is ample evidence for the bulk energization of nonrelativistic electrons whose interactions in the solar atmosphere produce hard X-rays. The energy contained in these electrons constitutes a major fraction of all the available flare energy. The acceleration of the gamma-ray producing ions in some flares is closely associated in time with this bulk energization. The energy content of these ions, while smaller than that of the nonrelativistic electrons, also constitutes a significant fraction of the total flare energy (Section 2.1).

The ions observed in interplanetary space are probably accelerated by shocks or turbulence. The correlation between Type II radio bursts and ion events provides the main observational support for this conclusion. Both stochastic and diffusive shock acceleration can explain the observed energy spectra, but the fact that these spectra do not vary much from one event to another (Section 2.2) seems to imply that one of these mechanisms is dominant. The relative importance of stochastic and shock acceleration depends on the injection thresholds, the relative magnitudes of the shock and turbulent speeds and on the magnitudes of the scattering mean free paths. The existence of suitable turbulence ahead of the shock is an essential requirement for shock acceleration (Section 3.2).

The chemical compositions of the energetic particles in space and their variability appears to be a problem of injection (Section 2.2). This follows from the fact that abundance variations are, generally, not accompanied by spectral variations. Both stochastic and shock acceleration mechanisms have injection thresholds, but the threshold of shock acceleration is expected to be higher than that of stochastic acceleration (Section 3). The abundance variations could be due to variations of the threshold energies, or of the preheating conditions. Preheating or pre-acceleration are required to produce particles above the thresholds.

Both stochastic and shock acceleration can be rapid enough to produce the observed rise times of the gamma-ray emission. We find that the necessary short scattering mean free paths are not inconsistent with other data.

The question of whether the ions responsible for gamma-ray production and those observed in interplanetary space are produced by the same mechanism remains unanswered. On the one hand, the energy spectra of these two populations appear to be similar, but on the other, the correlation between the number of particles seen in space and inferred at the Sun is quite poor. An important recent observation (Section 2.1) has been that of relativistic electrons seen in interplanetary space in correlation with gamma-ray emissions. Since stochastic and shock acceleration are not expected to accelerate relativistic electrons efficiently, this might be an indication for the acceleration of nuclei and electrons to essentially the same energies in electric fields (Section 3.3).

Both observational and theoretical work are required for future progress in the understanding of solar flare particle acceleration. On the observational side, much progress has recently been achieved by such novel investigations as gamma-ray and neutron observations, X-ray observations with high temporal resolution, and direct charged particle detections with good spectral elemental, isotopic, and ionic resolutions. These should be continued and correlated with each other as well as with other data on solar flares, such as shock parameters. For example, in energetic particle measurements there is a need for greater sensitivity. Some of the most bizarre and interesting flare events are the small ones; however, existing instrumentation cannot accumulate sufficient statistics to measure the composition in great detail. More sensitive gamma-ray observations are needed in order to determine whether prompt ion acceleration is present in all flares. Gamma-ray observations with high energy resolution are also needed to resolve the lines and measure Doppler shifts which would provide information on energetic particle beaming.

The theoretical and interpretative work making use of the existing observations is just beginning. We can foresee important advances from studies that will apply recent theoretical results on shock acceleration to solar flare acceleration, investigations of ion acceleration by electric fields leading to the beaming of the energetic particles which could have observable effects on gamma-ray spectra, and from a much more detailed study of acceleration theories which would consider the acceleration of particles and their effect on the accelerating agent in a self-consistent manner.

Acknowledgements

We wish to acknowledge W. I. Axford, G. A. Dulk, P. Evenson, D. C. Ellison, D. J. Forrest, M. L. Goldstein, J. R. Jokipii, R. P. Lin, R. E. McGuire, D. F. Smith, and M. A. I. Van Hollebeke for useful discussions, and D. V. Reames for providing the use of unpublished data. The research described in this paper was supported by NASA's Solar Terrestrial Theory Program. E. G. Zweibel wishes to thank the Aspen Center for Physics, where part of her research was carried out.

References

Abramowitz, M. and Stegun, I. A.: 1965, *Handbook of Mathematical Functions*, Dover Publications, New York.
Achterberg, A. and Norman, C. A.. 1980, *Astron. Astrophys.* 89, 353.
Achterberg, A.. 1981, *Astron. Astrophys.* 97, 259.
Anglin, J. D., Dietrich, W F., and Simpson, J. A.: 1977, *Proc. 15th Int. Cosmic Ray Conf., Plovdiv* 5, 43.

Asbridge, J. R., Bame, S. J., and Strong, I. B.: 1968, *J. Geophys. Res.* **73**, 5777.

Axford, W. I., Leer, E., and Skadron, G.: 1977, *Proc. 15th Int. Cosmic Ray Conf., Plovidiv* **11**, 132.

Axford, W. I.: 1982, in T. D. Tuyenne and T. Levy (eds), *Plasma Astrophysics*, ESA Publication SP-151.

Bai, T.: 1982, in R. E. Lingenfelter, H. S. Hudson, and D. M. Worrall (eds), *Gamma-Ray Transients and Related Astrophysical Phenomena*, Amer. Inst. of Physics, New York, p. 409.

Bai, T. and Ramaty, R.: 1976, *Solar Phys.* **49**, 343.

Bai, T. and Ramaty, R.: 1979, *Astrophys. J.* **227**, 1072.

Barbosa, D. D.: 1979, *Astrophys. J.* **233**, 383.

Barnes, G W. and Simpson, J. A.: 1976, *Astrophys. J.* **210**, L91.

Bertsch, D. L., Fichtel, C. E., and Reames, D. V.: 1969, *Astrophys. J.* **157**, L53.

Blandford, R. P. and Ostriker, J. P.: 1980, *Astrophys. J.* **237**, 793.

Brown, J. C.: 1971, *Solar Phys.* **18**, 489.

Brown, J. C. and Smith, D. F.: 1980, *Reports on Progress in Physics* **43**, 125.

Bryant, D. A., Cline, T. C., Desai, U. D., and McDonald, F. B.: 1965, *Astrophys. J.* **141**, 478.

Bulanov, S. V.: 1980, *Sov. Astron. Lett.* **6**, 206.

Bulanov, S. V. and Sasorov, P. V.: 1976, *Sov. Astron.* **19**, 464.

Burke, J. R. and Layzer, D.: 1969, *Astrophys. J.* **157**, 1169.

Canfield, R. C. and Cook, J. W.: 1978, *Astrophys. J.* **225**, 650.

Chen, G. and Armstrong, T. P.: 1975, *Proc. 14th Int. Cosmic Ray Conf., Munich* **5**, 1814.

Chubb, T. A.: 1970, in E. R. Dryer (ed.), 'Solar-Terrestrial Physics', *Proc. Int. Symp. on Solar-Terrestrial Physics, Leningrad*, Part 1, p. 99.

Chupp, E L.: 1982, in R. E. Lingenfelter, H. S. Hudson, and D. M. Worrall (eds), *Gamma-Ray Transients and Related Astrophysical Phenomena*, Amer. Inst. of Physics, New York, p. 363.

Chupp, E. L., Forrest, D. J., Higbie, P. R., Suri, A. N., Tsai, C., and Dunphy, P. P.: 1973, *Nature*, **241**, 333.

Chupp, E. L., Forrest, D. J., Ryan, J. M., Heslin, J., Reppin, C., Pinkau, K., Kanbach, G., Rieger, E., and Share, G. H.: 1982, *Astrophys, J.* **263**, L95.

Colgate, S. A.: 1978, *Astrophys. J.* **221**, 1068.

Colgate, S. A., Audouze, J., and Fowler, W. A.: 1977, *Astrophys. J.* **213**, 849.

Cook, W. R., Stone, E C., and Vogt, R. E.: 1980, *Astrophys. J.* **238**, L97.

Crannell, C. J., Frost, K. J., Matzler, C., Ohki, K., and Saba, J. L.: 1978, *Astrophys. J.* **223**, 670.

Datlowe, D.: 1971, *Solar Phys.* **17**, 436.

Datlowe, D. W. and Lin, R. P.: 1973, *Solar Phys.* **32**, 459.

Decker, R. B., Pesses, M. E., and Armstrong, T. P.: 1981, *Proc. 17th Int. Cosmic Ray Conf., Paris* **3**, 406.

Dietrich, W. F.: 1981, *Astrophys. J.* **245**, L41.

Dietrich, W. F. and Simpson, J. A.: 1979, *Astrophys. J.* **231**, L91.

Dietrich, W. F. and Simpson, J. A.: 1981, *Astrophys. J.* **245**, L41.

Drury, L. O'C., Axford, W. I., and Summers, D.: 1982, *Monthly Notices Roy. Astron. Soc.* **198**, 833.

Drury, L. O'C. and Volk, H. J.: 1981: *Astrophys. J.* **248**, 344.

Duggal, S. P.: 1979, *Reviews of Geophys. Space Phys.* **17**, 1021.

Dulk, G. A., Melrose, D. B., and Smerd, S. F.: 1978, *Proc. Astron. Soc. Australia* **3**, 243.

Eichler, D.: 1979, *Astrophys. J.* **229**, 419.

Eichler, D.: 1981, *Astrophys. J.* **244**, 711.

Ellison, D. C.: 1981a, *Geophys. Res. Lett.* **8**, 991.

Ellison, D. C.: 1981b, Ph.D. Thesis, Catholic Univ. of America, Washington, D.C.

Evenson, P., Meyer, P., and Yanagita, S.: 1981, *Proc. 17th Int. Cosmic Ray Conf., Paris* **3**, 32.

Fan, C. Y., Gloeckler, G., and Hovestadt, D.: 1975, *IAU Symp.* **68**, 411.

Fermi, E.: 1949, *Phys. Rev.* **75**, 1169.

Fichtel, C. E. and Guss, D. E.: 1961, *Phys. Rev. Lett* **6**, 495.

Fisk, L. A.: 1971, *J. Geophys. Res.* **76**, 1662.

Fisk, L. A.: 1978, *Astrophys. J.* **224**, 1048.

Fisk, L. A. and Axford, W. I.: 1968, *J. Geophys. Res.* **73**, 4396.

Fisk, L. A. and Lee, M. A.: 1980, *Astrophys. J.* **237**, 620.

Forman, M. A.: 1971, *J. Geophys. Res.* **76**, 759.

Forman, M. A.: 1981a, *Advances in Space Res.* 1, 41.

Forman, M. S.: 1981b, *Proc. 17th Int. Cosmic Ray Conf., Paris* 3, 467.

Forman, M. A. and Morfill, G.: 1979, *Proc. 16th Int. Cosmic Ray Conf., Kyoto* 5, 328.

Friedman, M.: 1969, *Phys. Rev.* 182, 1408.

Frost, K. J. and Dennis, B. R.: 1971, *Astrophys. J.* 165, 655.

Garrard, T. L. Stone, E. C., and Vogt, R. E.: 1973 in R. Ramaty and R. G. Stone (eds), *High Energy Phenomena on the Sun*, NASA SP-342, Washington, D. C., p. 341.

Geiss, J.: 1973, *Proc. 13th Int. Cosmic Ray Conf., Denver* 5, 3375.

Gloeckler, G.: 1979, *Rev. Geophys. Space Phys.* 17, 569.

Gloeckler, G., Sciambi, R. K., Fan, C. Y., and Hovestadt, D.: 1976, *Astrophys. J.* 209, L93.

Hasselmann, K. and Wibberenz, G.: 1968, *J. Geophys.* 34, 353.

Hayashi, T. and Sato, T.: 1978, *J. Geophys. Res.* 83, 217.

Heyvaerts, J.: 1981, in E. R. Priest (ed.), *Solar Flare Magnetohydrodynamics*, Gordon and Breach, New York, p. 429.

Holman, G. D., Ionson, J. A., and Scott, J. S.: 1979, *Astrophys. J.* 228, 576.

Hovestadt, D., Gloeckler, G., Höfner, H., Klecker, B., Ipavich, F. M., Fan, C. Y., Fisk, L. A., O'Gallagher, J. J., and Scholer, M.: 1981, *Astrophys. J.* 246, L81.

Hoyng, P., Brown, J. C., and Van Beek, H. F.: 1976, *Solar Phys.* 48, 197.

Hudson, H. S.: 1972, *Solar Phys.* 24, 414.

Hudson, H. S.: 1979, in J. Arons, C. McKee, and C. Max (eds), *Particle Acceleration Mechanisms in Astrophysics*, Amer. Inst. of Physics, New York, p. 115.

Hudson, H. S., Lin, R. P. and Stewart 1981

Ibragimov, I. A. and Kocharov, G. E.: 1977, *Sov. Astron. Lett.* 3 (5), 221.

Jokipii, J. R.: 1971, *Rev. Geophys. Space Phys.* 9, 27.

Kahler, S., Spicer, D., Uchida, Y., and Zirin, H.: 1980, in P. A. Sturrock (ed.), *Solar Flares*, Colorado Assoc. Univ. Press, Boulder, p. 83.

Kane, S. R., Crannell, C. J., Datlowe, D., Feldman, U., Gabriel, A., Hudson, H. S., Kundu, M. R., Mätzler, C., Neidig, D., Petrosian, V., and Sheelay, N. R. Jr,: 1980, in P. A. Sturrock (ed.), *Solar Flares*, Colorado Assoc. Univ. Press, Boulder, p. 187.

Kleckler, B., Hovestadt, D., Scholer, M., Gloeckler, G., and Ipavich, F. M.: 1982, *Trans. Amer. Geophys. Union* 63, 399.

Kocharov, G. E. and Kocharov, L. G.: 1978, *Proc. 10th Leningrad Symp. on Cosmic Physics*, A. F. Yoffe Physico-Technical Inst., Leningrad, p. 38.

Kozlovsky, B. and Ramaty, R.: 1977, *Astrophys. Lett.* 19, 19.

Kulsrud, R. M and Ferrari, A.: 1971, *Astrophys. Space Sci.* 12, 302.

Kurt, V. G., Logachev, Ju. I., Stolpovsky, V. G., and Daiborg, E. I.: 1981, *Proc. 17th Int Cosmic Ray Conf., Paris* 3, 69.

Lee, M. A.: 1982, *J. Geophys. Res.* 87, 5063.

Lee, M. A. and Fisk, L A.: 1981, *Proc. 17th Int. Cosmic Ray Conf., Paris* 3, 405 (and late papers).

Lee, M. A. and Fisk, L. A.: 1982, *Space Sci. Rev.* 32, 205.

Lin, R. P.: 1971, *Solar Phys.* 15, 453.

Lin, R P.: 1974, *Space Sci. Rev.* 16 189.

Lin, R P., Meng, C. I., and Anderson, K. A.: 1974, *J. Geophys. Res.* 79, 489.

Lin, R. P. and Hudson, H. S.: 1976, *Solar Phys.* 50, 153.

Lin, R. P., Mewaldt, R. A., and Van Hollebeke, M. A. I.: 1982, *Astrophys. J.* 253, 949.

Lingenfelter, R. E. and Ramaty, R.: 1967, in B. S. P. Shen (ed.), *High Energy Nuclear Reactions in Astrophysics*, Benjamin, New York, p. 99.

Levine, R. H.: 1974, *Astrophys. J.* 190, 447.

Luhmann, J. G .: 1976, *J. Geophys. Res.* 81, 2089.

Mason, G. M., Fisk, L. A., Hovestadt, D., and Gloeckler, G.: 1980, *Astrophys. J.* 239, 1070.

Ma Sung, L. S., Gloeckler, G., Fan, C. Y., and Hovestadt, D.: 1981, *Astrophys. J.* 245, L45.

McDonald, F. B.: 1981, *Proc. 17th Int. Cosmic Ray Conf., Paris* 13, 199.

McDonald, F. B., Fichtel, C. E., and Fisk, L. A.: 1974, in F. B. McDonald and C. E. Fichtel (eds), *High Energy Particles and Quanta in Astrophysics*, MIT Press, Cambridge, Mass, p. 212.

McDonald, F. B., Teegarden, B. J., Trainor, J. H., Von Rosenvinge, T. T., and Webber, W. R.: 1976, *Astrophys. J.* 203, L149.

McGuire, R. E., Von Rosenvinge, T. T., and McDonald, F. B.: 1979, *Proc. 16th Int. Cosmic Ray Conf., Kyoto* **5**, 61.

McGuire, R. E., Von Rosenvinge, T. T., and McDonald, F. B.: 1981, *Proc. 17th Int. Cosmic Ray Conf., Paris* **3**, 65.

Melrose, D. B.: 1974, *Solar Phys.* **37**, 353.

Melrose, D. B.: 1980, *Plasma Astrophysics,* Gordon and Breach, New York.

Mewaldt, R. A., Spalding, J. D., Stone, E. C., and Vogt, R. E.: 1979, *Astrophys. J.* **231**, L97.

Mewaldt, R. A., Spalding, J. D., Stone, E. C., and Vogt, R. E.: 1981, *Astrophys. J.* **243**, L163.

Meyer, J. P.: 1981, *Proc. 17th Int. Cosmic Ray Conf., Paris* **3**, 145, 149.

Mogro-Campero, A. and Simpson, J. A.: 1972, *Astrophys. J.* **171**, L5.

Mullan, D. J.: 1980, *Astrophys. J.* **237**, 244.

Mullan, D. J. and Levine, R. H.: 1981, *Astrophys. J. Suppl.* **47**, 87.

Ogilvie, K. W., Scudder, J. D., and Olbert, S.: 1978, *J. Geophys. Res.* **83**, 3776.

Palmer, I. A.: 1982, *Rev. Geophys. Space Phys.* **20**, 335.

Parker, E. N.: 1957, *J. Geophys. Res.* **62**, 509.

Parker, E. N.: 1979, *Cosmical Magnetic Fields,* Clarendon Press, Oxford.

Parker, E. N.: 1963, *Interplanetary Dynamical Processes,* Wiley, New York.

Parker, E. N. and Tidman, D. A.: 1958, *Phys. Rev.* **111**, 1206.

Pesses, M. E., Decker, R. B., and Armstrong, T. P.: 1982, *Space Science Rev.* **32**, 185.

Petschek, H. E.: 1964, in W. N. Hess (ed.), *Physics of Solar Flares,* NASA SP-50, Washington, D.C., p. 425.

Podosek, F. A.: 1978, *Ann. Rev. Astron. Astrophys.* **16**, 293.

Price, P. B., Hutcheon, I., Cowsik, R., and Barber, D. J.: 1971, *Phys. Rev. Lett.* **26**, 916.

Prince, T., Ling, J. C., Mahoney, W. A., Riegler, G. R., and Jacobson, A. S.: 1982, *Astrophys. J.* **255**, L81.

Ramaty, R.: 1969, *Astrophys. J.* **158**, 753.

Ramaty, R.: 1979, in J. Arons, C. McKee, and C. Max (eds), *Particle Acceleration Mechanisms in Astrophysics,* Amer. Inst. of Physics, New York, p. 135.

Ramaty, R. *et al.*: 1980, in P. A. Sturrock (ed.), *Solar Flares,* Colorado Assoc. Univ. Press, Boulder, p. 117.

Ramaty, R. and Crannell, R. C.: 1976, *Astrophys. J.* **203**, 766.

Ramaty, R. and Kozlovsky, B.: 1974, *Astrophys. J.* **193**, 729.

Ramaty, R., Kozlovsky, B., and Lingenfelter, R. E.: 1975, *Space Sci. Rev.* **18**, 341.

Ramaty, R., Kozlovsky, B., and Lingenfelter, R. E.: 1979, *Astrophys. J. Suppl.* **40**, 487.

Ramaty, R., Kozlovsky, B., and Suri, A. N.: 1977, *Astrophys. J.* **214**, 617.

Reames, D. V. and Von Rosenvinge, T. T.: 1981, *Proc. 17th Int. Cosmic Ray Conf., Paris* **3**, 162.

Richter, A. K. and Keppler, E.: 1977, *J. Geophys. Res.* **42**, 645.

Riechard, R. and Wibberenz, G.: 1974, *Solar Phys.* **36**, 473.

Rothwell, P. L.: 1976, *J. Geophys. Res.* **81**, 709.

Scholer, M.: 1976, *Astrophys. J.* **209**, L101.

Scholer, M.: Ipavich, F. M., Gloeckler, G., Hovestadt, D., and Klecker, B,: 1980, *Geophys. Res. Lett.* **7**, 73.

Sciambi, R. K., Gloeckler, G., Fan, C. Y., and Hovestadt, D.: 1977, *Astrophys. J.* **214**, 316.

Scudder, J. D. and Ogilvie, K. W.: 1979, *J. Geophys. Res.* **84**, 6603.

Shea, M. A. and Smart, D. F.: 1973, *Proc. 13th Int. Cosmic Ray Conf., Denver* **2**, 1548.

Simnett, G. M.: 1971, *Solar Phys.* **20**, 448.

Skilling, J. A.: 1975, *Monthly Notices Roy. Astron. Soc.* **172**, 557.

Smith, D. F.: 1979, in J. Arons, C. McKee, and C. Max (eds), *Particle Acceleration Mechanisms in Astrophysics,* Amer. Inst. of Physics, New York, p. 155.

Smith, D. F. and Auer, L. H.: 1980, *Astrophys. J.* **238**, 1126.

Sonnerup, B. U. O.: 1970, *J. Plasma Phys.* **4**, 161.

Sonnerup, B. U. O.: 1971, *J. Geophys. Res.* **76**, 8211.

Sonnerup, B. V. O.: 1973, in R. Ramaty and F. E. Stone (eds), *High Energy Phenomena on the Sun,* NASA SP-342, Washington, D.C., 357.

Speiser, T. W.: 1965, *J. Geophys. Res.* **70**, 4219.

Spicer, D. S.: 1982, *Space Sci. Revs.,* **31**, 351.

Stern, D. P. and Ness, N. F.: 1982, *Ann. Rev. Astron. Astrophys.* **20**, 139.

Sturrock, P. A.: 1980, in P. A. Sturrock (ed.), *Solar Flares*, Colorado Assoc. Univ. Press, Boulder, p. 411.

Suri, A. N., Chupp, E. L., Forrest, D. J., and Reppin, C.: 1975, *Solar Phys.* **43**, 414.

Svestka, Z. and Fritzova, L.: 1974, *Solar Phys.* **36**, 417.

Sweet, P. A.: 1958, *NOVO Cimento, Suppl.* **8**, 188.

Syrovatskii, S. I.: 1981, *Annual Revs. Astron. Astrophys.* **19**, 163.

Takakura, T.: 1960, *Publ. Astron. Soc. Japan* **12**, 325.

Teegarden, B. J., Von Rosenvinge, T. T., and McDonald, F. B.: 1973, *Astrophys. J.* **180**, 571.

Toptygin, I. N.: 1980, *Space Sci. Rev.* **26**, 157.

Tsurutani, B. T. and Rodriguez, P.: 1981, *J. Geophys. Res.* **86**, 4319.

Uchida, Y.: 1974, *Solar Phys.* **39**, 431.

Van Hollebecke, M. A. I.: 1979, *Rev. Geophys. Space Phys.* **17**, 545.

Van Hollebecke, M. A. I., MaSung, L. S., and McDonald, F. B.: 1975, *Solar Phys.* **41**, 189.

Vasyliunas, V. M.: 1975, *Rev. Geophys. Space Phys.* **13**, 308.

Von Rosenvinge, T. T., Ramaty, R., and Reames, D. V.: 1981, *Proc. 17th Int. Cosmic Ray Conf., Paris* **3**, p. 28.

Wentzel, D. G.: 1965, *J. Geophys. Res.* **70**, 2716.

Wentzel, D. G.: 1969, *Astrophys. J.* **157**, 545.

Wild, J. P., Smerd, S. F., and Weiss, A. A.: 1963, *Ann. Rev. Astron. Astrophys.* **1**, 291.

Zweibel, E. G. and Haber, D.: 1983, *Astrophys. J.* **264**, 648.

Zwickl, R. D., Krimigis, S. M., Carbary, J. F., Leath, E. P., Armstrong, T. P., Hamilton, D. C., and Gloeckler, G.: 1981, *J. Geophys. Res.* **86**, 8125.

M. A. Forman,
Dept of Earth and Space Sciences,
State University of New York,
Stony Brook, NY 11794,
U.S.A.

R. Ramaty,
Laboratory for High Energy
Astrophysics,
Goddard Space Flight Center,
Greenbelt, MD 20771,
U.S.A.

E. G. Zweibel,
Dept of Astro-Geophysics,
University of Colorado,
Boulder, CO 80309,
U.S.A.

CHAPTER 14

NUCLEAR PROCESSES IN SOLAR FLARES

R. RAMATY

1. Introduction

Nuclear reactions in solar flares take place between flare-accelerated protons and nuclei and the ambient solar atmosphere. Several reviews of the early work on this subject are available (Dolan and Fazio, 1965; Lingenfelter and Ramaty, 1967; Cheng, 1972). Using reasonably accurate and complete nuclear data, Lingenfelter and Ramaty (1967) have carried out a detailed calculation of the expected nuclear reaction rates in flares and predicted observable fluxes at Earth of the products of these reactions: gamma-ray lines, neutrons and nuclear fragments in the solar energetic particles.

Nuclear gamma-rays from solar flares were first observed by Chupp *et al.* (1973) with a NaI spectrometer flown on board the seventh Solar Orbiting Observatory (OSO-7). Gamma-ray lines at 0.51, 2.22, 4.44, and 6.13 MeV were observed from the August 4, 1972 flare. The lines at 0.51 and 2.22 MeV were also seen during the decay phase of the August 7, 1972 flare. These two flares were among the largest ever observed.

A considerable amount of theoretical and interpretative work has been carried out on the August 1972 observations (Ramaty and Lingenfelter, 1973; Reppin *et al.*, 1973; Wang and Ramaty, 1974; Forrest *et al.*, 1975; Ramaty, *et al.*, 1975; Wang, 1975; Kanbach *et al.*, 1975; Chupp, 1976; Lin and Hudson, 1976; Bai and Ramaty, 1976; Ramaty and Crannell, 1976; Crannell *et al.*, 1976; Kozlovsky and Ramaty, 1977; Ramaty *et al.*, 1977; Ibragimov and Kocharov, 1977; Lin and Ramaty, 1978; Crannell *et al.*, 1979; Ramaty, 1979; Ramaty *et al.*, 1980). These studies, together with additional gamma-ray data (Chupp *et al.*, 1975; Chupp, 1976; Suri *et al.*, 1975), energetic particle (Kohl *et al.*, 1973; Webber *et al.*, 1975), hard X-ray (Van Beek *et al.*, 1973), and microwave (Croom and Harris, 1973) data have led to the following broad outline of the origin and implications of solar gamma-rays.

Gamma-ray lines from solar flares result from the interaction of flare-accelerated protons and nuclei with the solar atmosphere. The emission measure of flare-heated material − i.e. the density squared of the hot gas times its volume which is determined from X-ray observations − is insufficient to produce any measurable amount of nuclear burning. The accelerated particles, on the other hand, produce neutrons, positrons, π mesons, radioactive nuclei, and excited nuclear levels, whose captures, annihilations, decays, and de-excitations lead to observable gamma-ray lines.

The strongest line in solar flares is at 2.223 MeV from neutron capture on hydrogen. The neutrons are produced mainly from the disintegration of ^{4}He and heavier nuclei and

Peter A. Sturrock (ed.), Physics of the Sun, Vol. II, pp. 291–323.

occasionally in proton–proton collisions resulting in high-energy neutrons and π mesons. High-energy neutrons could be directly detected near Earth (Lingenfelter *et al.*, 1965; Lingenfelter and Ramaty, 1967) and reports of their observation have just become available (Chupp *et al.* 1982).

The site of the nuclear reactions in the solar atmosphere is as yet unknown. Nevertheless, calculations (Wang and Ramaty, 1974) indicate that the bulk of neutrons with initial velocity vectors pointing towards the photosphere are thermalized in the photosphere and subsequently captured on either ^{1}H or ^{3}He. Capture on ^{1}H produces the 2.223 MeV gamma-ray line, but capture on ^{3}He results in tritium without emitting photons. The removal time of thermal neutrons from the photosphere, on the order of 1 min, can be measured by comparing the time dependence of the intensity of the 2.223 MeV line to that of a prompt nuclear line (see below). This removal time depends on the photospheric ^{3}He abundance, as does the capture probability on ^{1}H which determines the flux of the 2.223 MeV line. Gamma-ray line observations, therefore, can measure the abundance of ^{3}He in the photosphere. Because the 2.223 MeV line is formed at a larger depth in the solar atmosphere than the prompt gamma-ray lines, for flares close to the limb of the Sun, the 2.223 MeV line is substantially attenuated relative to the prompt emissions.

A variety of prompt gamma-ray lines are produced from nuclear de-excitations. The most important discrete lines (Ramaty *et al.*, 1975, 1979) are at 6.129 MeV from ^{16}O, 4.438 MeV from ^{12}C, 2.313 MeV from ^{14}N, 1.779 MeV from ^{28}Si, 1.634 MeV from ^{20}Ne, 1.369 MeV from ^{24}Mg, 1.238 MeV and 0.847 MeV from ^{56}Fe, all produced primarily by direct excitation of these nuclei, and at two lines, 0.478 MeV from ^{7}Li and at 0.431 MeV from ^{7}Be, which result from fusion reactions, ^{4}He$(\alpha, p)^7$Li* and ^{4}He$(\alpha, n)^7$Be*. The role of these fusion reactions for producing gamma-ray lines in astrophysics was first pointed out by Kozlovsky and Ramaty (1974). Nuclear de-excitations also produce Doppler broadened lines which, together with many unresolved lines, produce a significant gamma-ray continuum, in particular in the 4–7 MeV region (Ramaty *et al.*, 1977; Ibragimov and Kocharov, 1977).

Because of the short lifetimes of most excited nuclear levels, nuclear de-excitation radiation is an excellent tracer of the nuclear interaction rate of energetic particles in solar flares. This rate is directly proportional to the instantaneous number of accelerated particles in the interaction region which, in turn, is determined by the acceleration mechanism and the losses suffered by the particles. Through line shapes and Doppler shifts, prompt nuclear de-excitation lines also give information on the geometry of the interacting energetic particle beam (Ramaty and Crannell, 1976; Kozlovsky and Ramaty, 1977).

The interaction models of energetic protons and nuclei in solar flares can be crudely classified as thin- and thick-target models (e.g. Ramaty *et al.*, 1975). In the thin-target model the nuclear reactions are produced by energetic particles which escape from the interaction region at the Sun. These particles can be detected in the interplanetary medium. Furthermore, if sufficient thin-target nuclear reactions take place, their fragmentation products should also be detectable. On the other hand, in the thick-target model the nuclear reactions are produced by particles as they slow down in the solar atmosphere. These particles, and their spallation products, become thermalized and mixed back into the solar atmosphere. Gamma-ray lines and neutrons can, nevertheless, be seen from thick-target interactions.

The ratio of the flux in a prompt nuclear component (e.g. the 4.44 MeV line) to the flux in the 2.223 MeV line depends on the interaction model and on the energy spectrum of the accelerated particles. For a given energetic particle spectrum, this ratio is larger for the thin-target than for the thick-target model. Using this result and information on the energy spectrum derived from interplanetary particle observations (Van Hollebeke et al., 1975), Ramaty (1979) suggested that the gamma rays from the August 4, 1972 flare were produced predominantly in thick-target interactions. From the absolute fluxes of the gamma-ray lines, it is possible to deduce the energy deposited by the accelerated nuclei in the solar atmosphere. For the August 4, 1972 flare the energy deposited by the protons and nuclei amounts to several percent of the energy deposited by the electrons which make the impulsive hard X-rays (Lin and Hudson, 1976). Nevertheless, protons could deposit their energy in regions which are not accessible to electrons because they have a longer stopping range in the ambient medium than the <100 keV electrons.

Positrons in solar flares result from the decay of π^+ mesons and various radioactive nuclei produced by the nuclear reactions. The half-lives of the important positron emitters range from values less than 1 s to over 20 min and they produce positrons of energies from about 0.1 MeV to several tens of MeV. After their production, the positrons are decelerated to energies less than several hundred eV where they annihilate. The deceleration is due to interactions with the ambient solar atmosphere, and hence the deceleration time depends on the density and magnetic field of the medium in which the positrons annihilate. The positrons can annihilate with free electrons to produce two 0.511 MeV gamma-rays per annihilation, or they may form a positronium atom. This atom is similar to the hydrogen atom except that the proton is replaced by a positron. Positronium atoms also annihilate into gamma rays: 25% of the annihilations are from the singlet spin state producing two 0.511 MeV photons, and 75% of them are from the triplet state which anniliates into three photons of energies less than 0.511 MeV. Triplet positronium can annihilate before it is broken up by collisions only if the density of the ambient medium is less than about 10^{15} cm^{-3}. Observation of the characteristic 3-photon positronium continuum, therefore, would provide information on the density of the annihilation site (Crannell et al., 1976). Information on the temperature of this site would be obtained from the measurement of the width of the 0.511 MeV line. If the positrons annihilate below the transition layer — i.e. at temperatures less than 10^5 K — the width the 0.511 MeV line should be less than 3.5 keV.

Following the OSO-7 observations, solar gamma-ray lines were seen with the NaI spectrometer on the first High Energy Astrophysical Observatory (HEAO-1) (Hudson et al., 1980), with the NaI spectrometer on the Solar Maximum Mission (SMM) (Chupp et al., 1981; Chupp, 1982), and with the Ge spectrometer on HEAO-3 (Prince et al., 1982). The two HEAO detectors, designed to detect cosmic gamma-rays, observed solar gamma-ray lines through their shields and thus were sensitive to only the strongest one or two lines. The HEAO detectors are no longer operational. The SMM spectrometer has already seen line emission from many solar flares and thus demonstrated that nuclear reactions of flare-accelerated particles take place commonly in solar flares. At the time of this writing (early 1982) the detector is still operational and should continue to observe for several more years. The ten solar flares, from which gamma-ray lines were definitely seen so far, are discussed in the present paper.

In addition to gamma-ray lines and neutrons, nuclear reactions also produce energetic nuclear fragments (e.g. ^{2}H, ^{3}H, ^{3}He, Li, Be, B). The first attempt to measure the ^{3}He abundance in energetic solar particles was made by Schaeffer and Zahringer (1962). By using mass spectroscopy of material from the Discoverer 17 satellite, these authors found a ^{3}He/^{4}He ratio of ~0.2 for the November 12, 1960 flare. Subsequent measurements (Hsieh and Simpson, 1970; Anglin et al., 1973a; Dietrich, 1973; Garrard et al., 1973) have revealed the existence of a class of solar particle events in which the ^{3}He/^{4}He ratio is substantially larger than in the ambient solar atmosphere.

Enrichments of ^{3}He in energetic particle populations (for example, the galactic cosmic rays) have been generally attributed to nuclear reactions between the energetic particles and the ambient medium. But, as first pointed out by Garrard et al. (1973), this interpretation of the solar ^{3}He enrichments, in its simplest form, is inconsistent with much of the ^{3}He data. If the ^{3}He enrichments are due to nuclear reactions of the energetic particles, then they should be accompanied by similar enrichments in ^{2}H and, to a lesser degree, in ^{3}H. Such enrichments, however, are not observed.

Several schemes have been proposed to overcome this difficulty. These rely on the kinematics and angular distributions of the reaction products which favor the preferential escape of ^{3}He (Ramaty and Kozlovsky, 1974; Rothwell, 1976) and the thermonuclear destruction of ^{2}H and ^{3}H in a model in which the energetic products of the nuclear reactions are confined to thin filaments and interact with each other (Colgate et al., 1977). But, as proposed by Fisk (1978), the enhanced ^{3}He abundance in solar energetic particles could be due to preferential heating and acceleration of ambient atmospheric ^{3}He. The observation of energetic ^{3}He, therefore, cannot be used as indication for nuclear reactions unless the ^{3}He is accompanied by at least some other fragmentation product. No convincing observations of such products have yet been reported.

Nuclear reactions of accelerated particles could cause modifications of solar surface isotopic abundances. Thus, from the analysis of lunar surface material, Kerridge (1975) found a secular increase of the solar wind ^{15}N/^{14}N ratio, and Fireman et al. (1976) reported a measurable ^{14}C abundance on the lunar surface which they believe should be due to implantation by the solar wind. Even though solar surface nuclear reactions could, in principle, produce these isotopes, it is unlikely that this has indeed happened, because the necessary nuclear reaction rates on the ancient Sun would have to exceed the present rate (determined by the gamma-ray observations) by many orders of magnitude (Kerridge et al., 1977).

Thus, the only convincing evidence to-date for nuclear reactions in the solar atmosphere are the gamma-ray line and neutron observations. The observed line energies and line ratios are fully consistent with nuclear reactions produced by particles of energies in excess of several MeV. As already mentioned, thermonuclear burning makes no measurable contribution to these reactions. In Section 2 we review the interaction models for the production of gamma-rays in energetic particle reactions and present new calculations of the production rates. In Section 3 we review the observational data on gamma-ray lines, we compare them with theory and we discuss their implications. Because the neutron observations are still very preliminary, we defer their analysis to future studies. We summarize our conclusion in Section 4.

2. Nuclear Reactions in Solar Flares

In this section we first summarize the formalism of the thin- and thick-target interaction models used in calculating nuclear reaction yields in solar flares, and we discuss the energetic particle spectra and compositions that we use in these calculations. We then evaluate, for the various models, the neutron and 2.223 MeV photon productions, the positron and 0.511 MeV photon productions, and the productions of the various prompt nuclear de-excitation lines.

2.1. INTERACTION MODELS AND PROPERTIES OF THE ENERGETIC PARTICLES

We consider first the thin-target model. Here nuclear reactions take place between accelerated particles and a cold ambient medium in an interaction volume from which particles escape with negligible energy loss and with an escape probability which is energy independent and the same for all types of particles. Let $N_j(E, t) \, dE$ be the instantaneous number of energetic particles of type j in the volume having energies per nucleon in dE around E, and n_i the density of ambient particles of type i. The instantaneous reaction rate between accelerated particles of type j and ambient particles of type i is given by

$$q_{ij}(t) = \int_0^\infty c n_i N_j(E, t) \beta(E) \sigma_{ij}(E) \, dE, \qquad (1)$$

where $c\beta(E)$ is the particle velocity and $\sigma_{ij}(E)$ is the energy dependent cross-section of the reaction considered.

In the thin-target model the energetic particles that escape from the interaction region can, in principle, be observed in the interplanetary medium by detectors on spacecraft. The following relationship exists between the interplanetary particles, $N_{esc,j}(E)$, and the instantaneous number of particles in the thin-target interaction volume:

$$N_{esc,j}(E) = T^{-1} \int N_j(E, t) \, dt. \qquad (2)$$

Here T is the escape time from the volume and the time integral is over the duration of the nuclear interactions. By integrating Equation (1) over time, by substituting Equation (2), and by summing over all i and j that contribute to a particular reaction product (e.g. neutrons, positrons, excited levels), we obtain the time-integrated nuclear reaction yield of that product in the thin-target model:

$$Q = n_H T c \sum_{ij} (n_i/n_H) \int_0^\infty N_{esc,j}(E) \beta(E) \sigma_{ij}(E) \, dE. \qquad (3)$$

As opposed to the thin-target model, in which the particles escape from the interaction region and can be observed in space, the thick-target model is one in which the particles produce nuclear reactions as they slow down in the solar atmosphere. In this model, the ambient density in the interaction region is expected to be quite high (i.e. the region could be close to or even inside the photosphere). Since particles, in general, are not accelerated in high-density regions, it is reasonable to assume that in the thick-target model the acceleration takes place outside the interaction volume. Thus, let $\bar{N}_j(E) \, dE$ be the time-integrated number of particles of type j with E in dE incident on the interaction region. The time-integrated nuclear reaction yield can then be written as

$$Q = m_p^{-1} \sum_{ij} (n_i/n_H) \int_0^\infty \bar{N}_j(E)\, dE \int_0^E \sigma_{ij}(E') \left(\frac{dE'}{dx}\right)_j^{-1} dE', \qquad (4)$$

where m_p is the mass of the proton and $(dE/dx)_j$ is the energy loss rate per unit path length (measured in $(\text{MeV nucleon}^{-1})/(\text{g cm}^{-2})$) of accelerated particle j in the ambient solar atmosphere. In the present calculations we take $(dE/dx)_j$ equal to the energy loss rates of charged particles in a neutral medium,

$$\left(\frac{dE}{dx}\right)_j \simeq \left(\frac{dE}{dx}\right)_{j,H} \left[1 + \frac{n_{He}}{n_H} \frac{m_{He}}{m_p} \frac{(dE/dx)_{j,He}}{(dE/dx)_{j,H}}\right]. \qquad (5)$$

Here $(dE/dx)_{j,H}$ and $(dE/dx)_{j,He}$ are the energy loss rates of particle j in H and He, respectively, and m_{He} is the mass of ^4He. For the abundances of Table I, the term in the square brackets is approximately 1.13 and essentially independent of energy. Using the tabulations of Barkas and Berger (1964) and Northcliffe and Schilling (1970), $(dE/dX)_{j,H}$ can be approximated by

$$\left(\frac{dE}{dX}\right)_{j,H} \simeq (Z_{eff}^2/A)_j\, 630E^{-0.8}\ (\text{MeV nucleon}^{-1})/(\text{g cm}^{-2}), \qquad (6)$$

where (Pierce and Blann, 1968)

$$Z_{eff} = Z\,[1 - \exp(-137\beta/Z^{2/3})], \qquad (7)$$

and Z and A are the atomic and mass number of particle j. For computational purposes, it is convenient to invert the order of integration in Equation (4),

$$Q = m_p^{-1} \sum_{ij} (n_i/n_H) \int_0^\infty dE\, \sigma_{ij}(E)\, (dE/dx)_j^{-1} \int_E^\infty dE'\, \bar{N}_j(E'). \qquad (8)$$

A detailed discussion of solar energetic particle spectra and compositions based on interplanetary observation are given by Forman, Ramaty, and Zweibel in Chapter 13 of this work. Various forms of accelerated particle spectra have been used in previous treatments of nuclear reactions in solar flares (Lingenfelter and Ramaty, 1967; Ramaty et al., 1975; Ramaty, 1979). These are power laws in kinetic energy,

$$N_{esc,j}(E) \text{ or } \bar{N}_j(E) = \begin{cases} B_j E^{-s}, & E > E_c \\ B_j E_c^{-s}, & E < E_c \end{cases}, \qquad (9)$$

exponentials in rigidity,

$$N_{esc,j}(E) \text{ or } \bar{N}_j(E) = B_j \exp(-R_j/R_0)\, dR_j/dE, \qquad (10)$$

TABLE I

Elemental and isotopic abundances

Isotope (1)	Ambient particles (2)	Energetic particles (3)
^{1}H	1.00	1.00
^{4}He	0.07	0.15
^{12}C	4.15×10^{-4}	1.07×10^{-3}
^{13}C	4.64×10^{-6}	1.28×10^{-5}
^{14}N	9.00×10^{-5}	2.14×10^{-4}
^{15}N	3.46×10^{-7}	8.57×10^{-7}
^{16}O	6.92×10^{-4}	2.14×10^{-3}
^{18}O	1.38×10^{-6}	4.28×10^{-6}
^{20}Ne	9.00×10^{-5}	2.14×10^{-4}
^{22}Ne	1.00×10^{-5}	2.57×10^{-5}
^{23}Na	2.28×10^{-6}	4.28×10^{-5}
^{24}Mg	3.11×10^{-5}	6.42×10^{-4}
^{25}Mg	4.01×10^{-6}	8.14×10^{-5}
^{26}Mg	4.43×10^{-6}	8.49×10^{-5}
^{27}Al	3.18×10^{-6}	5.35×10^{-5}
^{28}Si	3.46×10^{-5}	6.42×10^{-4}
^{29}Si	1.80×10^{-6}	3.21×10^{-5}
^{30}Si	1.18×10^{-6}	2.14×10^{-5}
^{32}S	1.80×10^{-5}	1.07×10^{-4}
^{34}S	7.61×10^{-7}	4.71×10^{-6}
^{36}Ar	3.39×10^{-6}	2.14×10^{-5}
^{38}Ar	6.23×10^{-7}	4.28×10^{-6}
^{40}Ca	2.28×10^{-6}	4.28×10^{-5}
^{52}Cr	4.15×10^{-7}	2.14×10^{-5}
^{54}Fe	1.94×10^{-6}	6.85×10^{-5}
^{56}Fe	3.11×10^{-5}	1.07×10^{-3}
^{57}Fe	7.61×10^{-7}	2.57×10^{-5}
^{58}Ni	1.25×10^{-6}	2.14×10^{-5}
^{60}Ni	4.84×10^{-7}	8.57×10^{-6}

Note: Two sets of energetic particle abundances are discussed in the text:
EP1 – Energetic particle abundances given in column (3).
EP2 – Energetic particles with the same abundances as the ambient medium, column (2).

and Bessel functions,

$$N_{esc,j}(E) \text{ or } \bar{N}_j(E) = B_j K_2 \left[2(3p/(m_p c\alpha T))^{1/2} \right].$$ (11)

In these expressions the B_j are proportional to the abundances of energetic particles j, $R_j = (A/Z)_j p$ is particle rigidity, $p = \sqrt{E(E + 2m_p c^2)}$ is particle momentum per nucleon, and K_2 is the modified Bessel function of order 2 (e.g. Abramovitz and Stegun, 1966). The parameters, s and E_c for power laws, R_0 for the exponential in rigidity, and αT for the Bessel function, characterize the spectrum of the energetic particles.

Recent observations of energetic protons and α-particles on spacecraft (McGuire *et al.*, 1981) indicate that power laws cannot fit the observed energy spectra of these particles. This result is consistent with earlier studies (Freier and Webber, 1963) which have shown that exponentials in rigidity provide a better fit to the solar proton data than do power laws in kinetic energy. Therefore, we shall present new calculations of nuclear reaction yields only for the exponential spectrum of Equation (10) and the Bessel function spectrum of Equation (11). Calculations for power laws can be found in previous papers (Ramaty *et al.*, 1975, 1977).

In Figure 1 we show the energy spectra of protons and α-particles observed by detectors on the IMP-8 spacecraft from the June 7, 1980 flare (R. E. McGuire, private communication, 1981). This flare is one of the best studied gamma-ray flares observed by the SMM spectrometer (Chupp *et al.*, 1981). As can be seen, the curves in Figure 1, given by $dj/dE \propto \beta K_2 [2(3p/(m_p c\alpha T))^{1/2}]$, with $\alpha T = 0.015$, provide a very good fit to the data. The June 7, 1980 flare is the only gamma-ray flare observed to-date for which good interplanetary energetic particle spectra have been reported. But even for this flare it is not entirely clear that all the interplanetary particles shown in Figure 1 were produced by the same flare as the one which produced the gamma-ray lines because of evidence for multiple injection of particles from the Sun (T. von Rosenvinge, private communication, 1981).

Equation (11) was shown (M. Lee, private communication, 1978; Ramaty, 1979; see also Chapter 13 of this work) to be the solution of a Fokker–Planck equation for stochastic Fermi acceleration with acceleration efficiency coefficient, α, and escape time, T, which are independent of particle energy and particle charge. While there is no guarantee that such a simple acceleration model is appropriate for solar flares, we feel that the spectrum of Equation (11) can be adequately used for the analysis of the presently available gamma-ray data, particularly since this spectrum also provides a good fit to the observed interplanetary particle energy spectra, as can be seen from Figure 1 and from the more detailed analysis of McGuire *et al.* (1981).

The particle abundances that we use in the present calculation are given in Table I. The ambient medium abundances (Cameron, 1982) are listed in the second column. From interplanetary observations, it is well known that the energetic particle abundances vary substantially from one flare to another (e.g. Chapter 13 of this work). In the present calculations we use two sets of energetic particle abundances. The first set, denoted by EP1, is given in the third column of Table I, and the second set, EP2, is identical to the ambient medium abundances of the second column. The energetic particles, EP1, are significantly richer in heavy elements than EP2.

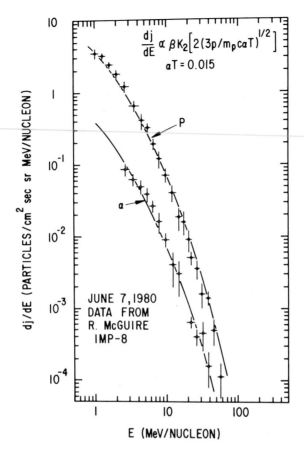

Fig. 1. Interplanetary protons and α-particles observed on June 7, 1980 (R. McGuire, private communication, 1980). The curves are from Equation (11) with the same αT for protons and α-particles.

2.2. NEUTRON AND 2.223 MeV PHOTON PRODUCTION

The strongest line observed in nearly all gamma-ray flares is that at 2.223 MeV from neutron capture on hydrogen, $^1H(n, \gamma)^2H$. Several theoretical studies have been made of neutron production in solar flares (Lingenfelter et al., 1965; Lingenfelter and Ramaty, 1967; Ramaty et al., 1975) and of 2.223 MeV photon production from the capture of these neutrons in the solar atmosphere (Wang and Ramaty, 1974; Kanbach et al., 1975). The neutron production cross-sections have been discussed in considerable detail by Ramaty et al. (1975). These cross-sections have recently been updated (B. Kozlovsky and R. E. Lingenfelter, private communication, 1981) with the addition of many new reactions that involve all the isotopes listed in Table I. These new cross-sections, to be published elsewhere, are used in the present calculations.

The calculated neutron production yields, Q_n, in the thin- and thick-target models are shown in Figure 2. These calculations are normalized such that the number of escaping protons of energies greater than 30 MeV, $N_{esc,p}(>30 \text{ MeV})$, and the number of protons

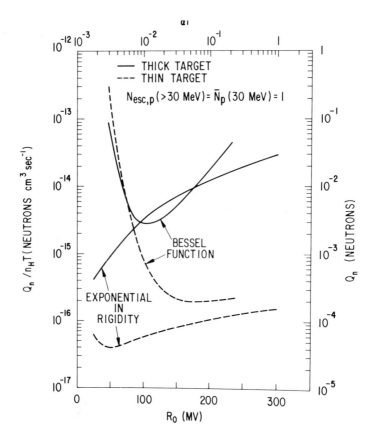

Fig. 2. Neutron yields in the thin-target and thick-target models calculated with the energetic-particle spectra of Equation (10) (exponential in rigidity) and Equation (11) (Bessel function). The ambient medium and energetic particle compositions are from Table I. $N_{esc, p}(>E)$ and $\bar{N}_p(>E)$ are, respectively, the numbers of protons of energies greater than E that escape from the thin-target region or are incident on the thick-target. $n_H T$ is the product of the hydrogen density in the thin-target region and the particle escape time from this region. In the thick-target calculations the particle energy loss rate is given by Equation (6), appropriate for a neutral medium.

incident on the thick target above the same energy, $\bar{N}_p(>30\ \mathrm{MeV})$, are both equal to unity. The energetic particle abundances are given by EP1 (see Table I). For relatively flat energetic particle spectra, corresponding to large values of R_0 or αT, the bulk of the neutrons are produced in reactions between protons and α-particles. For steep particle spectra, given by the Bessel function at small αT, the large neutron yields result from reactions between α-particles and heavy nuclei. These large neutron yields are absent for the rigidity spectra because particles with $Z \geq 2$ have lower energies per nucleon at the same rigidity and, hence, produce less nuclear reactions. These effects were discussed in more detail by Ramaty et al. (1975) in connection with the comparison of neutron production by particles with spectra that were either power laws in energy or exponentials in rigidity.

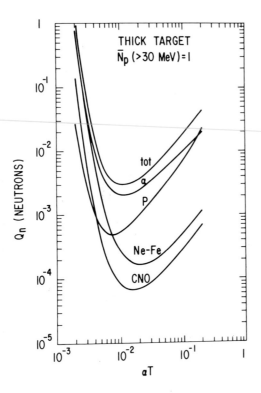

Fig. 3. Partial neutron production rates in the thick-target model for Bessel function energetic particle spectra. The various curves give the neutron production by energetic protons (P), α-particles (α), C, N, and O nuclei (CNO) and nuclei from Ne through Fe (Ne–Fe) interacting with the ambient medium. All the other parameters are as in Figure 2.

In Figure 3 we show partial neutron production rates in the thick-target model for energetic particle spectra given by Equation (11) and abundances again given by EP1. The curves labelled p, α, CNO and Ne–Fe represent, respectively, neutron yields of energetic protons, α-particles, C, N and O nuclei, and nuclei from Ne through Fe interacting with the ambient solar atmosphere. As can be seen, except for the very steep particle spectra (small αT) the neutrons result mostly from α-particles and protons. For very flat spectra, the neutron yield of protons includes an important contribution from the reaction $p + p \rightarrow (p + n) + \pi^+$.

In Figure 4 we show the energy in accelerated particles required to produce one neutron in the thick-target model. Here W is defined by

$$W = \sum_j A_j \int_E^\infty dE' \, E' \bar{N}_j(E'),$$ (12)

where $\bar{N}_j(E)$ is given by Equation (11) and the abundances are given by EP1. As can be seen from Figure 4, for $\alpha T = 0.015$, appropriate for the June 7, 1980 flare (see Figure 1), more than half of the accelerated particle energy is contained in particles of energies

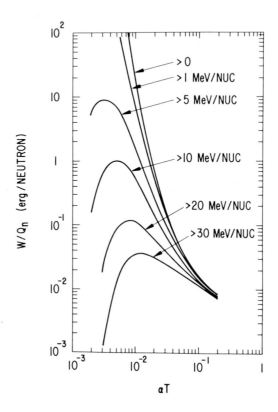

Fig. 4. Energy deposition per neutron produced in a thick-target model for Bessel function energetic particle spectra. The various curves give the energy deposited by particles of energies greater than the indicated values. The compositions of the ambient medium and energetic particles are given in Table I and the energetic particles are slowing down in a neutral medium.

greater than 1 MeV nucleon^{-1} and this fraction increases with increasing αT. A much lower fraction of the energy content, however, resides in particles of higher energies.

The values of W/Q_n given in Figure 4 should be considered as lower limits because the particle energy loss rate (Equation (6)) is valid only for slowing down of test particles in a neutral medium. In a fully ionized medium, the slowing-down rate of test particles is larger by about a factor of 3 (e.g. Ramaty, 1979). Furthermore, collective effects, such as the generation of an induced magnetic field by a beam of particles (Hoyng *et al.*, 1976; Colgate, 1978) would increase the energy required for the production of a given amount of neutrons.

The 2.223 MeV gamma-ray line is formed by neutron capture of ^{1}H in the photosphere. To study this line formation, Wang and Ramaty (1974) have carried out a detailed Monte Carlo simulation in which a distribution of neutrons was released above the photosphere, and the path of each neutron after its release was followed. For isotropic neutron release, any initially upward moving neutron escapes from the Sun. Some of the downward moving neutrons can also escape after being backscattered elastically by ambient protons, but most of these neutrons either are captured or decay at the Sun.

Because the probability for elastic scattering is much larger than the capture probability, the majority of the neutrons are thermalized before they get captured. Since the thermal speed in the photosphere (where most of the captures take place) is very much smaller than the speed of light, the energy of the gamma rays is almost exactly 2.223 MeV (Taylor et al., 1967), and the Doppler-broadened width of the line is very small (≤ 100 eV) (Ramaty et al., 1975). The energy of the line observed with the high-resolution Ge detector on board HEAO-3 from the November 9, 1979 flare is 2.225 ± 0.002 MeV (Prince et al., 1982).

The bulk of neutrons at the Sun are captured either on ^{1}H or on ^{3}He. Whereas capture on ^{1}H yields a 2.223 MeV photon, capture on ^{3}He proceeds via the radiationless reaction ^{3}He(n, p)^{3}H and hence produces no photons. The cross-sections for capture on ^{1}H and ^{3}H are $2.2 \times 10^{-30} \ \beta_n^{-1}$ cm^2 and $3.7 \times 10^{-26} \ \beta_n^{-1}$ cm^2, respectively, where $c\beta_n$ is the velocity of the neutron (for details see Wang and Ramaty, 1974). Thus, if the ^{3}He/H ratio in the photosphere is $\sim 5 \times 10^{-5}$, comparable to that observed in the solar wind (Geiss and Reeves, 1972) and in the chromosphere (Hall, 1975), nearly equal numbers of neutrons are captured on ^{3}He as on H.

The results of the Monte Carlo calculations of Wang and Ramaty (1974) are presented in Figure 5. In these calculations an isotropic distribution of monoenergetic neutrons of

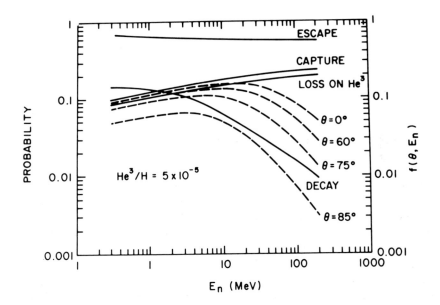

Fig. 5. Probabilities for neutron escape, decay, capture on protons and loss on ^{3}He in the solar atmosphere (solid lines); and photon yields per neutron (dashed lines). The parameter θ is the angle between the Earth–Sun line and the heliocentric radial direction through the flare. The ratio ^{3}He/H is the photospheric ^{3}He abundance, and E_n is the energy of the neutrons. The initial neutrons are assumed to be released isotropically above the photosphere. (From Wang and Ramaty, 1974.)

energy E_n is released above the photosphere. The solid lines are the probabilities for the various indicated processes. As can be seen, the capture and loss probabilities increase with increasing energy, because higher energy neutrons penetrate deeper into the photo-sphere. This reduces their escape probability and leads to a shorter capture time, thereby reducing the decay probability. The probability for loss on ^{3}He almost equals the capture probability on protons. The escape probability is greater than 0.5, because all initially upward moving neutrons were assumed to escape from the Sun. Note that the sum of all probabilities equals 1.

The dashed curves in Figure 5 are neutron-to-photon conversion coefficients evaluated (Wang and Ramaty, 1974) for specific emission angles θ between the Earth–Sun line and the vertical to the photosphere and given neutron energies. At low neutron energies and θ near zero, $f_{2.2}$ is close to the capture probability on protons. This means that gamma-rays from low-energy neutrons observed close to the vertical escape essentially unattenuated from the Sun. At higher energies and at larger angles, however, there is significant attenuation of the gamma-rays due to Compton scattering in the photosphere. Therefore, for flares close to the limb of the Sun, the 2.223 MeV line should be strongly attenuated in comparison with other nuclear de-excitation lines which are likely to be produced at higher altitudes in the solar atmosphere than the 2.223 MeV line. As we shall see in Section 3.1, this limb darkening is clearly seen in the SMM data.

The time integrated flux, or fluence, of 2.223 MeV photons at Earth resulting from neutron capture at the Sun can be written as

$$\phi(2.223 \text{ MeV}) = Q_n \bar{f}_{2.2}/(4\pi d^2), \qquad (13)$$

where $d = 1$ AU and $\bar{f}_{2.2}$ is the neutron-to-2.223 MeV photon conversion coefficient averaged over the neutron energy spectrum. For $\theta = 0$, Ramaty et al. (1975) find that $\bar{f}_{2.2}$ ranges from about 0.1 to 0.14, depending on the neutron energy spectrum. However, in the thick-target model, the neutron angular distribution is probably not isotropic and the neutrons could be produced in the photosphere (Kanbach et al., 1975). These effects should increase $\bar{f}_n$ by as much as a factor of 2. In our subsequent discussion we denote the conversion coefficient for vertical escape by $\bar{f}_{2.2}(0)$. The SMM observations now definitely justify new and more accurate calculations of neutron capture in the photosphere. These have not yet been carried out. Nevertheless, using the presently available calculations, we believe that $\bar{f}_{2.2}(0)$ should not exceed 0.3.

2.3. POSITRON AND 0.511 MeV PHOTON PRODUCTION

The 0.511 MeV gamma-ray line resulting from positron annihilation has been observed from several solar flares. Positrons in flares are produced in energetic particle interaction with the ambient solar atmosphere. A number of theoretical studies have been made of positron production in such interactions (Lingenfelter and Ramaty, 1967; Ramaty et al., 1975) and the positron slowing-down and annihilation (Crannell et al., 1976; Bussard et al., 1979). In the present paper we give the results of new calculations (B. Kozlovsky and R. E. Lingenfelter, private communication, 1981) of positron production based on a large number of β^+ emitters produced in nuclear reactions that involve all the isotopes listed in Table I. The results are given in Figure 6, where we show the ratio Q_+/Q_n for the thin- and thick-target models. Here Q_+ is calculated from Equations (3) and (8), respectively, with energetic particles given by Equation (11). The positron yields

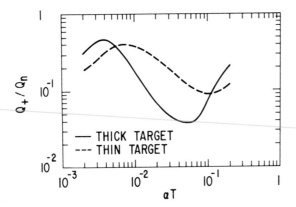

Fig. 6. The ratio of the positron yield, Q_+, to the neutron yield, Q_n, for Bessel function energetic particle spectra Equation (11). The composition of the ambient medium and energetic particles are given in Table I and the energetic particles are slowing down in a neutral medium.

shown in this figure represent total yields. Because of the finite half-lives of the various β^+ emitters, however, in a short observation time of a transient event, fewer positrons than indicated in Figure 6 are available for 0.511 MeV line production. This effect is shown in Figure 7, where dQ_+/dt is the instantaneous production rate of positrons from a burst (δ-function in time) of β^+ emitters produced at $t = 0$.

The positrons produced in nuclear reactions have initial energies ranging from about 0.1 MeV to tens of MeV, depending on the production mode. In a thick target, these positrons are rapidly slowed down to energies of tens of eV where the majority of them form positronium atoms. The slowing-down time, t_s, plus the positronium formation time, t_{pos}, are shown in Figure 8 as a function of initial positron energy for an ambient medium of temperature 10^4 K, degree of ionization η, and ambient density 10^{11} and 10^{13} cm^{-3} (from calculations of J. M. McKinley, private communication, 1981). The dependence of $(t_s + t_{pos})^{-1}$ on density is essentially linear.

The annihilation of positronium atoms is quite rapid (Heitler, 1954). If formed in the singlet state (25% of the time), positronium annihilates at a rate of 8×10^9 s^{-1} into two 0.511 MeV gamma rays. In the triplet state (formed 75% of the time) it annihilates at a rate of 7×10^6 s^{-1} into three photons of energies less than 0.511 MeV. This 3-photon continuum is, in principle, observable if the density of the ambient medium is less than $\sim 10^{15}$ cm^{-3}. At a higher density, collisions break up triplet positronium before it can annihilate (Crannell *et al.*, 1976).

We define a positron-to-0.511 MeV photon conversion coefficient, $\bar{f}_{0.51}$ analogous to $\bar{f}_{2.2}$ discussed above, such that the time-integrated flux, or fluence, of 0.511 MeV photons at Earth resulting from positron emitter production at the Sun is given by

$$\phi(0.511 \text{ MeV}) = Q_+ \bar{f}_{0.51}/(4\pi d^2). \tag{14}$$

Various effects influence the value of $\bar{f}_{0.51}$. If all the β^+ emitters produced in the flare decay during the observation period and if all the resultant positrons annihilate in this period, then $\bar{f}_{0.51}$ ranges from 0.5 to 2, depending on the fraction of positrons that

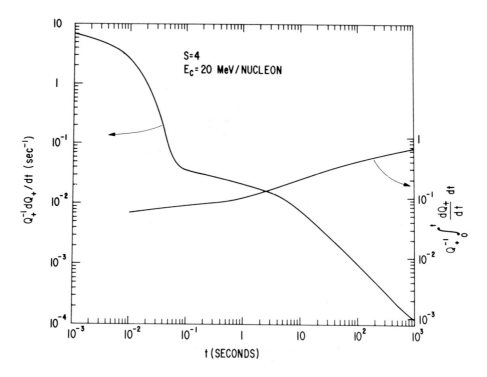

Fig. 7. The instantaneous fractional positron production rate and the time-integrated fractional positron yield in a thin-target model for a burst of β^+ emitter production (δ function) at $t = 0$. The energetic particle spectrum is given by Equation (9) and the compositions are close to those of Table I. The results of this figure do not depend strongly on the interaction model used.

annihilate via positronium. But if the observation period is shorter than the decay half-lives of the dominant β^+ emitters, or if some of the positrons escape from the Sun into low-density regions where their annihilation time is long, $\bar{f}_{0.51}$ can be substantially lower than the above values. For limb flares, there may be significant Compton scattering of the 0.511 MeV photons in the photosphere if the nuclear reactions themselves take place in the photosphere. This will also lower the value of $\bar{f}_{0.51}$.

Another observable of considerable interest is the width of the 0.511 MeV line. The dependence of this width on temperature, degree of ionization, and density was studied in considerable detail by Crannell et al. (1976) and Bussard et al. (1979). Using their results, we find that the 0.511 MeV line should be narrower than about 3.5 keV if the temperature of the annihilation site is less than 10^5 K. At higher temperatures the width should vary as $T^{1/2}$, with a full width at half maximum of about 11 keV at 10^6 K.

2.4. Prompt De-excitation Line Production

A variety of gamma-ray lines are produced in solar flares from de-excitation of nuclear levels. Figure 9 shows the spectrum from these de-excitations, calculated (Ramaty et al., 1979) by employing a Monte Carlo simulation for an energetic particle population

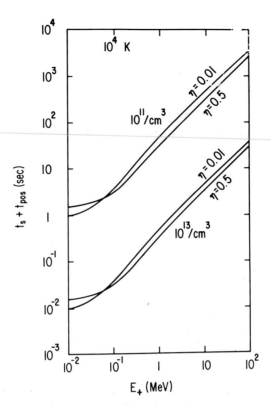

Fig. 8. The slowing-down time, t_s, plus positronium-formation time, t_{pos}, of positrons of initial energies E_+ in an ambient medium of temperature 10^4 K, degree of ionization η, and density 10^{11} or 10^{13} cm^{-3}. At the linear parts of the curves, $t_s \gg t_{pos}$; t_s is essentially independent of temperature and does not depend much on η; t_{pos}, however, depends strongly on these parameters (see Bussard *et al.*, 1979, for more details). The calculations of Figure 8 have been carried out by J. M. McKinley (private communication, 1981).

interacting with an ambient medium. The energetic particle spectrum is proportional to E^{-2} (Equation (9) with $E_c = 0$) and both the ambient medium and the energetic particles at the same E have a photospheric composition. The shapes of the lines are evaluated by taking into account nuclear kinematics and data on the differential cross-sections of the reactions. The results of the simulations are binned into energy intervals ranging from 2 to 5 keV (as indicated in the figure), consistent with the resolution of a Ge gamma-ray spectrometer.

Two line components can be distinguished in Figure 9. A narrow component resulting from the de-excitation of ambient, heavy nuclei excited by energetic protons and α-particles, and a broad component from the de-excitation of energetic heavy nuclei interacting with ambient H and He. The nuclei responsible for the emission of the strongest narrow lines are indicated in the figure.

In addition to these strong narrow lines, there are many other weaker narrow lines, which together with the broad component produced by heavy accelerated particles,

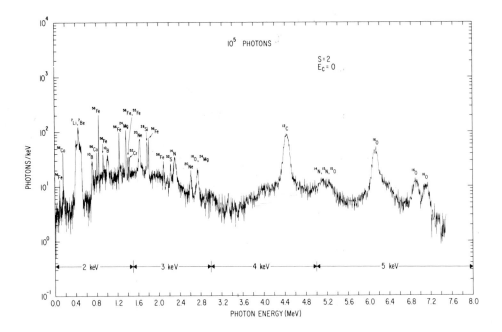

Fig. 9. Prompt nuclear gamma-ray spectrum from the interactions of energetic particles with the solar atmosphere in a thin-target model. The composition of the ambient medium is given in Table I. The energy spectrum of the energetic particles is given by Equation (9) and their composition is the same as that of the ambient medium (EP2). Not shown in this figure are the delayed lines, at 2.223 MeV from neutron capture and at 0.511 MeV from positron annihilation. In the August 4, 1972 flare, these lines were ~10 and 2 times more intense, respectively, than the ^{12}C line at 4.44 MeV. (From Ramaty *et al.*, 1980.)

merge into the underlying continuum. Above ~4 MeV most of the radiation is from C, N, and O, while below about 3 MeV the principal contributors are Mg, Si, and Fe. It has been shown (Ramaty *et al.*, 1977; Ibragimov and Kocharov, 1977) that the bulk of the gamma-ray flux observed between 4 and 7 MeV from the August 4, 1972 flare was of such nuclear origin rather than electron bremsstrahlung or other continuum emission processes. In addition, a substantial fraction of the photons in the 1–2 MeV band could be nuclear radiation resulting from an enhanced abundance of Ne, Mg, Si, and Fe in the energetic particles (Ramaty *et al.*, 1980, and Section 3.1). The 4–7 MeV energy band, referred to in the SMM observations as the 'main channel window', can provide a direct and sensitive measure of the interaction rate of protons and nuclei in solar flares (Sections 3.1 and 3.2).

Because the cross-sections for excitation of nuclear levels have different energy dependences from those of neutron and positron production, the calculated ratios of nuclear de-excitation line yields to the neutron yields depend strongly on the assumed spectra of the accelerated particles and on the interaction model. As an example, in Figure 10 we show the ratio $Q(4.44)/Q_n$ for the two interaction models as a function of αT for the Bessel function spectrum of Equation (11). Here $Q(4.44)$ is the yield of the

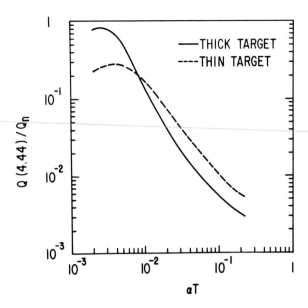

Fig. 10. The ratio of the narrow 4.44 MeV line yield to the neutron yield for energetic particle spectra given by Equation (11). The compositions of the ambient medium and energetic particles are given in Table I.

narrow 4.44 MeV line calculated from Equations (3) or (8), for thin or thick targets, respectively, with energetic particle abundances given by EP1 (see Table I). As can be seen, $Q(4.44)/Q_n$ generally decreases with increasing energetic particle spectral hardness, reflecting an increased neutron production and decreased 4.44 MeV photon production by particles of high energies. Likewise, this ratio is lower for thick targets than for thin targets (except for very steep particle spectra) because the energy losses harden the particle spectra in the thick target. For very steep (small αT) Bessel function spectra, the thick-target ratio exceeds that for thin targets. This results from the effect of the energy losses in the thick-target which suppress the heavy particle fluxes relative to the proton and α-particle fluxes. At low particle energies, the heavy particles contribute significantly to neutron production but not to the production of narrow 4.44 MeV photons.

As already mentioned, the photon energy band from 4 to 7 MeV is an important measure of the interaction rate of protons and nuclei in solar flares. In Figure 11 we show the ratio of the photon yield in this band, $Q(4-7)$, to the neutron yield Q_n. Here $Q(4-7)$ is calculated from Equation (3) or (8) for thin or thick targets, respectively, together with a Monte Carlo simulation similar to that employed in the evaluation of the gamma-ray spectrum of Figure 9. In this calculation we use both the EP1 and EP2 energetic particle abundances.

The same trends as in Figure 10 are also evident for $Q(4-7)/Q_n$ in Figure 11. $Q(4-7)/Q_n$ decreases with increasing αT and is larger in the thick-target than in the thin-target model. The variation of $Q(4-7)/Q_n$ with energetic particle composition,

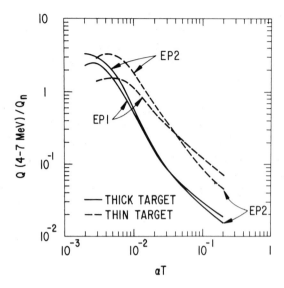

Fig. 11. The ratio of the photon yield in the 4–7 MeV channel to the neutron yield for energetic particle spectra given by Equation (11). The composition of the ambient medium is given in Table I. The curves EP1 are for energetic particles of composition as given in Table I while the curves EP2 are for an energetic particle composition which is the same as that of the ambient medium in Table I.

however, is not very large, because several particle species contribute simultaneously to both $Q(4-7)$ and Q_n.

The ratio $Q(4.44)/Q(4-7)$ is shown in Figure 12 for thin and thick targets and the

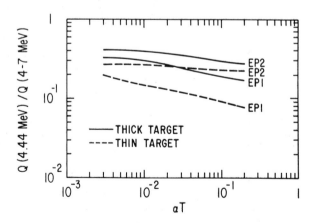

Fig. 12. The ratio of the narrow 4.44 MeV line yield to the 4–7 MeV channel yield. The rest of the parameters are the same as in Figure 11.

energetic particle abundances EP1 and EP2. As can be seen, this ratio does not depend strongly on energetic particle spectrum because of the similar energy dependences of all prompt line production cross-sections. However, $Q(4.44)/Q(4-7)$ is smaller for the thin target than for the thick target, because for the latter the contribution of the heavy particles is suppressed by the energy losses. Likewise $Q(4.44)/Q(4-7)$ is lower for EP1 than for EP2 because the former contains more heavy particles relative to protons than does the latter.

The shapes and peak energies of prompt nuclear gamma-ray lines are sensitive to anisotropies in the energetic particle angular distributions. Thus, Ramaty and Crannell (1976) have evaluated the shift in the peak of the 6.129 MeV line resulting from energetic particle beaming, while Ramaty et al. (1979) discussed the splitting of 4.438 MeV line that is observed (Kolata et al., 1967) when the excited ^{12}C nuclei are produced by a proton beam perpendicular to the direction of observation.

A unique test of energetic particle beaming was proposed by Kozlovsky and Ramaty (1977). This concerns the ^{7}Be* and ^{7}Li* de-excitation lines at 0.431 and 0.478 MeV produced in the reactions ^{4}He(α, n)^{7}Be* and ^{4}He(α, p)^{7}Li*, respectively. Here the stars indicate nuclei in excited states. The shapes of these lines are shown in Figure 13. As can

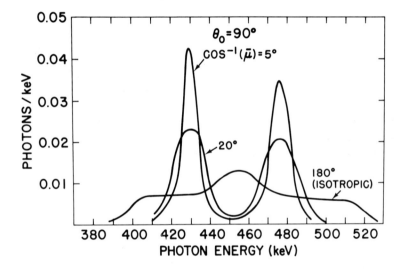

Fig. 13. Photon spectrum of the prompt ^{7}Li and ^{7}Be lines for an α-particle beam confined to a cone with half opening angle 5° or 20°, and for an isotopic distribution; θ_0 is the angle between the beam and the direction of observation. (From Kozlovsky and Ramaty, 1977.)

be seen, if the angular distribution of the energetic α-particles is isotropic, the Doppler broadening of the two lines is so large that they blend into a single feature that cannot, in general, be observed in the presence of a strong continuum. If, however, the α-particles are beamed, the line widths are much less than in the isotropic case and two discrete lines can be seen. In particular, if the direction of observations is perpendicular to the beam (as in Figure 13) the lines appear close to the rest energies of 0.431 and 0.478 MeV.

The fluence at Earth in the 4–7 MeV channel is given by

$$\phi(4-7 \text{ MeV}) = Q(4-7 \text{ MeV}) \bar{f}_{4-7}/(4\pi d^2), \tag{15}$$

where $\bar{f}_{4-7}$ is the conversion coefficient from nuclear de-excitations to photons observed. If the excited nuclei are produced isotropically and if there is no attenuation of the photons, $\bar{f}_{4-7} = 1$. However, if the protons and nuclei form a beam pointing away from the observed, then from the Monte Carlo calculations described above we find that $\bar{f}_{4-7} \simeq 0.8$. Thus, provided that there is not much attenuation, we expect that $0.8 \lesssim \bar{f}_{4-7} \lesssim 1$, depending on the geometry of the interacting particles.

3. Implications of Gamma-Ray Observations

In this section we consider the implications of the gamma-ray line observations on the nature of the interaction model (thin or thick target), on the accelerated particle spectrum, on the energy deposited by the particles and the number of particles that interact to produce the gamma-rays, on the timing of the acceleration, on the photospheric ^{3}He abundance, and on the beaming of the energetic particles.

The solar flares with observed gamma-ray lines are listed in Table II. The August 4 and

TABLE II

Gamma-ray flares

Flare	Fluences, ϕ(photons cm^{-2})					$\dfrac{\phi(4-7)}{\phi(2.22)}$	Location
	2.22 MeV	4–7 MeV	4.44 MeV	6.13 MeV	0.511 MeV		
Aug. 4, 1972	155 ± 12[1]	105 ± 11[2]	17 ± 5[1]	17 ± 5[1]	35 ± 11[1]	0.68 ± 0.09	E08 N14
Aug. 7, 1972	No data available during time of maximum emission					–	W38 N15
July 11, 1978	240 ± 70[3]	171[4]	43 ± 17[3]	–	–	0.71	E43 N18
Nov. 9, 1979	38 ± 9[5]	50 ± 4[5]	–	–	–	1.32 ± 0.33	E00 S16
June 7, 1980	6.6 ± 1[6]	11.5 ± 0.5[6]	–	–	<2	1.74 ± 0.27	W74 N12
June 21, 1980	No data available during time of maximum emission*					–	W91 N17
July 1, 1980	3.3 ± 0.5[6]	3.1 ± 0.4[6]	–	–	0.9 ± 0.4	0.94 ± 0.19	W37 S12
Nov. 6, 1980	10.3 ± 1.3[6]	14.8 ± 0.8[6]	–	–	<2	1.44 ± 0.2	E74 S12
Apr. 10, 1981	13.5 ± 1[6]	18.6 ± 1.6[6]	–	–	<6.6	1.38 ± 0.16	W37 N09
Apr. 27, 1981	11.7 ± 2[6]	118 ± 2[6]	–	–	–	10.1 ± 1.7	W90 N16

[1] Chupp (1976).
[2] Ramaty et al. (1977)
[3] Hudson et al. (1980)
[4] Theoretical
[5] T. Prince (private communication, 1981)
[6] Chupp (1982)

* Delayed 2.22 MeV and 0.511 MeV line, and high-energy neutrons were observed from this flare (Share et al., 1983; Chupp et al.,1982).

August 7, 1972, events were observed by Chupp et al. (1973) with the spectrometer on board OSO-7. The data given in Table II for the August 7, 1972 flare is from Chupp (1976), except for ϕ(4–7 MeV) which is from the analysis of Ramaty et al. (1977) We have multiplied the time-averaged fluxes given in these references by 553 s, the

observation time of gamma-rays from this flare (Chupp, 1976). Because of Earth occulta-
tion of the orbiting gamma-ray detector, however, the total duration of gamma-ray
emission from the August 4 flare was longer than this time interval. From the analysis
of Wang and Ramaty (1975), we estimate that all fluences in Table II for the August 4
flare should be increased by about a factor of 2, but without a significant modification
of $\phi(4-7)/\phi(2.22)$.

Only the 2.22 and 0.51 MeV lines were observed from the August 7, 1972 event
(Chupp, 1976) because the detector was behind the Earth during the flash phase of the
flare. These observations clearly demonstrate the delayed nature of these two lines:
at a time when all prompt emissions were very small, the 2.22 and 0.51 MeV lines were
still observable.

The data for the flares of July 11, 1978 and November 9, 1979 are, respectively,
from Hudson *et al.* (1980) and Prince *et al.* (1982).

The data for the other flares listed in Table II, still believed to be preliminary, were
observed with the gamma-ray spectrometer on SMM (Chupp, 1982). Because of the
very high intensity of the June 21, 1980 flare, the SMM detector saturated during the
flash phase of this event. Delayed 2.22 and 0.51 MeV lines, as well as high-energy neutrons (Chupp *et al.*, 1982), were observed from the June 21 flare.

The ratios of the photon fluence in the 4–7 MeV channel to that in the 2.22 MeV
line, shown in Table II, follow directly from the data except for the July 11, 1978
flare where $\phi(4-7)$ is determined using a theoretical ratio $Q(4.44)/Q(4-7) \simeq 0.25$. The
fact that this ratio is model dependent (Figure 12) leads to some uncertainty in the
determination of $\phi(4-7)$ for this flare.

3.1. INTERACTION MODEL, ENERGETIC PARTICLE SPECTRUM, NUMBER AND ENERGY CONTENT

We first consider the flare of June 7, 1980 for which there are both gamma-ray line
observations and interplanetary particle measurements. The combined analysis of these
data imply that for this flare the bulk of the gamma-ray line emission results from thick-
target interactions. Furthermore, most of the energetic protons and nuclei that produce
the gamma-ray lines remain trapped in the solar atmosphere and only a small fraction
of them escapes into the interplanetary medium.

The location of the June 7 flare at N12 W74 indicates that it was well connected
magnetically (e.g. Van Hollebeke *et al.*, 1975), so that particles escaping from the Sun
could be observed in interplanetary space near the orbit of the Earth. Indeed, several
observations of energetic particles have been reported (von Rosenvinge *et al.*, 1981;
Evenson *et al.*, 1981; Pesses *et al.*, 1981). Based on these, the number of protons of
energies greater than 10 MeV released into the interplanetary medium, $N_{esc,p}(>10 \text{ MeV})$,
has been estimated (von Rosenvinge *et al.*, 1981) to be $\sim 10^{31}$. As can be seen from
Figure 1, the spectrum of these protons fits well with the Bessel function of Equation
(11) with αT equal to 0.015 (R. E. McGuire, private communication, 1981). This spectral
form also fits the α-particle spectrum with essentially the same αT. For $\alpha T = 0.015$,
$N_{esc,p}(>10 \text{ MeV}) \simeq 10^{31}$ implies $N_{esc,p}(>30 \text{ MeV}) \simeq 5 \times 10^{29}$. If the gamma-rays were
produced by thin-target interactions, then from Equation (3), with the numerical results
of Figure 2 and $\bar{f}_{2.2} = 0.12$ (appropriate for a thin target), the observed 2.22 MeV line

fluence (Table II) implies that $n_H T \simeq 7.4 \times 10^{14}$ cm^3 s^{-1}. This is equivalent to a matter traversal for 30 MeV nucleon^{-1} particles of $\simeq 13$ g cm^{-2}. The large abundances of spallation products (^{2}H, ^{3}H, Li, Be B) that would result from such a long path length are not observed from solar flares (e.g. McGuire $et\ al.$, 1977). This indicates that the gamma-ray lines observed from the June 7, 1980 flare were probably not produced in thin-target interactions. In the thick-target models, on the other hand, the spallation products that accompany the production of gamma-ray lines, are slowed down in the solar atmosphere and hence are not expected to be seen in the interplanetary medium.

Analysis of the ratio of the 4–7 MeV fluence to the 2.223 MeV line fluence for the June 7, 1980 flare also suggests that the observed gamma-rays were produced in thick-target interactions, not thin-target. This can be seen as follows:

From Equations (13) and (15), with $\bar{f}_{4-7} = 1$, the neutron-to-2.223 MeV photon conversion coefficient, $\bar{f}_{2.2}$, can be written as

$$\bar{f}_{2.2} = \frac{Q(4\text{--}7\ \mathrm{MeV})/Q_n}{\phi(4\text{--}7\ \mathrm{MeV})/\phi(2.22\ \mathrm{MeV})} . \tag{16}$$

The numerator can be obtained from theory (see Figure 11) if αT is known. The interplanetary particle data suggests that $\alpha T \simeq 0.015$ (Figure 1) provided that all the observed particles were indeed produced in the gamma-ray flare and that the flare-particle spectrum is not greatly modified by the escape process and interplanetary propagation. The denominator in Equation (16) is from the gamma-ray data, $\phi(4\text{--}7)/\phi(2.22) \simeq 1.74$ (Table II). Then for thin-target interactions, $Q(4\text{--}7)/Q_n \simeq 0.7$, hence $\bar{f}_{2.2} \simeq 0.4$, while for thick-target interactions, $Q(4\text{--}7)/Q_n \simeq 0.25$, hence $\bar{f}_{2.2} \simeq 0.14$. From the calculations of Wang and Ramaty (1974) and Kanbach $et\ al.$ (1981) we estimate that for the location of the June 7, 1980 flare, $\bar{f}_{2.2}/\bar{f}_{2.2}(0) \simeq 0.6$. For this flare, therefore, $\bar{f}_{2.2}(0)$ would have to be $\sim$0.67, if the gamma-rays were made in thin-target interactions, and $\sim$0.23 if they were made in thick-target ones. The former value is clearly inconsistent with the range of values of $\bar{f}_{2.2}(0)$ obtained from the Monte Carlo simulations of neutron production in the solar atmosphere (Section 2.2). The latter value, for the thick target, is quite consistent with these calculations provided the neutrons are produced in the photosphere and/or their initial angular distribution is skewed downward towards the photosphere.

We proceed now to analyze the rest of the data listed in Table II. We note the relatively small variability of $\phi(4\text{--}7)/\phi(2.22)$ from one flare to another. The exception is the limb flare of April 27, 1981 for which the 2.22 MeV line is strongly attenuated by Compton scattering as the photons emerge from the photosphere at large angles to the local normal (Wang and Ramaty, 1974). We assume that in all of these flares, as in the June 7 flare, the gamma-rays are produced by thick-target interactions. This is justified because different interaction models for different flares would not be consistent with the relative constancy of $\phi(4\text{--}7)/\phi(2.22)$ shown in Table II. Furthermore, we assume the same neutron-to-2.223 photon conversion coefficient for all flares, except for the correction due to flare location. We use $\bar{f}_{2.2}(0) = 0.23$, a value consistent with both theory and the June 7 particle and gamma-ray data.

Table III lists the flares with available $\phi(4\text{--}7)/\phi(2.22)$ ratios and the April 27, 1981 flare. The second column gives an estimate of $\bar{f}_{2.2}/\bar{f}_{2.2}(0)$ obtained from the flare locations and the calculations of Wang and Ramaty (1974) and Kanbach $et\ al.$ (1981). The third column lists values of $Q(4\text{--}7)/Q_n$ deduced from Equation (16), while the fourth column

TABLE III

Spectral parameters, total energies and number of particles of gamma-ray flares

Flare	$\bar{f}_n/\bar{f}_n(0)$	$\dfrac{Q(4-7)}{Q_n}$	αT	$W(>1$ MeV$)$ (erg)	$\bar{N}_p(>10$ MeV$)$	$N^{obs}_{esc,\,p}(>10$ MeV$)$
Aug. 4, 1972	1	0.16	0.019	2.5×10^{30}	1.3×10^{34}	3×10^{35}
July 11, 1978	0.9	0.15	0.020	1.8×10^{30}	1.0×10^{34}	–
Nov. 9, 1979	1	0.30	0.014	9×10^{29}	3.4×10^{33}	–
June 7, 1980	0.6	0.24	0.015	2×10^{29}	8.5×10^{32}	10^{31}
July 1, 1980	0.95	0.21	0.016	5×10^{28}	2.3×10^{32}	–
Nov. 6, 1980	0.6	0.20	0.017	2×10^{29}	1.0×10^{33}	–
Apr. 10, 1981	0.95	0.30	0.014	3.5×10^{29}	1.3×10^{33}	–
Apr. 27, 1981	~0	–	0.019	1.4×10^{30}	7.3×10^{33}	–

provides the value of αT obtained from these ratios and Figure 11. We note the relatively small variability of αT from one flare to another, a direct consequence of the constancy of $\phi(4-7)/\phi(2.22)$. This result is consistent with the observations of McGuire *et al.* (1981) who find a range of αT's for protons and α-particles in interplanetary space which essentially overlaps that deduced for the gamma-ray flares. Particle acceleration evidently produces energy spectra that do not vary much from flare to flare. A similar conclusion has been obtained by Van Hollebeke *et al.* (1975).

We cannot deduce the αT for the limb flare of April 27, 1981 because of the strong attenuation of the 2.223 MeV line by Compton scattering in the photosphere. The small variability of αT, however, allows us to assume an αT for this flare. We take $\alpha T = 0.019$, equal to that for the August 4, 1972 flare, because both events have similar durations of gamma-ray emission and approximately equal fluences in the 4–7 MeV channel.

Using the αT's listed in column 4, we calculate, in the thick-target model, the number of particles that interact in the solar atmosphere and the energy deposited by them. The results, based on the numerical values of Figures 2 and 4, are given in columns 5 and 6 of Table III. Here, $W(>1$ MeV nucleon$^{-1})$ is the energy deposited by particles of energies greater than 1 MeV nucleon^{-1} and $\bar{N}_p(>10$ MeV$)$ is the number of protons of energies greater than 10 MeV incident on the thick target. For the June 7, 1980 flare, the number of protons of energies greater than 10 MeV that interacts at the Sun exceeds that observed in interplanetary space ($\sim 10^{31}$) by about two orders of magnitude. In contrast, for the August 4, 1972 flare, the number of protons above 10 MeV observed in the interplanetary medium ($\sim 3 \times 10^{35}$, Lin and Hudson, 1976) exceeds the number that interacts at the Sun (Table III) by more than an order of magnitude.

The very large number of interplanetary particles observed from the August 4, 1972 flare could have produced the observed gamma rays by thin-target interactions, as proposed by Lin and Hudson (1976). These authors have assumed a flatter energetic particle spectrum than that given by $\alpha T = 0.019$ in Table III. However, the interplanetary particle spectrum from the August 4 flare is only very poorly known because several interplanetary shocks were present at that time. Rather than using the interplanetary observations, we would now argue that the relative constancy of $\phi(4-7)/\phi(2.22)$ supports the same interaction model for all gamma-ray flares, and hence a thick-target model for the August 4, 1972 flare. This result is consistent with the average interplanetary proton spectrum observed (Webber *et al.* 1975) from August 2 to 11, which can be well fitted with Equation (11) with $\alpha T = 0.02$, in good agreement with the value of αT given in Table III for a thick target.

The energy depositions of the protons and nuclei given in Table III range from about 5×10^{28} to 2.5×10^{30} erg. However, as pointed out in Section 2.2, these should be considered as lower limits only, because of additional energy losses to the ionized component of the solar atmosphere and possible collective effects (e.g. Colgate, 1978). Nevertheless, we can compare the energy deposition of the nucleonic component with that of >25 keV electrons deduced from hard X-ray observations in a nonthermal model (Lin and Hudson, 1976). The energy deposition of the electrons ranges from about 2×10^{29} erg for small flares to $\sim 10^{32}$ erg for the August 4, 1972 flare (Lin and Hudson, 1976). We see that the nucleonic component could be responsible for the deposition of at least several percent of the total flare energy.

A final argument that supports the thick-target interaction model comes from the analysis of the 0.511 MeV line from positron annihilation. In Table IV we list three flares

TABLE IV

0.511 MeV line fluences

Flare	Duration s	αT	$Q_+/Q(4-7)$	$\phi(0.51)$ (photons cm^{-2}) Calculated	observed
Aug. 4, 1972	553	0.019	0.46	48 $\bar{f}_{0.51}$	35 ± 11
June 7, 1980	50	0.015	0.41	4.7 $\bar{f}_{0.51}$	<2
July 1, 1980	60	0.016	0.41	1.3 $\bar{f}_{0.51}$	0.9 ± 0.4

for which there is either an observation of, or an upper limit on, the fluence in this line. For the August 4, 1972 flare the data is from Chupp (1976); and for the June 7 and July 1, 1980 flares it is from Chupp (1982). Using the αT's of Table III and the results of Figures 6 and 11, we calculate the ratios $Q_+/Q(4-7$ MeV$)$. These are shown in column 4 of Table IV. In column 5 of this table we give the values of $\phi(0.51)$ calculated from the values of $\phi(4-7$ MeV$)$ given in Table II and with $\bar{f}_{0.51}$, the β^+ emitter-to-0.511 MeV photon conversion coefficient, a free parameter. The observed fluences or upper limits are given in column 6. By comparing the calculated and observed values of $\phi(0.51)$, we see that for the August 4 event there is good agreement if $\bar{f}_{0.51}$ is about 0.7 which is consistent with the theoretical expectation discussed in Section 2.3. A similar value of $\bar{f}_{0.51}$ could also account for the July 1, 1980 data. For the June 7, 1980 flare, $\bar{f}_{0.51}$ would have to be less than about 0.4, a value consistent with the short observation period (50 s) which does not allow the complete decay of all the positron emitters produced by the nuclear reactions (see Figure 7). Thus, for all the available data, the value of $\bar{f}_{0.511}$ is close to that expected if most of the positrons annihilate at the Sun and only a few of them escape. This result, if substantiated by further studies, should be a strong argument for the validity of the thick-target model.

3.2. TIME DEPENDENCES

The time dependences of the gamma-ray lines contain important information on a variety of questions in solar physics.

The time dependence of the strongest discrete line from flares, the 2.223 MeV line from neutron capture, is determined by the time history of the neutron production as well as by the removal rate of neutrons from the photosphere where they spend most of their time (~ 1 min) between production and radiative capture (Section 2.2).

The delay between neutron production and 2.223 MeV photon release has been unmistakably observed in several flares (Chupp et al., 1973; Hudson et al., 1980; Chupp et al., 1981; Prince et al., 1982). Such an observation entails the comparison of the time history of a prompt photon flux, for example that in the 4–7 MeV channel, with that of the 2.223 MeV line. This has been carried out in detail for the June 7, 1980 flare (Chupp et al., 1981). According to Chupp (1982), the characteristic neutron removal time from the photosphere, as deduced from the gamma-ray data, is on the order 50 s, consistent with theory (Section 2,2) and a photospheric ^{3}He/H ratio of $\sim 5 \times 10^{-5}$. As we shall see below (Section 3.3), the fact that the delay between neutron production and capture is not much shorter than 50 s can place an upper limit on the photospheric ^{3}He abundance.

The time dependence of the 0.511 MeV line is determined by the production rate of the β^+ emitters, the decay rate of these emitters (Figure 7) and the slowing down and annihilation time of the positrons (Figure 8). Considerable information on the annihilation site of the positrons should become available from the comparison of observable time dependences of the 0.511 MeV line with theory. No such comparison has yet been done with the SMM data.

Because of the delayed nature of both the 2.223 and 0.511 MeV lines, information on the timing of the acceleration of protons and nuclei can be best obtained from prompt nuclear de-excitation lines. The comparison of the time histories of such lines with those of hard X-rays of various energies can give information on the possible existence of multiple acceleration stages of energetic particles in solar flares. Because the 4–7 MeV channel is dominated by nuclear radiation (Section 3.1), the flux in this channel (the 'main channel window' in the SMM data) is an excellent diagnostic of the timing of acceleration of the nucleonic component in flares.

The time history of the 'main channel' flux was observed for the June 7, 1980 flare (Chupp, 1982). In particular, several gamma-ray spikes were seen in good correlation with the seven hard X-ray spikes. From the comparison of these two time histories it follows that, at least for the June 7 flare, the <100 keV electrons and >10 MeV nuclei were accelerated in very close time proximity (less than a few seconds). There is, nevertheless, a delay of approximately 2 s between the hard X-ray and gamma-ray peaks. Bai (1982) suggests that this delay is due to two-step acceleration, where the first step accelerates the <100 keV electrons and the second step accelerates the mildly relativistic electrons and the nuclei (see also Bai and Ramaty, 1979). On the other hand, Chupp (1982) attributes the delay to the difference in propagation time of ≥ 10 MeV per nucleon nuclei and ≤ 100 keV electrons along a magnetic arch of length $\sim 10^{10}$ cm. In any case, the rapid decay (~ 2 s) of the main channel emission for the June 7, 1980 flare requires a sufficiently high ambient density ($> 10^{13}$ cm^{-3}) and this provides additional support for the validity of the thick-target model for this flare.

The time dependences of the various hard photon emissions for the June 7, 1980 flare were quite different from those of the longer duration August 4, 1972 flare. For the latter, the X-ray continuum about 350 keV reached peak strength a few minutes later than the continuum above 30 keV and the 2.22 MeV line profile was better explained when the neutron production time profile was assumed to be similar to the time profile of the >350 keV continuum rather than the <100 keV continuum (Bai and Ramaty, 1976; Bai, 1982). These results, together with earlier X-ray observations (Frost and Dennis, 1971) and gamma-ray measurements (Hudson et al., 1980; Willet et al., 1982), demonstrate that not all energetic particle populations in flares are accelerated at the same time. The present status of the existence of multiple acceleration phases in flares has been reviewed by Bai (1982).

3.3. THE PHOTOSPHERIC ^{3}He ABUNDANCE

The abundance of ^{3}He in the solar atmosphere is of considerable astrophysical interest. Along with ^{2}H and ^{4}He, ^{3}He is formed by nucleosynthesis in the big bang (Wagoner, 1973). In addition, stellar evolution should increase the ^{3}He abundance. In particular, the Sun should have burned into ^{3}He any amount of deuterium it originally had, but the ^{3}He should not have been further burned into ^{4}He (Geiss and Reeves, 1972, and references therein). Thus, a measurement of the solar ^{3}He abundance provides an upper limit on the protosolar ^{2}H abundance which, in turn, provides information on nucleosynthesis in the big bang and on whether the Universe is open or closed (Gott et al, 1974).

^{3}He has been observed in the solar wind where the ^{3}He/^{4}He ratio is of the order of a few times 10^{-4} (Geiss and Reeves, 1972). Hall (1975) determined spectroscopically a ^{3}He/^{4}He ratio of $(4 \pm 2) \times 10^{-4}$ in a solar prominence. There is, however, no direct observation of ^{3}He in the photosphere. The solar gamma-ray observations can set limits on the photospheric ^{3}He abundance. This can be done in two ways. First, for a ^{3}He/H ratio much larger than 5×10^{-5}, the 2.22 MeV line fluence would be much lower relative to the 4–7 MeV fluence than observed. Second, the ^{3}He/H ratio must have an upper bound at a value not much higher than 5×10^{-5} because otherwise the delay between neutron production and 2.223 MeV photon release would be shorter than observed. There are as yet no firm values on the upper limits on the ^{3}He/H ratio from the SMM data, but a safe preliminary limit would be ^{3}He/H $< 2 \times 10^{-4}$.

In addition to setting an absolute upper bound on the photospheric ^{3}He/H ratio, the gamma-ray data also limit any possible variability of this ratio in time and with position on the solar surface. From the small variability of $\phi(4-7)/\phi(2.22)$ from flare to flare (Table II), it follows that ^{3}He/H should be constant to better than a factor of 2.

As discussed in Section 1, very high ^{3}He/^{4}He ratios have been observed in solar energetic particles (see Ramaty *et al.*, 1980, for a review of these observations). It is now believed that these enhancements are not of nuclear origin, but result from selective heating and acceleration (Fisk, 1978). Finite ^{2}H/^{1}H ratios in solar energetic particles averaged over several solar flares have been presented (Anglin *et al.*, 1973b; Hurford *et al.*, 1975). But the very large uncertainties in these measurements and the possibility of instrumental contamination preclude a definite conclusion regarding the positive detection of secondary nuclear products in solar energetic particles. It appears, nonetheless, that the bulk of the nuclear reactions are produced by the flare-accelerated particles that remain trapped at the Sun. The particles that escape from the Sun and are observed in the interplanetary medium are devoid of any measurable amount of nuclear spallation products.

3.4. BEAMING OF THE ENERGETIC PARTICLES

As we have seen in Section 2.4, gamma-ray line observations can give unique information on the beaming of the energetic particles. Shifts in the energies of narrow lines are indicative of such beaming (Ramaty and Crannell, 1976), but these effects are probably difficult to measure with low-resolution spectrometers. Another effect of beaming is the narrowing of the broad lines. These lines, produced by energetic heavy nuclei, are Doppler broadened by both the velocity spread and the angular distribution of the particles. In the case of a beam, however, the latter effect is greatly reduced and hence broad lines (Figure 9) can mimic narrow lines. For the same reason, the ^{7}Be and ^{7}Li lines shown in Figure 13 are much narrower if produced by α-particles in a beam than by such particles with an isotropic distribution.

The energies of these lines would then provide direct information on the angle between the beam and the direction of observation. A flare model in which the gamma-ray lines would be produced by a beam of energetic particles is that of Colgate (1978). Future gamma-ray line observations and more refined analysis of the SMM data should produce much new information on energetic particle beams in solar flares and hence on the flare model.

4. Summary

Gamma-ray lines are the most direct probe of nuclear processes in the solar atmosphere. The line observations from a number of flares, made with spectrometers on OSO-7, HEAO-1, HEAO-3, and SMM, are consistent with reactions produced by flare-accelerated particles of energies greater than several MeV nucleon^{-1}. These reactions involve the production of neutrons, β^+ emitters, π mesons, and excited nuclear levels, all of which lead to observable gamma-ray line emission.

The solar gamma-ray line observations can give information on the timing of the nucleonic component in flares, through measurements of the light curves of prompt lines, on the energy spectrum, number and energy content of these particles, through line ratios and line fluences, on the site of the nuclear reactions, through selective attenuation of lines from limb flares and the spectrum and time dependence of e^+-e^- annihilation radiation, on the geometry of particle beams, through line shapes and Doppler shifts, on the photospheric ^{3}He abundance, through the time dependence and fluence of the 2.223 MeV line, and on chemical compositions of both the ambient medium and the energetic particles, through gamma-ray line ratios.

At the time of this writing (early 1982), only limited portions of the SMM gamma-ray data are available for analysis and, hence, only a few hard conclusions can be drawn from them (see also Chupp, 1982).

The acceleration of protons and nuclei to energies above an MeV, in at least some flares, takes place in a time interval less than a few seconds. This sets important constraints on flare-accelerated mechanisms which have not yet been fully explored. For at least some flares, previous ideas on two phases of acceleration, which involve long delays ($>$1 min) between the acceleration of MeV nuclei and X-ray producing electrons, are not valid. But there is evidence that the $>$10 MeV protons are accelerated later than the $<$100 MeV electrons.

The 2.223 MeV line is strongly attenuated in limb flares. This provides direct observational confirmation for neutron capture in the photosphere. Further confirmation of this process comes from the precise measurement of the energy of this line (with the high-resolution Ge detector on HEAO-3) and from the observed delay between the 2.223 MeV flux and the flux of prompt nuclear radiation. The 2.223 MeV line observations also indicate that the photospheric ^{3}He abundance is about 5×10^{-5} relative to ^{1}H by number, and that it does not vary much with position on the Sun.

The gamma-ray line emission is produced in thick-target interactions, i.e. by energetic protons and nuclei which slow down in the solar atmosphere. The absence of nuclear fragments (^{2}H, ^{3}H, Li, Be, B) in the fluxes of the interplanetary particles indicates that thin-target interactions do not produce many gamma rays. This is probably the reason for the lack of correlation between the number of particles responsible for gamma-ray line production and the number observed in the interplanetary medium.

The ratio of the 4–7 MeV photon fluence to the fluence in the 2.223 MeV line is a strong function of the energy spectrum of the accelerated particles. This spectrum, as deduced from the gamma-ray observations, does not vary much from one flare to another, and is, within rather broad error ranges, similar to the particle energy spectra observed in the interplanetray medium. This argues for the same acceleration mechanism for both the gamma-ray producing particles and the interplanetary particles. In this case, the lack of correlation between the absolute numbers of particles in the two populations could be due to the varying escape conditions of energetic particles from solar flares. Multiple acceleration mechanisms, however, cannot be ruled out at the present time.

Much additional information on solar flares and on energetic particle acceleration therein is expected from the detailed analysis of the already available SMM data and from new data anticipated from the SMM spectrometer as well as from other spectrometers that hopefully will be flown during the next solar maximum towards the end of the 1980s. Of particular interest would be the observation of solar gamma-ray lines with high spectral resolution which could provide unique information on such questions as the beaming of the energetic particles, the temperature of the energetic particle interaction site, and the compositions of the ambient medium and the energetic particles.

Acknowledgements

The author wishes to acknowledge the contributions of B. Kozlovsky, R. E. Lingenfelter, J. M. McKinley and R. J. Murphy to both the calculations and interpretations presented in this paper and E. L. Chupp, D. Forrest, T. Prince, and R. McGuire for providing data prior to their publication. Part of the research described in this paper was supported by NASA's Solar Terrestrial Theory Program.

References

Abramovitz, M. and Stegun, I. A.: 1966, *Handbook of Mathematical Functions*, U.S. Government Printing Office, Washington, D. C.

Anglin, J. D., Dietrich, W. F., and Simpson, J. A.: 1973a, in R. Ramaty and R. G. Stone (eds) *High Energy Phenomena on the Sun*, NASA SP-342, Washington, D.C., p. 315.

Anglin, J. D., Dietrich, W. F., and Simpson, J. A.: 1973b, *Astrophys. J.* **186**, L41.

Bai, T.: 1982, in R. E. Lingenfelter, H. S. Hudson, and D. M. Worrall (eds), *Gamma Ray Transients and Related Astrophysics*, Amer. Inst. of Physics, New York, p. 409.

Bai, T. and Ramaty, R.: 1976, *Solar Phys.* **49**, 343.

Bai, T. and Ramaty, R.: 1979, *Astrophys. J.* **227**, 1072.

Barkas, W. H. and Berger, M. J.: 1964, *Tables of Energy Losses and Ranges of Heavy Charged Particles*, NASA SP-3013, Washington, D.C.

Bussard, R. W., Ramaty, R., and Drachman, R. J.: 1979, *Astrophys. J.* **228**, 928.

Cameron, A. G. W.: 1982 in C. Barnes, D. D. Clayton, and D. N. Schramm (eds), *Essays in Nuclear Astrophysics*, Combridge Univ. press, Cambridge, p. 23.

Cheng, C. C.: 1972, *Space Sci. Rev.* **13**, 3.

Chupp, E. L.: 1976, *Gamma Ray Astronomy*, D. Reidel, Dordrecht.

Chupp, E. L.: 1982, in R. E. Lingenfelter, H. S. Hudson, and D. M. Worrall (eds), *Gamma Ray Transients and Related Astrophysical Phenomena*, Amer. Inst. of Physics, New York, p. 363.

Chupp, E. L. *et al.*: 1973, *Nature* **241**, 333.

Chupp, E. L. *et al.*: 1981, *Astrophys. J.* **244**, L171.

Chupp, E. L. *et al.*: 1982, *Astrophys. J.* **263**, L95.

Chupp, E. L., Forrest, D. J., and Suri, A. N.: 1975, *IAU Symp.* **68**, 341.

Colgate, S. A.: 1978, *Astrophys. J.* **221**, 1068.

Colgate, S. A., Audouze, J., and Fowler, W. A.: 1977, *Astrophys. J.* **213**, 849.

Crannell, C. J., Crannell, H., and Ramaty, R.: 1979, *Astrophys. J.* **229**, 762.

Crannell, C. J., Joyce, G., Ramaty, R., and Werntz, C.: 1976, *Astrophys. J.* **210**, 582.

Croom, D. L. and Harris, L. D. J.: 1973, in H. E. Coffey (ed.), World Data Center Rept, UAG-28 Part I, *Collected Data Reports on August 1972 Solar-Terrestrial Events*, p. 210.

Dietrich, W. F.: 1973, *Astrophys. J.* **180**, 955.

Dolan, J. E. and Fazio, G. G.: 1965, *Rev. Geophys.* **3**, 319.

Evenson, P., Meyer, P., and Yanagita, S.: 1981, *Int. Cosmic Ray Conf. Papers, Paris* **3**, 32.

Fireman, E. L., DeFelice, J., and D'Amico, J.: 1976, *Proc. 7th Lunar Sci. Conf.*, p. 525.

Fisk, L. A.: 1978, *Astrophys. J.* **224**, 1048.

Forrest, D. J. Chupp, E. L., and Suri, A. N.: 1975, *Proc. Int. Conf. on X-Rays in Space*, Univ. of Calgary, Alberta, Canada, p. 341.

Freier, P. S. and Webber, W. R.: 1963, *J. Geophys. Res.* **68**, 1605.

Frost, K. J. and Dennis, B. R.: 1971, *Astrophys. J.* **165**, 655.

Garrard, T. L., Stone, E. C., and Vogt, R. E.: 1973, in R. Ramaty and R. G. Stone (eds), *High Energy Phenomena on the Sun*, NASA SP-342, Washington, D.C., p. 341.

Geiss, J. and Reeves, H.: 1972, *Astron. Astrophys.* **18**, 126.

Gott III, J. R., Gunn, J. E., Schramm, D. N., and Tinsley, B. M.: 1974, *Astrophys. J.* **194**, 543.

Hall, D. N. B.: 1975, *Astrophys. J.* **197**, 509.

Heitler, W.: 1954, *The Quantum Theory of Radiation*, Oxford Univ. Press, London.

Hoyng, P., Brown, J. C., and Van Beek, H. F.: 1976, *Solar Phys.* **48**, 197.

Hsieh, K. C. and Simpson, J. A.: 1970, *Astrophys. J. Lett.* **162**, L191.

Hudson, H. S., Bai, T., Gruber, D. E., Matteson, J. L., Nolan, P. L., and Peterson, L. E.: 1980, *Astrophys. J.* **236**, L91.

Hurford, G. F., Stone, E. C., and Vogt, R. E.: 1975, *14th Int. Cosmic Ray Conf. Papers, Munich*, **5**, 1624.

Ibragimov, I. A. and Kocharov, G. E.: 1977, *Sov. Astron. Lett.* **3**(5), 221.

Kanbach, G., Pinkau, K., Reppin, C., Rieger, E., Chupp, E. L., Forrest, D. J., Ryan, J. M., Share, G. H., and Kinzer, R. L.: 1981, *17th Int. Cosmic Ray Conf. Papers, Paris*, **10**, 9.

Kanbach, G., Reppin, C., Forrest, D. J., and Chupp, E. L.: 1975, *14th Int. Cosmic Ray Conf. Papers, Munich* **5**, 1644.

Kerridge, J. F.: 1975, *Science* **188**, 162.

Kerridge, J. F., Kaplan, I. R., Lingenfelter, R. E., and Boynton, W. V.: 1977, *Proc. 8th Lunar Sci. Conf.*, p. 3773.

Kohl, J. W., Bostrom, C. O., and Williams, D. J.: 1973, in H. E. Coffey (ed.), World Data Center Rept, UAG-28 Part II, *Collected Data Reports on August 1972 Solar-Terrestrial Events*, p. 330.

Kolata, J. J., Auble, R., and Galonsky, A.: 1967, *Phys. Rev.* **162**, 957.

Kozlovsky, B. and Ramaty, R.: 1974, *Astrophys. J.* **191**, L43.

Kozlovsky, B. and Ramaty, R.: 1977, *Astrophys. Lett.* **19**, 19.

Lin, R. P. and Hudson, H. S.: 1976, *Solar Phys.* **50**, 153.

Lin, R. P. and Ramaty, R.: 1978 in R. Ramaty and T. L. Cline (eds), *Gamma Ray Spectroscopy in Astrophysics*, NASA Tech. Mem. 79619, Washington, D.C., p. 76.

Lingenfelter, R. E., Flamm, E. J., Canfield, E. H., and Kellman, S.: 1965, *J. Geophys. Res.* **70**, 4077 and 4087.

Lingenfelter, R. E. and Ramaty, R.: 1967, in B. S. P. Shen (ed.), *High Energy Nuclear Reactions in Astrophysics*, Benjamin, New York, p. 99.

McGuire, R. E., von Rosenvinge, T. T., and McDonald, F. B.: 1977, *15th Int. Cosmic Ray Conf. Papers, Plovdiv* **5**, 54.

McGuire, R. E., von Rosenvinge, T. T., and McDonald, F. B.: 1981, *17th Int. Cosmic Ray Conf. Papers, Paris* **3**, 65.

Northcliffe, L. C. and Schilling, R. F.: 1970, *Nuclear Data Tables* **A7**, 233.

Pesses, M. E., Klecker, B., Gloeckler, G., and Hovestadt, D.: 1981, *17th Int. Cosmic Ray Conf. Papers, Paris* **3**, 36.

Pierce, T. E. and Blann, M.: 1968, *Phys. Rev.* **173**, 390.

Prince, T., Ling, J. C., Mahoney, W. A., Riegler, G. R., and Jacobson, A. S.: 1982, *Astrophys. J.*, **255**, L81.

Ramaty, R.: 1979, in J. Arons, C. Max, and C. McKee (eds), *Particle Acceleration Mechanisms in Astrophysics*, Amer. Inst. of Physics, New York, p. 135.

Ramaty, R., *et al.*: 1980, in P. A. Sturrock (ed.), *Solar Flares*, Colorado Assoc. Univ. Press, Boulder, p. 117.

Ramaty, R. and Crannell, C. J.: 1976, *Astrophys. J.* **203**, 766.

Ramaty, R. and Kozlovsky, B.: 1974, *Astrophys. J.* **193**, 729.

Ramaty, R., Kozlovsky, B., and Lingenfelter, R. E.: 1975, *Space Sci. Rev.* **18**, 341.

Ramaty, R., Kozlovsky, B., and Lingenfelter, R. E.: 1979, *Astrophys. J. (Suppl.)* **40**, 487.

Ramaty, R., Kozlovsky, B., and Suri, A. N.: 1977, *Astrophys. J.* **214**, 617.

Ramaty, R. and Lingenfelter, R. E.: 1973, in R. Ramaty and R. G. Stone (eds), *High Energy Phenomena on the Sun*, NASA SP-342, Washington, D.C., p. 301.

Reppin, C., Chupp, E. L., Forrest, D. J., and Suri, A. N.: 1973, *13th Int. Cosmic Ray Conf. Papers, Denver, Colorado*, p. 1577.

Rothwell, P. L.: 1976, *J. Geophys. Res.* **81**, 709.

Schaeffer, O. A. and Zahringer, J.: 1962, *Phys. Rev. Lett.* **8**, 389.

Share, G. H. *et al.*: 1983, in M. L. Burns, A. K. Harding, and R. Ramaty (eds), *Electron-Positron Paris in Astrophysics*, Amer. Inst. of Physics, New York, p. 15.

Suri, A. N., Chupp, E. L., Forrest, D. J., and Reppin, C.: 1975, *Solar Phys.* **43**, 414.

Taylor, H. W., Neff, N., and King: 1967, *Phys. Rev. Lett.* **24B**, 659.

Van Beek, H. F., Hoyng, P., and Stevens, G. A.: 1973, in H. E. Coffey (ed.), World Data Center Rept UAG-28 Part II, *Collected Data Reports on August 1972 Solar-Terrestrial Events*, p. 319.

Van Hollebeke, M. A. I., MaSung, L. S., and McDonald, F. B.: 1975, *Solar Phys.* **41**, 189.

von Rosenvinge, T. T., Ramaty, R., and Reames, D. V.: 1981, *17th Int. Cosmic Ray Conf. Papers, Paris* **3**, 28.

Wagoner, R. V.: 1973, *Astrophys. J.* **179**, 343.

Wang, H. T.: 1975, '2.2 MeV and 0.51 MeV Gamma-Ray Line Emissions from Solar Flares', Ph.D. Thesis, Univ. of Maryland.

Wang, H. T. and Ramaty, R.: 1974, *Solar Phys.* **36**, 129.

Wang, H. T. and Ramaty, R.: 1975, *Astrophys. J.* **202**, 532.

Webber, W. R., Roelof, E. C., McDonald, F. B., Teegarden, B. J., and Trainor, J.: 1975, *Astrophys. J.* **199**, 482.

Willet, J. B., Ling, J. C., Mahoney, W. A., Riegler, G. R., and Jacobson, A. S.: 1982, in R. E., Lingenfelter, H. S. Hudson, and D. M. Worrall (eds), *Gamma Ray Transients and Related Astrophysics*, Amer. Inst. of Physics, New York, p. 401.

Laboratory for High Energy Astrophysics,
NASA/Goddard Space Flight Center,
Greenbelt, MD 20771,
U.S.A.

Additional information on gamma rays and neutrons from solar flares, obtained after the completion of the present article, can be found in R. Ramaty *et al.* (1983, *Solar Phys.* **86**, 395), D. J. Forrest (1983, *Positron-Electron Pairs in Astrophysics*, Amer. Inst. of Physics, New York, p. 3.), E. L. Chupp (1984, *Ann. Rev. Astron. Astrophys.* **22**, 359), R. J. Murphy and R. Ramaty (1985, *Advances in Space Research* **4**, 127), and M. Yoshimori (1985, *J. Phys. Soc. Japan,* **54**, 487). The last reference presents data from the Japanese satellite HINOTORI which was launched in 1981.

CHAPTER 15

SOLAR RADIO EMISSION

MARTIN V. GOLDMAN AND DEAN F. SMITH

1. Introduction

The subject of solar radio emission is quite broad and has been reviewed repeatedly in articles and books (Wild *et al.*, 1963; Kundu, 1965; Zheleznyakov, 1970; Wild and Smerd, 1972; Rosenberg, 1976; Smerd, 1976a; Kruger, 1979; Melrose, 1980a). In addition, recent conference proceedings include issue 9 of *Radiophysics and Quantum Electronics* **20** (1977) and 'Radio Physics of the Sun', *IAU Symposium* **86** (Reidel, Dordrecht, 1980). Entire books exist on specialized topics, and we shall refer to them as necessary.

Because of this wealth of material, no attempt will be made in this review at completeness, either concerning subject matter or references. Rather, we have selected a number of topics which are present active areas of both observational and theoretical research. New observational and theoretical material is forcing the re-examination of present ideas and in many cases the development of new theories. This sensitive interplay between observations and theory is vital to the further development of the field. We have chosen some of the topics where rapid progress is being made in our physical understanding of the phenomenon or could be made in the near future.

We begin with a brief review of the range of phenomena in the field, as shown schematically in Figure 1. This is a dynamic spectrum or frequency versus time plot. Assuming that the frequency is related to the electron plasma frequency $\omega_p = (4\pi n e^2/m)^{1/2}$, where n is the electron density, the dynamic spectrum can also be converted into a height versus time plot as shown on the right-hand side of Figure 1. This is the reason that decreasing frequency is plotted on the left.

The classification of radio emissions from the Sun are based on the characteristic signatures shown in Figure 1. Type I noise storms in the range 40–400 MHz and low-frequency Type III bursts are shown on the left of the figure. Superimposed on the continuuim of the Type I noise storms are brighter Type I bursts (see, e.g., Fig. 5 of Smerd, 1976a) which are narrow in bandwidth as indicated by the dashed lines in the center of Figure 1 which would have been better labelled 'I storm bursts'.

When a large flare occurs, a continuum microwave burst is produced at the flashphase (the phase in which most emissions increase most rapidly), which may be followed by a microwave Type IV burst which lasts 30 min to 1 h. A flare-associated Type III burst is also produced at the flashphase which usually extends to higher frequencies than low-frequency Type III storm bursts and is more intense at meter wavelengths. It may have a continuum attached to it, which is not shown in Figure 1 and is known as a Type V

Peter A. Sturrock (ed.), Physics of the Sun, Vol. II, pp. 325–376.
© 1986 *by D. Reidel Publishing Company.*

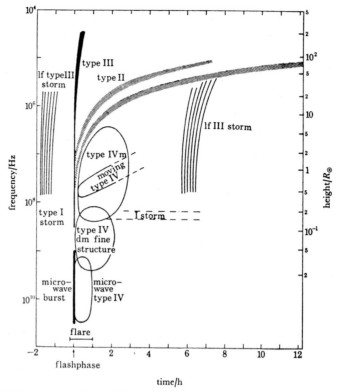

Fig. 1. A schematic representation of the radio spectrum during and after a large flare. The low-frequency Type III and Type I storms preceding and following the flare are not necessarily ingredients. Only one Type III burst has been drawn although a group of approximately ten occurs at the flash phase. Only the envelopes of the respective Type IV bursts have been drawn and usually on parts of them are filled. The height scale on the right-hand side corresponds to the plasma level of the frequency scale on the left-hand side. (After Rosenberg, 1976.)

burst. Type III bursts are also known as fast-drift bursts because of the rapid rate at which the bursts drift from high to low frequencies.

The Type II burst also starts near the flashphase and is made up of a pair of bands, as are a few high-frequency Type III bursts; the lower frequency one is called the 'fundamental' because of radiation near ω_p, and the upper frequency one is called the 'second harmonic' because of radiation near $2\omega_p$. Type II bursts have a slow drift rate.

Several minutes after the flashphase, a metric Type IVm burst, which is a continuum burst, may develop. It can be connected to the microwave Type IV burst by a decimetre Type IVdm burst which is a combination of continuum with fine structure. These continuum bursts are all stationary, but when viewed with a radio interferometer or heliograph, a separate continuum source, the moving Type IV burst, may break off from the stationary Type IVm and move out into the corona with a drift rate somewhat slower than for a Type II burst. After this, a Type I noise storm may continue for hours or days and have low-frequency Type III bursts associated with it. More complete accounts of the observations of microwave, Type I, Type II, Type III, and moving Type IV bursts are given in Section 2.

We now consider the physical mechanisms which give rise to these bursts. The reader is referred to Wild *et al.* (1963) and Smerd (1976a) for arguments for these choices. Microwave bursts are caused by $\lesssim 100$ keV electrons trapped in a magnetic arch and will be treated in Section 5, along with other bursts caused by trapped electrons. Type III bursts are caused by dilute streams of mildly relativistic electrons (≈ 10–100 keV). As such they have one of the least perturbing sources of the corona and thus represent one of the simplest phenomena. For this reason, the physics of Type III bursts will be considered first in Section 3. Type II bursts are caused by collisionless shock waves which are a more perturbative phenomenon since the density increases behind the shock. A similar region of density enhancement is a current sheet between two oppositely directed magnetic fields. The physics of Type II bursts and their related shocks is considered in Section 4. Some Type I bursts may be associated with current sheets. Alternatively, Type I bursts may be caused by electrons trapped in a magnetic arch or loop which is a region of higher density than the normal corona. Occasionally, part of this loop is blown off by reconnection processes and we see a white-light coronal transient. This is the most perturbing type of source and is often associated with a moving Type IV burst. Sometimes these bursts appear to come from their own self-contained plasmoid which is a trap for the emitting electrons. The physics of both stationary and moving traps and the associated microwave, Type I, and moving Type IV bursts are considered in Section 5.

In Section 6 we consider the status of the field of solar radio emission, where it is going, and what will be needed to insure its future health. We have not considered how the radiating electrons are accelerated since this is covered in a separate review in this work.

2. Observational Results

2.1. TYPE III BURSTS

2.1.1. *Ground-Based Observations* (above ~ 8 MHz)

The discussion of Type III radio bursts (Wild, 1950a, b; Wild and McCready, 1950) was enabled by the dynamic radiospectrograph (Sheridan, 1967; Dulk and Suzuki, 1980), which records contours of equal flux (intensity per frequency interval) on a frequency-time plot. The drift rate, defined as the time rate of change of frequency, f, is given roughly by $f^{1.85}/(100\ \text{s})$, for Type III bursts, with f in MHz. This formula holds from 10^3 MHz, all the way down to 0.1 MHz (Alvarez and Haddock, 1973), corresponding to electron streams with velocity between $0.2c$ and $0.6c$. The lifetime of a time profile at $f \approx 80$ MHz is 3–5 s. A typical time profile is characterized by a rapid rise, and a slower, approximately exponential decay of form $\exp(-10^{-8} f\, t)$ (Wild, 1950a). Recent measurements yield flux densities of the order of a few times 10^{-19} W m^{-2} Hz^{-1} (Dulk and Suzuki, 1980). For a fully resolved source (Melrose, 1980b), this corresponds to average brightness temperatures of 10^{10} K at 80 MHz and 3×10^{10} K at 43 MHz (Dulk and Suzuki, 1980).

Bursts commonly occur in groups of ten or more, with a separation of seconds (Wild *et al.*, 1963; McLean, 1971). This is probably a manifestation of an interrupted stream of electrons in the lower corona. It seems quite certain (Smerd, 1976a) that fundamental-harmonic pairs can be distinguished from groups of bursts. In Figure 2 we see a number of such pairs, recorded in 1954. The existence of the harmonic lends strong support to the early hypothesis that resonant electron plasma (Langmuir) waves are involved in Type III emissions. At first, such pairs were only found in about 10% of bursts (Wild *et al.*, 1954), but recent measuresments show that they may comprise a significant fraction of Type III bursts in the frequency range 30–210 MHz (Dulk and Suzuki, 1980). The fundamental usually begins below 100 MHz, while the harmonic component can begin from a frequency as large as 500 MHz (Dulk and Suzuki, 1980). The wide frequency bandwidth at any given time is due to the wide range of plasma frequencies encountered by the electron stream at that instant, and the relatively slow decay times of frequencies excited earlier. The frequency ratio between the two bands is generally of the order of 1.85, and never exceeds 2. This has been attributed to a 'chopping-off' of the lowest frequencies, f, of the fundamental band, due to reflection by small density irregularities, whose plasma frequency, f_p, rises above f (Roberts, 1959; Riddle, 1972, 1974; Stewart, 1974).

In Figures 2(d) and (e), the drift to lower frequencies slows to a halt. In other cases (not shown), the drift can reverse towards higher frequencies after stopping. Due to the appearance of such bursts on spectrograms in which time is on the horizontal axis, and the frequency is plotted vertically, they are called J and U bursts (Maxell and Swarup, 1958; Stewart, 1975). A likely explanation for such bursts is that they are generated by electron streams which follow *closed* magnetic field lines, and sample higher densities first, then lower densities higher in the corona, and finally higher densities lower in the corona. Such events have even been seen, from satellites, at frequencies as low as 800 kHz, corresponding to closed magnetic loops of size 35 $R_\odot$ (Fainberg and Stone, 1974).

For normal Type III bursts associated with streams which travel along open field lines, the drift to lower frequencies continues indefinitely, due to the stream front encountering progressively lower densities as it travels along an open magnetic field line.

In order to determine whether the emission corresponds to a local plasma frequency or twice the local plasma frequency, it is necessary to identify the true source height in the corona and to associate the correct electron plasma density with that height. This has proved to be a difficult and elusive task both at high ($f > 20$ MHz) and at low (20 MHz $> f$) frequencies.

At 43, 80, and 160 MHz the Culgoora radioheliograph (Wild, 1967; Sheridan *et al.*, 1973) has been used to record the apparent position, shape, and polarization of Type III bursts. In Figure 3 we see the apparent source regions for a fundamental–harmonic pair in a Type III burst at the limb, for these three frequencies. It is reassuring that the lower frequencies appear to emanate from higher altitudes, but the apparent heights probably do not coincide with true source heights due to refraction and scattering effects in the propagation of the emission. Refraction shifts the apparent fundamental position outward, and the harmonic inward (McLean, 1971; Riddle, 1972). A simple correction for this effect (Stewart, 1976) still yields coronal electron densities which need to be an order of magnitude larger than quiet Sun values at solar minimum (Saito, 1970) in order for the plasma frequencies of the true source heights to correspond correctly to the

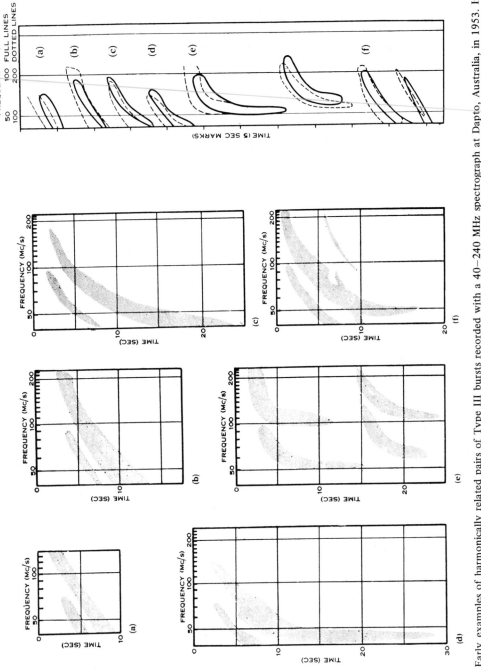

Fig. 2. Early examples of harmonically related pairs of Type III bursts recorded with a 40–240 MHz spectrograph at Dapto, Australia, in 1953. In the right-hand column the seven bursts are replotted with the harmonic band shifted 2 : 1 in frequency. The displacement, mostly leftward, of the harmonic bands indicates harmonic ratios <2 (Wild *et al.*, 1954).

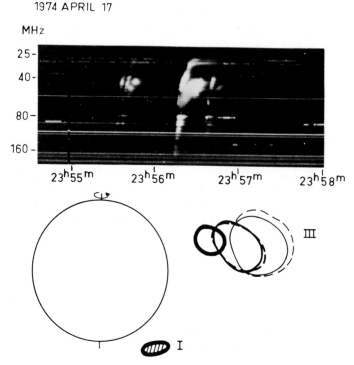

Fig. 3. Dynamic spectrum and half-peak brightness contours from the Culgoora radioheliograph, for a fundamental-harmonic Type III burst at the limb. The heavy solid line is interpreted as emission at $2f_p$ (harmonic emission at 160 MHz. The medium solid line is $2f_p$ emission at 80 MHz, and the medium dashed line is f_p (fundamental) emission at 80 MHz. The light lines show the $2f_p$ (solid) and the f_p (dashed) emission at 43 MHz (Stewart, 1976).

observed frequency for fundamental emissions and to one-half the observed frequency for harmonic emissions. To some extent, the radio bursts are likely to be generated in dense coronal structures such as loops and streamers. A ducting mechanism, proposed by Duncan (1979), helps to bring observed frequency–height correlations more into line with the quiet Sun density profile, but enhanced streamer densities are probably still necessary.

It is fairly certain that the apparent source sizes are larger than the true source sizes. The discrepancy is probably due in part to scattering of the emission from small-scale density irregularities (Riddle, 1972), but this cannot be the whole story. A close examination of Girue 3 reveals a number of surprises. First, the apparent source size rapidly increases as the frequency decreases (see also Dulk and Suzuki, 1980). Second, at a given frequency (say 80 MHz) the apparent height and size of the fundamental source (dashed medium line) is nearly identical to the apparent height of the harmonic source (solid medium line) which arrives later (Smerd et al., 1962; Bourgeret et al., 1970; McLean, 1971). Third, at a given time, the source of the fundamental (at say 80 MHz) is observed at a greater apparent height than the source of the harmonic (at 160 MHz), even though the two frequencies should be emitted from the same volume of space. All of these features are explained by Duncan's (1979) mechanism of radiowave *ducting* by radially elongated density inhomogeneities, together with the assumption of extremely divergent open magnetic field lines (Dulk *et al.*, 1979). This latter assumption also fits in nicely with satellite observations of sources of low-frequency emission near the Earth, in which a source size of 1 AU is observed (Fainberg and Stone, 1971).

The radioheliograph and spectropolarimeter (Suzuki, 1974) have provided valuable information concerning the polarization of Type III bursts in the range 24–220 MHz. At the fundamental, the average degree of circular polarization is 35% (Dulk and Suzuki, 1980; Suzuki and Sheridan, 1977), although at lower frequencies, almost complete polarization has occasionally been observed (Hanasz, *et al.*, 1980). The sense of polarization is consistent with emission in the O-mode of magneto-ionic theory (Melrose, 1980a). At the second harmonic, the average degree of polarization is 11%, and the sense is the same; however, if the second harmonic lasts longer than 1 min, then the sense of polarization reverses, possibly due to emission in the X-mode (Dulk *et al.*, 1980). For any given harmonic–fundamental pair, the degree of polarization is always greater for the fundamental. Another important affect is that, on the average, the degree of circular polarization decreases for *all* burst from center to limb (Dulk, *et al.*, 1979; Dulk and Suzuki, 1980).

A small percentage of Type III bursts are followed by Type V emission. The Type V emission has some of the character of Type III. It is thought to be second harmonic plasma emission from an abnormally slowly propagating or widely dispersed beam (Melrose, 1974b) since it exhibits such a slow drift. Its most unusual characteristic is that it has the opposite polarization (but the same degree) as the second harmonic in Type III bursts (Dulk *et al.*, 1980).

2.1.2. *Spacecraft Observations* (below 1 MHz)

The study of general properties of bursts below about 5–10 MHz is not usually possible from Earth-based radio receivers, due to the maximum electron density in the ionospheric F-layer, which causes total reflection away from the Earth. Since about 1964, but especially in the last decade, important satellite measurements have been made at a variety of radial distances from the Sun, between 0.3 AU (Helios 1 and 2) and 1.2 AU, and beyond (Voyager 1 and 2).

In order to observe the solar radio emissions, the spacecraft would be equipped typically with a dipole antenna (of length 30–120 m) and multichannel spectrum analyzer which, together, comprise a satellite radiospectrograph. In Figure 4 we see the characteristic drift from high to low frequencies, here laid out as a sequence of time profiles. Each is again characterized by a rapid rise and a slower, approximately exponential decay. At the lower frequencies the burst extends over minutes or hours, instead of the seconds associated with ground-based observations. These data come from an in-depth study and summary of early Earth-orbiting satellite results by Fainberg and Stone (1974). The observation made by Wild that the decay in time of any frequency component is exponential over many decades also holds true for the low-frequency bursts, and his decay formula need only be slightly modified to cover the frequency range from 2.8 MHz down to 67 kHz. Evans *et al.* (1973) find an approximate decay as $\exp[-(0.5 \times 10^{-8} f^{1.09} t)]$, with f in hertz. Attempts to explain the decay in terms of collisional damping (free–free absorption) of the emission by a thermal background plasma do not seem to give the

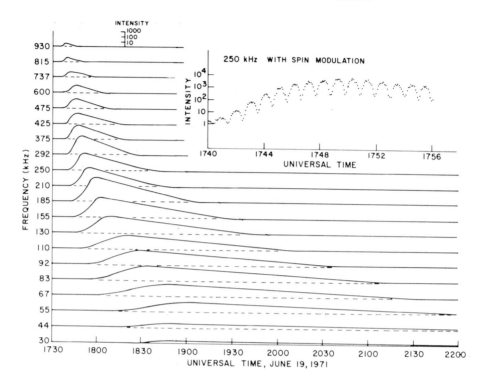

Fig. 4. A Type III burst observed between 1 MHz and 30 kHz by the IMP-6 satellite experiment. The insert figure illustrates the observed spin modulation at a frequency of 250 kHz, while, for the main figure, only the burst envelopes are shown for clarity (Fainberg *et al.*, 1972). (© 1972 A.A.A.S.)

correct temperature near 1 AU (Evans *et al.*, 1973; Haddock and Graedel, 1970; Fainberg and Stone, 1974). The excitation time, t_e, from burst onset to maximum also can be fitted (Evans *et al.*, 1973) by $t_e = 4 \times 10^8/f^{1.08}$, with f in hertz.

Due to antenna rotation on a spin-stabilized spacecraft, the received radiation shows a definite modulation pattern (Figure 4). Hence, one can determine the direction of arrival of a given frequency component of a burst, in the plane defined by the rotating dipole. This is known as the spin modulation technique, and it can be used to help determine the source location of the low-frequency bursts (Slysh, 1967; Fainberg and Stone, 1974).

There are a number of problems involved in locating the source for each frequency, associating a local density and plasma frequency with that location, and deciding whether the corresponding emission is fundamental or second harmonic. However, if these determinations can be made, one can, in principle, construct the dynamical trajectory of the exciting electron stream. The stream is guided by open solar magnetic field lines, so the trajectory of sources for a sequence of decreasing frequencies can be expected to follow the Archimedes spiral (Parker, 1958) of the field line. Fainberg *et al.* 1972) and Fainberg and Stone (1974) constructed source trajectories for a number of Type III bursts. Spin modulation gave the direction of arrival projected onto the ecliptic plane. This defined for each frequency a line of possible source locations in the ecliptic plane, emanating from the spacecraft. The location of the source on that line was determined by a model for the average emission frequency as a function of radial distance from the Sun:

$$f_{obs} = 66.8 \, R^{-1.315} \tag{2.1.1}$$

where f is in megahertz and R is distance to the Sun in solar radii, $R_\odot$. The resulting source trajectory closely matched the expected Archimedean spiral form of the magnetic field in the solar wind. Using a very crude model for the electron density profile as a function of radius, they were able to show that the associated profile of plasma frequencies was only slightly lower than the profile of *half*-frequencies determined from the observed frequency profile. From this they concluded that the observed burst was second harmonic emission from about 400 kHz down to 30 kHz. This, together with other evidence based on Type II bursts and also on the absence of spin modulation at twice the local plasma frequency at the site of the IMP-6 satellite, led them to the conclusion that most Type III bursts below 1 MHz are second harmonic.

It is important to note that, in many theousands of observed low-frequency Type III bursts, they found *no* cases in which fundamental and harmonic components were both observed.

Haddock and Alvarez (1973) interpreted most complex Type III events observed on OGO-5 as making a transition from predominantly fundamental to predominantly harmonic, below about 1 MHz. They identified the emission as fundamental or second harmonic by the time delay in its arrival. Using the same analysis for Type III bursts observed from the IMP-6 satellite, Alvarez *et al.* (1974) found the transition from fundamental to harmonic occurred at 230 kHz.

Recently, Kellogg (1980) found a transition from second harmonic to fundamental as slow as 50 kHz in one burst observed from the solar orbiting Helios 2 spacecraft. This work was based on direct observation of the Langmuir wave instability as well as the electromagnetic emissions. Hence, some low-frequency Type III bursts may be fundamental, rather than second harmonic.

Gurnett *et al.* (1978a, b) used the solar orbiting Helios 1 and 2, together with the IMP-8 and Hawkeye 1 satellites, to measure the source locations and (a few days later) the *in situ* densities and plasma frequencies along the source trajectory. This enabled a comparison to be made between the *measured* plasma frequencies and the observed emission frequencies. It appeared that the burst was second harmonic rather than fundamental at the measured frequencies, thus supporting the conclusions of Fainberg and Stone (1974).

The peak intensity of a given low-frequency component of a burst can be quite high. Evans *et al.* (1971) found the brightness temperatures for the 1 MHz component of one Type III burst was in excess of 10^{15} K (fluxes in excess of 2×10^{-15} W m^{-2} Hz^{-1}). However, Melrose (1980a) quotes a typical brightness temperature of 10^{11} K for the 50 kHz (harmonic) component at 1 AU. In Figure 4 the peak intensity is largest at 185 kHz, corresponding to about 50 $R_\odot$.

A more important intensity measure is the volume emissivity, J, which is the power emitted per unit volume per unit solid angle: $J \equiv (\Delta P / \Delta V \, \Delta \Omega)$ W m^{-3} ster^{-1}. Recently Tokar and Gurnett (1980) – see also Gurnett *et al.* (1980) – complied the volume emissivity at a number of frequencies in each of 36 low-frequency Type III events observed by IMP-8 and ISEE-1 satellites. By using Equation (2.1.1), they associated a heliocentric radial distance with each emissivity. The result is shown in Figure 5. Variations in emissivity of ever five orders of magnitude are evident at some radii, but the best power law fit is $J = J_0 R^{-6.0}$, where $J_0 = 1.5 \times 10^{-24}$ W m^{-3} ster^{-1}.

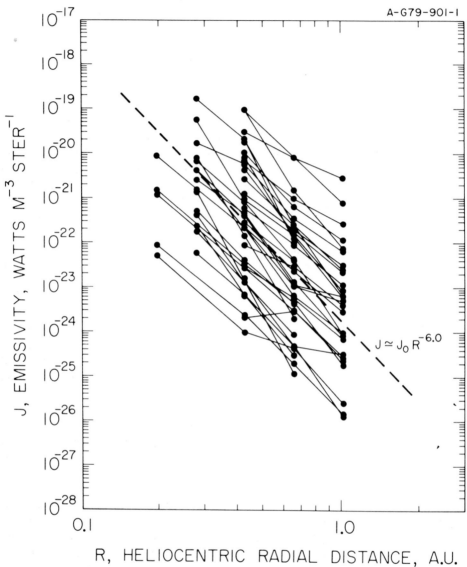

Fig. 5. The volume emissivity as a function of radial distance from the Sun, determined form 36 Type III radio bursts detected by IMP-8 and ISEE-1. Frequency components common to one event are linked by straight lines. (From Gurnett *et al.*, 1980.)

2.1.3. *Langmuir Waves and Electron Streams*

A fundamental question concerning the origin of Type III bursts is how well they correlate with the electron streams and Langmuir waves which are supposed to produce the observed emissions. Spacecraft measurements over the past decade have verified the existence of both the electron stream and Langmuir waves.

It has been established fairly definitely that stream electrons (presumably from solar flares or other activity) are primarily responsible for the Type III radio emission (Lin, 1970, 1974; Alvarez et al., 1972; Frank and Curnett, 1972; Lin et al., 1973, 1981; Gurnett and Frank, 1975). The streams have the proper speed to account for the observed drifts of the Type III bursts ($0.2c$ to $0.6c$). Typical beam to background density ratios are 10^{-7}, or, at most, 10^{-6}, at energies around $\sim$25 keV.

An interesting correlation between radio flux and the flux of electrons with energies in excess of 18 keV has been demonstrated by Fitzenreiter et al. (1976) and is illustrated in Figure 6. They examined a number of Type III events at 1 AU from the IMP-6 spacecraft at frequencies believed to be local (because of the absence of spin modulation). The onset of >18 keV electrons coincided with the onset of local emission. (Lower energy electrons arrived after the emission had peaked.) As both flux of radio emission F_R, and the flux of high energy electrons, F_E, increased in turn, a sequence of values of F_R were plotted against F_E. Two kinds of power laws, $F_R \propto (F_E)^\alpha$, were found. At flux values less than 50 per cm^2 s sr, $\alpha = 1$, whereas for flux values greater than 50 per cm^2 s sr, $\alpha = 2.4$. In Figures 6(h) and (i), the transition is visible. They argue that the existence of two distinct regimes of radio emission implies a fundamental change in the emission mechanism of Type III bursts when the electron flux reaches a critical level.

Very recently, more detailed studies of the electron streams were presented, based on data collected from the ISEE-3 spacecraft, 259 R_e upstream from the Earth (Lin et al., 1981). In Figure 7(b) we see the spin-averaged electron fluxes in different energy intervals, as a function of time. In Figure 7(a), the measured electric field strength are plotted simultaneously, in various wide band frequency channels. The time profiles at 100 and 56.2 kHz are interpreted as harmonic emission, and the 56.2 kHz component is identified as local because of the absence of spin modulation. (It is at roughly twice the local plasma frequency.) Two temporally consecutive bursts are evident, but only the second is of concern here. Its onset coincides with the arrival only of the electrons with energy above 200 keV, at a very low flux level. Lower energy electrons arrive progressively later due to velocity dispersion in the stream. The electric field profiles in the 31.1 and 17.8 kHz channels in Figure 7(a) are interpreted as electron plasma (Langmuir) waves. At energies of 20 keV and above, the pitch-angle distribution (not shown) is flat-topped and sharp-sided, whereas it is more beam-like below 10 keV.

By assuming the electron distribution function is symmetric about the magnetic field, Lin et al. (1981) were able, for the first time, to plot the electron distribution function for parallel velocities, V (integrated over all $V_\perp$). The resulting one-dimensional distribution function is shown in Figure 8 in a sequence of (displaced) plots over 5-min intervals. More detailed analysis shows the distribution function first develops a tiny, short-lived positive slope, $f' > 0$, at 1953 UT. This rises by two orders of magnitude around 2000 UT, and remains positive until about 2045 UT. The region of $f' > 0$ begins at $V = 1.3 \times 10^{10}$ cm s^{-1}, and moves down to $V = 3 \times 10^9$ cm s^{-1}, with a typical range of $\Delta V = 0.3$ V. There are several important conclusions which can be drawn from Figure 8. First, there is a surprising fact to note about the ambient background electron distribution (black dots in Figure 8(a)). An enhanced tail of nonthermal electrons is observed at all times. This tail may be fitted approximately with an exponential distribution, having an effective temperature of around 10 keV. Second, the authors note that the onset times of large positive slopes, f', correspond very well with the onset times of Langmuir waves at 2000 UT (see Figure 7(a)). This establishes for the first time the theoretically expected

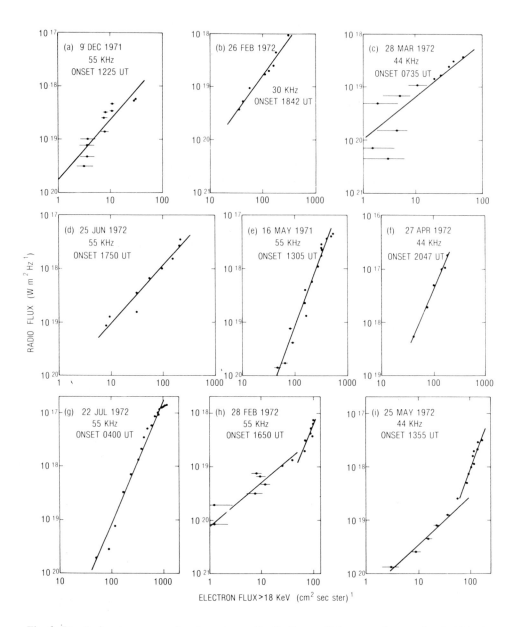

Fig. 6. Events showing a power law dependence of radio flux on high-energy electron flux (>18 kev). The slopes of the fitted straight lines (equal to the power law index, α) fall into two distinct groups. Events (a)–(d) have $\alpha \sim 1$; events (e)–(g) have $\alpha = 2.4$; events (h) and (i) show an abrupt transition from $\alpha \sim 1$ to $\alpha \sim 2.4$ (Fitzenreiter *et al.*, 1976).

causal relation between the $\gtrsim 25$ keV part of the electron stream and (Cerenkov-emitted) unstable Langmuir waves.

However, the Langmuir waves, and the $\gtrsim 25$ keV electrons which produce them, are both observed about 20 min after the onset of the local harmonic emission, so their

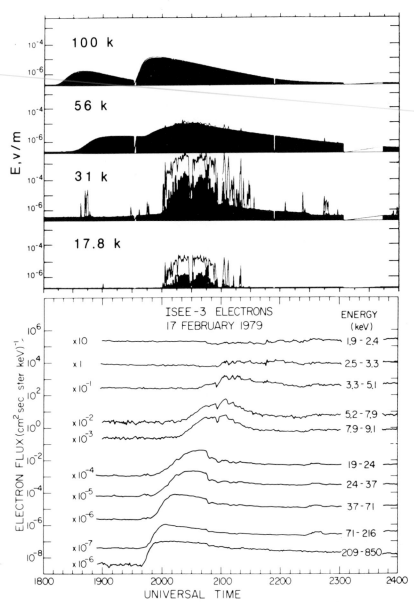

Fig. 7. Electric field intensities and electron fluxes versus time for an event studies from the ISEE-3 satellite. (a) Top panel shows intensity, measured in four broad frequency bands. Black areas show 64 s averages, solid lines give peak intensity, measures every 0.5 s. Smothly varying profiles in 100 and 56.2 kHz channels show two Type III bursts, but only the second is of interest here. Impulsive emission in the 31.1 and 17.8 kHz channels are electron plasma (Langmuir) waves. (b) Omnidirectional electron fluxes from 2 to >200 keV, showing velocity dispersion. No significant change in flux is observed below ~2.5 keV (Lin *et al.*, 1981). (Reprinted countesy of the authors and *The Astrophysical Journal*, publ. by Univ. Chicago Press; © 1981 Am. Astron. Soc.)

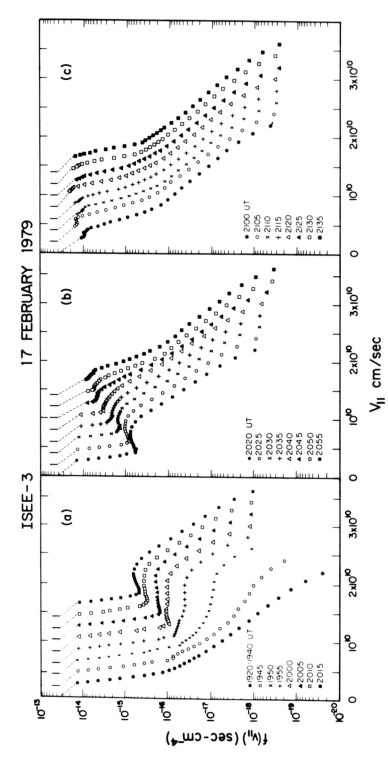

Fig. 8. Synthesis of data to construct a 1-D velocity distribution function of the electrons as a function of time. Each succeeding distribution within a panel is shifted to the right in velocity by 2×10^9 cm s^{-1}. The distribution averaged over 20 min prior to the event onset is indicated by the sold dots in panel (a). 64 s measurements of the distribution during the event are shown every 5 min thereafter (Lin et al., 1981).

theoretically expected causative role in the second harmonic emission (Section 3) would seem to be ruled out! The delay in the appearance of Langmuir waves until the harmonic emission is well underway is not unique to this event or to observations at 1 AU. It seems to be a pervasive feature of all measurements which detect both the local second harmonic emission and the local plasma waves, including those observed *in situ* near 0.5 AU (Gurnett and Anderson, 1977; Gurnett *et al.*, 1978b; Kellogg, 1980).

This raises a serious theoretical challenge. The most upsetting explanation would be that Langmuir waves do not play a role in harmonic emission, although a viable alternative mechanism is unknown. Another explanation would be that the emission interpreted as local second harmonic is really fundamental, coming from much closer to the Sun, but scattered significantly from density inhomogeneities to account for the observed absence of spin modulation. However, this seems contrary to most (but not all) of the observational evidence we have reviewed earlier in this section. It appears that the only other possibility is that the spacecraft somehow consistently misses the early Langmuir waves, although in this case one needs a causative agent for the Langmuir waves other than the usual <25 keV electrons.

Since the Langmuir waves are usually deemed essential for the emission process, we conclude by reviewing the observational evidence concerning them. The frequency of the observed spiky electric field structures, such as those in Figure 7(a) at 31.1 kHz corresponds satisfactorily to the local plasma frequency based on the measured local electron density. It seems to be *inferred* that these fields are electrostatic (i.e. longitudinal), since a dipole antenna cannot distinguish polarization. Spin modulation has sometimes shown the fields to be closely aligned with the solar wind magnetic field (Gurnett and Anderson, 1977).

The spiky structure is characteristic. Since the solar wind is sweeping the plasma waves past the spacecraft at $\sim$600 km s^{-1}, and the instrumental time resolution of the electric field is 50 ms, spatial structures smaller than about 30 km cannot be resolved. However, many of the observed Langmuir field spikes do tend to be associated with this size, indicating that even smaller unresolved spatial structures cannot be ruled out.

Gurnett *et al.* (1980) have grouped together all of the 90 electron plasma oscillation events which have been identified to date in conjunction with Type III bursts. The data, taken from Helios 1 and 2, Voyager 1 and 2, and IMP-8, are summarized in Figure 9. The maximum electric field strength is plotted versus the heliocentric radial distance at which the waves are observed. Although a wide spread is evident at any radius, a power law fit shows a decrease of field with radius, roughly with the dependence $E \simeq E_0 R^{-1.4}$, where $E_0 = 0.5$ mV m^{-1}. The fall-off with radius is consistent with the fall-off of the volume emissivity with radius (Figure 5), and suggests a causal relationship. The most intense electron plasma oscillations (field strengths from 1 to 10 mV m^{-1}) are usually detected close to the Sun, at heliocentric radial distances less than 0.5 AU. It is noteworthy that the majority of Type III bursts occurring near 0.5 AU are *not* accompanied by measurable Langmuir waves on resolvable scales (Gurnett and Anderson, 1977).

2.2. MICROWAVE BURSTS

Microwave bursts are a type of continuum burst, so-called because they extend over a broad range of frequencies from a few tens of gigahertz to several hundred megahertz

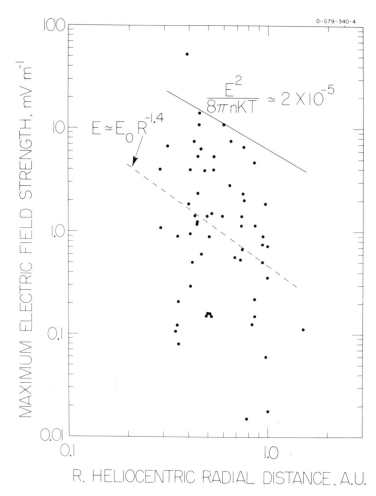

Fig. 9. A plot of the peak electric field strength for all of the plasma oscillation events associated with Type III bursts (detected to date) as a function of radial distance from the Sun.

without pronounced spectral structure (Figure 1). Microwave bursts can be classified into impulsive bursts, gradual bursts, and microwave Type IV bursts (Wild *et al.*, 1963). Impulsive bursts have a time-scale of 1–5 min and brightness temperatures up to 10^9 K. (The brightness temperature of solar radiation is the equivalent temperature which a black body would have which emitted radiation of the same intensity at the same frequency.) Gradual bursts have a time-scale of tens of minutes and brightness temperatures up to 10^6 K. Microwave Type IV bursts have a time-scale of 10 to 30 min and brightness temperatures up to 10^9 K.

The impulsive microwave bursts are closely correlated with hard X-ray bursts and the intensity profiles usually track each other, but with some time delay of the order of 1 s for the microwaves when seen with subsecond time resolution. When seen with 20 ms time resolution, the X-rays have 80 ms spikes which are absent in the microwaves as though the microwaves were a smoothed out version of the hard X-rays (Lin *et al.*, 1980).

Thus, while microwaves and hard X-rays come from related electron populations, they clearly do not come entirely from the same population. This is born out by the VLA maps at 15 and 22.5 GHz with arcsec resolution (Marsh and Hurford, 1980) and the hard X-ray images taken with the Solar Maximum Mission (SMM) with 8″ resolution (Hoyng et al., 1981). The microwaves come from the tops of loops and the 16–30 keV X-rays and Hα emission come mostly from the footpoints of the loops.

The accepted radiation mechanism for the microwave bursts is gyrosynchrotron radiation due to electrons with energies greater than about 100 keV spiraling in a magnetic field (Ramaty, 1969; Trulsen and Fejer, 1970). However, occasional fine structure in microwave Type IV bursts at wavelengths $\gtrsim 10$ cm observed with a time constant of 20 ms have brightness temperatures greater than 10^{13} K (Slottje, 1980) and can only be explained by plasma radiation (Smith and Spicer, 1979). The major problem in interpreting microwave bursts is that several factors affect their intensities and spectra, and we have no independent handle on many of them. The most important of these are non-uniformity of the magnetic field and various low-frequency absorption mechanisms such as synchrotron self-absorption. Because of these uncertainties, we shall not treat the theory of microwave bursts in detail, but shall consider the related problems of Type I and moving Type IV bursts in Section 5.

2.3. TYPE II BURSTS

Type II bursts consist of two slow drifting bands near the fundamental and second harmonic of the plasma frequency at metre wavelengths (Figure 1). When the drift rate is converted into an effective radial velocity, a velocity in the range 800–2000 km s^{-1} is obtained which was identified with a collisionless magnetohydrodynamic (MHD) shock wave ascending through the corona. This identification was confirmed when a Type II burst was observed down to 30 kHz with the IMP-6 satellite (Malitson et al., 1973). The last observation was made just before a sudden commencement geomagnetic storm which is known to occur when an interplanetary shock wave impinges on the magnetosphere of the Earth. Several Type II bursts have recently been observed below 1.3 MHz with the Voyager spacecraft (Boishot et al., 1980). The relatively frequent occurrence of these bursts at large distances from the Sun would favor the hypothesis of shock propagating parallel to the ambient magnetic field. However, Boishot et al. found that the observed spectral characteristics showed that the source of emission was restricted to only a small portion of the shock which could well be a region where the magnetic field is locally perpendicular to the shock. Unfortunately, no measurements have been reported to date where the properties of the shock, the energetic electrons and the radio emission have all been measured together with the detail which we have for Type III bursts.

It is important for discussing the theory to determine whether radio emission is produced for shock propagation primarily parallel or perpendicular to the ambient magnetic field. For one Type II burst at metre wavelengths, Smerd (1970) concluded that a better case could be made for parallel than perpendicular propagation. However, it is also true at metre wavelengths that only a part of the shock emits at any one time since when seen with the Culgoora radioheliograph, one part of the source brightens and fades, and then another part brightens (Wild and Smerd, 1972). This indicates that some special condition must be satisfied for radio emission which may well be related to the mode of propagation of the shock. The brightness temperatures of Type II bursts reach about 10^{11} K for both the fundamental and the harmonic in the bright parts of the bursts.

Type II bursts are rich in structure (Wild *et al.*, 1963; Wild and Smerd, 1972). Among the most important of these are:

(1) *Band-splitting*. Occasionally, each harmonic band is split into two or more components separated by about 10% of the midfrequency (Figure 13 of Wild and Smerd (1972)). These split bands clearly emanate from different spatial locations at 80 MHz which led Smerd *et al.* (1975) to postulate that they arise from emission from ahead of and behind the shock. This is consistent with the fact exemplified in Figure 13 of Wild and Smerd (1972) that detailed spectral features are sometimes duplicated in the two components of a split band.

(2) *Herringbone structure*. In about 10% of the bursts the harmonic bands consist of a succession of short-lived, broad-band elements which have fast frequency drifts of both positive and negative signs like mini-Type III bursts. Sometimes these diverge from a narrow-band feature and sometimes this feature is absent.

It can be seen immediately in comparing the observations of Type II and Type III bursts that we are dealing with a more complex phenomenon with Type II bursts. On the other hand, because of their much slower drift, they provide the clearest example of plasma emission at the fundamental and second harmonic and other fine structure clues whose interpretation we shall discuss in Section 4.

2.4. MOVING TYPE IV BURSTS

Type IV bursts are a very complex type of continuum radiation which typically occurs after Type II bursts in large flares (Figure 1). Some of the radiation must be produced near the plasma frequency due to its high brightness temperature and some of it must be synchrotron radiation due to the high degree of circular polarization (Wild and Smerd, 1972). A part of the Type IV burst moves progressively outward through the corona when viewed with the Culgoora radioheliograph to heights as large as 6 $R_\odot$ with velocities in the range 20–1400 km s^{-1}. This is a moving Type IV burst and is often associated with white-light coronal transients indicating the ejection of material (Stewart *et al.*, 1974a, b). Moving Type IV bursts have been further classified into three types (Smerd and Dulk, 1971).

(1) *Advancing shock front*. This appears as a wide irregular arc on the heliograph record some minutes after a Type II source has occurred. The arc gradually expands outward and can be explained as synchrotron radiation at 80 MHz as the source attains a height of $\gtrsim 1 R_\odot$. All varieties of moving Type IV bursts have the characteristic of a late first appearance at a height $\gtrsim 1 R_\odot$ at 80 MHz which can sometimes be explained by suppression of synchrotron radiation by the medium (Boishot and Clavelier, 1967).

(2) *Expanding magnetic arch*. This second variety is due to electrons trapped in a magnetic arch which expands with time. This variety is often associated with an activated filament seen in Hα. The arch progressively expands at a velocity ≈ 300 km s^{-1} and develops strong circular polarization of opposite senses at its two feet as though electrons are mirroring near these positions. Often after some

expansion as a whole arch, the source condenses into several discrete sources, but still arranged along a loop of increasing dimensions. The emission from the footpoints is best explained as plasma emission for this type.

(3) *Ejected plasmoid.* The last variety is characterized by uniform radial motion of the source to very great heights occasionally, but more often to $2-3\ R_\odot$. These sources often break up into two sources which are circularly polarized in opposite senses. Although the original interpretation of this type was that a plasma with its own magnetic field is being ejected from the corona (i.e. a plasmoid), the most recent observations with the Culgoora radioheliograph operating at three frequencies (Duncan *et al.*, 1981; Duncan, 1981) show that this type is remarkably similar to the expanding arch in a corona whose density has increased by a large factor due to the ejection of a transient. The only difference between these two types is that in the expanding arch the accompanying density enhancement causes a bulging out of the corona whereas in the ejected plasmoid the accompanying density enhancement actually loses its solar attachment and becomes a transient.

The latest observations (Duncan *et al.*, 1981; Duncan, 1981) also show brightness temperatures up to 5×10^{12} K which can only be explained by plasma emission. The observed degree of polarization of up to 100% implies that the emitting electrons cannot have energies much above 100 keV on the gyrosynchrotron hypothesis. However, 100 keV electrons cannot give gyrosynchrotron brightness temperatures above 10^9 K (Duncan, 1981). The observed sense of polarization is O-mode consistent with plasma emission and inconsistent with the X-mode sense expected from gyrosynchrotron emission. It should be noted that until these recent observations plasma emission had been rejected for moving Type IV bursts (Dulk *et al.*, 1978) because it was thought that electron densities at the heights of moving Type IV sources were too small and, on the evidence of one-dimensional interferometers, moving Type IV sources showed no dispersion of source position with observing frequency as would be expected for plasma emission. With the Skylab observations of coronal transients, cases were observed with densities as high as 1.5×10^9 cm^{-3} at heights of $3\ R_\odot$, corresponding to a plasma frequency of 270 MHz (Schmahl and Hildner, 1977). The material was confined in threads with steep density gradients so that source dispersion at different frequencies should be different than in the normal corona traversed by a Type III burst. The three frequency two-dimensional interferometer observations with the Culgoora heliograph have shown that there is source dispersion at different frequencies in moving Type IV sources (Duncan, 1981). Thus we have a case where the use of powerful new observing techniques has forced a complete rethinking of the interpretation of these bursts. This interpretation for advancing front and ejected plasmoid type sources for which we now have a rich database will be considered in Section 5.

2.5. TYPE I NOISE STORMS

Type I noise storms are the most persistent form of solar activity at metre wavelengths and are not associated with flares, but occur continuously in active regions. They have been reviewed by Elgaroy (1977). The storms consist of Type I continuum in the 40–400 MHz range which has a slow rise time, long duration of hours to days, and a relative bandwidth of about 100%, and Type I bursts which have a rise time ≈ 0.1 s, a duration of

0.1–10 s, and a relative bandwidth of a few percent. The emission of both continuum and bursts is predominately polarized in the sense of the O-mode and often reaches nearly 100%, consistent with fundamental plasma emission. As shown schematically in Figure 1, low-frequency Type III storms have approximately the same starting frequency as the lowest frequency of Type I emission which lends some support to the hypothesis that the frequency of Type I emission is related to the plasma frequency.

However, the directivity properties of Type I and Type III emissions are quite different. As discussed in Section 2.2, Type III bursts have broad cones of emission. Observations made from the Earth and from a spacecraft at 169 MHz to give a stereo capability have shown that the beamwidth of individual Type I bursts is less than 25° (Steinberg et al., 1974) and sometimes tilted 60° away from the local solar vertical (Bougeret and Steinberg, 1980). The observations of individual Type I bursts with high spatial (3'.4) and temporal (0.1 s) resolution have shown the existence of bursts whose peak intensity moves during their lifetime which can only be explained by propagation effects that take place very close to the primary source (Bourgeret and Steinberg, 1977). The high directivity of individual bursts argues against much isotropic angular scattering of the radiation far from the source. On this basis, Bougeret and Steinberg (1977) have developed a model in which the radiation is produced in bunches of overdense fibers and suffers multiple reflections off these fibers. For the model to work they need emission directed along the fibers as would be expected for gyrosynchrotron emission for fibers aligned along the magnetic field.

Heliography has also shown the persistence of a given spatial–temporal shape at the same position which means a broad noise storm center can be divided into a few distinct and fixed sources where bursts of constant characteristics are emitted. This indicates that the burst sources are very well localized and connected with very stationary structures in the corona. In the context of the fiber model, many loops consisting of a number of fibers with different densities and orientations can make up an active region. Thus close-by sources with different beam orientation would be expected and are sometimes observed (Bougeret and Steinberg, 1980). The noise storm center then consists of many sources that cannot be observed simultaneously from a given direction, and the spatial–temporal shape is obtained by strong scattering close to the source inside the fibrous medium. Other observations show that Type I sources are located over regions with soft X-ray loop structures (Stewart and Vorpahl, 1977) and it is possible that there are many more loops which are too cool to be observed.

The directivity of Type I emission leads to a very marked center-to-limb effect for their observability (Elgaroy, 1977). Since noise storms are associated with loop structures, but can be up to 100% polarized, the emission must be confined primarily to one leg of the loop. Within the fiber bunch model, the center-to-limb variation and degree of polarization is hypothesized to occur as follows (Bougeret and Steinberg, 1980):

(1) The radiation is most often oriented in the direction of the fibers with less frequent orientation at large angles.

(2) Propagation transverse to the direction of the fibers results in a depolarization due to multiple reflections. Propagation along the fibers results in very few reflections and little depolarization.

(3) Burst sources are more frequently located in regions where the radiating part of the arch (bunch of fibers) is close to the solar radial which will produce the observed center-to-limb distribution.

The Type I noise storm phenomenon is our last example of a quite complex process which is rich in details which should give many clues for a theory. It is fair to say that no theory to date has been able to explain all the observations. Type I noise storms are related to moving Type IV bursts in that they both arise from trapped electron populations, but without the very rapid movement of the traps possible during flare conditions. In fact, Type I bursts cluster in 'drifting chains' which drift to lower frequency 70% of the time at 1 MHz s^{-1} at 150 MHz which leads to an average velocity of 90 km s^{-1} assuming fundamental plasma emission. At lower frequencies the drift rates and derived speeds are considerably smaller, and consistent with the observed speeds of non-flare-associated rising loops observed in white light (Golsing *et al.*, 1976). Thus we shall consider the theory for Type I noise storms along with that for moving Type IV bursts in Section 5.

3. Theory of Type III Radio Bursts (Radio Emission from Electron Streams)

3.1. OVERVIEW

We shall only discuss those theories which regard electron-stream-excited Langmuir waves as the source of observed electromagnetic emission, and in which only a small fraction of the Langmuir wave energy is lost by electromagnetic emission processes. The evolution of the Langmuir wave spectrum can therefore be studied independently of any coupling to transverse fields. There are two clases of nonlinear mechanisms which govern the evolution of the Langmuir wave spectrum. 'Quasilinear theory' studies the interaction between the waves and the electron stream, and their mutual evolution. 'Mode-coupling (or wave–wave interaction) theory' involves coupling between Langmuir waves in different parts of k-space, and includes induced scatter, as well as nonlinear refractive effects. Finally, there may be linear refractive effects due to small density irregularities, which can affect the Langmuir wave spectrum.

3.2. QUASILINEAR THEORY

Quasilinear theory is a statistical theory, in which the ensemble-averaged spectral energy density of Langmuir waves, $P(k)$, grows due to the free energy in a 'bump-on-tail' electron distribution function, $F(v)$, which simultaneously evolves because of diffusion in velocity space as electron orbits are perturbed by the Langmuir waves. Most theories are one-dimensional. For a homogeneous beam, a plateau eventually forms in velocity space, and Langmuir wave growth stops. (On a much longer time-scale, both the beam and the waves will thermalize, due to collisions.)

However, we know from observation (Figures 7 and 8) that the electron streams associated with Type III bursts are not spatially homogeneous. Due to velocity dispersion, the fast electrons arrive before the slow ones Figure 7(b). As shown in Figure 8, the peak of the 'bump-on-tail' distribution moves from higher to lower speeds. Waves emitted at early times with phase velocities matched to electron velocities, v, for which the 'bump' in $F(v)$ has positive slope, should therefore be reabsorbed at later times, when the phase velocities correspond to a (displaced) bump with negative slope. In this way, the stream can propagate over long distances.

'Inhomogeneous' quasilinear theory is required to treat the process quantitatively. Early analytical predictions by Ryutov and Sagdeev (1970) have been developed and confirmed in the context of Type III burst streams (Zheleznyakov and Zaitsev, 1970a, b; Zaitsev *et al.*, 1972; Grognard, 1975). The most complete calculations require extensive numerical work (Takakura and Shibahashi, 1976; Magelssen and Smith, 1977; Grognard, 1980; de Genouillac and Escande, 1981). The one-dimensional inhomogeneous quasilinear equations solved by Magelssen and Smith had the following form:

$$\frac{\partial P}{\partial t} = \left[av^2 \frac{\partial F}{\partial v} - \gamma_b \right] P + a' \, vF + \beta, \tag{3.2.1a}$$

$$\frac{\partial F}{\partial t} + v \frac{\partial F}{\partial x} = -\tau F + b \left[\frac{\partial F}{\partial v} + \frac{\partial}{\partial v} \left(\frac{P}{v} \frac{\partial F}{\partial v} \right) \right], \tag{3.2.1b}$$

where a, a', and b are constants, γ_b and β represent wave damping and spontaneous emission by the background plasma, and τ represents (slow) collisional relaxation of the beam. In (3.2.1a) $P(k)$ is driven unstable at wavenumbers k where the slope of the bump, $\partial F/\partial v$, is positive, and where the Cerenkov condition, $v = \omega_p/k$, is satisfied. Spontaneous emission is included in the a' term. In (3.2.1b) the advective term, $v(\partial F/\partial x)$, is the essential new feature in the inhomogeneous theory. Equations (3.2.1a, b) were solved, subject to the boundary condition of an assumed stream with $F \propto v^{-5} \exp[-(t - t_0)^2/T^2]$, generated at the injection point, low in the solar corona. Typically, t_0 and T were chosen on the order of seconds. In Figure 10, we see a time-ordered sequence of profiles of F and of P, at a distance 2×10^{11} cm from the injection point. Both velocity dispersion and reabsorption are evident, as well as plateau formation.

A central result of any nonlinear calculation is the predicted total energy density in Langmuir waves at a given spatial point. We define the dimensionless Langmuir energy density as

$$W \equiv \frac{\langle |E|^2 \rangle}{4\pi \, n_e k_B T_e} = \int \frac{dk}{2\pi} \frac{P(k)}{n_e k_B T_e}, \tag{3.2.2}$$

where n_e is the electron density and $k_B T_e$ the electron thermal energy. For the quasilinear calculations the peak energy density is $W \approx 10^{-5}$ typically, for beam to background density ratio of 10^{-7}. Grognard (1980) used the one-dimensional distribution function found by Lin *et al.* (1981; see Figure 8) as a boundary condition for integrating the quasilinear equations forward to a spatial point downstream. An initially low level of waves was found numerically to grow to $W \approx 10^{-5}$, in agreement with the peak Langmuir electric fields of several millivolts per metre measured by Lin.

Finally, it is worth remarking that inhomogeneous quasilinear theory predicts, at a given spatial point (and thus a given plasma frequency), a temporal build-up of W^2 to a maximum and then a temporal relaxation which is in accord with the temporal emission profiles in Figure 4. This was first noted by Zaitsev *et al.* (1972), and confirmed by Smith and Magelsson. It is significant because the theory of second harmonic emissions gives

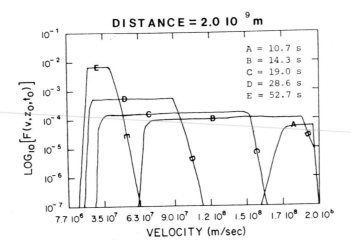

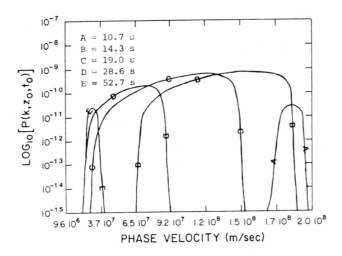

Fig. 10. Inhomogeneous 1-D quasilinear calculation of simultaneous stream and Langmuir spectral evolution. The electron stream distribution function, and the corresponding Langmuir wave distribution are shown at five different times at a single spatial point, 2×10^9 m above the Sun. The Langmuir distribution is plotted as a function of the phase velocity, ω_p/k. Both velocity dispersion and re-absorption and evident (Magelssen and Smith, 1977).

volume emissivities quadratic in the Langmuir wave spectral energy density (see Section 3.5). If the emission is assumed to be fundamental, a somewhat less adequate fit is still possible (Zaitsev *et al.*, 1972).

3 3. INDUCED SCATTER OFF IONS

Quasilinear phenomena and the process of induced Langmuir scatter off the polarization clouds of ions (or off ion-acoustic waves, when $T_e \gg T_i$) together comprise the subject matter of 'weak turbulence theory', provided the scatter is treated statistically (using the random phase approximation). Kaplan and Tsytovich (1967, 1968, 1973) considered induced scatter off ions independently of quasilinear theory, as a stabilization mechanism which removes Langmuir wave energy from resonance with the beam. Zheleznyakov and Zaitsev (1970a) treated induced scatter off ions together with quasilinear theory for the lower corona (100 MHz $< f_p < $ 200 MHz), and concluded that the former was negligible. Latter work seemed to support this conclusion, by showing that the energy density in Langmuir waves had to be comparable to the energy in the electron stream for induced scatter to become important (Smith and Fung, 1971; Heyvaerts and be Genouillac, 1974).

More recent testimate (Smith, 1977), based on spacecraft observations at lower frequencies, show that it *cannot* be neglected in comparison with quasilinear effects for the value $W \approx 10^{-5}$ measured in intense bursts at 0.45 AU. It is estimated that the time for the scattered wave energy to equal the beam-resonant wave energy is about 40 s, which is an order of magnitude shorter than the duration of the Langmuir waves. Induced scatter is therefore a fast process.

There is some difference of opinion over whether induced scatter is principally a 1-D or a 3-D process. In 2-D numerical calculations of induced scatter off ion-acoustic quasi-modes, it is found to occur as a 1-D backward or forward scatter (Bardwell and Goldman, 1976; Nicholson and Goldman, 1978; Hafizi *et al.*, 1982). Other authors claim it is essentially 3-D (Heyvaerts and de Genouillac, 1974; Perkins *et al.*, (1974).

If one considers the 1-D scatter in the dimension defined by the wave vector, $\mathbf{k}_0$, of a beam-resonant 'pump' Langmuir wave, then the scattered wave vector is determined from the kinematical condition $\omega_L(\mathbf{k}_0) = \omega_L(\mathbf{k}_s) + v_i |\mathbf{k}_0 - \mathbf{k}_s|$, where $\omega_L(k) = (\omega_p^2 + 3v_e^2 k^2)^{1/2}$ is the Langmuir wave dispersion relation and v_i the ion thermal velocity. This leads to a scattered Langmuir wave vector, $k_s/k_D = \alpha - k_0/k_D$, where $\alpha = \frac{2}{3}(m/M)^{1/2} = 1.55 \times 10^{-2}$, in a plasma with equal electron and ion temperatures, and k_D is the Debye wavenumber, $(4\pi n e^2/k_B T_e)^{1/2}$. Since the beam-resonant wave number is $k_0/k_D = v_e/v_b$, the character of the scatter depends on the size of v_e/v_b relative to α. If it is assumed that $v_e/v_b \gg \alpha$, as in most of the early work near 100 MHz, then the scattering is in the backward direction until the scattered mode builds up enough energy to act as a pump for a secondary scatter into the forward direction. During each scatter, the wavenumber is reduced by α (see Nicholson and Goldman, 1978, for a graphic illustration) until the cascade of multiple scatterings builds up a distribution of Langmuir waves in both directions, between $\mathbf{k}_0$ and $-\mathbf{k}_0$. In the 3–D version of this, an isotropic distribution is created. However, the *measured* beam near 1 AU (Figure 8) has the property that k_0/k_D is of the same order as α, so that only a *single* scatter should occur, bringing Langmuir waves to essentially zero wavenumber. In preliminary calculations, using the data in Figure 8, Grognard (1981) has studied quasilinear and induced scatter effects near 1 AU simultaneously, and observed an important role played by the latter, leading to a build-up of scattered energy at wavenumbers much less than beam-resonant wavenumbers. The resulting spectrum may be unstable to modulational instability and spatial self-focusing as discussed in the next section.

3.4. WAVE–WAVE EFFECTS OF THE NONLINEAR REFRACTIVE AND SELF-FOCUSING VARIETY

The fundamental equations underlying most discussions of this wider class of wave–wave interactions is the pair of beam-driven Zakharov equations for the slowly varying complex envelope, $\mathbf{E}(\mathbf{r}, t)$ of electrostatic electric field oscillations near the plasma frequency [of form $E(\mathbf{r}, t) = \text{Re } \mathbf{E}(\mathbf{r}, t) \exp(-i\omega_p t)$], and the slowly varying density perturbation $\delta n_2(\mathbf{r}, t)$ which is second order in $\mathbf{E}$. With time in units of ω_p^{-1}, and distances in units of $\sqrt{3} \, k_D^{-1}$, these equations assume the following general form (Zakharov, 1972; Nicholson et al., 1978; Goldman et al., 1980a)

$$i \, \partial_t \, \mathbf{E} + \frac{1}{2} \, \nabla^2 \, \mathbf{E} - \frac{\delta n_2}{2n_0} \, \mathbf{E} - i \, \hat{\gamma} \, \mathbf{E} = 0, \qquad (3.4.1a)$$

$$[\partial_t^2 + \hat{\nu}_i \, \partial_t - C_i^2 \, \nabla^2] \, \delta n_2 = \frac{n}{2\pi M \omega_p^2} \, \nabla^2 \, |\mathbf{E}|^2. \qquad (3.4.1b)$$

In the first equation, the term $\delta n_2 \mathbf{E}$ contains nonlinear refractive effects arising from density changes, δn_2. In the second equation δn_2 is driven by the divergence of the pondermotive force, $-\nabla(e^2 \, |\mathbf{E}|^2 / m \, \omega_p^2)$, and responds through the linear ion-acoustic (quasi-) mode operator in square brackets on the left. The velocity is the ion (sound) speed in these special units. The term $\hat{\gamma} \mathbf{E}$ in the first equation has the spatial Fourier transform $\gamma_k \mathbf{E}_k$, where γ_k is the growth rate of Langmuir waves due to an idealized electron stream, with no velocity dispersion, which propagates unaffected by the waves. The growth rate γ_k is a maximum of order $(\pi/8e)^{1/2} \, (n_b/n_0) \, (v_b/\Delta v)^2$ in a small region of k-space near $\mathbf{k}_0 = \omega_p \hat{\mathbf{v}}_b/v_b$, where resonant waves are driven by the beam. The term $\hat{\nu}_i \, \delta n_2$ in the second equation has the spatial Fourier transform $\nu_{ik} \, \delta n_k$, where ν_{ik} is the damping rate of ion-acoustic waves, usually assumed to be of the order of their frequency, $c_i k$, because of heavy ion Landau damping of ion-acoustic (quasi-) modes in a plasma with $T_e = T_i$.

The Zakharov equations (3.4.1) are dynamical, rather than statistical in nature, unlike the quasilinear equations (3.2.1). One solves for the amplitude and phase of the electric field envelope $\mathbf{E}(\mathbf{r}, t)$ or its transform, $\mathbf{E}(\mathbf{k}, t)$, rather than for a spectral energy density, $P(k) = \int d^3 (\mathbf{r}_1 - \mathbf{r}_2) \langle \mathbf{E}(\mathbf{r}_1) \cdot \mathbf{E}(\mathbf{r}_2) \rangle \exp i\mathbf{k} \cdot (\mathbf{r}_1 - \mathbf{r}_2)$, where $\langle \rangle$ denotes an ensemble average. The Zakharov equations do not allow for nonlinear modifications of the beam distribution function, whereas the pure quasilinear equations do not allow for any wave–wave interactions. Moreover, the Zakharov equations are three-dimensional, and contain induced scatter off ions, as well as important effects such as self-focusing which require at least two spatial dimensions (Goldman and Nicholson, 1978; Goldman et al., 1980b; Hafizi et al., 1982), whereas the quasilinear equations are usually solved in one dimension.

In Equation (3.4.1a) it is assumed that the transverse electric field and its sources are much smaller than the longitudinal field and its sources. With this approximation, the coupled equations have been solved numerically in a two-dimensional cell with periodic boundary conditions (Nicholson et al., 1978; Goldman et al., 1981). The most recent results, for parameters appropriate to 0.5 AU, are shown in Figure 11 (Hafizi et al., 1982) in which contours of equal $|\mathbf{E_k}|$ are plotted in k-space. Initially all modes are at

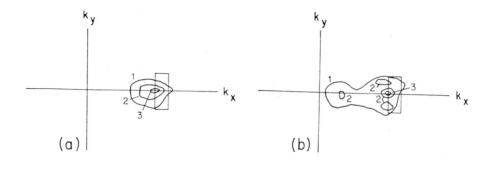

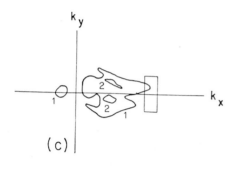

Fig. 11. Solutions to the stream-driven Zak-harov equation in two dimensions, relevant to a Type III burst at 0.5 AU. Contours of equal $|E_k|$ are plotted in k-space. The beam-driven modes lie in the rectangle, and are randomly phased with respect to one another. The central wavenumber is $k_0/k_D = 0.011$. (a) Time t_1, (b) time, t_2, showing induced formed scatter off ions to lower wavenumbers, and off axis modulational instability. (c) is at time t_3, after collapse is underway. (From Hafizi et al., 1982.)

a low level, and randomly phased with respect to one another. The beam-resonant or 'pump' modes in the rectangle grow up temporally, to a level $W_p = 10^{-5}$. At time t_1, energy is beginning to go to lower wavenumbers (Figure 11(a)). At time t_2, the three contours labelled by 2 in Figure 11(b) represent linear wave instabilities pumped by the beam-resonant modes. The two contours off the axis represent a modulational instability, in the geometry predicted by Bardwell and Goldman (1976). This differs from the one-dimensional geometry predicted by Papadopoulos et al. (1974). The contour near the origin represents induced scatter off ions, which is contained in the dynamical Zakharov equations as well as in weak turbulence theory. At the slightly later time, t_3, shown in Figure 11(c), there is very little Langmuir energy left in resonance with the pump, and the rate of energy injection into all modes has therefore slowed considerably. As viewed in coordinate space around this time, Langmuir wave packets are seen to begin to collapse spatially to smaller dimensions. The background solar magnetic field may be incorporated into the calculation, and tends to make the wave packets slightly pancake-shaped at the early stage of collapse (Goldman et al., 1981), but does not change the scenario significantly.

The physical origin of this spatial self-focusing is as follows: Pondermotive force causes a local reduction in density and increase in the 'index of refraction' seen by the Langmuir waves. In two or more dimensions (Goldman and Nicholson, 1978; Goldman et al., 1980b) this refraction cannot be compensated by dispersion, and the packet collapses unstably until it is dissipated by background electrons at a scale size of several Debye lengths, or broken up by scattering off density cavities.

Unfortunately, the Zakharov equations cannot be solved accurately for times much later than t_3, so it is not known how long the collapse continues. Presumably a steady state evolves, and the electromagnetic emission may come from either the small-scale or larger scale structures, or both. Kruchina *et al.* (1980) argue heuristically that this steady state is dominated by Langmuir scatter off density cavities. They construct a crude statistical theory in which field correlation times are controlled by phase shifts associated with this scatter. They argue that the large-scale structures are principally responsible for the emission, and derive emissivities which are consistent with the data of Figure 5.

Goldman *et al.* (1980a) have calculated the emission from a single collapsing (self-similar) wave packet in the late stage, when small spatial scales are reached. However, a volume emissivity requires, in addition, a knowledge of the density of such collapsing packets, which cannot be found reliably without a knowledge of the steady state.

The study of the implications of the Zakharov equations for the Langmuir turbulence associated with Type III bursts has been frought with difficulties. It was originally thought that beam-resonant wave packets collapsed directly, before any induced scatter off ions could occur (Nicholson *et al.*, 1978; Goldman and Nicholson, 1978). Other authors also neglected induced scatter off ions, in heuristic 1-D statistical theories in which direct modulational instability of the beam modes led to a transfer of energy to higher wavenumbers and eventual stabilization to a steady state (Smith *et al.*, 1979; Goldstein *et al.*, 1979; for a more rigorous theory of 1-D strong Langmuir turbulence, also see Dubois and Rose, 1981). Such 1-D theories probably do not describe the rapid collapse stage properly (Rowland *et al.*, 1981; Hafizi *et al.*, 1982) and the one-dimensional nature of the modulational instability has been criticized. However, in the context of the Type III problem, all of these attempts give peak energy densities between $W = 10^{-5}$ and $W = 10^{-4}$, and all quickly remove Langmuir energy density from resonance with the beam. This is because the thresholds for induced scatter off ions, modulational instability, and direct collapse are all very close to each other, and all involve relatively fast transfers of energy in k-space, compared to the time-scale for quasilinear plateau formation.

The 'spiky' spectrum of Langmuir waves measured between 0.5 and 1 AU (Figure 7) cannot be taken as evidence for collapsed wave packets, because the minimum resolvable distance of ~ 25 km is much larger than the size of a collapsed packet of ten Debye lengths (~ 50 m) or even the wavelength of the beam-resonant Langmuir modes (~ 3 km). It is possible the small k condensate of Figure 11(c) is observable, but we do not know theoretically what fraction of the energy of an eventual steady-state Langmuir spectrum lies at such small wavenumbers.

In Figure 9 the line corresponding to $W = 4 \times 10^{-5}$ lies near the upper limit of the measured Langmuir field strengths. It is important to note that this value of W, although common to most theoretical treatments, is associated with much shorter scale length Langmuir turbulence than the measured fields with scale lengths $\gtrsim 25$ km.

3.5. SECOND HARMONIC EMISSION FROM LANGMUIR WAVES

There are no plausible mechanisms for production of second harmonic emission other than the coalescence of two Langmuir waves, first proposed by Ginzburg and Zheleznyakov (1958). Even this requires the beam-resonant Langmuir modes to undergo spectral modification, by wave–wave interaction or by scattering off density irregularities. The reason has to do with the necessary kinematical matching restriction, $\omega_L(\mathbf{k}_1) + \omega_L(\mathbf{k}_2) = (\omega_p^2 + c^2 k_T^2)^{1/2}$. Since the two Langmuir frequencies are very close to the plasma frequency, the transverse wavenumber of the harmonic emission must be $k_T \approx \sqrt{3}\,\omega_p/c$, regardless of the Langmuir wave vectors $\mathbf{k}_1$, $\mathbf{k}_2$. If we take the magnitude of these wave vectors to be of the order ω_p/v_b, assuming them to be resonant with the beam, then the momentum matching condition, $\mathbf{k}_1 + \mathbf{k}_2 = \mathbf{k}_T$, tells us that we must have $v_b \approx (2/\sqrt{3})c$, which is impossible. The stream velocity is known to be $c/2$ or less, so the wave vectors of the beam-resonant waves are too large to add up properly. Induced scatter off ions into the backward direction may lead to a 'correct' spectrum for second harmonic emissions. Also, the real space collapse of this condensate may eventually create a broad spectrum of backward and forward waves which would be kinematically suitable for second harmonic emission.

The current which produces second harmonic emission is second order in the Langmuir field: $j_2 \sim \delta n_1 E$, where $\delta n_1 \propto \nabla \cdot \mathbf{E}$, by Poisson's equation. The emissivity is proportional to $|j_2|^2$ and so depends on E to the fourth power. In weak turbulence statistical theories of the volume emissivity the random phase approximation is employed and the emissivity goes as P^2. This may not be the case, however, for statistical theories of strong turbulence even with random phases (Kruchina et al., 1981; Goldman et al., 1980a; Papadopoulos and Freund, 1978). Moreover, there can be important phase effects, in the dynamical evolution of the Langmuir fields, which are neglected if collapse plays a role (Hafizi and Goldman, 1981).

There are a number of different estimates for the dependence on W of the volume emissivity at the second harmonic. Smith's (1977) estimate, combining rigorous inhomogeneous quasilinear theory results with heuristic arguments concerning induced scatter off ions gives a volume emissivity of $J_{2\omega_p} = 10^{-13}\,E^4/\sqrt{n}$ W m^{-3} ster^{-1}, which is large enough (Gurnett et al., 1980) to account for most of the observations at 1 AU (Figure 8). However, his treatment of the scatter is very heuristic and Smith's own claim that the induced scatter is very fast, and goes towards small forward wavenumbers during the rise of the burst, suggests that the estimate may not be self-consistent. Grognard's preliminary self-consistent calculations of the two effects together, shows the Langmuir spectrum builds to a large 'condensate' at small k. This spectrum would not radiate at the second harmonic at all because the kinematics cannot be satisfied. Moreover, the spectrum seems to be above the collapse threshold, and would therefore be expected to spread to higher wavenumbers.

Papadopoulos et al.'s (1974) estimate assumes a 1-D isotropic spectrum peaked at $k \sim 0.1 k_D$, where they claim the modulational instability has the maximum growth rate. Gurnett et al. (1980) obtain from this the volume emissivity $J_{2\omega_p} = 5.8 \times 10^{-15} (T/T_0)^{3/2} E^4/\sqrt{n}$ W m^{-3} ster^{-1}, where T_0 is the electron temperature at the Earth. This accounts for many fewer of the events observed at 1 AU. Both the one-dimensional nature of this calculation, and the related failure to describe collapse have been criticized (Goldman et al., 1980a; Hafizi et al., 1982; Rowland et al., 1981). Gurnett et al. (1980), using their measured radial variation of the emissivity, however, show that both the above estimates are consistent with the measured radial variation of the Langmuir field (although we must point out once again the disparity of theoretical and measured Langmuir wave scale sizes).

Papadopoulos and Freud (1978) calculate the emission from stationary solitons, assumed to arise from a balance of growth in beam modes, γ_0, against a transfer rate γ_{NL}

associated with 1-D modulational instability. They find the volume emissivity for second harmonic generation is proportional to the *first* power of W rather than the second. In addition, they argue that they can account for the two regimes of dependence of the emissivity on the electron flux J_E, observed by Fitzenreiter *et al.* (1979), and shown in Figure 6. The argument is nominally based on a transition between two forms of the modulational instability, the subsonic (in which the first two terms on the left in Equation (3.4.1b) are negligible) and the supersonic (in which they are dominant). We believe this argument to be unconvincing for a number of reasons. The most serious objection is that it requires a beam to background density ratio, $n_b/n_e \approx 3 \times 10^{-4}$, which is three orders of magnitude higher than is usually assumed reasonable. In addition, it is difficult to justify the stable solitons they assume, or the closely packed density of such solitons (Goldman *et al.*, 1980a).

Goldman *et al.* (1980a) calculate the second harmonic (and fundamental) emissivity from a *single* collapsing soliton, and find both an upper limit which is quadratic in W, and a lower estimate, based on a supersonic self-similar solution, which is independent of W. A volume emissivity is then obtained from a crude statistical model which gives the density of solitons by balancing the power flow into beam modes against the power transfer to collapsing solitons. However, this model contains an undetermined parameter, F, relating to the lack of knowledge concerning the amount of Langmuir energy in resonance with the beam. The parameter F cannot be determined without the (2-D) steady-state strongly turbulent spectrum of Langmuir waves. At present, this spectrum has not been determined numerically. In addition, the assumption was made that the collapse was direct, i.e. that it proceeded from wave packets whose size and shape was determined by the width and location in k-space of the beam-resonant mode spectrum. The latest, more detailed, results (Figure 11) show the collapse is not direct, but proceeds from a condensate near zero wavenumber, which forms after induced scatter off ions has occurred (Hafizi *et al.*, 1982). This will affect both the dynamical and statistical assumptions which enter into the calculations of emission from collapsing solitons, particularly because the beam modes are severely depleted up to the latest times which can be followed (Figure 11).

Kruchina *et al.* (1981) consider the emissions from a strongly turbulent Langmuir state in which Langmuir wave packets are nonlinearly phase de-correlated due to scatter off density cavities. They find an emissivity for second harmonic emission which is linear in W, and also claim good agreement with the data of Figure 5. Their model, however, is one-dimensional and based on a scenario for steady-state Langmuir turbulence which has never been obtained numerically even in 1-D for the streams associated with Type III bursts (although there is some evidence for such a 1-D spectrum in laser-driven plasmas). In addition, one might expect much lower cross-sections for scatter of wave packets from density cavities in two or three dimensions.

These theoretical studies usually interpret local low-frequency Type III emissions as second harmonic, in which case, none is able to account for the characteristic 10–20-min delay between the onset of the local emission and the onset of both the observed Langmuir waves *and* the positive slopes part of the parallel electron distribution function. Since Lin *et al.* (1981) are now calling into question the interpretation of the low-frequency emissions as second harmonic, more careful consideration ought to be given to the predictions from strong turbulence theory of *fundamental* emission at low frequencies.

3.6. FUNDAMENTAL EMISSION FROM LANGMUIR WAVES

There is no way a Langmuir wave can convert into fundamental radiation near the plasma frequency without some agent to take up momentum. This follows from the frequency and momentum conservation laws. In the presence of local density gradients, the inverse scale length of the gradient can act as the required momentum, and so-called direct emission can occur. This effect cannot be dismissed out of hand (Melrose, 1981, 1980b).

In the more traditional scenario, the extra momentum is supplied by ions (through their surrounding polarization clouds). This process was first proposed by Ginzburg and Zheleznyakov (1958, 1959). It has been shown (Smith, 1970; Melrose, 1977a, b) that unless this conversion off ions is *induced*, there is a problem in accounting quantitatively for fundamental emission at high frequencies.

Another possibility is that the conversion is off very low-frequency waves, which may be present at turbulent levels. Melrose (1980b) has considered this process in connection with emission above 60 MHz, although no theory for the Langmuir turbulence is provided. If the low-frequency waves have W_{low} greater than about 10^{-9}, he finds the fundamental can be as bright as the Langmuir waves. Melrose (1980b) also considers the usual co-alescence process for second harmonic emission. He shows that the fundamental, second harmonic, and Langmuir waves can all have the same brightness temperature (in agreement with observation) provided that the Langmuir wave brightness temperature is as large as 10^{15} K. The argument is advanced that the observed brightness temperatures (of 10^{12} K for $f_p > 40$ MHz) should be much less than the actual brightness temperature of the emission if the actual size of the source is much less than the observed size (due, for example, to coronal scattering, or to clumpy Langmuir waves).

At lower frequencies, there has been less work done on fundamental emission. Kruchina *et al.* (1981) find that the *fundamental* emissivity goes as W^2 in their version of strong turbulence theory. At 0.5 AU they find equal emissivity for fundamental and harmonic emission, and general agreement with the data of Figure 5. Goldman *et al.* (1980a) also find comparable fundamental and harmonic emission from a self-similar supersonic collapsing soliton.

3.7. DENSITY IRREGULARITIES AND ION-ACOUSTIC WAVES

The existence and significance of static density irregularities and low-frequency wavelike density structures, in relation to Type III bursts, can no longer be ignored. Scattering and ducting of high-frequency emissions off density irregularities have been invoked to explain the differences between apparent and true source sizes and heights in the lower corona (Section 2.1.1). Density irregularities can play a role in fundamental emission, either by allowing direct emission (Melrose, 1980b), or by reducing absorption in induced conversion off ions (Smith and Riddle, 1975; Melrose, 1980b). Irregularities have also been called upon to explain the 1.85 ratio of harmonic to fundamental as due to truncation of the lowest fundamental frequencies due to reflection off irregular density peaks.

Spacecraft proton density measurements show considerable variation in the average density with position and time (Gurnett *et al.*, 1978a). There is also direct evidence for density irregularities from interplanetary scintillations in the emission from radio stars. Coles and Harmon (1978) have studied the spectrum of scintillations at distances

$\gtrsim 0.5$ AU from the Sun. It has been shown that the amplitude, $\delta n/n$, of density irregularities of size between 50 and 200 km is $\delta n/n = 10^{-3}$, to within an order of magnitude. Smith and Sime (1979) have found that linear beam-driven Langmuir rays in such a (weakly) clumpy plasma are strongly refracted into a random pattern of intense and weak spots. The random pattern of Langmuir intensity is consistent with the typically spiky Langmuir fields found by all observers (see Figure 9) and provides a better explanation than the collapse scenario, because the spatial scales are correct. In addition, the isotropization of Langmuir rays as they wander through the density irregularities may help enable the kinematical condition for second harmonic emission to be satisfied.

Ion-acoustic turbulence has been found experimentally to be a permanent feature of the solar wind (Gurnett *et al.*, 1979). Low-frequency energy densities $W_{\text{low}} = E^2/4\pi n\Theta$ of the order of 10^{-8} are common between 0.6 and 1 AU. This corresponds to values of $\delta n/n = (k_{\text{D}e}/k) W_{\text{low}}^{1/2}$ which can be as large as 10^{-4} or more, if the waves are assumed to be ion-acoustic. The observed high levels of ion-acoustic turbulence would also be sufficient to enable Melrose's process of fundamental amplification by induced conversion of Langmuir waves off ion-acoustic waves to occur. The analogue of this process, induced scatter of Langmuir waves off ion-acoustic waves into other Langmuir waves, should be equally fast and efficient. (Ion-acoustic turbulence in the vicinity of Jupiter's bow shock has been found by Gurnett *et al.* (1981), and seems to have a profound effect on Langmuir turbulence, shifting it up to high wavenumbers by a cascade of scatters.) The effect of ion-acoustic turbulence on collapse has not been considered, but inhibition would not be surprising. Finally, we should mention that there are processes which enable electron streams to radiate fundamental emission directly, in the presence of ion-acoustic turbulence, without the need for any excitation of Langmuir waves (Melrose, 1982).

4. Radio Emission from Shock Waves and Current Sheets

We move from the relatively simple case of a low-density beam which acts as a nonperturbing exciting agent to shock waves which act as a perturbing exciting agent. A perpendicular magnetosonic collisionless shock wave (defined in Section 4.1) is of necessity a current sheet because the magnetic field changes across the shock and its curl gives rise to a current. A stationary current sheet in the corona can always be thought of as two colliding, and thus stationary, shock waves of this type. We shall not treat current sheets in detail, but having noted the relation between current sheets and shock waves, it should be clear how the results for shock waves could be extended to current sheets. Some results on radio emission from current sheets are given in Smith and Spicer (1979).

4.1. SHOCK CONFIGURATION

By a collisionless shock wave we mean a propagating transition layer that causes a change of state in which the primary dissipation mechanism is not Coulomb collisions between particles. The shock is, at least on the average, stationary in time in its rest frame. Since ordinary Coulomb collisions are not important, the time-scales of all relevant processes must be much less than the collision time. The change in state which occurs when the shock traverses a plasma comes from the collective interaction between particles and electric and magnetic fields. The fields can be of two kinds: (1) constant in time, produced by charge separation, currents, or external sources; (2) fluctuating in time, produced by plasma instabilities. The first case is called 'laminar' and the second 'turbulent'. The turbulence can be either microturbulence generated by short-wavelength instabilities inside a laminar shock layer or large-scale turbulence associated with the dominant mode of the shock structure. Collisionless shock waves have been treated in detail by Tidman and Krall (1971) and reviewed by Galeev (1976). Hereafter only shock waves of this type will be considered.

The main paramaters required to describe the state of a shock wave are the Alfvén Mach number,

$$M_A = \frac{u}{V_A},$$
(4.1.1)

where u is the shock speed, and the angle between the unshocked or upstream magnetic field B and u. We are restricting the discussion to a low beta plasma, where $\beta = 8\pi n_1 K(T_{e1} + T_{i1})/B_1{}^2$ is the ratio of plasma to magnetic pressure. Extension to the case of arbitrary β can be found in Tidman and Krall (1971). Subscript 1 refers to upstream variables and subscript 2 refers to downstream or shocked variables. Shocks with $\psi_1 = 0$ and 90° are called 'parallel' and 'perpendicular' shocks' repectively. Shocks with other values of ψ_1 are called 'oblique' shocks.

The important property of shock waves for our purposes is the manner in which they can lead to the generation of Langmuir waves. This can occur in two ways: (1) A relative drift of two distinct groups of electrons with a drift velocity $v_D > (3)^{1/2} v_e$ occurs in the shock structure. (2) The shock accelerates electrons to velocities much larger than v_e in the ambient plasma upstream or downstream from the shock. The resulting beams of electrons produce Langmuir waves upstream and/or downstream from the shock. We note in passing that it is insufficient to heat electrons in the shock and let them interact with the cooler electrons of the ambient corona. As shown by Melrose (1980a), no nonrelativistic isotropic distribution of electrons can lead to plasma waves with an effective temperature greater than 3×10^9 K. Since the brightness temperature of Type II emission reaches 10^{11} K, this type of process, which was proposed by Krall and Smith (1975) and Zaitsev (1977), can be dismissed. Heating in the shock plus a selection mechanism which only allows fast electrons to escape along field lines as in Smith (1971) is a viable mechanism because a beam is formed. Since fundamental and second harmonic emission in Type II bursts are often of comparable brightness temperature, the fundamental radiation must be amplified (cf. Section 4.3). The amplification requires regions of thickness much larger than the thicknesses of any of the above shocks. These regions can occur upstream or downstream of the shock if the shock can either (1) selectively accelerate electrons to velocities greater than $(3)^{1/2} v_e$ in these regions so that beams are formed, or (2) inject beams into these regions by heating electrons together with a selection mechanism.

It might be thought that Langmuir waves excited in the shock could propagate upstream or downstream from the shock. However, the frequencies of these waves in the upstream or downstream plasma are always considerably less than the local plasma frequency ω_{pe} and they can neither propagate nor lead to radiation which can propagate.

This was shown by Smith and Krall (1974) for perpendicular shocks and can easily be generalized to oblique shocks which are rarefaction shocks (cf. Tidman and Krall, 1971). Parallel shocks do not give rise to drifts v_D approaching v_e (Zaitsev and Ledenev, 1976) and thus there is no possibility of exciting Langmuir waves in the shock itself. The physical reason is that most of the current in the shock is carried by electrons and Langmuir waves exist in the frame of these electrons. The Langmuir waves can be excited by an ion beam or an electron beam associated with a small group of suprathermal electrons associated with the ∇B drift (Pesses et al., 1982, §4.2). In either case, for Langmuir waves to be excited the effective velocity of the current v_D must be a significant fraction of v_e. Thus, in the coordinate system of Figure 12, the wavenumber k_y

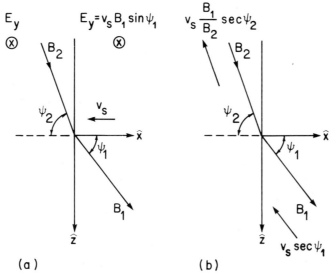

(a)   (b)

Fig. 12. (a) Shock rest frame with the coordinate system used indicated. E_y at the back of the shock has the same value as E_y at the front of the shock. (b) E = 0 frame.

of the excited plasma waves must be of order ω_p/v_e and $k_y v_D$ must be a significant fraction of ω_p. The frequency of the waves in the rest frame of the ions or shock frame, is $\omega_p' - k_y v_D \approx \omega_p - \delta\omega_p \approx \alpha\omega_p$, where ω_p' is the frequency of plasma waves in the current-carrying electrons rest frame, δ is a number of order unity, and α is a number of order zero. Thus these waves are of low frequency in the frame of the upstream or downstream plasma and cannot propagate.

Zaitsev (1977) proposed that the frequencies of these waves could be boosted by induced scattering on electrons moving with velocities $v_s > v_e$. The problem with this process is that the amplification distances required are much larger than the thicknesses of any of the above shocks for any energy density in Langmuir waves $W_p < nKT_e$ (Smith, 1972c). Zaitsev does not consider whether the Langmuir waves can be amplified, but only if they can be isotropized. The formulae for isotropizing the Langmuir waves which are low frequency in the ion frame and for amplifying Langmuir waves which have been scattered to high frequency in this frame are quite different. In fact, a significant energy density in Langmuir waves scattered to high frequency in the ion frame does not occur.

This means that plasma emission from shock waves must be associated with the acceleration or injection of beams of electrons in the upstream or downstream plasma.

As a prelude to discussing beam formation properties in Section 4.2, we state the thicknesses of fast-mode magnetosonic shocks. Fast-mode shocks are shocks in which the basic flow is decelerated and $B_{t2} > B_{t1}$, where B_t refers to the transverse component of the magnetic field (Tidman and Krall, 1971). These are the only shocks capable of effectively accelerating electrons. A laminar perpendicular shock has a characteristic thickness of c/ω_{pe}. When this shock becomes turbulent for Alfvén Mach numbers $M_A \gtrsim 2$, the width increases to $\approx 10\ c/\omega_p$ (Smith, 1971). High Alfvén Mach number laminar oblique shocks also have a width of several c/ω_{pe} while for low Alfvén Mach numbers ($M_A \lesssim 2$) the characteristic thickness is c/ω_{pi}, where ω_{pi} is the ion plasma frequency (Tidman and Krall, 1971). Turbulent oblique shocks have not been studied in detail. For typical coronal parameters ($n = 10^8\ \mathrm{cm}^{-3}$) $c/\omega_{pi} = 2.3 \times 10^3$ cm, which shows how thin even the thickest of these structures are. We do not consider parallel shocks because possibilities for producing beams with them are poor.

4.2. GENERATION OF ELECTRON STREAMS BY SHOCK WAVES

There are two main possibilities for this process: (1) heating of electrons in the shock coupled with a selection mechanism; (2) multiple encounters of an electron with the shock causing direct acceleration. An example of the first process was given by Smith (1972b). An almost perpendicular turbulent shock with $2.0 \lesssim M_A \lesssim 2.9$ heats electrons preferentially through the ion-acoustic instability. For $86° < \tilde{\psi}_1 \lesssim 90°$ only fast electrons can run upstream of the shock and form a beam. The potential drop associated with the gradient of the electric field E_y in Figure 12 was not taken into account in these calculations and they need to be refined. While this will change the selection criterion somewhat because the electrons have to climb up a potential hill, it will still remain a viable mechanism. However, the quite restrictive range of ψ_1 required is unlikely to be fulfilled over large areas of the shock and we proceed to examine the possibilities for acceleration in the upstream or downstream plasma.

A large amount of work has been done in the past few years on acceleration in interplanetary and interstellar shock waves (Armstrong *et al.*, 1977; Axford *et al.*, 1977; Bell, 1978; Pesses *et al.*, 1982). It is the last of the referenced papers and extensions in progress (Pesses, private communication) which are the main basis of our review. For fast-mode nonparallel ($\psi \neq 0$) magnetosonic shocks the incoming plasma is decelerated and heated in the shock front over a distance of a few thermal ion gyroradii which is comparable to the c/ω_{pi} estimate at the end of Section 4.1. The large gradients in $|\mathbf{B}|$ and the plasma bulk velocity combined with the induced electric field $\mathbf{E}_y$ that exists in the shock rest frame (Figure 12(a)) are responsible for the acceleration of energetic electrons. By energetic we mean electrons whose kinetic energy is much larger than the mean thermal energy. The acceleration mechanisms are independent of whether the shock is laminar or turbulent as long as the waves associated with the turbulence do not affect the energetic electrons.

The two acceleration mechanisms are the shock drift and compression mechanisms. The shock drift mechanism occurs because in the rest frame of nonparallel magnetosonic shock waves there is a $\mathbf{V} \times \mathbf{B}$ electric field $\mathbf{E}_y$ due to the motion of the upstream and downstream plasma, where $\mathbf{V}$ is the fluid velocity. For the shock geometry in Figure 12, $\mathbf{E}_y = \hat{y}|\mathbf{V}_s||\mathbf{B}_1|\sin\psi_1$, where $\mathbf{V}_s$ is the shock velocity in the upstream plasma rest frame. As pointed out in Armstrong *et al.* (1977), the $\nabla|\mathbf{B}|$ guiding center drift of electrons interacting with the shock is antiparallel to $\mathbf{E}_y$. Thus, the $\nabla|\mathbf{B}|$ drives a current $\mathbf{J}$ and the electrons comprising the current experience a $\mathbf{J} \cdot \mathbf{E}_y$ energy gain. The compression mechanism works because of the difference in the plasma bulk flow velocity across the shock front (Axford *et al.*, 1977; Bell, 1978). Electrons that diffuse back and forth across the shock in its frame are accelerated by reflection of approaching upstream scattering centers and decelerated by reflection off receding downstream scattering centers. These postulated scattering centers are convected by the bulk plasma motion so that the approaching centers move faster than the receding ones. Thus electrons gain a net energy by being effectively compressed between upstream and downstream scattering centers.

To find the post-shock energies and pitch angles we go to an inertial frame in which both the upstream and downstream $\partial\mathbf{B}/\partial t$ and $\mathbf{V}\times\mathbf{B}$ electric fields are simultaneously zero ($\mathbf{E} = 0$ frame) shown in Figure 12(b). In this frame, the plasma bulk velocity is along $\mathbf{B}$ and electrons gain no energy from the shock. For the shock waves that are planar in the x–z plane of Figure 12 and for which $\mathbf{V}_s$, ψ_1 and ψ_2 (angle between $\hat{n}$ and $\mathbf{B}_z$) do not vary in space or time, Pesses *et al.* (1981) have calculated the energy gains and pitch-angle changes that result from the shock drift and compression mechanisms analytically given the post-shock pitch angle in the $\mathbf{E} = 0$ frame. The pitch angle α is the polar angle between the particle velocity $\mathbf{v}$ and $\mathbf{B}$. The effects of the shock drift mechanism are calculated by transformations between the shock rest frame and the $\mathbf{E} = 0$ frame. The effects of the shock compression mechanism are calculated by transformations between the pre- and post-shock plasma rest frame of the electron and it is assumed that the scattering centers are at rest.

The reader is referred to Pesses *et al.* (1982) for the detailed results. We shall merely note the general properties. Electrons reflected at fast-mode shocks gain energy only in the component of their velocity parallel to the magnetic field. The minimum number of times an upstream electron must reflect off the shock front, N, for its speed to exceed $(3)^{1/2}v_e$ in the upstream plasma is (Pesses, private communication)

$$N > 1.6 \, v_{ni} \sec\psi_1/v_e, \tag{4.21}$$

where v_{ni} is the normal component of the upstream bulk velocity in the shock rest frame. It is possible that counterstreaming electrons become unstable and excite Langmuir waves for a smaller number of reflections. Particles transmitted at fast-mode shocks gain energy primarily in the component of their velocity perpendicular to the magnetic field. There are instabilites which can convert such an anisotropic distribution into electron cyclotron waves and these could nonlinearly couple to Langmuir waves. However, the details remain to be worked out. Both the magnitude of $\mathbf{E}_y$ and the velocity of the $\nabla|\mathbf{B}|$ drift increase with increasing values of ψ_1 and thus also the gain in electron energy per encounter.

In summary, it appears that our knowledge of collisionless shock waves is now sufficiently complete so that viable mechanisms for producing electron beams and Langmuir waves can be constructed. However, to date the only quantitative calculation is that of

Smith (1972b) and this needs to be repeated taking into account the potential drop across the shock. In general the upstream and downstream beam producing potential of a shock increases with ψ_1 and, since this should be reflected in its plasma emission potential, an explanation for the patchy nature of Type II emission (cf. Section 2.3) appears possible.

4.3. RADIATION MECHANISMS

The question of the radiation mechanisms in this case is frought with all the problems for Type III bursts (Sections 3.5 and 3.6) together with the uncertainty in the beam characteristics and density. Counteracting this problem is the relative slowness with which the burst occurs (cf. Figure 1) and the richnss of structure (Section 2.3). We know that in the majority of type II bursts the fundamental and harmonic are of comparable magnitude, and brightness temperatures reach 10^{11} K. It was shown by Smith (1972b) that, without amplification, the fundamental at metre wavelengths would be more than five orders of magnitude less intense than the second harmonic. In any case, without amplification, the brightness temperature of the fundamental cannot exceed 3×10^9 K (Smith, 1970). Thus, amplification of the fundamental is essential.

The problem of amplifying fundamental radiation in a plasma with random density inhomogeneities was analyzed by Smith and Riddle (1975). They showed that inhomogeneities of scale 35 km and strength $\epsilon = (\delta n^2)^{1/2}/n = 0.016$ at the 80 MHz plasma level would not allow amplification. Thus, a mode with homogeneous Langmuir waves over scale sizes of 100 km or more, as in Smith (1972b), is unlikely to be applicable. Smith and Sime (1979) studied the amplification of Langmuir waves in a plasma with random density inhomogeneities and showed that inhomogeneities of scale size $\sim$50 km at 0.5 AU with $\epsilon = 4.8 \times 10^{-3}$ would allow amplification only in certain clumps or spikes where the energy density in Langmuir waves could reach high levels. The value of ϵ in their analysis was determined from interplanetary scintillation data. There is no method of directly measuring density inhomogeneities at the 80 MHz plasma level, but, since these inhomogeneities presumably originate at the Sun and are only smoothed out by the solar wind, they are only likely to be stronger at this level. The area downstream from a turbulent shock is likely to be turbulent and also have a variety of low-frequency waves present.

Thus, a general picture of emission from Type II shocks is as follows. The emission mechanism is plasma emission near the fundamental and second harmonic (Smith, 1970, 1972a; Melrose, 1980a). Langmuir waves are produced by beams produced in the shock or accelerated between the shock and scattering centers both upstream and downstream from the shock. Because of the presence of density inhomogeneities and/or low-frequency waves, the beams relax and produce Langmuir waves only in spatially localized clumps of scale $\sim$35 km at the 80 MHz plasma level. The Langmuir waves in these clumps are at a sufficiently high level to allow the fundamental to be amplified up to the same level as the second harmonic. In the region downstream from the shock some fundamental emission may also be produced by coalescence of Langmuir waves with low-frequency waves (cf. Section 5.4). The Langmuir waves are likely to have a more or less isotropic distribution if produced by counterstreaming beams or due to induced scattering (Smith, 1970) and/or because of scattering off of the density inhomogeneities. As explained in Smith (1972a) and Section 3.5, this will greatly facilitate production of the second harmonic. The spontaneous scattering considered in Smith (1972a) is unlikely to be important.

Normally the scattering centers and area of beam relaxation occur close to the shock and we see normal Type II emission. When the regions upstream and downstream from the shock are well differentiated in density, split bands occur (cf. Section 2.3). When the scattering centers are sufficiently far from the shock and the shock is sufficiently oblique, we see herringbone structure due to the much larger beam relaxation regions. If there is enhanced beam relaxation near the shock, a 'backbone' is present in the herringbone structure. This whole scenario is quite speculative at present and needs to be worked out in detail. However, it appears that our knowledge of the plasma physics involved is, or soon will be, sufficiently complete to make such an endeavor worth the effort.

5. Radio Emission from Moving Plasmoids and Other Traps

We move now to a greyer area of solar radio emission with some overlap with the current sheets of Section 4 since such sheets may be embedded in the configurations of this section. This area is greyer because the emission takes place in regions which are atypically dense and/or have an atypically high magnetic field so that we are sampling a special region of the corona, in some sense, which often is seen only by its radio emission. Thus, independent means of establishing density and/or magnetic field, etc., are often absent and the interpretation of the observations is necessarily more speculative.

5.1. PLASMOID AND OTHER TRAPPING CONFIGURATIONS

This subsection could be subtitled 'detached and rooted coronal loops', including the possibility that detached loops close on themselves and form complete toruses. The quantitative analysis of loops and their evolution is in its infancy since their importance in flare physics was only brought out by the Skylab observations (Sturrock *et al.*, 1980). All the quantitative analyses assume a form for the magnetic field since it has not been measured directly. Most of the analyses of the density and temperature structure of loops have been carried out on large, long-lived loops (e.g. Foukal, 1978) which may be related to Type I emission, but are most likely unrelated to moving Type IV emission. Thus the basic configurations for the magnetic field can only be inferred from the observations and modeling and, except in a few cases, the same is true for the density and temperature. Modeling has been performed by Altschuler *et al.* (1968) for photospheric conditions. These results have been used by Dulk and Altschuler (1971) in the 'cartoon approximation' for plasmoid type, moving Type IV bursts, as shown in Figure 13. Here a toroidal current ring with its own poloidal magnetic field moves outward through a decreasing background poloidal field. While this configuration is able to explain many of the features of plasmoid type moving Type IV bursts, it is not clear how it might evolve from an expanding magnetic arch which would favor development of a toroidal magnetic ring.

A toroidal ring could explain the occurrence of double sources in moving Type IV bursts which are oppositely circularly polarized. It could also evolve naturally from an expanding magnetic arch which is ejected from the Sun simply by reconnection of the

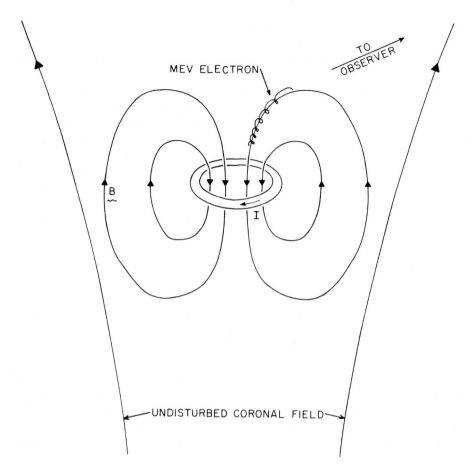

Fig. 13. Schematic diagram of a possible plasmoid configuration at one stage of its motion along the diverging coronal magnetic field. (After Dulk and Altschuler, 1971.)

feet of the loop which might be squeezed together in the ejection process. Some modeling work on toroidal magnetic rings has been done by Lillequist *et al.* (1971) and Altschuler *et al.* (1973) who included the effect of the Hall term in Ohm's law which is important for densities $n < 10^7$ cm^{-3} and scale lengths less than 2×10^4 cm. However, no work has been done modeling large toroidal rings for coronal conditions which is the case possibly relevant to moving Type IV bursts. A problem with toroidal rings is that moving Type IV sources sometimes switch from a single source with O-mode polarization to a double source with O-mode and X-mode polarizations during their evolution. They can also break up into as many as four sources, all with the same sense of polarization. A toroidal ring, when small, would be expected to have little net polarization because polarizations from different regions would tend to cancel for most viewing angles. A poloidal field, as in Figure 13 where there is a concentration of field in one direction, can much more easily explain a single sense of polarization although why this would always be O-mode is a difficulty for gyrosynchrotron emission. A further difficulty is why an observer should be always looking within $30°$ of the direction of the strong field which is needed to explain strong polarization (Wild and Smerd, 1972). This problem becomes especially severe for sources near the limb of the Sun which are often observed to have high degrees of polarization in the late phases of events.

In summary, plasmoid type configurations of a hybrid variety with both toroidal and poloidal components appear to offer the best possibility for explaining the observations, but even then there are difficulties with pure gyrosynchrotron emission. As pointed out by Robinson (1977), a poloidal field is inevitably accompanied by a toroidal field in any case if the plasmoid is to be force-free and thus stable. Any gradient in the poloidal field produces a toroidal field under the force-free condition. How this mix of poloidal–toroidal field is created remains unclear. A plasmoid with its own magnetic field still remains a likely candidate for some moving Type IV bursts.

The alternative basic configuration is a loop with some attachment, however loose, to the background solar field. The magnetic field is primarily toroidal and runs along the loop. Any current running along the loop will create a poloidal field component. The main problem with this type of configuration is containment of fast electrons. There are two possible ways in which this could occur:

(1) The loop is rooted in stronger magnetic fields so that electrons moving to higher fields with finite pitch angles mirror successively. We then have a magnetic trap.

(2) Electrons with velocities $v > 43v_A$ excite whistler waves (Melrose, 1974a) which scatter these electrons and keep them confined to the top or end of the loop. This process is called resonant scattering.

Just how long and under what conditions the first alternative will work in the presence of losses is discussed in Section 5.3. The quantitative analysis of the second alternative is just being worked out (Dulk, private communication) since it may have special relevance to the observed propensity of microwave bursts to be confined to the tops of loops (cf. Section 2.1). In terms of energy the criterion $v > 43v_A$ becomes (Melrose and Brown, 1976)

$$E > E_{min} \approx 5.2(v_A/10^8 \text{ cm s}^{-1})^2 \text{ keV} \tag{5.1.1}$$

and the Alfvén speed

$$v_A = 2.2 \times 10^{11} Bn^{-1/2} \text{ cm s}^{-1}. \tag{5.1.2.}$$

For typical loop densities and magnetic fields, v_A lies in the range $1-4 \times 10^8$ cm s^{-1} and thus $E_{min} \approx 6-85$ keV. Taking the lower limit, which is probably more applicable to Type I and moving Type IV emissions, it is possible in principle to contain electrons which could radiate by plasma radiation.

There are many facets of the loop configuration which remain to be worked out. For example, what is the effect of a steady current on the resonant scattering process? Does counterstreaming of fast electrons mirroring in a trap lead to instability? About all that can be done now is to take model distribution functions, put them in a loop, and see how they evolve in the manner of Melrose and Brown (1976). Nevertheless as a model for microwave, Type I, and some moving Type IV bursts, a loop with containment is a strong candidate for the basic configuration.

5.2. SOURCES OF ELECTRONS IN PLASMOIDS AND TRAPS

To be able to study emission mechanisms in the configurations of Section 5.1 we need to know the energization and loss processes for electrons which are considered in the following two sections. As stated in Section 1, the initial energization of electrons will not be considered although some of the mechanisms for the continuing energization of electrons are the same. The main possibilities are direct electric field acceleration, as in a current-carrying loop with tearing mode instabilities, and acceleration by hydromagnetic turbulence generated as the configuration moves through the surrounding medium. There are also currents associated with contained plasmoid configurations, but since the exact configuration remains unknown, it is difficult to analyze this possibility for tearing instabilities in detail. Thus we shall only consider tearing instabilities in a loop in a general way which could be applied to plasmoid configurations. Similarly, although the motion of plasmoids and loops through the surrounding medium is almost certain to generate hydromagnetic turbulence, the exact spectrum of the turbulence has not been worked out for specific motions. Hence, we shall only consider acceleration by hydromagnetic turbulence for an arbitrary level of the turbulence.

The theory of particle acceleration and heating due to fast tearing modes has been considered by Van Hoven (1979) and Spicer (1981) and is also considered in the article by Spicer in this volume. The basic characteristic of interest is the rate at which electrons in a current-carrying loop gain energy. The actual energy gain occurs in very small regions in a very inhomogeneous manner, but since we are only interested in the emission averaged over the loop resulting from the spatially averaged distribution function, this is of little concern. The pitch-angle distribution is of importance and we shall take the result of Spicer *et al.* (1981) in agreement with the results of Smith (1980) that tearing modes result primarily in heating rather than acceleration. Thus we shall take the initial pitch-angle distribution as isotropic. Both Van Hoven (1979) and Spicer (1981) only consider a single tearing region. The rate of energy release in the volume of the loop is given by (Spicer, 1977)

$$\frac{dE}{dt} \approx \frac{\gamma B_p^2 \, \Delta V}{4\pi} \, , \tag{5.2.1}$$

where γ is the growth rate, B_p is the poloidal field, and ΔV is the incremental volume of the loop in which the tearing mode occurs. A rough estimate for the growth rate for fast tearing modes is (Spicer, 1981)

$$\gamma \approx S^{2/3}/\tau_R, \tag{5.2.2}$$

where $S = \tau_R/\tau_A$ is the magnetic Reynolds number and the resistive diffusion time

$$\tau_R = \frac{4\pi(\delta l)^2}{\eta c^2} \, . \tag{5.2.3}$$

Here δl is the characteristic gradient scale length and η is the resistivity. The Alfvén transit time $\tau_A = \delta l/v_A$.

Further analysis requires the specification of the parameters of a loop and the tearing modes occurring in the loop. We simply state some typical rates for coronal parameters,

$n = 10^8$ cm^{-3}, $T_e = 1.6 \times 10^6$ K, $B_p = 1$ G and $\Delta V = 10^{26}$ cm^3. Typical growth rates are 10^3 s^{-1} so that, from Equation (5.2.1), $dE/dt \approx 8 \times 10^{27}$ erg s^{-1} or 8×10^{-7} erg s^{-1} per electron. It would take 0.2 s to heat an electron to 100 keV, but since the total volume involved is of the order of 10^{32} cm^3, only some fraction of the electrons will be heated as the tearing volume moves within this region. Slower tearing rates and accelera-tion times are also possible

Acceleration by hydromagnetic turbulence has been considered by Melrose (1974b) using the model of Kulsrud and Ferrara (1971) together with scattering by whistler waves. The basic idea is that electrons interact with low-frequency, large-amplitude hydromagnetic (HM) waves. The HM waves can cause changes in the pitch-angle distri-bution due to conservation of the adiabatic invariant $E_\perp/B$. The perpendicular energy increases with the magnetic field strength during the first half of the wave cycle and would be returned to the field during the second half-cycle in the absence of scattering. Scattering of the electrons transfers some of the gained perpendicular energy to the parallel component which is unaffected by magnetic field variations and the intensity of the HM waves decreases with an increase in electron energy. The transfer of energy is produced by the scattering and its rate is thus largely determined by the scattering rate ν. Clearly this must be larger than the frequency of the turbulence ω. It is argued by Melrose (1974b) that wave–particle interactions are the only known process capable of giving the required scattering rate. The resonant waves are generated by the electrons themselves, undergoing induced emission during the compression phase (increasing B) and reabsorption during the rarefaction. Since the distribution function changes due to the scattering, not all the whistler waves will be reabsorbed during the second half-cycle. Thus, there is a gradual build-up of the wave intensity and scattering effectiveness with a corresponding increase in the acceleration rate. They reach a constant level when the energy density in whistlers is so large that the anisotropy driving the whistlers is removed in one wave period or less.

The acceleration rate after whistlers reach their saturation level is (Melrose, 1974b)

$$\nu_A = \frac{\pi}{4} \omega \epsilon^2 \frac{\beta_A}{\beta}, \tag{5.2.4}$$

where $\epsilon = B_t/B_0$ is the relative amplitude of the turbulence with B_t the turbulent field strength at maximum and B_0 the background magnetic field. The electron velocity $v = \beta c$ and $\beta_A c = v_A$. The restrictions on (5.2.4) are:
(a) the mean free path is greater than the wavelength of the turbulence;
(b) the scattering rate $\nu \gg \nu_A$;
(c) the electrons can resonate with the whistlers (Equation (5.1.1));
(d) the whistlers have a sufficient time to grow, which means that the density of electrons n_1 being accelerated must satisfy

$$\frac{n_1}{n} \gg \frac{\omega}{\Omega_i} \left(\frac{\gamma\beta}{43\beta_0} \right)^2, \tag{5.2.5}$$

where Ω_i is the ion gyrofrequency, $\gamma = (1-\beta^2)^{-1/2}$, and $\beta_0 = 43\beta_A$.
Further application requires specification of the turbulence, and acceleration to $\sim$200 keV will occur on a time-scale of a few minutes from turbulence with periods in the

range 0.1–10 s which is indicated by pulsations of metre-wave continuum radiation (McLean *et al.*, 1971).

5.3. LOSS CONE AND COLLISIONAL ELECTRON LOSSES

We consider how electrons mirroring in a loop or other trap lose energy. The pitch angle α is the angle between the direction of electron motion and the magnetic field. It is convenient to write distributions in terms of cos $\alpha = \mu$. A loss cone distribution is flat out to the value $\pm\mu_0$ and rapidly falls to zero for larger $|\mu|$. Electrons which are scattered into the loss cone by wave-particle or collisional scattering are lost to the system. The resulting anisotropic pitch-angle distribution can lead to the growth of Langmuir waves under certain conditions (Stepanov, 1973; Kuijpers, 1974). The energy relaxation rate due to collisions is ν_e and the deflection rate is ν_D, which is the rate at which pitch-angle diffusion occurs. For Coulomb collisions, $\nu_E/\nu_D = 0.5$ since fast electrons are scattered at the same rate by both thermal electrons and ions, but only the interactions with thermal electrons cause energy changes (Trubnikov, 1965). The value

$$\nu_D \approx 10^{-8} \, nE^{-3/2} \, s^{-1} \tag{5.3.1}$$

where E is the electron energy in keV (Melrose and Brown, 1976). There are two simple approximations to the precipitation rate ν_p which is the rate at which energy is lost from the trap. The pitch angle at the edge of the loss cone is α_0 and the bounce rate in the trap is ν_b. The two approximations which Kennel (1969) called the weak and strong diffusion limits are:

$$\nu_p \approx \begin{cases} \nu_D, & \nu_D \ll \frac{1}{2} \alpha_0^2 \nu_b \\ \frac{1}{2} \alpha_0^2 \nu_b, & \nu_D \gg \frac{1}{2} \alpha_0^2 \nu_b. \end{cases} \tag{5.3.2}$$

It has been assumed that $\alpha_0^2 \ll 1$. These two limits represent approximately empty and full loss cones, respectively. For coronal traps, the weak diffusion limit will usually apply.

Without a loss cone as in a closed plasmoid, the only losses are collisional losses between fast electrons and the electrons of the background distribution since collisions with ions only lead to deflections, to a good approximation. The loss rate in this case is just $\nu_E = 0.5\nu_D$, with ν_D given by Equation (5.3.1).

5.4. RADIATION MECHANISMS

The main candidates for the radiation mechanism for moving Type IV, Type I, and microwave bursts are plasma radiation and gyrosynchrotron radiation. Since we have already reviewed plasma radiation in relation to Type II and Type III bursts (cf. Sections 3.5 and 4.3), we shall only consider applying these results to the bursts mentioned except for Type I bursts where coalescence of Langmuir and ion-acoustic waves will be treated. The alternative possibility of gyrosynchrotron radiation will then be reviewed.

In general, plasma emission can be broken into three parts:

(1) Wave source.
(2) Transformation of Langmuir waves into radiation.
(3) Propagation of radiation from the source to the observer.

We consider this scenario for each of the burst types of this section. Because of the difficulties noted in Sections 3.5 and 3.6 with radiation in the strong turbulence regime, we limit the discussion to weak turbulence emission processes which is consistent with all the published results on these bursts. As noted in Section 2.2, we do not intend to treat the theory of microwave bursts in detail. The only microwave emission which requires plasma emission at present is the fine structure in microwave Type IV bursts although some microwave bursts recently observed are pushing the gyrosynchrotron theory to its limits. The theory for the microwave Type IV fine structure would be similar to plasma emission for either moving Type IV or Type I bursts.

We begin with plasma emission from a moving Type IV burst for either the expanding arch or plasmoid type source. Because of the rather open nature of the basic configuration (cf. Section 5.1) and the consequent open nature of the electron population (cf. Section 5.2), the source of Langmuir waves is fairly arbitrary. It should be able to sustain the losses due to Langmuir waves and those of Section 5.3 if continuous acceleration is not required. The most likely types of electron distribution for an arch or a plasmoid are a 'gap' or plateau, as shown in Figure 14. A 'gap' distribution (Melrose, 1975) is one which

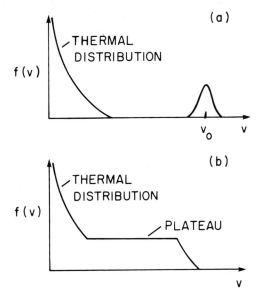

Fig. 14. (a) A gap electron distribution. (b) A plateau electron distribution.

peaks at a velocity $v = v_0$ and rapidly falls for $v < v_0$ with a region of velocity space between the thermal distribution and the nonthermal peak in which $n_1 \lambda_D^3 < 1$, i.e. the density n_1 is less than one particle per cubic Debye length, λ_D. When the distribution is unidirectional, we have the beam treated in Sections 3 and 4. The number density of particles with $v \approx v_0$ must be sufficiently high to dominate over the thermal particles in the emission and absorption of Langmuir waves with $v_{ph} \approx v_0$, where v_{ph} is the phase velocity. In the case of an arch, the most likely distribution is an anisotropic gap distribution with the anisotropy caused by loss of particles through the loss cone and the gap created by resonant scattering and/or collisional losses. In a plasmoid there is no loss cone and collisions tend to create a gap, but quasilinear relaxation (Section 3.2) tends to fill in the gap with a plateau, resulting in a distribution intermediate between Figures 14(a) and (b). For times that are long compared to relaxation and collision times, but still short compared to the lifetime of the plasmoid, the distribution should be isotropic.

The plasma emission from an arch depends on the ratio of suprathermal to background plasma densities n_1/n. The growth rate for Langmuir waves due to the loss cone instability was estimated by Melrose (1980a) as

$$\gamma(k, \theta) \approx \pi \frac{n_1}{n} \omega_{pe} G(\theta), \tag{5.4.1}$$

where $G(\theta)$ is a function depending on the details of the distribution function f. The main point is that this growth rate is large compared to collision times so that the instability should saturate, giving an energy density in Langmuir waves comparable with the energy density in the trapped electrons. For example, for $n_1/n = 10^{-3}$ and 30 keV electrons, the Langmuir waves will have an effective temperature $T_p \approx 10^{14}$ K. Since there will be counterstreaming in the source the Langmuir wave distribution should be approximately isotropic resulting in second harmonic emission with $T_b < T_p$. Although there is some controversy over the expected polarization characteristics, the degree of polarization should be weak. As in Section 4.3, unless the fundamental is amplified, it will be much weaker than the second harmonic. It could be suppressed by density inhomogeneities or lack of sufficient optical depth. This would be consistent with observations since moving Type IV sources generally appear to be well above the plasma level. However, because of scattering the fundamental source can also appear much higher than its true position. Since the bursts only have high brightness temperature with low degrees of polarization (Duncan, 1981), there is no compelling need for fundamental emission.

The plasma emission from a plasmoid with an isotropic gap f is limited to $T_b < 3 \times 10^9$ K for nonrelativistic electron energies (Melrose, 1980a) and if the gap is partially filled in, even less. Thus, without some continuous acceleration process, a plasmoid with plasma emission is a possible model only for the late phases of many moving Type IV bursts. Only fundamental emission could lead to the observed high degrees of circular polarization.

Proceeding to Type I bursts and continuum, we note that because of the strong O-mode polarization up to 100% (Section 2.5), plasma radiation could only be near the fundamental. Thus the problem here is how to obtain fundamental $T_b > 10^{10}$ K without observable second harmonic emission. The recent approaches to this problem have invoked the coalescence of Langmuir waves with ion-acoustic waves (Melrose, 1980c; Benz and Wentzel, 1981). The source of Langmuir waves, according to Melrose (1980c), is a loss cone instability which is just strong enough to overcome collisional losses of Langmuir waves and maintain the waves in a marginally steady state. The plasma waves then build up over a time of order $\gamma c^{-1} \approx 0.1-1$ s, where γc is the collisional damping rate. For a 5-keV beam with characteristic velocity $v_0 \approx 10^9$ cm s^{-1}, the waves build up over a distance $v_0 \gamma_c^{-1} \approx 10^3 - 10^4$ km and it is assumed that Langmuir waves occur over such time and distance scales. According to Benz and Wentzel (1981) the source of Langmuir waves may be trapped nonthermal electrons from previous burst sources. The idea is that current instability in an unspecified manner accelerates electrons. They become trapped by an anomalous-cyclotron instability (Papadopoulos and Palmadesso, 1976). The current instability is also the direct source of ion-acoustic waves which thus have an effective temperature $T_s \gg T_p$.

The emission process is the coalescence of a Langmuir and an ion-acoustic or other low-frequency wave. For simplicity, we confine the discussion to ion-acoustic waves. Benz and Wentzel have argued that these are the most likely candidates because they saturate at a relatively high level. This, in turn, allows an optical depth of order unity to be reached for a source thickness of 0.1 km. Melrose argues for a minimum source size of 10^3 km, but he does not consider in detail the excitation condition for the ion-acoustic waves. Benz and Wentzel do consider the excitation condition and conclude that the ion-acoustic waves must be confined to a thin sheet. In Melrose, the emission is controlled by the Langmuir waves. In Benz and Wentzel, T_b is controlled by the Langmuir waves but the optical depth is controlled by the ion-acoustic waves. This allows them to obtain $T_b \leqslant 4 \times 10^{13}$ K without a detectable harmonic ($<10^{-20}$ erg cm^{-2} s^{-1} Hz^{-1}). Because Melrose has a source size of 10^3 km, he expects a detectable harmonic (10^8 K) for a bright Type I burst with $T_b = 10^{10}$ K. Of course, Benz and Wentzel will not have a detectable harmonic only to the extent that the Langmuir waves are confined to thin sheets as for the ion-acoustic waves, and they have not given any mechanism for such confinement. The continuum, according to Melrose, is due to coalescence of stable Langmuir and low-frequency waves in a large source. Benz and Wentzel are more specific: they suggest that low-frequency waves with small k decay into low-frequency waves with sufficiently high k to combine with Langmuir waves to produce radiation. Since the k of the radiation is small, small-k low-frequency waves are of no use without some further process. Benz and Wentzel suggest that whistlers and lower hybrid waves are likely candidates for the low-frequency waves.

Moving to gyrosynchrotron radiation, we begin again with moving Type IV bursts. As noted in Section 2.4, until recently this was the accepted emission mechanism for these bursts and is still possibly the best explanation for the late phases of these events. Gyrosynchrotron emission requires electrons of 100 keV similar to microwave bursts. Dulk (1973) developed the theory of this emission for moving Type IV sources and argued that the relatively sharp cutoff at low frequencies is due to synchrotron self-absorption; in other words, when the source becomes optically thick, as it does at low frequencies, only a fraction of the radiation escapes. Robinson (1974) extended Dulk's calculations to the case of inhomogeneous magnetic field configurations. The results are that $T_b \leqslant 10^9$ K and the radiation is polarized in the X-mode up to 100%, for an optically thin source, but depends sensitively on the viewing angle. The primary success of the gyrosynchrotron hypothesis is an explanation why the degree of emission increases as the source moves out, which is applicable mainly to plasmoid type sources. With self-absorption both the ordinary and extraordinary modes are comparable and there is little polarization. As the source moves out, it become less self-absorbed and the X-mode becomes dominant, leading to high degrees of polarization. There is no analogous explanation with plasma emission. Counteracting this success of gyrosynchrotron emission is the sensitivity of this result to the viewing angle of the observer.

A combined model in which plasma emission dominates for sources close to the Sun and gyrosynchrotron emission dominates for sources further out would seem the best choice at present, but clearly more work needs to be done on plasma emission in this context.

Moving to Type I emission, gyroemission at low harmonic numbers excited by a beam with large perpendicular velocity has been considered by Mangency and Veltri (1976). They showed that coupling of unstable whistler and X-mode waves with low-frequency MHD waves would stabilize these modes, but that the O-mode would remain unstable. This leads to O-mode radiation of relative bandwidth $3-4 \times 10^{-2}$ and high directivity. The average opening angle of the radiation is, at most, $12°$. This is the principal success of this model. There is no explanation of such a high degree of directivity of Type I emission with plasma radiation. Counteracting this success is an explanation of how the beams are accelerated. Mangeney and Veltri's model applies to bursts, but offers no explanation for the continuum.

At the present time it is difficult to judge whether plasma or gyroemission is a better model for Type I bursts. Excitation of Langmuir waves by a loss cone instability seems plausible and less *ad hoc* than the beams of Mangeney and Veltri (1976) and a tie-in with the escape of electrons from traps as a source for storm Type III bursts also seems natural. The high directivity of gyroemission, on the other hand, is also attractive. It is difficult to see how such directivity, which is required by the observations (Section 2.5), could come from plasma emission except through a propagation effect. This is an area of research which is bound to be active in the next few years.

6. Conclusions and Recommendations

Solar radio astronomy is now at a stage where observations, particularly from satellites, are beginning to provide enough pertinent data to enable theory to be put to the test. This has been true especially for the theory of Type III bursts, one of the most exhaustively studied and supposedly well-understood of solar radio emissions. The results are surprising and provocative. Even the very cornerstones of the theory of Type III bursts have been challenged. A vigorous period of re-examination of fundamental processes seems to be in order. This should represent a healthy stage in the development of the underlying physics, in which reasonable hypothesis is bridled by reality, and the dominant factors governing emission are finally identified from among speculative alternatives.

We separately present our conclusions and recommendations for the various kinds of radio emissions, beginning with the Type III burst.

6.1. TYPE III EMISSIONS

Above about 10 MHz, it is known that the source heights observed by radioheliograph cannot be the true heights of fundamental and second harmonic emission. Scattering and ducting of the emissions off density irregularities have been invoked to construct a reasonable picture, but this is somewhat *ad hoc*, and even with these processes, the frequencies do not relate properly to the quiescent local plasma frequencies, and density enhancement seems to be required. The observed polarizations at these frequencies also have not been explained by theory (Melrose, 1980b; Melrose and Sy, 1972).

Emissions below 1 MHz are usually identified as second harmonic, although this appears to create irreconcilable difficulties for the causative role of either Langmuir waves or the (high flux) 5–30 keV electron streams which drive them. Some of these difficulties disappear if the emissions can be re-identified as fundamental, but then, once more, it is necessary to postulate strong scattering off density irregularities. The possibility that the observed source positions are not the true source positions ought to be explored seriously for emissions below 1 MHz.

Clearly, what is needed is a positive identification of the emissions below 1 MHz as fundamental or second harmonic, and a careful, comprehensive study of density irregularities in the solar wind. It would be especially helpful if the region between 1 and 20 MHz could be probed to determine which member of the high-frequency fundamental–harmonic pair disappears towards lower frequencies. Density irregularities could be studied more systematically by scintillation techniques, or, in the lower corona, by VHF scattering techniques. Microwave scattering from the lower corona has already been attempted as a means of remotely probing Langmuir waves (Benz and Fitze, 1980; Fitze and Benz 1981), and may be further exploited.

It is likely that part of the difficulty in connecting theory with observation lies in an oversimplified picture of the ambient solar wind. We are now beginning to appreciate that the wind has permanent features like a nonthermal ($\sim$10 keV) component of the electron velocity distribution, and low-frequency turbulence which may correspond to ion-acoustic waves (and may even be the source of density irregularities). Such features can have important implications for the underlying plasma physics during radio emission events, and can allow the occurrence of stronger incoherent and coherent emission, or even coherent beam emission in the absence of Langmuir waves. These phenomena ought now to be studied theoretically with more confidence and input from observation.

The most recent spacecraft experiments seem to confirm that the electron streams associated with Type III bursts do drive Langmuir waves unstable, even if their relation to the radio emissions remains in question because of their low intensity, occasional absence, and consistently delayed arrival. The theory of the nonlinear saturation of Langmuir waves is complicated by the fact that a number of unrelated nonlinear mechanisms such as plateau formation, induced scatter, and self-focusing, come into play simultaneously at wave energy densities about 10^{-5} times the background electron energy density. The interplay of all these effects against the background of a realistic corona and solar wind will be one of the tasks of theory in the coming years. Until the properties of Langmuir turbulence, which underlies all plasma emission processes, are well understood, there can be no complete and self-consistent calculations of emissivities and polarizations of the related bursts.

6.2. TYPE II BURSTS

Here, it is desirable to have spacecraft observations of the electron distributions, plasma waves, and radiation of the same variety and quality as for Type III bursts, and, in addition, to gather direct data concerning the related shocks. Theoretically, shock heating

and beam selection mechanisms should be reconsidered for interplanetary shock parameters, and beam acceleration and relaxation properties should be studied further. The
effect of large- and small-scale density inhomogeneities should be taken into account for
both plasma waves and amplified fundamental emission. Execution of this program would
result in the same confrontation of theory and observations for Type II bursts as outlined
in Section 6.1 for Type III bursts.

6.3. MOVING TYPE IV BURSTS

On the observational side, more multifrequency interferometer measurements are necessary to determine source structures better, and possible correlations with coronal transients need to be examined in more depth. Theoretically, the MHD stability of moving
arches and plasmoids should be established, and continuous acceleration processes studied.
Controversy over the polarization characteristics of plasma emission in a magnetic field
should be resolved (for the Type III burst problem as well).

6.4. TYPE I BURSTS

Very high time resolution (better than 10 ms) observations would enable a determination
of any short time-scale structure within bursts to be made. Multifrequency interferometer
observations would help determine the relationship of Type I storms to storm Type III
bursts more precisely. Theoretically, MHD studies of localized current channels in solar
arches would be informative. Wave production and electron acceleration in the arches
should also be treated.

Acknowledgements

One of us (M.V.G.) gratefully acknowledges the help and encouragement of the CSIRO
Solar Radiophysics group, and the Department of Theoretical Physics at the University
of Sydney during the writing of this review. We are also indebted to G. Dulk, R. Lin,
K. Sheridan, R. Grognard, D. Melrose, M. Kundu, M. Pesses, and R. Stewart for helpful
discussions. This work was supported by National Aeronautics and Space Administration
Grant No. NAGW-91, National Science Foundation Grants No. ATM-7916837 and
ATM-8020426, and Air Force Office of Scientific Research Grant No. 80-0022.

References

Altschuler, M. D., Nakagawa, Y., and Lilliequist, C. G.: 1968, *Solar Phys.* **3**, 466.
Altschuler, M. D., Smith, D. F., Swarztrauber, P. N., and Priest, E. R.: 1973, *Solar Phys.* **32**, 153.
Alvarez, H. and Haddock, F. T.: 1973, *Solar Phys.* **29**, 197.
Alvarez, H., Haddock, F. T., and Lin, R. P.: 1972, *Solar Phys.* **26**, 468.
Alvarez, H., Haddock, F. T., and Potter, W. H.: 1974, *Solar Phys.* **34**, 413.
Armstrong, T., Chen, G., Sarris, E., and Krimigis, S.: 1977, in M. Shea, D. Smart, and S. Wu (eds),
    Study of Travelling Interplanetary Phenomena, D. Reidel, Dordrecht, p. 367.
Axford, W., Leer, E., and Skadron, G.: 1977, *Proc. 15th Int. Cosmic Ray Conf.*, Vol. II, p. 132.

Bardwell, S. and Goldman, M. V.: 1976, *Astrophys. J.* **209**, 912.

Bell. A.: 1978, *Monthly Notices Roy. Astron. Soc.* **182**, 147.

Benz, A. O. and Fitze, H. R.: 1980, *IAU Symp.* **86**, 247.

Benz, A. O. and Wentzel, D. G.: 1981, *Astron. Astrophys.* **94**, 100.

Boishot, A. and Clavelier, B.: 1967, *Astrophys. Lett.* **1**, 7.

Boishot, A., Riddle, A. C., Pearce, J. B., and Warwick, J. W.: 1980, *Solar Phys.* **65**, 397.

Bougeret, J. L. and Steinberg, J. L.: 1977, *Astron. Astrophys.* **61**, 777.

Bougeret, J. L. and Steinberg, J. L.: 1980, *IAU Symp.* **86**, 401.

Bourgeret, J. L., Caroubalos, C., Mercier, C., and Pick, M.: 1970, *Astron. Astrophys.* **6**, 406.

Coles, W. A. and Harmon, J. K.: 1978, *J. Geophys. Res.* **83**, 1413.

de Genouillac, G. V. and Escande, D. F.: 1981, *Astron. Astrophys.* **94**, 219.

DuBois, D. F. and Rose, H. A.: 1981, *Phys. Rev.* **A24**, 1476.

Dulk, G. A.: 1973, *Solar Phys.* **32**, 491.

Dulk, G. A. and Altschuler, M. D.: 1971, *Solar Phys.* **20**, 438.

Dulk, G. A. and Suzuki, S.: 1980, *Astron. Astrophys.* **88**, 203.

Dulk, G. A., Melrose, D. B., and Smerd, S. F.: 1978, *Proc. Astron. Soc. Austral.* **3**, 243.

Dulk, G. A., Melrose, D. B., and Suzuki, S.: 1979, *Proc. Astron. Soc. Austral.* **3**, 375.

Dulk, G. A., Suzuki, S., and Gary, D. E.: 1980, *Astron. Astrophys.* **88**, 218.

Duncan, R. A.: 1979, *Solar Phys.* **63**, 389.

Duncan, R. A.: 1981, *Solar Phys.* **73**, 191.

Duncan, R. A., Stewart, R. T., and Nelson, G. J.: 1981, *IAU Symp.* **91**.

Elgaroy, O.: 1977, *Solar Noise Storms*, Pergamon Press, Oxford.

Evans, L. G., Fainberg, J., and Stone, R. G.: 1971, *Solar Phys.* **21**, 198.

Evans, L. G., Fainberg, J., and Stone, R. G.: 1973, *Solar Phys.* **31**, 501.

Fainberg, J. and Stone, R. G.: 1971, *Astrophys. J.* **164**, 123.

Fainberg, J. and Stone, R. G.: 1974, *Space Sci. Rev.* **16**, 145.

Fainberg, J., Evans, L. G., and Stone, R. G.: 1972, *Science* **178**, 743.

Fitze, H. R. and Benz, A. O.: 1981, *Astrophys. J.* **250**, 782.

Fitzenreiter, R. J., Evans, L. G., and Lin, R. P.: 1976, *Solar Phys.* **46**, 437.

Foukal, P.: 1978, *Astrophys. J.* **223**, 1046.

Frank, L. A. and Gurnett, D. A.: 1972, *Solar Phys.* **27**, 446.

Freund, H. P. and Papadopoulos, K.: 1980, *Phys. Fluids* **23**, 139.

Galeev, A. A.: 1976, in D. J. Williams (ed), *Physics of Solar Planetary Environments*, A.G.U., Washington, p. 464.

Ginzburg, V. L. and Zheleznyakov, V. V.: 1958, *Astron. Zh.* **35**, 694; *Soviet Astron. – AJ* **2**, 653.

Ginzburg, V. L. and Zheleznyakov, V. V.: 1959, *Astron. Zh.* **36**, 233; *Soviet Astron. – AJ* **3**, 235.

Goldman, M. V and Nicholson, D. R.: 1978, *Phys. Rev. Lett.* **41**, 406.

Goldman, M. V., Reiter, F. G., and Nicholson, D. R.: 1980a, *Phys. Fluids* **23**, 388.

Goldman, M. V., Rypdal, K., and Hafizi, B.: 1980b, *Phys. Fluids* **23**, 975.

Goldman, M. V., Weatherall, J. C., and NIcholson, D. R.: 1981, *Phys. Fluids* **24**, 668.

Goldstein, M. L., Smith, R. A., and Papadopoulos, K.: 1979, *Astrophys. J.* **237**, 683.

Gosling, J. T., Hildner, E., MacQueen, R. M., Munru, R. H., Poland, A. I., and Ross, C. L.: 1976, *Solar Phys.* **48**, 389.

Grognard, R. J.-M.: 1975, *Austral. J. Phys.* **28**, 731.

Grognard, R. J.-M.: 1980, *IAU Symp.* **86**, 303.

Grognard, R. J.-M.: 1982, *Solar Phys.* **81**, 173.

Gurnett, D. A. and Anderson, R. R.: 1977, *J. Geophys. Res.* **82**, 632.

Gurnett, D. A. and Frank, L. A.: 1975, *Solar Phys.* **45**, 477.

Gurnett, D. A., Anderson, R. R., and Tokar, R. L.: 1980, *IAU Symp.* **86**, 369.

Gurnett, D. A., Anderson, R. R., Scarf, F. L., and Kurth, W. S.: 1978a, *J. Geophys, Res.* **83**, 4147.

Gurnett, D. A., Baumback, M. M., and Rosenbauer, H.: 1978b, *J. Geophys. Res.* **83**, 616.

Gurnett, D. A., Maggs, J. E., Gallagher, D. L., Kurth, W. S., and Scarf, F. L.: 1981, *J. Geophys. Res.* **86**, 8833.

Gurnett, D. A., Marsch, E., Pilipp, W., Schwenn, R., and Rosenbauer, H.: 1979, *J. Geophys. Res.* **84**, 2029.

Haddock, F. T. and Alvarez, H.: 1973, *Solar Phys.* **29**, 183.
Haddock, F. T. and Graedel, T. E.: 1970, *Astrophys. J.* **160**, 293.
Hafizi, B. and Goldman, M. V.: 1981, *Phys. Fluids* **24**, 145.
Hafizi, B., Weatherall, J. C., Goldman, M. V., and Nicholson, D.: 1982, *Phys. Fluids* **25**, 392.
Hanasz, J., Schreiber, R., and Aksenov, V. I.: 1980, *Astron. Astrophys* **91**, 311.
Heyvaerts, J. and de Genouillac, G. V.: 1974, *Astron. Astrophys.* **30**, 211.
Hoyng, P., Duijveman, A., Machado, M. E., Rust, D. M. Svestka, A., Boelee, A., de Jager, C., Frost, K. J., Lafleur, H., Simnett, G. M., Van Beek, H. F., and Woodgate, B. E.: 1981, *Astrophys. J.* (submitted).
Kaplan, S. A. and Tsytovich, V. N.: 1967, *Astron. Zh.* **44**, 1036.
Kaplan, S. A. and Tsytovich, V. N.: 1968, *Sov. Astron. – AJ* **11**, 834.
Kaplan, S. A., and Tsytovich, V. N.: 1973, *Plasma Astrophysics*, Pergamon Press, Oxford.
Kellogg, P. J.: 1980, *Astrophys. J.* **236**, 696.
Kennel, C. F.: 1969, *Rev. Geophys.* **7**, 379.
Krall, N. A. and Smith, D. F.: 1975, *Astrophys. J.* **199**, 500.
Kruchina, E. N., Sagdeev, R. Z., and Shapiro, V. D.: 1980, *Pis'mn V. Zh. Eksp. Teor. Fiz. (USSR)*, **32**, 443 (*JETP Letters*).
Kruger, A.: 1979, *Introduction to Solar Radio Astronomy and Radio Physics*, D. Reidel, Dordrecht.
Kuijpers, J.: 1974, *Solar Phys.* **36**, 157.
Kulsrud, R. M. and Ferrara, A.: 1971, *Astrophys. Space Sci.* **12**, 302.
Kundu, M. R.: 1965, *Solar Radio Astronomy*, Interscience, New York.
Lilliequist, C. G., Altschuler, M. D., and Nakagawa, Y.: 1971, *Solar Phys.* **20**, 348.
Lin, A. T., Kaw, P. K., and Dawson, J. M.: 1973, *Phys. Rev.* **A8**, 2618.
Lin, R. P.: 1970, *Solar Phys.* **12**, 266.
Lin, R. P.: 1974, *Space Sci. Rev.* **16**, 189.
Lin, R. P., Evans, L. G., and Fainberg, J.: 1973, *Astrophys. Lett.* **14**, 191.
Lin, R. P., Potter, D. W., Gurnett, D. A., and Scarf, F. L.: 1981, *Astrophys. J.* **251**, 364.
Lin, R. P., Schwartz, R., Pelling, M., and Hurford, G.: 1980, *Bull. Amer. Astron. Soc.* **12**, 892.
Magelssen, G. R. and Smith, D. F.: 1977, *Solar Phys.* **55**, 211.
Malitson, H. H., Fainberg, J., and Stone, R. G.: 1973, *Astrophys. J.* **183**, L35.
Mangeney, A. and Veltri, P.: 1976, *Astron. Astrophys.* **47**, 165 and 181.
Marsh, K. A. and Hurford, G. J.: 1980, *Astrophys. J.* **240**, L111.
Maxell, A. and Swarup, G.: 1958, *Nature* **181**, 36.
McLean, D. J.: 1971, *Austral. J. Phys.* **24**, 201.
McLean, D. J., Sheridan, K. V., Stewart, R. T., and Wild, J. P.: 1971, *Nature* **234**, 140.
Melrose, D. B.: 1974a, *Proc. Astron. Soc. Austral* **2**, 261.
Melrose, D. B.: 1974b, *Solar Phys.* **37**, 353.
Melrose, D. B.; 1975, *Solar Phys.* **43**, 211.
Melrose, D. B.: 1977a, *Izv. Vyssh. Uchebn. Zaved., Radiofiz.* **20**, 1369.
Melrose, D. B.: 1977b, *Radiophys. Quantum Electron.* **20**, 945.
Melrose, D. B.: 1980a, *Plasma Astrophysics*, Vols 1 and 2, Gordon and Breach, New York.
Melrose, D. B.: 1980b, *Space Sci. Rev.* **26**, 3.
Melrose, D. B.: 1980c, *Solar Phys.* **67**, 357.
Melrose, D. B.: 1982, *Austral. J. Phys.* **35**, 67.
Melrose, D. B. and Brown, J. C.: 1976, *Monthly Notices Roy. Astron. Soc.* **176**, 15.
Melrose, D. B. and Sy, W. N.: 1972, *Austral. J. Phys.* **25**, 387.
Melrose, D. B., Dulk, G. A., and Smerd, S. F.: 1978, *Astron. Astrophys.* **66**, 315.
Nicholson, D. R. and Goldman, M. V.: 1978, *Phys. Fluids* **21**, 1766.
Nicholson, D. R., Goldman, M. V., Hoyng, P., and Weatherall, J. C.: 1978, *Astrophys. J.* **223**, 605.
Papadopoulos, K., and Freund, H. P.: 1978, *Geophys. Res. Lett.* **5**, 881.
Papadopoulos, K., Goldstein, M. L., and Smith, R. A.: 1974, *Astrophys. J.* **190**, 175.
Papadopoulos, K. and Palmadesso, P. J.: 1976, *Phys. Fluids* **19**, 605.
Parker, E. N.: 1958, *Astrophys. J.* **128**, 664.
Perkins, F. W., Oberman, C., and Valeo, E. J.: 1974, *J. Geophys. Res.* **79**, 1478.
Pesses, M. E., Decker, R. B., and Armstrong, T. P.: 1982, *Space Sci. Rev.* **32**, 185.

Ramaty, R.: 1969, *Astrophys. J.* **158**, 753.
Riddle, A. C.: 1972, *Proc. Astron. Soc. Austral.* **2**, 148.
Riddle, A. C.: 1974, *Solar Phys.* **35**, 153.
Roberts, J. A.: 1959, *Aust. J. Phys.* **12**, 327.
Robinson, R. D.: 1974, *Proc. Astron. Soc. Austral.* **2**, 258.
Robinson, R. D.: 1977, Ph.D. Thesis, Univ. of Colorado, Boulder.
Rosenberg, H.: 1976, *Phil. Trans. Roy. Soc.* **A281**, 461.
Rowland, H. L., Lyon, J. G., and Papadopoulos, K.: 1981, *Phys. Rev. Lett.* **46**, 346.
Ryutov, D. D. and Sagdeev, R. Z.: 1970, *Sov. Phys. – JETP* **31**, 396.
Saito, K.: 1970, *Ann. Tokyo Astron. Obs.* **12**, 53.
Schmahl, E. and Hildner, E.: 1977, *Solar Phys.* **55**, 473.
Sheridan, K. V.: 1967, *Proc. Astron. Soc. Austral.* **1**, 58.
Sheridan, K. V., Labrum, N. R., and Paton, W. J.: 1973, *Proc. IEEE* **61**, 1312.
Slottje, C.: 1980, *IAU Symp.* **86**, 195.
Slysh, V. I.: 1967, *Kosm. Issled.* **5**, 897; *Cosmic Res.* **5**, 759.
Smerd, S. F.: 1970, *Proc. Astron. Soc. Austral.* **1**, 305.
Smerd, S. F.: 1976a, in D. J. Williams (ed.), *Physics of Solar Planetary Environments*, A.G.U., Washington, Vol. 1 p. 193.
Smerd, S. F.: 1976b, *Solar Phys.* **46**, 493.
Smerd, S. F. and Dulk, G. A.: 1971, in R. Howard (ed.), *Solar Magnetic Fields*, D. Reidel, Dordrecht, p. 616.
Smerd, S. F., Sheridan, K. V., and Stewart, R. T.: 1975, *Astrophys. Lett.* **16**, 23.
Smerd, S. F., Wild, J. P., and Sheridan, K. V.: 1962, *Aust. J. Phys.* **15**, 180.
Smith, D. F.: 1970, *Adv. Astron. Astrophys.* **7**, 147.
Smith, D. F.: 1971, *Astrophys. J.* **170**, 559.
Smith, D. F.: 1972a, *Astrophys. J.* **174**, 121.
Smith, D. F.: 1972b, *Astrophys. J.* **174**, 643.
Smith, D. F.: 1972c, *Astron. Astrophys.* **18**, 403.
Smith. D. F.: 1977, *Astrophys. J.* **216**, L53.
Smith, D. F.: 1980, *Solar Phys.* **66**, 135.
Smith, D. F. and Fung, P. C. W.: 1971, *J. Plasma Phys.* **5**, 1.
Smith, D. F. and Krall, N. A.: 1974, *Astrophys. J.* **194**, L163.
Smith, D. F. and Riddle, A. C.: 1975, *Solar Phys.* **44**, 471.
Smith, D. F. and Sime, D.: 1979, *Astrophys. J.* **233**, 998.
Smith, D. F. and Spicer, D. S.: 1979, *Solar Phys.* **62**, 359.
Smith, R. A., Goldstein, M. L., and Papadopoulos, K.: 1979, *Astrophys. J.* **234**, 348.
Spicer, D. S.: 1977, *Solar Phys.* **53**, 305.
Spicer, D. S.: 1981, *Solar Phys.* **71**, 115.
Steinberg, J. L. Caroubalos, C., and Bougeret, J. L.: 1974, *Astron. Astrophys.* **37**, 109.
Stepanov, A. V.: 1973, *Soviet Astron. -AJ* **17**, 781.
Stewart, R. T.: 1974, *Solar Phys.* **39**, 451.
Stewart, R. T.: 1975, *Solar Phys.* **40**, 417.
Stewart, R. T.: 1976, *Solar Phys.* **50**, 437.
Stewart, R. T. and Vorpahl, J.: 1977, *Solar Phys.* **55**, 111.
Stewart, R. T., Howard, R. A., Hansen, F., Gergely, T., and Kundu, M.: 1974b, *Solar Phys.* **36**, 219.
Stewart, R. T., McCabe, M. K., Koomen, M. J., Hansen, R. T., and Dulk, G. A.: 1974a, *Solar Phys.* **36**, 203.
Sturrock, P. A. *et al.*: 1980, *Solar Flares*, Colorado Assoc. Univ. Press, Boulder.
Suzuki, S.: 1974, *Solar Phys.* **38**, 1.
Suzuki, S. and Sheridan, K. V.: 1977, *Radiofizica* **20**, 1432.
Takakura, T. and Shibahashi, H.: 1976, *Solar Phys.* **46**, 323.
Tidman, D. A. and Krall, N. A.: 1971, *Shock Waves in Collisionless Plasmas*, Interscience, New York.
Tokar, R. L. and Gurnett, D.: 1980, *J. Geophys. Res.* **85**, 2353.
Trubnikov, B. A.: 1965, *Rev. Plasma Physics* **1**, Consultants Bureau, New York, p. 205.
Trulsen, T. and Fejer, J. A.: 1970, *J. Plasma Phys.* **4**, 825.

Van Hoven, G.: 1979, *Astrophys. J.* **232**, 572.
Weatherall, J. C., Goldman, M. V., and Nicholson, D. R.: 1981, *Astrophys. J.* **246**, 306.
Wild, J. P.: 1950a, *Austral. J. Sci. Res.* **A3**, 541.
Wild, J. P.: 1950b, *Austral. J. Sci. Res.* **A3**, 399.
Wild, J. P. (ed.): 1967, *Proc. Inst. Radio Electron Engrs. Austral.* **28** (Special issue).
Wild, J. P.: 1974, *Records of Austral. Acad. Sci.* **3**, 93.
Wild, J. P. and McCready, L. L.: 1950, *Austral J. Sci. Res.* **A3**, 387.
Wild, J. P. and Smerd, S. F.: 1972, *Ann. Rev. Astron. Astrophys.* **10**, 159.
Wild, J. P., Murray, J. D., and Rowe, W. C.: 1954, *Austral. J. Phys.* **7**, 439.
Wild, J. P., Smerd, S. F., and Weiss, A. A.: 1963, *Ann. Rev. Astron. Astrophys.* **1**, 291.
Zaitsev, V. V.: 1977, *Radiophys. Quantum Electron.* **20**, 952.
Zaitsev, V. V. and Ledenev, V. G.: 1976, *Sov. Astron. Lett.* **2**, 443 (Russian original page number).
Zaitsev, V. V., Mityakov, N. A., and Rapoport, V. O.: 1972, *Solar Phys.* **24**, 444.
Zakharov, V. E.: 1972, *Zh. Eksp. Teor. Fiz.* **62**, 1745; *Sov. Phys. – JETP* **35**, 908.
Zheleznyakov, V. V.: 1970, *Radio Emission of the Sun and Planets*, Pergamon Press, Oxford.
Zheleznyakov, V. V. and Zaitsev, V. V.: 1970a, *Sov. Astron. – A.J.* **14**, 47.
Zheleznyakov, V. V. and Zaitsev, V. V.: 1970b, *Sov. Astron. – A.J.* **14**, 250.

Dept of Astro-Geophysics,
University of Colorado,
Boulder, CO 80309,
U.S.A.

INDEX

377